30

Springer Series in Chemical Physics

Edited by J. Peter Toennies

Springer Series in Chemical Physics

Editors: V. I. Goldanskii R. Gomer F. P. Schäfer J. P. Toennies

1 **Atomic Spectra and Radiative Transitions**
By I. I. Sobelman
2 **Surface Crystallography by LEED**
Theory, Computation and Structural Results. By M. A. Van Hove, S. Y. Tong
3 **Advances in Laser Chemistry**
Editor: A. H. Zewail
4 **Picosecond Phenomena**
Editors: C. V. Shank, E. P. Ippen, S. L. Shapiro
5 **Laser Spectroscopy**
Basic Concepts and Instrumentation
By W. Demtröder 2nd Printing
6 **Laser-Induced Processes in Molecules** Physics and Chemistry
Editors: K. L. Kompa, S. D. Smith
7 **Excitation of Atoms and Broadening of Spectral Lines** By I. I. Sobelman, L. A. Vainshtein, E. A. Yukov
8 **Spin Exchange**
Principles and Applications in Chemistry and Biology
By Yu. N. Molin, K. M. Salikhov, K. I. Zamaraev
9 **Secondary Ion Mass Spectrometry SIMS II** Editors: A. Benninghoven, C. A. Evans, Jr., R. A. Powell, R. Shimizu, H. A. Storms
10 **Lasers and Chemical Change**
By A. Ben-Shaul, Y. Haas, K. L. Kompa, R. D. Levine
11 **Liquid Crystals of One- and Two-Dimensional Order**
Editors: W. Helfrich, G. Heppke
12 **Gasdynamic Laser** By S. A. Losev
13 **Atomic Many-Body Theory**
By I. Lindgren, J. Morrison
14 **Picosecond Phenomena II**
Editors: R. M. Hochstrasser, W. Kaiser, C. V. Shank
15 **Vibrational Spectroscopy of Adsorbates** Editor: R. F. Willis
16 **Spectroscopy of Molecular Excitions**
By V. L. Broude, E. I. Rashba, E. F. Sheka
17 **Inelastic Particle-Surface Collisions**
Editors: E. Taglauer, W. Heiland
18 **Modelling of Chemical Reaction Systems** Editors: K. H. Ebert, P. Deuflhard, W. Jäger
19 **Secondary Ion Mass Spectrometry SIMS III**
Editors: A. Benninghoven, J. Giber, J. László, M. Riedel, H. W. Werner
20 **Chemistry and Physics of Solid Surfaces IV** Editors: R. Vanselow, R. Howe
21 **Dynamics of Gas-Surface Interaction**
Editors: G. Benedek, U. Valbusa
22 **Nonlinear Laser Chemistry**
Multiple-Photon Excitation
By V. S. Letokhov
23 **Picosecond Phenomena III**
Editors: K. B. Eisenthal, R. M. Hochstrasser, W. Kaiser, A. Laubereau
24 **Desorption Induced by Electronic Transitions DIET I** Editors: N. H. Tolk, M. M. Traum, J. C. Tully, T. E. Madey
25 **Ion Formation from Organic Solids**
Editor: A. Benninghoven
26 **Semiclassical Theories of Molecular Scattering** By B. C. Eu
27 **EXAFS and Near Edge Structures**
Editors: A. Bianconi, L. Incoccia, S. Stipcich
28 **Atoms in Strong Light Fields**
By N. B. Delone, V. P. Krainov
29 **Gas Flow in Nozzles**
By U. Pirumov, G. Roslyakov
30 **Theory of Slow Atomic Collisions**
By E. E. Nikitin, S. Ya. Umanskii
31 **Reference Data on Atoms, Molecules, and Ions** By A. A. Radzig, B. M. Smirnov
32 **Adsorption Processes on Semiconductor and Dielectric Surfaces I**
By V. F. Kiselev, O. V. Krylov
33 **Surface Studies with Lasers**
Editors: F.R. Aussenegg, A. Leitner, M.E. Lippitsch
34 **Inert Gases**
Potentials, Dynamics, and Energy Transfer in Doped Crystals. Editor: M. L. Klein
35 **Chemistry and Physics of Solid Surfaces V** Editors: R. Vanselow, R. Howe
36 **Secondary Ion Mass Spectrometry, SIMS IV** Editors: A. Benninghoven, J. Okano, R. Shimizu, H. W. Werner
37 **X-Ray Spectra and Chemical Binding**
By A. Meisel, G. Leonhardt, R. Szargan
38 **Ultrafast Phenomena IV**
By D. H. Auston, K. B. Eisenthal
39 **Laser Processing and Diagnostics**
Editor: D. Bäuerle

E. E. Nikitin S. Ya. Umanskii

Theory of Slow Atomic Collisions

With 92 Figures

Springer-Verlag
Berlin Heidelberg New York Tokyo 1984

Professor Evgenii E. Nikitin
Stanislav Ya. Umanskii

Institute of Chemical Physics, Academy of Sciences of the USSR, Kosygin Street 4
Moscow V-334, USSR

Title of the original Russian edition:
Neadiabaticheskiye perekhody pri medlennykh atomnykh stolknoveniyakh

---

ISBN 3-540-12414-4 Springer-Verlag Berlin Heidelberg New York Tokyo
ISBN 0-387-12414-4 Springer-Verlag New York Heidelberg Berlin Tokyo

Library of Congress Cataloging in Publication Data. Nikitin, E. E. (Evgeniĭ Evgen'evich), 1933-. Theory of slow atomic collisions. (Springer series in chemical physics; v. 30). 1. Collisions (Nuclear physics) 2. Scattering (Physics) I. Umanskiĭ, Stanislav I͡Akovlevich. II. Title. III. Series. QC794.6.C6N545 1984 539.7'54 84-1345

Typesetting: Daten- und Lichtsatz-Service, Würzburg
Offsetprinting and bookbinding: Graphischer Betrieb Konrad Triltsch, Würzburg
2153/3020-543210

Preface

The theory of atom-molecule collisions is one of the basic fields in chemical physics. Its most challenging part – the dynamics of chemical reactions – is as yet unresolved, but is developing very quickly. It is here a great help to have an analysis of those parts of collision theory which are already complete, a good example being the theory of atomic collisions in processes specific to chemical physics.

It has long been observed that many notions of this theory can also be applied successfully to reactive and unreactive molecular collisions. Moreover, atomic collisions often represent a touchstone in testing approaches proposed for the solution of more complicated problems. Research on the theory of slow atomic collisions carried out at the Moscow Institute of Chemical Physics has been based on just these ideas. A general viewpoint concerning the setting up and representation of the theory came out of these studies, and appeared to be useful in studying complicated systems as well. It underlies the representation of the theory of slow atomic collisions in this book.

Analytical approaches to the theory of slow atomic collisions, including the calculation of electronic states of diatoms, nonadiabatic coupling models, calculation of differential and total cross sections, are discussed here. These approaches are based on the exact and approximate symmetry properties of an electronic subsystem, and on simplifications provided by slow quasi-classical nuclear motion. A transparent correlation between interatomic interactions and the characteristic features of scattering emerges from these approaches.

The theory presented in this book is self-contained. It can be applied to the interpretation of various processes occurring in atomic collisions over a relatively wide energy range, from thermal energies to hundreds of eV.

The authors wish to thank Dr. E. I. Dashevskaya, Dr. G. K. Ivanov, Dr. M. Ya. Ovchinnikova and Dr. A. I. Reznikov for their valuable advice, as well as colleagues at the Moscow Institute of Chemical Physics for very fruitful discussions. Thanks are also due to Mrs V. D. Grammatchikova for her help in preparing the text.

Moscow, January 1984 *E. E. Nikitin · S. Ya. Umanskii*

Contents

1. Introduction

The theory of atomic collisions was given a fresh incentive towards the end of the nineteen twenties, immediately after the advent of quantum mechanics. The first edition of the celebrated book by *Mott* and *Massey*, *The Theory of Atomic Collisions* appeared in 1933 [1.1]. Since then, the theory has greatly advanced. Yet the only part which is relatively complete (at least in principle) seems to be that dealing with slow collisions in which the velocity of the atoms is substantially smaller than the velocity of the electrons in the atoms. This condition makes it possible to formulate the collision problem correctly, and since the scattering equations are well known, the solution of such a problem with a computer would cause no great difficulty. However, a solely numerical solution of any problem leaves a certain discontent, since the mechanism of the process studied remains obscure. Therefore, it appears to be desirable to obtain more analytical results although this is possible only under additional drastic simplifications.

One of the most significant simplifications stems from the quasi-classical nature of the motion of the colliding atoms. Alongside the low-velocity condition, this makes the theory relatively simple and, which is very important, illustrative. Its basic pattern is the adiabatic motion of the atoms over the electronic terms of a quasi-molecule, accompanied by nonadiabatic transitions. As a result, there emerges a transparent correlation between the interatomic interaction and the characteristic features of scattering. The theory thus formulated can be applied to the interpretation of various processes occurring in atomic collisions in gases, low-temperature plasmas, and electronic and ionic beams, over a relatively wide energy range from thermal energies to hundreds of electron volts.

In this way, theoretical analysis of slow quasi-classical atomic collisions involves three stages: (i) calculation of the adiabatic electronic terms and couplings between relevant states; (ii) calculation of the probability amplitudes of nonadiabatic transitions between these states, induced by the relative motion of the atoms; (iii) calculation of differential and total cross sections, using the probability amplitudes of nonadiabatic transitions. The books dealing with the general theory of scattering [1.1–8] do not pay much attention to these problems taken together. More specific books on atom-molecule collisions [1.9–14] consider separate problems related to the second or third stage only, but the models of nonadiabatic coupling and the quasi-classical theory of differential cross sections for inelastic collisions are not treated in detail. Yet these problems are now of particular importance in connection with the present experimental possibility of high angular and energy resolution.

All the three stages of analysis are treated in two books [1.15, 16], but only for processes occurring at large distances. The calculation of diatomic terms is considered in many books (see references cited in Chap. 4), but mostly from the molecular-spectroscopy point of view.

Taking all this into account, it was thought expedient to write a book which would treat all aspects of the theory of slow quasi-classical collisions – interatomic interaction, interference phenomena in scattering, and nonadiabatic transitions – from a unique standpoint. The text is arranged as follows.

Chapter 2 gives a general formulation of the problem of slow atomic collisions under quasi-classical conditions with the aim of describing an inelastic event by several "classical" trajectories corresponding to various channels, the probability amplitudes being associated with these trajectories.

Chapters 3 and 4 are concerned with the electronic states of diatoms. Chapter 3 deals with the exact and approximate symmetry properties of these states and discusses the notion of diabatic states widely invoked in the theory of atomic collisions. Chapter 4 describes the approximate methods of calculating electronic adiabatic wave functions and diatomic terms at different internuclear seperations. Attention is mainly paid here to simple approaches developed for specific problems connected with atomic collisions.

The elastic collisions are discussed in Chap. 5 for the quasi-classical case only, mostly in order to introduce and illustrate the notions and approaches used for approximate analysis of inelastic processes. This is, first of all, the matching method of solving the radial problem for quasi-classical conditions violated in localized regions, as well as a discussion of the scattering amplitude in terms of the interference of waves scattered through the same angle by different parts of the potential.

Chapter 6 is concerned with the derivation and analysis of various methods of constructing the quasi-classical scattering matrix for the multichannel case. These are the common-trajectory method, the matching method using the theory of adiabatic and sudden perturbations, the distorted-wave method. The matching method seems to be quite general for evaluation of the quasi-classical scattering matrix. This method implies that the propagation of waves in various channels is associated with the adiabatic motion of atoms over relatively extended regions of the internuclear distance, accompanied by nonadiabatic transitions within regions of small extent. The localization of these regions means that relatively simple models described by comparison equations can be introduced. These models can be used in various combinations for the construction of a specific scattering matrix. Most frequently one has to account for strong coupling of two states only. Therefore the next three chapters are devoted to discussion of two-state models.

Chapter 7 deals with the classification and the general features of two-state models under quasi-classical conditions, from the standpoint of calculating both the scattering matrix and the differential and integral cross sections.

Specific models of strong nonadiabatic coupling of two states are discussed in Chap. 8 (the linear model representing a generalization of the familiar Landau-Zener model), and in Chap. 9 (various nonlinear models).

Chapter 10 treats certain models of strong coupling between many states, including interaction with the continuum.

Finally, the theory presented – from the calculation of terms to the calculation of cross sections – is illustrated in Chap. 11 by an extensively studied processes of intramultiplet mixing in the collision of an alkali atom in the first excited state ^2P_j with a noble gas atom. This example was chosen for two reasons. First, one of the authors (Nikitin) has long been concerned with intramultiplet mixing. Second, this process depends, over a wide energy range, on the interaction of a small number of states (three), and the determination of adiabatic terms and nonadiabatic coupling can be sufficiently reliable. This enables a relatively simple, but at the same time nontrivial, treatment demonstrating the specific features of multichannel scattering under quasi-classical conditions.

In treating the general theory it was assumed for definiteness that the colliding atoms were neutral. However, over the energy range considered, this theory can also be applied, with obvious modifications, to processes involving ions.

To make reading easier the book is provided with an appendix devoted to the quantum theory of angular momentum, in which all the notions of this theory encountered in the text are defined. It is not referred to explicitly, but the reader may resort to it, when needed.

The notation used is as close as possible to that generally used, and coincides with that in two previous books of this series [1.17, 18]. The Clebsch-Gordan coefficients are the only essential exception here. Use is made of the seemingly most compact notation involving the quantum numbers of all momenta and their projections and containing neither subscripts nor superscripts. The book [1.19] is followed in this respect.

Since both the structure of atoms and molecules and the dynamics of atomic collisions are discussed, the same letter is sometimes used to denote different quantities, in particular for wave functions and quantum numbers of angular momenta and their projections. This cannot, however, result in confusion, since these different quantities are only met in separate chapters. When some physical property is conserved in a collision, the appropriate sub- or superscript is written only once, or not at all if it is not essential to the problem discussed.

The set of quantum numbers characteristic of the state of a free atom or diatom is denoted by one letter, a. If certain quantum numbers $p, \ldots, r$ must be singled out of this set, the total set will be denoted by $a, p, \ldots, r$. In this latter designation, a does not stand for the complete set of quantum numbers, but rather for a truncated set – without the quantum numbers $p, \ldots, r$. This notation greatly simplifies the writing of expressions.

Finally, atomic units are used throughout the text, unless stated otherwise.

2. General Formulation of Scattering Problem Under Quasi-classical Conditions

Slow collisions are those for which we can neglect all effects due to the additional translational momenta of the electrons produced in their motion together with the atomic nuclei. Since the additional electron velocity δv_{el} corresponding to this translational motion coincides with that of the nuclei v there is an upper bound on the velocity of the nuclei, $v \ll v_{el}$, where v_{el} is the characteristic velocity of the electrons in an atom at rest. If v_{el} is taken to be the velocity of the electrons in the valence shell of the atom ($v_{el} \simeq 1$ a.u. $\simeq 10^8$ cm/s), the above conditions represent the most stringent upper bound on the relative energy. However, for transitions of inner-shell atomic electrons, many results of the theory discussed below are also valid for nuclear velocities comparable with 1 a.u.

The quasi-classical conditions impose a lower bound on the relative energy and are expressed as the relation between the wavelength $\lambda\!\!\!^{-}$ for relative atomic motion, and the size of the interaction region R^*,

$$\lambda\!\!\!^{-} \ll R^*.$$

This condition permits the use of various approximations generally known as quasi-classical or semiclassical approximations.

It will readily be seen that both conditions are fulfilled simultaneously over a wide range of relative energies, from tenths (and even hundredths) to hundreds of eV.

For simplicity this chapter mainly considers collisions of an atom A with a structureless atom B, i.e., with an atom in an unchanging spherically symmetric state $^1S_0$. Such a restriction does not obscure the basic features of slow collisions under quasi-classical conditions, and at the same time enables a substantial simplification of all expressions. If needed, these expressions can readily be generalized for the case of an arbitrary atom B (Sect. 2.3).

In general, the internal state of an atom A (or an ensemble of atoms) is specified by the density matrix $\varrho_{a_1 a_2}$. If the transition frequencies $\Omega_{a_1 a_2}$ between the states a_1 and a_2 are substantially larger than the inverse times $1/\tau$ characteristic of the evolution of the ensemble of atoms, $\varrho_{a_1 a_2}$ may often be taken as diagonal in the indices a_1 and a_2. Although fulfillment of the inequality $\Omega_{a_1 a_2} \tau \gg 1$ depends on the precise meaning of τ, it is clear that this inequality is never satisfied for degenerate states and for some states with very small energy splittings.

States of the first type differ only in the quantum number of the total angular momentum projection on a certain axis fixed in space. The second type of states are the atomic hyperfine structure states for which $\Omega_{a_1 a_2}$ is of the order of $10^8-10^9\ \mathrm{s}^{-1}$. Since the hyperfine interaction is very small on the scale of typical atomic interactions, it does not influence the collision dynamics. Therefore the corresponding quantum numbers will not be included hereafter within the set a[1] that specifies the state of the free atom at rest. Thus, this set involves the following quantum numbers: j – the total electronic angular momentum, m – its projection on the Z axis of a certain coordinate frame XYZ fixed in space, w – the parity, as well as additional quantum numbers γ. The physical meaning of the latter will be discussed in detail in Sect. 3.1.

For simplicity we shall assume in Sects. 2.1–3 that the atomic state before the collision corresponds to a definite value of $m\,(\varrho_{a_1 m_1; a_2 m_2} = \delta_{a_1 a_2}\delta_{m_1 m}\delta_{m_2 m})$ and shall calculate the amplitudes and cross sections for transitions to states with fixed m'. Under these conditions the internal state of an atom can be described by a wave function $|\gamma j m\rangle_{\mathrm{A}}$, rather than by a density matrix. This restricts the problem to a certain extent, since the scattering of an atom A in the state $|\gamma j m\rangle_{\mathrm{A}}$ leads, generally speaking, to a coherent superposition of states with different m described by a density matrix non-diagonal in m . This incompleteness is partially removed in Sect. 2.4 where isotropic collisions are dealt with. The high symmetry of such collisions makes the number of linearly independent cross sections rather small.

2.1 Scattering Amplitudes and Cross Sections

2.1.1 Representations of Amplitudes and Cross Sections

The most general description of the process

$$\mathrm{A}\,(a m) + \mathrm{B} \rightarrow \mathrm{A}\,(a' m') + \mathrm{B} \tag{2.1}$$

is given by the differential cross section $q_{a'm';am}(\omega)$, i.e., by the ratio of the flux of atoms $\mathrm{A}\,(a'm')$ per unit solid angle in the direction $\omega = (\theta, \varphi)$ to the incident flux density of atoms $\mathrm{A}\,(a, m)$. Here and throughout the following text, scattering is discussed for the center-of-mass space-fixed coordinate

[1] If needed, transitions between hyperfine structure levels may be taken into account in accordance with the results obtained for the collisions of atoms with zero nuclear spin (Sect. 2.4).

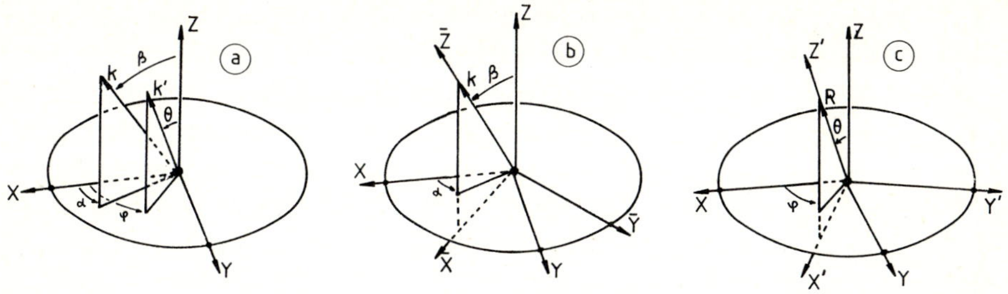

Fig. 2.1 a–c. Coordinate systems used in quantum treatment of atomic collisions. (**a**) Arbitrary space-fixed coordinate system XYZ. The vector $\boldsymbol{k}$ is characterized by the azimuthal angle α and the polar angle β and the vector $\boldsymbol{k}'$ by the azimuthal angle φ and the polar angle θ. (**b**) Standard space-fixed coordinate system $\bar{Y}\bar{Y}\bar{Z}$ obtained from the system XYZ by rotation through Euler angles α, β, γ ($\gamma = 0$). (**c**) Body-fixed (molecular) coordinate system obtained from the system XYZ by rotation through Euler angles φ, θ, γ ($\gamma = 0$). Note that for $R \to \infty$ the relation $\hat{\boldsymbol{k}}' = \hat{\boldsymbol{R}}$ holds

frame XYZ, and the angles θ and φ are defined by the direction of the relative velocity vector $\boldsymbol{v}'$ in this frame after the collision. The initial velocity $\boldsymbol{v}$ in the XYZ frame is characterized by the polar angle β and the azimuthal angle α (Fig. 2.1 a).

The differential cross section is found by solving the Schrödinger equation for the system considered,

$$\mathscr{H}\Psi = \mathscr{E}\Psi, \tag{2.2}$$

where $\mathscr{H}$ is the total Hamiltonian of the system, Ψ its wave function, $\mathscr{E}$ the total energy. In the asymptotic region when A and B are separated by a distance R sufficiently large for their interaction to be disregarded, the state of the system is described by a superposition of plane [incident atom A (a, m)] and scattered [scattered atom A $(a'm')$] waves. Each such wave can be represented as the product of an electronic wave function $|am\rangle_A$ of an atom at rest with the wave function of relative nuclear motion.[2] For $R \to \infty$ the latter is the sum of a plane wave and an outgoing spherical wave. Thus, the boundary conditions

$$\Psi \to \exp(\mathrm{i}\boldsymbol{k}_a\boldsymbol{R})|am\rangle_A + \sum_{a',m'} f_{a'm';am}(\hat{\boldsymbol{k}}_{a'}, \hat{\boldsymbol{k}}_a) \times \frac{\exp(\mathrm{i}k_{a'}R)}{R}|a'm'\rangle_A \tag{2.3}$$

[2] Strictly speaking, R must be the distance between B and the center of mass of A, and the electronic wave function of the free atom A must be referred to its center of mass. However, when B is a structureless particle, a rigorous formulation results in a minor (of the order of m_{el}/M_{nuc}) redefinition of the potential energy of the A–B interaction, and in an isotope shift of the levels of the free atom A. When needed, these corrections can be readily accounted for. However, the real difficulties, due to the additional electron momentum connected with the motion of the atom as a whole, arise only when taking the indistinguishability of the electrons in atoms A and B into account. This, will be discussed in more detail in Sect. 2.3.

are imposed on the wave function Ψ. Here $\boldsymbol{k}_a$ and $\boldsymbol{k}_{a'}$ are the initial and final wave vectors, and $\hat{\boldsymbol{k}}_a$, $\hat{\boldsymbol{k}}_{a'}$ are unit vectors.[3] The values k_a in different channels are connected by the energy conservation law

$$k_{a'}^2/2\mu + \varepsilon_{a'} = k_a^2/2\mu + \varepsilon_a, \tag{2.4}$$

where μ is the reduced mass of the colliding atoms and ε_a is the internal energy of A (a, m). From this point on we shall omit the subscripts on the wave vectors, and shall use $\boldsymbol{k}$ and $\boldsymbol{k}'$ instead of $\boldsymbol{k}_a$ and $\boldsymbol{k}_{a'}$. The differential cross section of process (2.1) is known to be expressed via the amplitude of an outgoing spherical wave,

$$q_{a'm';am} = \frac{k'}{k} |f_{a'm';am}(\hat{\boldsymbol{k}}', \hat{\boldsymbol{k}})|^2 . \tag{2.5}$$

The solution of (2.2) can be simplified if Ψ is expanded over the eigenfunctions of operators representing the integrals of motion. In the present case these are the total angular momentum squared $\boldsymbol{J}^2$ and its projection J_Z on the space-fixed axis Z. Let

$$\Psi = \sum_{JM} \Psi^{JM} \tag{2.6}$$

where J and M are quantum numbers of $\boldsymbol{J}^2$ and J_Z, respectively. Then for each component Ψ^{JM} of the expansion we get

$$\mathscr{H} \Psi^{JM} = \mathscr{E} \Psi^{JM} . \tag{2.7}$$

The functions Ψ^{JM} ensuring the boundary conditions (2.3) for Ψ can be expressed in the form

$$\begin{aligned} \Psi^{JM} = & \frac{1}{R} \sum_{a'm'} \sum_{ll'} \langle \hat{\boldsymbol{R}}, l' a' m' | \mathrm{Y}^J | \hat{\boldsymbol{k}}, l a m \rangle \\ & \times 2\pi \mathrm{i} k^{-1/2} |a' m'\rangle_{\mathrm{A}} \, \mathrm{i}^l \chi^{+J}_{a'l';al}(R) . \end{aligned} \tag{2.8}$$

Here

$$\begin{aligned} & \langle \hat{\boldsymbol{R}}, l' a' m' | \mathrm{Y}^J | \hat{\boldsymbol{k}}, l a m \rangle \\ & = \sum_{n'n} Y_{l'}^{n'}(\hat{\boldsymbol{R}}) \, Y_l^{n*}(\hat{\boldsymbol{k}}) \begin{bmatrix} l' & j' & J \\ n' & m' & M \end{bmatrix} \begin{bmatrix} l & j & J \\ n & m & M \end{bmatrix}, \end{aligned} \tag{2.9}$$

$\begin{bmatrix} l & j & J \\ n & m & M \end{bmatrix}$ and $\begin{bmatrix} l' & j' & J' \\ n' & m' & M' \end{bmatrix}$ are Clebsch-Gordan coefficients and $Y_{l'}^{n'}(\hat{\boldsymbol{R}})$, $Y_l^n(\hat{\boldsymbol{k}})$ are spherical functions; the functions $\chi^{+J}_{a'j';al}$ stand for the radial wave

---

[3] Here and throughout the following text $\hat{\boldsymbol{A}} = \boldsymbol{A}/A$ is a unit vector in the direction $\boldsymbol{A}$.

functions of relative motion, with the asymptotic behavior

$$\chi^{+J}_{a'l';al}|_{R\to\infty} = (k')^{-1/2}\left\{\exp\left[-\mathrm{i}\left(kR - \frac{l\pi}{2}\right)\right]\delta_{ll'}\delta_{aa'} - S^{J}_{a'l';al}\exp\left[\mathrm{i}\left(k'R - \frac{l'\pi}{2}\right)\right]\right\}, \tag{2.10}$$

where $S^{J}_{a'l';al}$ stands for the elements of the scattering matrix (S matrix) in the total angular momentum representation referred to as the J representation.

In the J representation the S matrix is diagonal in J and M and independent of M. This is a manifestation of the invariance of a collision event with respect to different choices of the space-fixed frame. By the principle of probability conservation, the S matrix is unitary:

$$\sum_{a''l''} S^{J*}_{a''l'';a'l'} S^{J}_{a''l'';al} = \delta_{a'a}\delta_{l'l}. \tag{2.11}$$

Additional restrictions upon the S matrix in the J representation are imposed by inversion and time-reversal invariances of the total Hamiltonian $\mathscr{H}$. The state $|a\,w\,l\,J\,M\rangle$ is an eigenstate of the total inversion operator $\hat{I}$ with the eigenvalue

$$I = w(-1)^{l}. \tag{2.12}$$

Hence inversion invariance requires that

$$S^{J}_{a'l';al} = 0 \quad \text{if} \quad w\,w'(-1)^{l-l'} = -1. \tag{2.13}$$

When considering time-reversal invariance special care is needed in choosing the phases of the wave functions. The problem of phase conventions in connection with time reversal has been discussed in detail [2.1]. The following phase conventions are used in this book. The spherical functions Y^{m}_{l} are defined according to *Condon* and *Shortley* [2.2] and transform under time reversal $\hat{T}$ as

$$\hat{T}Y^{m}_{l} = Y^{m*}_{l} = (-1)^{-m}Y^{-m}_{l}. \tag{2.14}$$

The spin wave functions $\chi^{1/2}_{m}$ are such that

$$\hat{T}\chi^{1/2}_{m} = (-1)^{(1/2)-m}\chi^{1/2}_{-m}. \tag{2.15}$$

With these phase conditions, it can be readily seen that the atomic wave functions $|a\,w\,j\,m\rangle$ transform under time reversal in the following way:

$$\hat{T}|a\,w\,j\,m\rangle = w(-1)^{j-m}|a\,w\,j\,-m\rangle. \tag{2.16}$$

General collision theory (see, e.g., [2.3]) commonly uses modified spherical functions

$$\tilde{Y}_l^m = \mathrm{i}^l Y_l^m, \quad \hat{T}\tilde{Y}_l^m = (-1)^{l-m}\tilde{Y}_l^{-m} \tag{2.17}$$

for relative motion and target states $|\widetilde{awjm}\rangle$ such that

$$\hat{T}|\widetilde{awjm}\rangle = (-1)^{j-m}|\widetilde{awj-m}\rangle . \tag{2.18}$$

The advantage of the latter phase convention is that the wave function

$$|\widetilde{j_1 j_2 jm}\rangle = \sum_{m_1 m_2} \begin{bmatrix} j_1 & j_2 & j \\ m_1 & m_2 & m \end{bmatrix} |\widetilde{j_1 m_1}\rangle |\widetilde{j_2 m_2}\rangle \tag{2.19}$$

transforms in the same way as do the functions $|\widetilde{j_1 m_1}\rangle$ and $|\widetilde{j_2 m_2}\rangle$:

$$\hat{T}|\widetilde{j_1 j_2 jm}\rangle = (-1)^{j-m}|\widetilde{j_1 j_2 j-m}\rangle . \tag{2.20}$$

The invariance of the transformation properties under angular momentum addition with this phase convention allows one to prove, solely on the basis of time-reversal invariance, that the corresponding S matrix in the J representation is symmetric [2.1, 3, 4]:

$$\tilde{S}^J_{a'l';al} = \tilde{S}^J_{al;a'l'} . \tag{2.21}$$

On the other hand, it is easy to show using (2.14, 16–18) that

$$S^J_{a'l';al} = \frac{\sqrt{w'}}{\sqrt{w}}\,\mathrm{i}^{l-l'}\,\tilde{S}^J_{a'l';al} . \tag{2.22}$$

It follows from (2.22) that the symmetry of $\tilde{S}^J_{a'l';al}$ and the parity conservation expressed by (2.13) leads to the symmetry of $S^J_{a'l';al}$:

$$S^J_{a'l';al} = S^J_{al;a'l'} . \tag{2.23}$$

The scattering amplitude is completely controlled by the S matrix, or the T matrix defined as

$$T = 1 - S . \tag{2.24}$$

Then

$$f_{a'm';am}(\hat{\boldsymbol{k}}', \hat{\boldsymbol{k}}) = \frac{2\pi\mathrm{i}}{\sqrt{k'k}} \sum_{l'n'} \sum_{ln} \mathrm{i}^{l-l'}\, Y^{n'}_{l'}(\hat{\boldsymbol{k}}')\, Y^{n*}_l(\hat{\boldsymbol{k}})\, T^{m'n';mn}_{a'l';al} \tag{2.25}$$

where $T^{m'n';mn}_{a'l';al}$ stands for the matrix elements of the T matrix in the $jmln$ representation (referred to as the j representation). The connection between the j and J representations is given by

$$T^{m'n';mn}_{a'l';al} = \sum_{JM} \begin{bmatrix} l' & j' & J \\ n' & m' & M \end{bmatrix} T^J_{a'l';al} \begin{bmatrix} l & j & J \\ n & m & M \end{bmatrix} . \tag{2.26}$$

The direction of the quantization axis in (2.25) is arbitrary, but is the same for the initial and final states. For the space-fixed coordinate frame

$\bar{X}\,\bar{Y}\bar{Z}$ with axis $\bar{Z}$ directed along $\boldsymbol{k}$ (Fig. 2.1 b),

$$Y_l^n(\hat{\boldsymbol{k}}) = \left(\frac{2l+1}{4\pi}\right)^{1/2} \delta_{n0}\,, \tag{2.27}$$

and (2.25) for $f_{a'\bar{m}';a\bar{m}}$ (here and throughout the following text $\bar{m}$, $\bar{m}'$ denote quantum numbers of the projection of the atomic angular momentum on the direction of the initial velocity) becomes

$$f_{a'\bar{m}';a\bar{m}}(\theta,\varphi) = \mathrm{i}\left(\frac{\pi}{k\,k'}\right)^{1/2} \sum_{ll'} \mathrm{i}^{l-l'}(2l+1)^{1/2} \times Y_{l'}^{\bar{m}-\bar{m}'}(\theta,\varphi)\, T_{a'l';al}^{\bar{m}'\bar{m}-\bar{m}';\bar{m}0}\,. \tag{2.28}$$

In general, the differential cross section $q_{a'\bar{m}';a\bar{m}}$ given by (2.5) is taken as the four-fold sum over the quantum numbers of the orbital angular momenta when the j representation is used, or even higher sums for the J representation. Though these expressions can be simplified (see, e.g., [2.1]), they will not be written here. The relevant expressions for quasi-classical conditions will be given later. We give here only rather simple expressions for the integral cross sections obtained by integrating the differential cross sections over all scattering angles:

$$\sigma_{a'\bar{m}';a\bar{m}} = \frac{\pi}{k^2} \sum_{l'l_1l_2} \mathrm{i}^{l_1-l_2}(2l_1+1)^{1/2}(2l_2+1)^{1/2} \times T_{a'l';al_1}^{\bar{m}'\bar{m}-\bar{m}';\bar{m}0}\left(T_{a'l';al_2}^{\bar{m}'\bar{m}-\bar{m}';\bar{m}0}\right)^*, \tag{2.29}$$

$$\sigma_{a';a} = \sum_{\bar{m}\bar{m}'} \frac{1}{(2j+1)}\sigma_{a'\bar{m}';a\bar{m}} = \frac{\pi}{k^2}\sum_{Jll'}\frac{(2J+1)}{(2j+1)} \times |T_{a'l';al}^J|^2\,, \tag{2.30}$$

$$\sigma_{a\bar{m}}^{\mathrm{tot}} = \sum_{a'\bar{m}'} \sigma_{a'\bar{m}';a\bar{m}} = \frac{2\pi}{k^2}\sum_{Jll'} \mathrm{i}^{l-l'}(2J+1) \times \begin{bmatrix} j & J & l \\ -\bar{m} & \bar{m} & 0 \end{bmatrix} \begin{bmatrix} j & J & l' \\ -\bar{m} & \bar{m} & 0 \end{bmatrix} \mathrm{Re}\{T_{al';al}^J\}\,, \tag{2.31}$$

$$\sigma_a^{\mathrm{tot}} = \frac{1}{2j+1}\sum_{\bar{m}}\sigma_{a\bar{m}}^{\mathrm{tot}} = \frac{2\pi}{k^2}\sum_{Jl}\frac{2J+1}{2j+1}\,\mathrm{Re}\{T_{al;al}^J\}\,. \tag{2.32}$$

Sometimes it is convenient to quantize the initial and final states on different axes, namely, in the initial state the qunatization axis is assumed to be directed towards the incident atom (this is opposite to the direction of $\boldsymbol{k}$), and in the final state towards the receding atom (this coincides with the direction of $\boldsymbol{k}'$). Such a choice of quantization axes is convenient for two reasons. First, the projections of j on $-\hat{\boldsymbol{k}}$ (specified by the quantum number $\tilde{\Omega}$) and on $\hat{\boldsymbol{k}}'$ (specified by the quantum number $\tilde{\Omega}'$) are conserved simultaneously along with both the linear momentum and the total angular momentum [2.5]. Second, as will be seen later, the quantum number $\tilde{\Omega}$ specifies the adiabatic electronic term of the diatom AB (Sect. 3.2) in the initial state, and the same is true for $\tilde{\Omega}'$ in the final state. Consequently, often only the transitions involving $\tilde{\Omega}' = \tilde{\Omega}$ occur with appreciable cross sections. In the literature on general scattering theory [2.3–5] the angular momentum is usually quantized on $\hat{\boldsymbol{k}}$ and $\hat{\boldsymbol{k}}'$. The relevant representation of the S matrix, known as the helicity (sometimes P-helicity) representation, is simply connected with that obtained by quantization on $-\hat{\boldsymbol{k}}$ and $\hat{\boldsymbol{k}}'$. We will call the latter the R-helicity representation.

The wave function Ψ corresponding to the quantization directions for the initial and final states described above represents the solution to (2.2) and has the asymptotic form

$$\Psi\,|_{R\to\infty} \to \exp(\mathrm{i}\,\boldsymbol{k}\,\boldsymbol{R})\,|a\,\tilde{\Omega}\rangle_{\mathrm{A}} + \sum_{a'\tilde{\Omega}'} f_{a'\tilde{\Omega}';a\tilde{\Omega}}(\hat{\boldsymbol{k}}',\hat{\boldsymbol{k}})\frac{\exp(\mathrm{i}\,k'R)}{R}|a'\tilde{\Omega}'\rangle_{\mathrm{A}}. \quad (2.33)$$

Here the atomic wave functions $|a\,\tilde{\Omega}\rangle_{\mathrm{A}}$ and $|a'\tilde{\Omega}'\rangle_{\mathrm{A}}$ are related to the functions $|am\rangle_{\mathrm{A}}$ and $|a'\,m'\rangle_{\mathrm{A}}$:

$$|a\,\tilde{\Omega}\rangle_{\mathrm{A}} = \sum_m D^i_{m\tilde{\Omega}}(\alpha+\pi,\pi-\beta,0)\,|a\,m\rangle_{\mathrm{A}}, \quad (2.34)$$

$$|a'\,\tilde{\Omega}'\rangle_{\mathrm{A}} = \sum_{m'} D^{j'}_{m'\tilde{\Omega}'}(\varphi,\theta,0)\,|a'\,m'\rangle_{\mathrm{A}}, \quad (2.35)$$

where $D^j_{m\tilde{\Omega}}$, $D^{j'}_{m'\tilde{\Omega}'}$ are the rotation matrices, α and β are the azimuthal and polar angles of $\boldsymbol{k}$, φ and θ are the azimuthal and polar angles of $\boldsymbol{k}'$. Note that since for $R\to\infty$ the vectors $\hat{\boldsymbol{R}}$ and $\hat{\boldsymbol{k}}'$ coincide, θ and φ can be considered as polar and azimuthal angles of $\boldsymbol{R}$ at finite R and this will be implied everywhere below (see Fig. 2.1 c). The plane wave expansion over states with definite J and M in the R-helicity representation is

$$\exp(\mathrm{i}\,\boldsymbol{k}\,\boldsymbol{R})\,|a\,\tilde{\Omega}\rangle_{\mathrm{A}} = \frac{1}{R}\sum_{JM}\sum_{a'\tilde{\Omega}'}\frac{2J+1}{4\pi}D^{J*}_{M\tilde{\Omega}'}(\varphi,\theta,0)$$
$$\times D^J_{M-\tilde{\Omega}}(\alpha,\beta,0)\,|a'\,\tilde{\Omega}'\rangle_{\mathrm{A}}\,\mathrm{i}^J\,\chi^{0J}_{a'\tilde{\Omega}';a\tilde{\Omega}}, \quad (2.36)$$

where

$$\chi^{0J}_{a'\tilde{\Omega}';a\tilde{\Omega}} = \mathrm{i}^{-J}(-1)^j\sum_{ll'}\mathrm{i}^l\begin{bmatrix} l' & j' & J \\ 0 & \tilde{\Omega}' & \tilde{\Omega}'\end{bmatrix}\left(\frac{2l'+1}{2J+1}\right)^{1/2} 4\pi R\,\mathrm{j}_l(kR)$$
$$\times\left(\frac{2l+1}{2J+1}\right)^{1/2}\begin{bmatrix} l & j & J \\ 0 & -\tilde{\Omega} & -\tilde{\Omega}\end{bmatrix}\delta_{ll'}\,\delta_{aa'}\,|_{R\to\infty} \to 4\pi\frac{\mathrm{i}}{2(k\,k')^{1/2}}$$
$$\times\left\{\exp\left[-\mathrm{i}\left(kR-\frac{J\pi}{2}\right)\right]\delta_{\tilde{\Omega}\tilde{\Omega}'} - \delta_{-\tilde{\Omega}\tilde{\Omega}'}\exp\left[\mathrm{i}\left(kR-\frac{J\pi}{2}\right)\right]\right\}\delta_{aa'} \quad (2.37)$$

and $\mathrm{j}_l(kR)$ is a spherical Bessel function.

Consequently, the expansion of Ψ over states with fixed J and M will be of the form

$$\Psi = 2\pi \mathrm{i} k^{-1/2} \frac{1}{R} \sum_{JM} \sum_{a'\tilde{\Omega}'} \mathrm{i}^J \left(\frac{2J+1}{4\pi}\right) D^{J*}_{M\tilde{\Omega}'}(\varphi,\theta,0)$$
$$\times D^J_{M-\tilde{\Omega}}(\alpha,\beta,0)\, \chi^{+J}_{a'\tilde{\Omega}';a\tilde{\Omega}}(R)\, |a'\tilde{\Omega}'\rangle \,. \qquad (2.38)$$

Here the functions $|a'\tilde{\Omega}'\rangle$ depend on the spatial and spin coordinates of electrons in the rotating (body-fixed) coordinate frame $X'Y'Z'$ obtained from the initial coordinate frame XYZ by rotation through the Euler angles φ, θ, 0. In the general case the $|a'\tilde{\Omega}'\rangle$ depend on R and in particular they may coincide with the molecular functions (see Sect. 2.2.2). These functions are linear combinations of $|a'\tilde{\Omega}'\rangle_{\mathrm{A}}$ with various a'. For $R \to \infty$ the functions $|a'\tilde{\Omega}'\rangle$ will convert to definite functions $|\gamma j'\tilde{\Omega}'\rangle_{\mathrm{A}}$, so that the quantum numbers $\gamma' j'$ completely specify the functions $|a'\tilde{\Omega}'\rangle$. The asymptotic expression of the functions $\chi^{+J}_{a'\tilde{\Omega}';a\tilde{\Omega}}(R)$ has the form

$$\chi^{+J}_{a'\tilde{\Omega}';a\tilde{\Omega}}|_{R\to\infty} = (k')^{-1/2}\left\{\exp\left[-\mathrm{i}\left(kR - \frac{J\pi}{2}\right)\right]\delta_{aa'}\,\delta_{\tilde{\Omega}\tilde{\Omega}'}\right.$$
$$\left. - S^J_{a'\tilde{\Omega}';a\tilde{\Omega}} \exp\left[\mathrm{i}\left(kR - \frac{J\pi}{2}\right)\right]\right\}, \qquad (2.39)$$

where the $S^J_{a'\tilde{\Omega}';a\tilde{\Omega}}$ are the S matrix elements in the R-helicity representation. It follows from (2.37, 10) that $S^J_{a'\tilde{\Omega}';a\tilde{\Omega}}$ are related to $S^J_{a'l';al}$ as

$$S^J_{a'\tilde{\Omega}';a\tilde{\Omega}} = (-1)^j \sum_{ll'} \mathrm{i}^{l-l'} \left(\frac{2l'+1}{2J+1}\right)^{1/2} \begin{bmatrix} l' & j' & J \\ 0 & \tilde{\Omega}' & \tilde{\Omega}' \end{bmatrix}$$
$$\times S^J_{a'l';al} \begin{bmatrix} l & j & J \\ 0 & -\tilde{\Omega} & -\Omega \end{bmatrix} \left(\frac{2l+1}{2J+1}\right)^{1/2} . \qquad (2.40)$$

Note that $S^J_{a'\tilde{\Omega}';a\tilde{\Omega}}$ defined by (2.40) is not the S matrix in the generally accepted sense, since when the interaction becomes zero,

$$S^{J(0)}_{a'\tilde{\Omega}';a\tilde{\Omega}} = (-1)^j \,\delta_{a'a}\,\delta_{\tilde{\Omega}';-\tilde{\Omega}} \,. \qquad (2.41)$$

This comes from the fact that $S^J_{a'\tilde{\Omega}';a\tilde{\Omega}}$ is the S matrix in the rotating coordinate frame where the Coriolis coupling always exists. For zero interaction the effect of Coriolis coupling is equivalent to a change in the quantization direction from $-\boldsymbol{k}$ to $\boldsymbol{k}$. This yields the factor $\delta_{\tilde{\Omega}';-\tilde{\Omega}}$ in (2.41). The phase factor $(-1)^j$ appears when inversion of the quantization direction is achieved by a rotation through the angle π about the initial direction of the axis X'. This choice of the rotation axis follows naturally from the semiclassical picture of scattering (Sect. 6.1).

It is seen from (2.40), using (2.23), that $S^J_{a'\tilde{\Omega}';a\tilde{\Omega}}$ is symmetric:

$$S^J_{a'\tilde{\Omega}';a\tilde{\Omega}} = S^J_{a\tilde{\Omega};a'\tilde{\Omega}'}\,. \tag{2.42}$$

The states with definite J and R-helicity are not eigenstates of $\hat{I}$. Therefore, parity conservation does not impose selection rules on the elements $S^J_{a'\tilde{\Omega}';a\tilde{\Omega}}$. However it does lead to relations between these elements which are readily obtained from (2.40) using the selection rule (2.13):

$$S^J_{a'\tilde{\Omega}';a\tilde{\Omega}} = w\,w'(-1)^{j-j'}\,S^J_{a'-\tilde{\Omega}';a-\tilde{\Omega}}\,. \tag{2.43}$$

In practical close-coupling calculations it is often useful to introduce linear combinations of states with different R-helicities. These linear combinations possess definite parity thus decomposing the set of coupled equations into two uncoupled blocks [2.6]. We shall consider this point in more detail in the next section.

The expression for the scattering amplitude $f_{a'\tilde{\Omega}';a\tilde{\Omega}}(\hat{\boldsymbol{k}}',\hat{\boldsymbol{k}})$ via the elements $S^J_{a'\tilde{\Omega}';a\tilde{\Omega}}$ is obtained by comparing the asymptotic of expressions for Ψ, defined by (2.38, 39), with (2.33) while allowing for (2.37). Taking into account that $\hat{\boldsymbol{k}}' = \hat{\boldsymbol{R}}$, such a comparison yields

$$\begin{aligned} f_{a'\tilde{\Omega}';a\tilde{\Omega}}(\hat{\boldsymbol{k}}',\hat{\boldsymbol{k}}) = {} & \frac{\mathrm{i}}{2\sqrt{k\,k'}}\sum_{JM}(2J+1)\,D^{J*}_{M\tilde{\Omega}'}(\varphi,\theta,0) \\ & \times D^J_{M-\tilde{\Omega}}(\alpha,\beta,0)\,[(-1)^j\,\delta_{aa'}\,\delta_{-\tilde{\Omega}\tilde{\Omega}'} - S^J_{a'\tilde{\Omega}';a\tilde{\Omega}}]\,. \end{aligned} \tag{2.44}$$

Note that due to the possibility of the introduction of the R-helicity quantum number $\tilde{\Omega}$ together with either the linear momentum $\boldsymbol{k}$ or the angular momentum quantum numbers J, M for the system of free atoms [2.4], (2.44) is much simpler than the similar equation (2.25). For instance, the integral scattering cross section $\sigma_{a'a}$ in the R-helicity representation is

$$\begin{aligned} \sigma_{a'a} &= \frac{k'}{k}\int_0^{2\pi} d\varphi \int_0^{\pi}\sin\theta\,d\theta \sum_{\tilde{\Omega}\tilde{\Omega}'}\frac{1}{2j+1}|f_{a'\tilde{\Omega}';a\tilde{\Omega}}|^2 \\ &= \frac{\pi}{k^2}\sum_J(2J+1)\sum_{\tilde{\Omega}\tilde{\Omega}'}\frac{1}{2j+1}|\delta_{aa'}\,\delta_{\tilde{\Omega}';-\tilde{\Omega}}(-1)^j - S^J_{a'\tilde{\Omega}';a\tilde{\Omega}}|^2\,. \end{aligned} \tag{2.45}$$

2.1.2 Scattering Amplitudes and Cross Sections Under Quasi-classical Conditions

Consider now the simplifications that could be attained in (2.28, 29) under those conditions whereby the values l^*, l'^* and J^* of the quantum numbers l, l' and J, essential for the sums, are high.[4] In this case instead of the

[4] It will be assumed hereafter that the values j and j' do not exceed a few units, so that the conditions $l, l', J \gg j', j$ and $l, l', J \gg 1$ are virtually the same.

Clebsch-Gordan coefficients in (2.26) for the matrix elements $T^{m'n';mn}_{a'l';al}$, use can be made of their asymptotic expressions in terms of the d-functions (Appendix A.1) leading to

$$f_{a'\bar{m}';a\bar{m}}(\theta,\varphi) = \mathrm{i}\left(\frac{\pi}{k k'}\right)^{1/2} \sum_{l l' J} \mathrm{i}^{l-l'}(2l+1)^{1/2} \times Y_{l'}^{\bar{m}-\bar{m}'}(\theta,\varphi)\, d^{j'}_{\bar{m}';J-l'}(\beta)\, T^{J}_{a'l';al}\, d^{j}_{\bar{m},J-l}(\beta) \tag{2.46}$$

where $\cos\beta = \bar{m}/J$. Since under our conditions $J^* \gg |\bar{m}|, |J-l|, |J-l'|$, it can be assumed that $\beta = \pi/2$ [5] and terms of the order $1/J^*$ can be neglected, so that ultimately

$$f_{a'\bar{m}';a\bar{m}}(\theta,\varphi) = \mathrm{i}\left(\frac{\pi}{k k'}\right)^{1/2} \sum_{J\nu\nu'} (2J)^{1/2}\, Y^{\bar{m}-\bar{m}}_{J-\nu'}(\theta,\varphi) \times \exp(-\mathrm{i}\pi\nu')\, \Delta^{j}_{\bar{m}'\nu'}\, T^{J}_{a'\nu';a\nu}\, \Delta^{j}_{\bar{m}\nu}, \tag{2.47}$$

where $\nu, \nu' = J-l,\ J-l'$, $\Delta^{j}_{\bar{m}\nu} = d^{j}_{\bar{m}\nu}(\pi/2)$ and

$$T^{J}_{a'\nu';a\nu} = \exp\left[\frac{\mathrm{i}\pi}{2}(3\nu-\nu')\right] T^{J}_{a'l';\alpha l}\Big|_{\substack{l'=J-\nu'\\ l=J-\nu}}. \tag{2.48}$$

The phase factor in (2.48) is introduced to coordinate the notation with that introduced in the next section, where this factor appears quite naturally.

Further simplification of (2.47) is possible under additional assumptions concerning the scattering angles or the form of the matrix $T^{J}_{a'\nu';a\nu}$.

Consider first *scattering through small angles*. Here the spherical functions will be represented via the Bessel functions

$$Y_l^n(\theta,\varphi) = \sqrt{\frac{l}{2\pi}}\exp(\mathrm{i}n\varphi)\,\mathrm{J}_{-n}(l\theta) \tag{2.49}$$

Passing from the summation over J to an integral in (2.47) and neglecting terms of the order of j/J, we obtain

$$f_{a'\bar{m}';a\bar{m}} = \mathrm{i}\left(\frac{\pi}{k k'}\right)^{1/2} \frac{1}{(2\pi)^{1/2}} \exp[\mathrm{i}(\bar{m}-\bar{m}')\varphi] \times \int_0^\infty \mathrm{J}_{\bar{m}'-\bar{m}}(J\theta)\, T^{J}_{a'\bar{m}';a\bar{m}}(2)^{1/2} J\, dJ, \tag{2.50}$$

where

$$T^{J}_{a'\bar{m}';a\bar{m}} = \sum_{\nu\nu'} \exp(-\mathrm{i}\pi\nu')\, \Delta^{j'}_{\bar{m}'\nu'}\, T^{J}_{a'\nu';a\nu}\, \Delta^{j}_{\bar{m}\nu}. \tag{2.51}$$

---

[5] Indeed, the characteristic scale $\delta\beta$ of the angular change of the function $d^{j}_{\lambda\lambda'}(\beta)$ is $1/j$. At the same time the deviation $\Delta\beta$ of the angle β from $\pi/2$ is of the order j/J^*. Consequently the approximation adopted is valid for $\Delta\beta \ll \delta\beta$, i.e., for $J^* \gg j$.

If $f_{a'\bar{m}';a\bar{m}}$ rapidly tends to zero (for instance exponentially) for $\theta > 1/J^*$, the approximation considered can also be used for calculations of the integral scattering cross section. This would require integrating $(k'/k)\,|f_{a'\bar{m}';a\bar{m}}|^2$ over $d\varphi\, d\theta$ with a weighting factor θ. Taking into account the orthogonality of the Bessel functions this yields

$$\sigma_{a'\bar{m}';a\bar{m}} = \frac{\pi}{k^2}\int_0^\infty |T^J_{a'\bar{m}';a\bar{m}}|^2\, 2J\, dJ\,. \tag{2.52}$$

Since (2.52) is valid only when the kinetic energy E is much larger than the interactions in the system, it can be given a simple semiclassical meaning. To this end an impact parameter b is introduced instead of J according to the relation $b = J/k$. Since E is high, $b \simeq J/k'$. Then (2.52) can be rewritten as

$$\sigma_{a'\bar{m}';a\bar{m}} = 2\pi\int_0^\infty |T_{a'\bar{m}';a\bar{m}}(b)|^2\, b\, db\,, \tag{2.53}$$

where the quantity

$$|T_{a'\bar{m}';a\bar{m}}(b)|^2 = |T^J_{a'\bar{m}';a\bar{m}}|^2, \quad b = \frac{J}{k} \simeq \frac{J}{k'} \tag{2.54}$$

can be treated as the probability of the transition $a\bar{m} \to a'\bar{m}'$ induced by the relative motion of the collision partners along a rectilinear trajectory (the main contribution to the cross section occurs for $\theta \ll 1$) defined by the impact parameter b and the initial relative velocity v.

Consider now another limiting case $\theta \gg 1/J^*$ which, by analogy with elastic scattering (see Chap. 5), can be called *scattering through classical angles*. When $J\theta \gg 1$, the spherical functions in (2.47) can be represented asymptotically as

$$\begin{aligned} Y^{\bar{m}-\bar{m}'}_{J-\nu'} = \frac{\exp[\mathrm{i}(\bar{m}-\bar{m}')\varphi]}{2(\sin\theta)^{1/2}} \Big\{ &\exp(-\mathrm{i}\nu'\theta) \\ &\times \exp\left[\mathrm{i}(J+1/2)\theta + \frac{\mathrm{i}(\bar{m}-\bar{m}'-1)\pi}{2}\right] \\ &+ \exp(\mathrm{i}\nu'\theta)\exp\left[-\mathrm{i}(J+1/2)\theta - \frac{\mathrm{i}(\bar{m}-\bar{m}'-1/2)\pi}{2}\right]\Big\}. \end{aligned} \tag{2.55}$$

Then, passing from the summation over J to an integral, and neglecting the terms of the order of j/J, (2.47) can be rewritten as

$$\begin{aligned} f_{a'\bar{m}';a\bar{m}}(\theta,\varphi) = &-\mathrm{i}\,\frac{\exp[\mathrm{i}(\bar{m}-\bar{m}')\varphi]}{2(\pi k k'\sin\theta)^{1/2}} \\ &\times \int_0^\infty (2J)^{1/2}\Big\{\exp\left[\mathrm{i}(J+1/2)\theta + \mathrm{i}(\bar{m}-\bar{m}'-1/2)\frac{\pi}{2}\right] A^J_{a'\bar{m}';a\bar{m}}(\theta) \\ &+ \exp\left[-\mathrm{i}(J+1/2) - \mathrm{i}(\bar{m}-\bar{m}'-1/2)\frac{\pi}{2}\right] A^J_{a'\bar{m}';a\bar{m}}(-\theta)\Big\}\, dJ\,, \end{aligned} \tag{2.56}$$

where

$$A^J_{a'\bar{m}';a\bar{m}}(\theta) = \sum_{\nu\nu'} \exp[-\mathrm{i}(\pi\nu + \theta\nu')]\,\Delta^{j'}_{\bar{m}'\nu'}\,S^J_{a'\nu';a\nu}\,\Delta^j_{\bar{m}\nu}\,. \tag{2.57}$$

It should be noted that the S matrix is used here instead of the T matrix. This is justified since the contribution of the unit matrix [see (2.24)] to the sum (2.47) is proportional to $\delta(1 - \cos\theta)$, and thus the unit matrix does not contribute to the scattering amplitude for $\theta \gg 1/J^*$.

Further simplification of (2.56) takes into account the fact that the main contribution to scattering through classical angles is from those values of J upon which the phase factors contained within the S matrix are strongly dependent. Generally speaking, this dependence can be found only after solving the scattering equations. However, in many cases the S matrix can be represented as

$$S^J_{a'\nu';a\nu} = \sum_{\Lambda} S^{\Lambda}_{a'\nu';a\nu}(J)\exp[2\,\mathrm{i}\,\delta^{\Lambda}_{a'a}(J)]\,, \tag{2.58}$$

where the dependence of $S^{\Lambda}_{a'\nu';a\nu}(J)$ on J is substantially weaker than that of the phase factors which have been singled out explicitly. It is essential for the treatment below that the phases $\delta^{\Lambda}_{a'a}$ be assumed independent of ν and ν'. The S matrix representing in the form (2.58) can always be obtained by retaining a sufficiently large number of terms with different Λ. It should be noted however that this representation naturally arises when the scattering of two atoms obeys some rigorous selection rules over certain electronic quantum numbers (parity, spin, etc.). In this case the coefficients $S^{\Lambda}_{a'\nu';a\nu}$ (or some of these) are determined simply by the rule used to convert the initial free-atom representation to that of the united system.

After substituting (2.58) into (2.56) the integration over J can be carried out by the stationary phase method. All stationary phase points are obtained using the equation

$$\pm\,\theta = \eta^{\Lambda}_{a'a}(J) + 2\pi M\,, \qquad M \text{ is an integer}, \tag{2.59 a}$$

where

$$\eta^{\Lambda}_{a'a}(J) = 2\,\partial\delta^{\Lambda}_{a'a}(J)/\partial J\,. \tag{2.59 b}$$

Now, in the case of one-channel scattering, which is necessarily elastic, (2.59 a) specifies classical trajectories which are characterized by their deflection function $\eta(J)$. This will be discussed in detail in Sect. 5.2. We use the above analogy to name $\eta^{\Lambda}_{a'a}(J)$, defined by (2.59 b), the deflection function for "trajectories" Λ connecting the channels a and a'. Several solutions to (2.59 a) are possible for fixed values of a, a' and Λ (for instance, because of the different signs on the left, or for some other reasons). Let these be denoted by the index χ. Accordingly, the roots of (2.59 a) will be denoted by $J^{\Lambda\chi}_{a'a} = J^{\Lambda\chi}_{a'a}(\theta)$. The "trajectories" Λ, χ connecting the channels a and a' correspond to these roots.

Assume now that the separation of stationary phase points is such that the differences in the values of $\delta^{\Lambda}_{a'a}(J^{\Lambda\chi}_{a'a})$ corresponding to different sets of indices a, a', Λ, χ are large. Then the scattering amplitude will be represented as the sum of contributions[6]

$$f_{a'\bar{m}';a\bar{m}}(\theta,\varphi) = \sum_{\Lambda\chi} f^{\Lambda\chi}_{a'\bar{m}';a\bar{m}}(\theta,\varphi)\,, \tag{2.60}$$

where

$$f^{\Lambda\chi}_{a'\bar{m}';a\bar{m}} = -\,\mathrm{i}\,\frac{\exp\left[\mathrm{i}\,(\bar{m}-\bar{m}')\,\varphi\right]}{(k\,k'\sin\theta)^{1/2}}$$

$$\times\exp\left\{-\,\mathrm{i}\,w^{\Lambda\chi}_{a'a}\left[(J+1/2)\,\theta+\frac{(m-m'-1/2)\,\pi}{2}\right]+\mathrm{i}\,v^{\Lambda\chi}_{a'a}\,\pi/4\right\}$$

$$\times\exp\left[2\,\mathrm{i}\,\delta_{a'a}(J)\right]B^{\Lambda\chi}_{a'\bar{m}';a\bar{m}}(J,\theta)\,(J/|\partial\eta^{\Lambda}_{a'a}/\partial J|)^{1/2}\,|_{J=J^{\Lambda\chi}_{a'a}}\,. \tag{2.61}$$

Here the factor $w^{\Lambda\chi}_{a'a} = \pm 1$ corresponds to the sign in front of θ on the lhs of (2.59), $v^{\Lambda\chi}_{a'a} = \pm 1$ according to the sign of $\partial\eta^{\Lambda}_{a'a}/\partial J\,|_{J=J^{\Lambda\chi}_{a'a}}$ and

$$B^{\Lambda\chi}_{a'\bar{m}';a\bar{m}}(J,\theta) = \sum_{vv'}\exp\left[-\,\mathrm{i}\,(\pi\,v' - w^{\Lambda\chi}_{a'a}\,\theta\,v')\right]$$

$$\times\,\Delta^{j'}_{\bar{m}'v'}\,S^{\Lambda}_{a'v';av}(J)\,\Delta^{j}_{mv}. \tag{2.62}$$

Following the analogy with elastic scattering, we introduce the classical scattering cross section for the trajectory Λ, χ connecting channels a and a':

$${}^{\mathrm{cl}}q^{\Lambda\chi}_{a'a}(\theta) = \frac{1}{k^2\sin\theta}\,J\,|\partial\eta^{\Lambda}_{a'a}(J)/\partial J|^{-1}\,|_{J=J^{\Lambda\chi}_{a'a}}\,. \tag{2.63}$$

In this way every classical "trajectory" Λ, χ connecting channels a and a' defines the partial amplitude $f^{\Lambda\chi}_{a'\bar{m}';a\bar{m}}$ in terms of three factors,

$$f^{\Lambda\chi}_{a'\bar{m}';a\bar{m}}(\theta,\varphi) = \left[\sqrt{k/k'}\,B^{\Lambda\chi}_{a'\bar{m}';a\bar{m}}(J^{\Lambda\chi}_{a'a},\theta)\right]$$

$$\times\exp\left(\mathrm{i}\left\{(\bar{m}-\bar{m})\,\varphi-\frac{\pi}{2}+v^{\Lambda\chi}_{a'a}\frac{\pi}{4}-w^{\Lambda\chi}_{a'a}\left[(J_{a'a}+1/2)\,\theta\right.\right.\right.$$

$$\left.\left.\left.+\,\pi\,(\bar{m}-\bar{m}'-1/2)/2\right]\right\}+2\,\mathrm{i}\,\delta^{\Lambda}_{a'a}(J^{\Lambda\chi}_{a'a})\right)\left[{}^{\mathrm{cl}}q^{\Lambda\chi}_{a'a}(\theta)\right]^{1/2}. \tag{2.64}$$

The first depends on slowly varying parameters of the S matrix, the second rapidly oscillates, and the third is controlled by the cross section for classical scattering specified by the deflection function $\eta^{\Lambda}_{a'a}(J)$.

---

6 Similar expressions for the scattering amplitudes (as sums of conribiutions from individual trajectories) valid at large phase differences have been previously reported [2.7–18]. Here we shall emphasize only the possibility of a unique approach to the description of elastic and inelastic scattering.

In accordance with the generally accepted terminology [2.19–22], the approximation defined by (2.60) can be called a primitive semiclassical approximation. In this approximation, the differential cross section $q_{a'\bar{m}';a\bar{m}}$ is given by the sum over the differential cross sections corresponding to the trajectories Λ, χ, and also by the additional contribution (the interference contribution) given by the sum over the amplitude products corresponding to different Λ, χ and Λ', χ'. Each term of this contribution is proportional to the oscillating factor involving the phase differences $\delta^{\Lambda}_{a'a}(J^{\Lambda\chi}_{a'a}) - \delta^{\Lambda'}_{a'a}(J^{\Lambda'\chi'}_{a'a})$ between various stationary-phase points.

The next simplifying step, which will be referred to as the transition to the classical approximation, consists in neglecting the interference contribution. This is often justified, since the rapidly oscillating terms it contains tend to zero when averaged over small ranges of angles $\Delta\theta$ or energies ΔE. The extent to which this is justified depends on the relations between $\Delta\theta$, ΔE and the rates of change in the phase differences induced by variations in θ and E. In the classical approximation,

$$q_{a'\bar{m}';a\bar{m}} = \sum_{\Lambda\chi} P^{\Lambda\chi}_{a'\bar{m}';a\bar{m}}(\theta)\,{}^{\mathrm{cl}}q^{\Lambda\chi}_{a'a}(\theta)\,, \tag{2.65}$$

where

$$P^{\Lambda\chi}_{a'\bar{m}';a\bar{m}}(\theta) = |B^{\Lambda\chi}_{a'\bar{m}';a\bar{m}}(J^{\Lambda\chi}_{a'a},\theta)|^2\,. \tag{2.66}$$

As is consistent with the natural interpretation of classical scattering, the values $P^{\Lambda\chi}_{a'\bar{m}';a\bar{m}}(\theta)$ can be interpreted as the probabilities of particular processes in classical scattering specified by the deflection function $\eta^{\Lambda}_{a'a}(J)$.

The integral cross section for inelastic scattering ($a'\bar{m}' \neq a\bar{m}$) in the classical approximation can be obtained readily in the following way. Substitute into the integral over angles which leads to (2.29), the expressions (2.47) for $f_{a'\bar{m}';a\bar{m}}(\theta,\varphi)$ in which the summation over J is replaced by an integral, and use the representation (2.58) for the S matrix elements. Then taking into account the orthogonality of the spherical functions and neglecting the rapidly oscillating terms (classical approximation) we have

$$\sigma_{a'\bar{m}';a\bar{m}} = \frac{\pi}{k^2}\int_0^\infty P^{J}_{a'\bar{m}';a\bar{m}}\,2J\,dJ\,,$$

$$P^{J}_{a'\bar{m}';a\bar{m}} = \sum_{\Lambda} P^{\Lambda J}_{a'\bar{m}';a\bar{m}}\,, \tag{2.67}$$

where

$$\begin{aligned} P^{\Lambda J}_{a'\bar{m}';a\bar{m}} &= \sum_{\nu\nu'}\sum_{\bar{\nu}\bar{\nu}'} \exp\{2\mathrm{i}[\delta^{\Lambda}_{a'a}(J) - \delta^{\Lambda}_{a'a}(J-\nu'+\bar{\nu}')]\} \\ &\quad\times \exp(-\mathrm{i}\pi\nu')\,\Delta^{j'}_{\bar{m}'\nu'}\,S^{\Lambda}_{a'\nu';a\nu}(J)\,\Delta^{j}_{\bar{m}\nu}\exp(\mathrm{i}\pi\bar{\nu}') \\ &\quad\times \Delta^{j'}_{\bar{m}'\bar{\nu}'}\,S^{\Lambda*}_{a'\bar{\nu}';a\bar{\nu}}(J-\nu'+\bar{\nu}')\,\Delta^{j}_{\bar{m}\bar{\nu}}\,. \end{aligned} \tag{2.68}$$

Since under quasi-classical conditions the main contribution to the integral (2.67) is from large J, i.e., $J \gg v, v', \bar{v}, \bar{v}'$, (2.68) may be simplified. First, it can be assumed for the slowly varying function $S^{A}_{a'v';av}(J)$ that

$$S^{A*}_{a'\bar{v}';a\bar{v}}(J - v' + \bar{v}') \simeq S^{A*}_{a'\bar{v}';a\bar{v}}(J). \tag{2.69}$$

Second, the phase $\delta^{A}_{a'a}(J - v' + \bar{v}')$ can be expressed as

$$\begin{aligned}\delta^{A}_{a'a}(J - v' + \bar{v}') &= \delta^{A}_{a'a}(J) + (\bar{v}' - v')\,\partial\delta^{A}_{a'a}(J)/\partial J \\ &= \delta^{A}_{a'a}(J) + (\bar{v}' - v')\,\eta^{A}_{a'a}(J)/2.\end{aligned} \tag{2.70}$$

Consequently, the ultimate expression for $P^{AJ}_{a'\bar{m};a\bar{m}}$ will be

$$P^{AJ}_{a'\bar{m}';a\bar{m}} = |\sum_{vv'} \exp[\mathrm{i}\,v'(-\pi + \eta^{A}_{a'a})]\,\Delta^{j'}_{\bar{m}'v'}\,S^{A}_{a'v';av}(J)\,\Delta^{j}_{\bar{m}v}|^2. \tag{2.71}$$

The quantities $P^{AJ}_{a'\bar{m}';a\bar{m}}$ are to be interpreted as classical probabilities for motion along classical "trajectories" A connecting channels a and a' and corresponding to definite J values.

Note that the differential cross section (2.65) is represented as the sum over contributions from trajectories with different J, each corresponding to scattering through the same angle θ, while the total cross section (2.67) is expressed by values for trajectories with the same J, but with different deflection angles $\eta^{A}_{a'a}(J)$. In this sense any inelastic scattering corresponds to the relative motion of particles along several different trajectories. If these trajectories are sufficiently close, so that

$$|\eta^{A}_{a'a}(J) - \eta^{A'}_{a'a}(J)| \ll 1, \tag{2.72}$$

they can be considered to coincide, i.e., $\eta^{A}_{a'a}(J) = \eta^{A'}_{a'a}(J)$ (the common-trajectory approximation, see Sect. 6.1). It should be noted that condition (2.72) does not contradict the condition for large phase differences $|\delta^{A}_{a'a}(J) - \delta^{A'}_{a'a}(J)|$, since, crudely speaking, under quasi-classical conditions the ratio of these two differences is of the order of the small quantity $1/J^*$.

Equations (2.66, 67, 71) enable a simple physical interpretation of the partial elements $S^{A}_{a'v';av}(J)$ to be made. To this end we introduce the coordinate frame $\bar{X}_1\,\bar{Y}_1\,\bar{Z}_1$ with the axis $\bar{Z}_1$ normal to the plane defined by the vectors $\boldsymbol{k}$ and $\boldsymbol{k}'$ and the axis $\bar{X}_1$ directed along $-\boldsymbol{k}$ (Fig. 2.2). This frame is obtained from the frame $\bar{X}\bar{Y}\bar{Z}$ by rotation through the Euler angles $\alpha = \varphi - (\pi/2)$, $\beta = \pi/2$ and $\gamma = 0$. Denoting the quantum numbers of the j projection along $\bar{Z}_1$ by v, v', we get

$$\begin{aligned}f_{a\bar{m}';a\bar{m}}(\theta,\varphi) &= \exp\left[\mathrm{i}(\bar{m} - \bar{m}')\left(\varphi - \frac{\pi}{2}\right)\right] \\ &\quad \times \sum_{vv'} \Delta_{\bar{m}'v'}\,f_{a'v';av}(\theta)\,\Delta_{\bar{m}v},\end{aligned} \tag{2.73}$$

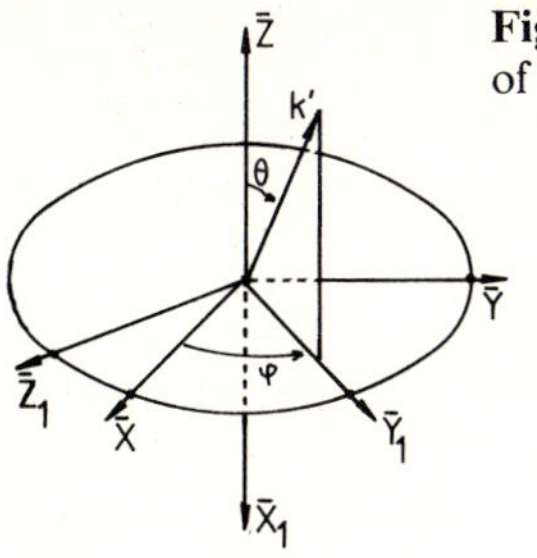

Fig. 2.2. Space-fixed coordinate system $\bar{X}_1\,\bar{Y}_1\,\bar{Z}_1$ used for definition of the scattering amplitude $f_{a'v';av}(\theta)$

where $f_{a'v';av}(\theta)$ is the scattering amplitude in the new coordinate frame. Comparing this expression with (2.62) and (2.64) and taking into account (2.65, 66), it can be readily seen that, except for the phase factor independent of v, v', the quantity

$$\exp[-\mathrm{i}(\pi v' - w^{\Lambda\chi}_{a'a}\,\theta\, v')]\, S^{\Lambda}_{a'v';av}(J^{\Lambda\chi}_{a'a})$$

coincides with the amplitude of the transition probability $a\,v \to a'\,v'$ for motion along the classical trajectory $\Lambda\,\chi$ connecting channels a and a' and corresponding to scattering through an angle θ. In a quite similar way, on the basis of (2.67, 71), the quantity

$$\exp\{-\mathrm{i}[\pi v' - \eta^{\Lambda}_{a'a}(J)\, v']\}\, S^{\Lambda}_{a'v';av}(J)$$

coincides with the amplitude of the transition probability for motion along a trajectory with the deflection function $\eta^{\Lambda}_{a'a}(J)$.

In the quasi-classical limit, when J^*, $l^* \gg 1$, it can be considered that to a good degree of accuracy the motion of the colliding atoms occurs in a fixed plane (collision plane) normal to the vector $\boldsymbol{J}$. This is because the collisional change in l is $\Delta l \sim j \ll l^*, J^*$. Thus, the quantum numbers v, v' can be taken as the quantum numbers of the $\boldsymbol{j}$ projection on the axis normal to the collision plane, and this enables a direct comparison of the amplitudes $S^{\Lambda}_{a'v';av}(J)$ obtained by a quantum calculation with the semiclassical transition amplitudes. Such a comparison is performed in detail in Chapter 6 which deals with the accurate reduction of the quantum to semiclassical scattering equations. This reduction shows that the transition amplitudes $S^{\Lambda}_{a'v';av}$ correspond to semiclassical transition amplitudes in the rotating coordinate frame. Clearly, since $f_{a'v';av}(\theta)$ is defined in the space-fixed frame, the appearance of additional phase factors $\exp[-\mathrm{i}(\pi v' - w^{\Lambda\chi}_{a'a}\,\theta\, v')]$ or $\exp\{-\mathrm{i}[\pi v' - \eta^{\Lambda}_{a'a}(J)\, v']\}$ is due (as is shown explicitly in Sect. 6.1) to the transformation of the semiclassical transition amplitudes from the rotating to the space-fixed frame.

2.2 Scattering Equations

2.2.1 Atomic Basis

To find the S matrix one has to solve the Schrödinger equation (2.2) with the boundary condition (2.3). As a rule, we make use of the expansion of the wave function Ψ^{JM} over some complete basis set. We now consider the collision of an atom A with a structureless particle B and choose as such a basis the set of functions

$$\Psi_{la}^{JM} = \sum_{nm} \begin{bmatrix} l & j & J \\ n & m & M \end{bmatrix} Y_l^n(\hat{\boldsymbol{R}})|a\,m\rangle_{\mathrm{A}}. \tag{2.74}$$

The convenience of this set lies in the fact that the appropriate system of coupled equations directly yields the desired matrix elements $S^J_{a'l';al}$. The basis constructed from functions (2.74) is referred to as the *atomic basis*.

In the present case the Hamiltonian $\mathscr{H}$ in the center-of-mass system is

$$\mathscr{H} = \boldsymbol{p}_{\mathrm{R}}^2/2\mu + H_{\mathrm{A}} + V_{\mathrm{AB}}, \tag{2.75}$$

where $\boldsymbol{p}_{\mathrm{R}}$ is the momentum of relative motion, H_{A} is the electronic Hamiltonian of A, and V_{AB} is the interaction potential between A and B. Since the collisions discussed here are slow, the center of mass of atom A can be taken to coincide with the A nucleus (see Footnote 2 and Sect. 2.3.1). Then V_{AB} will be

$$V_{\mathrm{AB}} = \sum_{i=1}^{N_{\mathrm{A}}} V_{\mathrm{B}}(|\boldsymbol{r}_i - \boldsymbol{R}|) + V_{\mathrm{AB}}^0(R), \tag{2.76}$$

where $\boldsymbol{r}_i$ is the radius vector of the ith electron referred to the A nucleus, N_{A} is the number of electrons in atom A, V_{B} is the interaction potential of the electron with B and V_{AB}^0 is the interaction potential of the A core with B. It will be convenient for further treatment to use the expansion of V_{B} over multipoles,

$$V_{\mathrm{B}}(|\boldsymbol{r} - \boldsymbol{R}|) = \sum_{\varkappa} \frac{4\pi}{2\varkappa + 1} V_{\mathrm{B}\varkappa}(r, R) \sum_q (-1)^q Y_\varkappa^q(\hat{\boldsymbol{R}})\, Y_\varkappa^{-q}(\hat{\boldsymbol{r}}). \tag{2.77}$$

which allows V_{AB} to be represented in the form

$$V_{\mathrm{AB}} = \sum_{\varkappa} \sum_q (-1)^q Y_\varkappa^q(\hat{\boldsymbol{R}})\, W_\varkappa^{-q}(\boldsymbol{r}, \ldots, \boldsymbol{r}_{N_{\mathrm{A}}}, R)\,(4\pi/2\varkappa + 1)^{1/2}, \tag{2.78}$$

where

$$W_\varkappa^{-q} = \sum_{i=1}^{N_{\mathrm{A}}} [V_{\mathrm{B}\varkappa}(r_i, R) + \delta_{\varkappa 0} V_{\mathrm{AB}}^0(R)]\,(4\pi/2\varkappa + 1)^{1/2}\, Y_\varkappa^{-q}(\hat{\boldsymbol{r}}_i) \tag{2.79}$$

is an irreducible tensor operator of rank $\varkappa$ with respect to rotations of the electronic coordinates.

The Hamiltonian $\mathscr{H}$ defined by (2.75, 76) can be taken to be a good approximation for the collision of an atom with an ion with a closed electronic shell, or with an inert gas atom in the ground state at energies lower than the excitation threshold for particles with closed shells. The problem of the calculation of V_B and V_{AB} in these cases will be considered in detail in Sects. 4.3, 4.

Multiplying (2.7) throughout by the functions $\Psi_{l'a'}^{JM*}$, substituting Ψ^{JM} in the form of expansion (2.8), integrating over $\hat{\boldsymbol{R}}$ and the electronic spatial variables $\boldsymbol{r}_i$, and summing over the electron spin variables σ_i, we obtain for the functions $\chi_{a'l';al}^{+J}$,

$$\sum_{a''l''}\left[\left(-\frac{1}{2\mu}\frac{d^2}{dR^2}+\frac{l'(l'+1)}{2\mu R^2}+\varepsilon_{a'}-\mathscr{E}\right)\delta_{l'l''}\delta_{a'a''}\right.$$

$$\left.+\langle l'a'JM|V_{AB}|l''a''JM\rangle\right]\chi_{a''l'';al}^{+J}=0, \tag{2.80}$$

where

$$\langle l'a'JM|V_{AB}|l''a''JM\rangle$$

$$=\sum_{\sigma_1\ldots\sigma_{N_A}}\int\Psi_{l'a'}^{JM*}V_{AB}\Psi_{l''a''}^{JM}d^2\hat{\boldsymbol{R}}\,d^3\boldsymbol{r}_1\ldots d^3\boldsymbol{r}_{N_A}. \tag{2.81}$$

The equations (2.80) must be solved with the boundary conditions (2.10) and

$$\chi_{a'l';al}^{+J}\xrightarrow[R\to 0]{}0. \tag{2.82}$$

To simplify (2.80), the quasi-classical nature of the collisions must be taken into account. Under quasi-classical conditions the values of interest of J, l and l' are large. Just as in Sect. 2.1, we first introduce the values $\nu=J-l$ and $\nu'=J-l'$ instead of l and l'. Since, as follows from (2.74), $|\nu|, |\nu'|\sim j, j'$, the centrifugal potential $l'(l'+1)/2\mu R^2$ can be replaced by $(J+1/2)^2/2\mu R^2-\nu'(J+1/2)/\mu R^2$ with an accuracy up terms of order $1/J$. Making use of (2.78) for V_{AB} and of the known expression for the matrix element of the scalar product of irreducible tensor operators, the matrix elements (2.82) will be

$$\langle l'a'JM|V_{AB}|l''a''JM\rangle|_{\substack{l'=J-\nu'\\ l''=J-\nu''}}=(-1)^{\nu'-\nu''}$$

$$\times\langle a'\nu'J|V_{AB}|a''\nu''J\rangle=\sum_{\varkappa}(-1)^{2J+j'-\nu''}\begin{Bmatrix}\varkappa & j'' & j'\\ J & J-\nu' & J-\nu''\end{Bmatrix}$$

$$\times[2(J-\nu'')+1]^{1/2}\begin{bmatrix}J-\nu'' & \varkappa & J-\nu'\\ 0 & 0 & 0\end{bmatrix}\langle a'\|W_{\varkappa}\|a''\rangle. \tag{2.83}$$

Here $\begin{Bmatrix} \varkappa & j'' & j' \\ J & J-v' & J-v'' \end{Bmatrix}$ is a 6-j symbol and $\langle a' \| W_\varkappa \| a'' \rangle$ is a reduced matrix element of the one-electron tensor operator (2.79). Taking into account that, as follows from the properties of 6-j symbols, only the terms with $\varkappa \leqq j' + j''$ contribute to the sum over $\varkappa$, and using the asymptotic expressions for the Clebsch-Gordan coefficients and the 6-j symbols, we obtain the following expression for the matrix elements $\langle a' v' J | V_{AB} | a'' v'' J \rangle$ in the quasi-classical limit (this expression is no longer dependent on J and consequently the index J will be omitted below)

$$\langle a' v' | V_{AB} | a'' v'' \rangle = \sum_\varkappa \Delta^\varkappa_{v'-v'';0} (2j'+1)^{-1/2} \begin{bmatrix} j'' & \varkappa & j' \\ v'' & v'-v'' & v' \end{bmatrix} \langle a' \| W_\varkappa \| a'' \rangle . \tag{2.84}$$

The radial equations for large J for the wave functions

$$\chi^{+J}_{a'v';av} = \exp\{i\pi[(3v/2) - v']\} \chi^{+J}_{a'l';al} \Big|_{\substack{l'=J-v' \\ l=J-v}}$$

ultimately become

$$\left[-\frac{1}{2\mu} \frac{d^2}{dR^2} + \frac{(J+1/2)^2}{2\mu R^2} + \varepsilon_{a'} - \mathscr{E} \right] \chi^{+J}_{a'v';av} = -\sum_{a''v''} \left(\langle a' v' | V_{AB} | a'' v'' \rangle - \frac{(J+1/2)v''}{\mu R^2} \delta_{v'v''} \delta_{a'a''} \right) \chi^{+J}_{a''v'';av} . \tag{2.85}$$

The functions $\chi^{+J}_{a'v';av}$ here will satisfy the boundary condition for $R \to \infty$:

$$\chi^{+J}_{a'v';av} \to (k')^{-1/2} \left\{ \exp\left[-i\left(kR - J\frac{\pi}{2}\right)\right] \delta_{aa'} \delta_{vv'} - S^J_{a'v';av} \exp\left[i\left(k'R - J\frac{\pi}{2}\right)\right] \right\}, \tag{2.86}$$

where

$$S^J_{a'v';av} = \exp\left[\frac{i\pi}{2}(3v - v')\right] S^J_{a'l';al} \Big|_{\substack{l'=J-v' \\ l=J-v}} . \tag{2.87}$$

The set (2.85) contains the same number of coupled equations as does the initial set (2.80) and therefore it is not simpler from a mathematical point of view. However, it is particularly suitable for further simplifications based on the quasi-classical nature of the collision. Such simplifications will be considered in detail in Chap. 6. Here we shall comment on the general properties of the amplitudes $S^J_{a'v';av}$ and the equations (2.85).

First of all, note that in spite of the nonsymmetric form of (2.87), it is actually symmetric because $\exp(2i\pi v) \equiv \exp(2i\pi v')$. The parity-conservation selection rule (2.13) now takes the form

$$S^J_{a'\nu';a\nu} = 0 \quad \text{if} \quad ww'(-1)^{\nu-\nu'} = -1. \tag{2.88}$$

This selection rule can be obtained directly from (2.84) and (2.85). Indeed,

$$\langle a' \| W_\varkappa \| a'' \rangle = 0 \quad \text{if} \quad w'w''(-1)^\varkappa = -1. \tag{2.89}$$

This means that (2.84) contains only even values of $\varkappa$ for transitions between states with the same parity, and only odd values for transitions between states with different parity. Furthermore, the symmetry properties of the rotation matrices imply that

$$\Delta^\varkappa_{\nu'-\nu'';0} = 0 \quad \text{if} \quad (-1)^{\varkappa+\nu'-\nu''} = -1. \tag{2.90}$$

Taking into account (2.89, 90), the set of coupled equations (2.85) actually decomposes to two noninteracting blocks such that in each block the equality $w'w''(-1)^{\nu'-\nu''} = 1$ is fulfilled. This considerably reduces the number of coupled equations to be solved and leads to the selection rule (2.88) for the S matrix elements.

The second comment to be made deals with the fact that, when the interaction V_{AB} becomes zero,

$$S^{J(0)}_{a'\nu';a\nu} = \exp(i\pi\nu)\, \delta_{a'a}\delta_{\nu'\nu} \tag{2.91}$$

rather than $\delta_{a'a}\delta_{\nu'\nu}$. This is due to the additional term $\delta_{a'a''}\delta_{\nu'\nu''}$ $\nu'(J+1/2)/\mu R^2$ in the effective interaction matrix in (2.85) which does not vanish together with V_{AB}. This additional term, as is shown in Sect. 2.2.2, is nothing other than the Coriolis coupling matrix when the quantization axis is normal to the collision plane. This implies that the quantum S matrix elements $S^J_{a'\nu';a\nu}$ correspond to the semiclassical ones calculated in the rotating coordinate frame. The explicit character of the correspondence is considered in Sect. 6.1.

Although, as stated at the beginning of this section, the atomic basis is the most natural one for treating the problem of atomic collisions, it has certain drawbacks. The most serious of these is that the couplings between channels in this basis are, as a rule, of the same order as the diagonal interactions. Of great importance here is the coupling with closed channels which is, for example, solely responsible for the van der Waals interaction. Furthermore, the matrix elements of interaction in the atomic basis become very complicated in passing over to collisions of two complex atoms, for which the interaction V_{AB} defined by (2.76) cannot be used and the effects of the indistinguishability of electrons belonging to different atoms must be taken into account.

At the same time, however, extensive information is available about the adiabatic potential curves and the electronic wave functions of diatoms with indistinguishability effects and also interactions with excited states properly taken into account. The interactions between various adiabatic electronic states induced by atomic motion is usually small at low atomic velocities. Thus, it can be expected that the above-mentioned drawbacks of the atomic basis will be absent when the adiabatic electronic wave functions of diatoms are used as the basis set.

2.2.2 Molecular Basis

Since the ratio m_{el}/M_{nuc} is small, the motion of electrons follows that of the nuclei in slow atomic collisions. Obviously, in the limit $v \to 0$ transitions between nondegenerate atomic states will not occur. In this limit the scattering problem degenerates to calculating the energy levels $U_{a\tilde{\Omega}}$ and the wave functions $|a\tilde{\Omega}\rangle$ of electrons in the field of nuclei at rest. The relevant Schrödinger equation for the case under consideration of an atom interacting with a structureless particle is

$$H\,|a\tilde{\Omega}\rangle = (H_A + V_{AB})\,|a\tilde{\Omega}\rangle = U_{a\tilde{\Omega}}\,|a\,\tilde{\Omega}\rangle. \tag{2.92}$$

The quantum number $\tilde{\Omega}$ corresponding to the projection of the total electronic angular momentum on the internuclear axis is singled out explicitly. It is a good quantum number for this problem because H is invariant under rotations about this axis. Moreover H is invariant under reflection in any plane containing the internuclear axis. The degeneracy of the energy levels $U_{a\tilde{\Omega}}$ and $U_{a-\tilde{\Omega}}$ is an implication of this invariance. The wave functions $|a\tilde{\Omega}\rangle$ will be referred to as *molecular electronic wave functions* (or simply molecular functions), and the energy levels $U_{a\tilde{\Omega}}$ as *electronic molecular terms* (or simply molecular terms). The definitions *adiabatic electronic wave functions* and *adiabatic electronic terms* are also often used. The symmetry properties and the methods of calculating molecular functions and terms are discussed in Chaps. 3 and 4.

In deriving scattering equations in the molecular basis, we take advantage of the freedom of choice of the basis set and use the wave functions $|a\tilde{\Omega}\rangle_0$ satisfying the equations

$$(H_A + V_{AB}^{(0)})\,|a\tilde{\Omega}\rangle_0 = U_{a\tilde{\Omega}}^{(0)}\,|a\tilde{\Omega}\rangle_0 \tag{2.93}$$

rather than $|a\tilde{\Omega}\rangle$. Here

$$V_{AB}^{(0)} = V_{AB} - V_{AB}^{(1)} \tag{2.94}$$

is chosen in each specific case from the standpoint of convenience. The only requirement is that $V_{AB}^{(0)}$ be axially symmetric and tend to zero when $R \to \infty$. In particular, $V_{AB}^{(0)}$ may be equal to V_{AB} or to zero, which would lead to an exact molecular basis, or to an atomic basis, respectively. Further, to simplify the notation, we shall omit the index 0 in the wave functions of the molecular basis.

In order to write the set of radial wave equations in the molecular basis, one has first to transform the kinetic-energy operator of relative nuclear motion $\boldsymbol{p}_{\mathrm{R}}^2/2\mu$. For a space-fixed coordinate frame

$$\boldsymbol{p}_{\mathrm{R}}^2/2\mu = -\frac{1}{2\mu R^2}\frac{\partial}{\partial R}\left(R^2\frac{\partial}{\partial R}\right) + \frac{1}{2\mu R^2}\boldsymbol{l}^2, \tag{2.95}$$

where the operator of the relative angular momentum squared,

$$\boldsymbol{l}^2 = -\left[\frac{1}{\sin^2\theta}\frac{\partial^2}{\partial\varphi^2} + \frac{1}{\sin\theta}\frac{\partial}{\partial\theta}\left(\sin\theta\frac{\partial}{\partial\theta}\right)\right], \tag{2.96}$$

and θ and φ are the polar and azimuthal angles of the internuclear axis. The operators $\partial/\partial\theta$ and $\partial/\partial\varphi$ specify differentiation at constant $\boldsymbol{r}_i$, σ_i in a space-fixed coordinate frame. Since the molecular wave functions $|a\tilde{\Omega}\rangle$ are defined as functions of the variables $\boldsymbol{r}_i'$, σ_i' in the coordinate frame $X'Y'Z'$ moving together with the nuclei (Fig. 2.1 c), the operators $\partial/\partial\theta$ and $\partial/\partial\varphi$ must be transformed in such a way that differentiation with respect to θ and φ is performed at constant $\boldsymbol{r}_i'$ and σ_i'. Such a transformation (see, e.g., [2.23–27]) yields

$$\boldsymbol{l}^2 = \boldsymbol{J}^2 - (\boldsymbol{J}\cdot\boldsymbol{j} + \boldsymbol{j}\cdot\boldsymbol{J}) + \boldsymbol{j}^2, \tag{2.97}$$

where

$$\boldsymbol{J}^2 = -\left(\frac{\partial'^2}{\partial\theta^2} + \cot\theta\frac{\partial'}{\partial\theta} + \frac{1}{\sin^2\theta}\left[\frac{\partial'^2}{\partial\varphi^2} - 2\mathrm{i}\cos\theta\frac{\partial'}{\partial\varphi}j_{Z'} - j_{Z'}^2\right]\right), \tag{2.98}$$

$$\begin{aligned}\boldsymbol{J}\cdot\boldsymbol{j} + \boldsymbol{j}\cdot\boldsymbol{J} = &\left(\frac{\mathrm{i}}{\sin\theta}\frac{\partial'}{\partial\varphi} + j_{Z'}\cot\theta\right)j_{X'}\\ &+ j_{X'}\left(\frac{\mathrm{i}}{\sin\theta}\frac{\partial'}{\partial\varphi} + j_{Z'}\cot\theta\right) - j_{Y'}2\mathrm{i}\frac{\partial'}{\partial\theta} + 2j_{Z'}^2.\end{aligned} \tag{2.99}$$

In (2.98, 99) $\partial'/\partial\theta$ and $\partial'/\partial\varphi$ are differentiation operators at constant $\boldsymbol{r}_i'$, σ_i', and $j_{X'}, j_{Y'}, j_{Z'}$ are operators of the $\boldsymbol{j}$ projections on the axes X', Y', Z'. The operators $j_{X'}, j_{Y'}$ and $j_{Z'}$ act in the usual way on the functions $|a\tilde{\Omega}\rangle$ which depend on the variables $\boldsymbol{r}_i'$, σ_i' (to avoid misunderstanding it is emphasized that here $\boldsymbol{r}_i'$, just as $\boldsymbol{r}_i$, is referred to the nucleus of atom A). Using the definition of rotation matrices, the following expressions will be readily obtained:

$$\boldsymbol{J}^2 D^{J*}_{M\tilde{\Omega}'}(\varphi,\theta,0)\,|a'\tilde{\Omega}'\rangle = J(J+1)\,D^{J*}_{M\tilde{\Omega}'}(\varphi,\theta,0)\,|a'\tilde{\Omega}'\rangle, \tag{2.100}$$

$$\int_0^{2\pi} d\varphi \int_0^{\pi} \sin\theta \, d\theta \, D^{J'}_{M'\tilde{\Omega}'}(\varphi,\theta,0) \langle a'\tilde{\Omega}'| \boldsymbol{Jj} + \boldsymbol{jJ} |a''\tilde{\Omega}''\rangle$$

$$\times D^{J*}_{M\tilde{\Omega}''}(\varphi,\theta,0) = \frac{4\pi}{2J+1} \delta_{JJ'}\delta_{MM'} 2[\delta_{\tilde{\Omega}''\tilde{\Omega}'}\delta_{a''a'}\tilde{\Omega}'^2$$

$$+ \lambda^-(J) \langle a'\tilde{\Omega}'| j_{X'} |a''\tilde{\Omega}' - 1\rangle \delta_{\tilde{\Omega}'-1;\tilde{\Omega}''} + \lambda^+(J)$$

$$\times \langle a'\tilde{\Omega}'| j_{X'} |a''\tilde{\Omega}' + 1\rangle \delta_{\tilde{\Omega}'+1;\tilde{\Omega}''}], \tag{2.101}$$

where

$$\lambda^{\pm}(J) = [(J \mp \tilde{\Omega}')(J \pm \tilde{\Omega}' + 1)]^{1/2}. \tag{2.102}$$

We now multiply (2.7) throughout by $D^J_{M\tilde{\Omega}'}\langle a'\tilde{\Omega}'|$, substitute Ψ^{JM} in the form of expansion (2.38), integrate over $\hat{\boldsymbol{R}}, \boldsymbol{r}'_i$, and sum over σ'_i. Then taking into account (2.95, 97, 100, 101, 93) we obtain a set of equations for the functions $\chi^{+J}_{a'\tilde{\Omega}';a\tilde{\Omega}}$:

$$\left(-\frac{1}{2\mu}\frac{d^2}{dR^2} + \frac{J(J+1) - \tilde{\Omega}'^2}{2\mu R^2} + U^0_{a'\tilde{\Omega}'} - \mathscr{E}\right)\chi^{+J}_{a'\tilde{\Omega}';a\tilde{\Omega}}$$

$$= -\sum_{a''\tilde{\Omega}''} (\langle a'\tilde{\Omega}'| V_{\mathrm{R}} |a''\tilde{\Omega}''\rangle + \langle a'\tilde{\Omega}'| V_{\mathrm{Cor}} |a''\tilde{\Omega}''\rangle$$

$$+ \langle a'\tilde{\Omega}'| V^{(1)}_{\mathrm{AB}} |a''\tilde{\Omega}''\rangle)\chi^{+J}_{a''\tilde{\Omega}'';a\tilde{\Omega}}. \tag{2.103}$$

Here the interaction

$$\langle a'\tilde{\Omega}'| V_{\mathrm{R}} |a''\tilde{\Omega}''\rangle = \delta_{\tilde{\Omega}'\tilde{\Omega}''}\left(-\frac{1}{\mu}\langle a'\tilde{\Omega}'|\frac{\partial}{\partial R}|a''\tilde{\Omega}''\rangle\frac{d}{dR}\right.$$

$$\left. - \frac{1}{2\mu}\langle a'\tilde{\Omega}'|\frac{\partial^2}{\partial R^2}|a''\tilde{\Omega}'\rangle + \frac{1}{2\mu R^2}\langle a'\tilde{\Omega}'| j^2_{X'} + j^2_{Y'} |a''\tilde{\Omega}'\rangle\right) \tag{2.104}$$

is mostly due to the radial motion of the nuclei and results in transitions only between molecular states with the same $\tilde{\Omega}$. This coupling of the molecular states is usually referred to as radial nonadiabatic coupling. It will be noted that since normalization of the electronic wave function is conserved,

$$\langle a'\tilde{\Omega}'|\frac{\partial}{\partial R}|a'\tilde{\Omega}'\rangle = 0. \tag{2.105}$$

The matrix elements $\langle a'\tilde{\Omega}'| \partial/\partial R |a''\tilde{\Omega}'\rangle$ for $a' \neq a''$ and $\langle a'\tilde{\Omega}'| \partial^2/\partial R^2 |a''\tilde{\Omega}'\rangle$ can be expressed using (2.93) via the matrix elements of the interaction potential $V^{(0)}_{\mathrm{AB}}$ [2.28] as

$$\langle a'\tilde{\Omega}'|\frac{\partial}{\partial R}|a''\tilde{\Omega}'\rangle$$

$$= -\langle a'\tilde{\Omega}'| \partial V^{(0)}_{\mathrm{AB}}/\partial R |a''\tilde{\Omega}'\rangle / (U^{(0)}_{a'\tilde{\Omega}'} - U^{(0)}_{a''\tilde{\Omega}'}), \tag{2.106}$$

$$\langle a'\tilde{\Omega}'|\,\partial^2/\partial R^2\,|a''\tilde{\Omega}'\rangle|_{a''\neq a'}$$
$$= -\langle a'\tilde{\Omega}'|\,\partial^2 V_{AB}^{(0)}/\partial R^2\,|a''\tilde{\Omega}'\rangle/(U_{a'\tilde{\Omega}'}^{(0)} - U_{a''\tilde{\Omega}'}^{(0)})$$
$$+ 2\frac{\langle a'\tilde{\Omega}'|\,\partial V_{AB}^{(0)}/\partial R\,|a''\tilde{\Omega}'\rangle}{(U_{a'\tilde{\Omega}'}^{(0)} - U_{a''\tilde{\Omega}'}^{(0)})^2}\frac{d}{dR}(U_{a'\tilde{\Omega}'}^{(0)} - U_{a''\tilde{\Omega}'}^{(0)})$$
$$+ 2\sum_{\bar{a}\neq a', a''}\frac{\langle a'\tilde{\Omega}'|\,\partial V_{AB}^{(0)}/\partial R\,|\bar{a}\tilde{\Omega}'\rangle\langle\bar{a}\tilde{\Omega}'|\,\partial V_{AB}^{(0)}/\partial R\,|a''\tilde{\Omega}'\rangle}{(U_{a'\tilde{\Omega}'}^{(0)} - U_{a''\tilde{\Omega}'}^{(0)})\,(U_{\bar{a}\tilde{\Omega}'}^{(0)} - U_{a''\tilde{\Omega}'}^{(0)})}, \tag{2.107a}$$

$$\langle a'\tilde{\Omega}'|\,\partial^2/\partial R^2\,|a'\tilde{\Omega}'\rangle$$
$$= \sum_{\bar{a}\neq a'}\frac{\langle a'\tilde{\Omega}'|\,\partial V_{AB}^{(0)}/\partial R\,|\bar{a}\tilde{\Omega}'\rangle\,\langle\bar{a}\tilde{\Omega}'|\,\partial V_{AB}^{(0)}/\partial R\,|a'\tilde{\Omega}'\rangle}{(U_{a'\tilde{\Omega}'}^{(0)} - U_{\bar{a}\tilde{\Omega}'}^{(0)})^2}. \tag{2.107b}$$

The interaction

$$\langle a'\tilde{\Omega}'|\,V_{\mathrm{Cor}}\,|a''\tilde{\Omega}''\rangle = -\frac{1}{\mu R^2}[\lambda^+(J)$$
$$\times\langle a'\tilde{\Omega}'|\,j_{X'}\,|a''\tilde{\Omega}'+1\rangle\,\delta_{\tilde{\Omega}'';\tilde{\Omega}'+1} + \lambda^-(J)$$
$$\times\langle a'\tilde{\Omega}'|\,j_{X'}\,|a''\tilde{\Omega}'-1\rangle\,\delta_{\tilde{\Omega}'';\tilde{\Omega}'-1}] \tag{2.108}$$

is due to the correlation between the motion of the electrons and the rotation of the internuclear axis. This interaction is known as rotational nonadiabatic coupling or Coriolis coupling.

The solution of (2.103) with the boundary conditions (2.39) and

$$\chi_{a'\tilde{\Omega}';a\tilde{\Omega}}^{+J}|_{R\to 0}\to 0 \tag{2.109}$$

enables one to obtain a set of elements of the S matrix in the R-helicity representation.

As opposed to the scattering equations in the atomic basis, accounting for the condition $J \gg 1$ virtually does not modify (2.103). It only amounts to substituting $(J+1/2)^2$ for $[J(J+1)+\tilde{\Omega}'^2]$ and $J+1/2$ for $\lambda^\pm(J)$. Essential simplifications of the radial coupling can be obtained only in the regions where the radial motion is quasi-classical. In these regions

$$\lambda\!\!\!^{-}\alpha \ll 1, \tag{2.110}$$

where $\lambda\!\!\!^{-}$ is the wavelength of the radial motion and $1/\alpha$ is the distance at which the interaction between the atoms changes substantially. Taking into account (2.106, 107), it is possible to obtain for $a''\neq a'$ the following crude estimates:

$$\left|\frac{1}{\mu}\langle a'\tilde{\Omega}'|\,\partial/\partial R\,|a''\tilde{\Omega}'\rangle\frac{\partial}{\partial R}\chi_{a''\tilde{\Omega}';a\tilde{\Omega}}^{+J}\right|$$
$$\simeq\frac{\alpha}{\mu\lambda\!\!\!^{-}\,|U_{a'\tilde{\Omega}'}^{(0)} - U_{a''\tilde{\Omega}'}^{(0)}|}|\langle a'\tilde{\Omega}'|\,V_{AB}^{(0)}\,|a''\tilde{\Omega}'\rangle\,\chi_{a''\tilde{\Omega}';a\tilde{\Omega}}^{+J}|, \tag{2.111}$$

$$\left|\frac{1}{2\mu}\langle a'\tilde{\Omega}'|\,\partial^2/\partial R^2\,|a''\tilde{\Omega}'\rangle\right| \simeq (\alpha\lambda\!\!\!^{-})\frac{\alpha}{|U^{(0)}_{a''\tilde{\Omega}'}-U^{(0)}_{a''\tilde{\Omega}'}|}(\mu\lambda\!\!\!^{-})^{-1} \times |\langle a'\tilde{\Omega}'|\,V^{(0)}_{\mathrm{AB}}\,|a''\tilde{\Omega}'\rangle|, \tag{2.112}$$

$$\left|\frac{1}{2\mu R^2}\langle a'\tilde{\Omega}'|\,j^2_{X'}+j^2_{Y'}\,|a''\tilde{\Omega}'\rangle\right| \simeq (\alpha\lambda\!\!\!^{-})\frac{\alpha}{\mu\lambda\!\!\!^{-}\,|U^{(0)}_{a'\tilde{\Omega}'}-U^{(0)}_{a''\tilde{\Omega}'}|}\,\frac{|U^{(0)}_{a'\tilde{\Omega}'}-U^{(0)}_{a''\tilde{\Omega}'}|}{(\alpha R)^2}. \tag{2.113}$$

It follows from these estimates that the two last terms on the rhs of (2.104) are small compared with the first term, if the inequality (2.110) is satisfied. The diagonal matrix elements $\langle a'\tilde{\Omega}'|\,V_{\mathrm{R}}\,|a'\tilde{\Omega}'\rangle$ proportional to the small value $1/\mu$, can always be neglected compared to $U^{(0)}_{a'\tilde{\Omega}'}+\langle a'\tilde{\Omega}'|\,V^{(1)}_{\mathrm{AB}}\,|a'\tilde{\Omega}'\rangle$. Thus, in the regions where radial motion is quasi-classical, (2.103) reduce to

$$\left(-\frac{1}{2\mu}\frac{d^2}{dR^2}+\frac{(J+1/2)^2}{2\mu R^2}+U^{(0)}_{a'\tilde{\Omega}'}-\mathscr{E}\right)\chi^{+J}_{a'\tilde{\Omega}';a\tilde{\Omega}}$$
$$=-\sum_{a''\tilde{\Omega}''}\Bigg[\delta_{\tilde{\Omega}'\tilde{\Omega}''}\left(\langle a'\tilde{\Omega}'|\,V^{(1)}_{\mathrm{AB}}\,|a''\tilde{\Omega}'\rangle-\frac{1}{\mu}\langle a'\tilde{\Omega}'|\,\partial/\partial R\,|a''\tilde{\Omega}'\rangle\frac{d}{dR}\right)$$
$$-\frac{J+1/2}{\mu R^2}\big(\delta_{\tilde{\Omega}'';\tilde{\Omega}'+1}\langle a'\tilde{\Omega}'|\,j_{X'}\,|a''\tilde{\Omega}'+1\rangle$$
$$+\,\delta_{\tilde{\Omega}'';\tilde{\Omega}'-1}\langle a'\tilde{\Omega}'|\,j_{X'}\,|a''\tilde{\Omega}'-1\rangle\big)\Bigg]\chi^{+J}_{a''\tilde{\Omega}'';a\tilde{\Omega}}. \tag{2.114}$$

Since we have used the inequality (2.110) to simplify the equations, there is, strictly speaking, little sense in writing the second-order equations: in this limit they must be reduced to first-order ones. This will be done in Sect. 6.1. Here the second derivative is retained in order to consider the connection with the scattering equations in the atomic basis. This connection is expected to be very close when $V^{(1)}_{\mathrm{AB}}=V_{\mathrm{AB}}$, $V^{(0)}_{\mathrm{AB}}=0$. But first consider the exact molecular basis corresponding to $V^{(1)}_{\mathrm{AB}}=0$, $V^{(0)}_{\mathrm{AB}}=V_{\mathrm{AB}}$.

In this case, inelastic transitions are induced solely by nonadiabatic coupling. If the radial motion is quasi-classical, it follows from estimate (2.111) that the radial nonadiabatic coupling is of the order $\langle a'\tilde{\Omega}'|\,V_{\mathrm{AB}}\,|a''\tilde{\Omega}'\rangle/\xi_{a'\tilde{\Omega}';a''\tilde{\Omega}'}$, where

$$\xi_{a'\tilde{\Omega}';a''\tilde{\Omega}'}=\mu\lambda\!\!\!^{-}\,|U_{a'\tilde{\Omega}'}-U_{a''\tilde{\Omega}'}|/\alpha=|U_{a'\tilde{\Omega}'}-U_{a''\tilde{\Omega}'}|/\alpha v \tag{2.115}$$

is the *Massey parameter* which characterizes the adiabaticity of the given inelastic transitions. In most cases of slow atomic collisions $\xi_{a'\tilde{\Omega}';a''\tilde{\Omega}'}\gg 1$ for many $a''\neq a'$. The above estimate shows that the radial nonadiabatic coupling is small compared to

$$\mathscr{E} - \frac{(J + 1/2)^2}{2\mu R^2} - U_{a'\tilde{\Omega}'} = \mathscr{K}^2_{a'\tilde{\Omega}'}/2\mu. \tag{2.116}$$

Note that in the quasi-classical regions (far from the turning points) $\mathscr{K}^2_{a'\tilde{\Omega}'}/2\mu \simeq \mu v^2/2 = E$. The Massey parameter can become small for two reasons, either due to the low value of $U_{a'\tilde{\Omega}'} - U_{a''\tilde{\Omega}'}$ for some pairs of terms, or due to the high relative velocity. In the first case, as follows from the noncrossing rule (Sect. 3.3), $\langle a'\tilde{\Omega}'|\, V_{\mathrm{AB}}\, |a''\tilde{\Omega}'\rangle \simeq |U_{a'\tilde{\Omega}'} - U_{a''\tilde{\Omega}'}|$ and the radial nonadiabatic coupling for not very low value of E is usually still lower than $\mathscr{K}^2_{a'\tilde{\Omega}'}/2\mu$. In the second case, at very high energies only, the Massey parameter drops off as $1/v$, $\mathscr{K}^2_{a'\tilde{\Omega}'}/2\mu$ increases as v^2 and the radial nonadiabatic coupling is again smaller than $\mathscr{K}^2_{a'\tilde{\Omega}'}/2\mu$. The Coriolis coupling is equal to within an order of magnitude to $(E/\mu)^{1/2}(R^*/R^2)$, where R^* is the characteristic atomic radius, and obviously is always smaller than $\mathscr{K}^2_{a'\tilde{\Omega}'}/2\mu$. The smallness of the nonadiabatic coupling compared to $\mathscr{K}^2_{a'\tilde{\Omega}'}/2\mu$, ultimately due to the small ratio $m_{\mathrm{el}}/M_{\mathrm{nuc}}$, actually makes the molecular basis particularly convenient in treating the scattering of slow atoms, since it enables the second-order scattering equations to be reduced to first-order equations (Sect. 6.1).

The set (2.114), as well as (2.103), has the disadvantage [compared to the sets (2.85) or (2.80)] that parity conservation does not decompose these equations in two noninteracting blocks. However, such a decomposition can be readily achieved by taking into account the fact that the invariance of the nonadiabatic interaction and of $V^{(1)}_{\mathrm{AB}}$ under reflection in the $Y'Z'$ plane leads to the relation

$$\langle a'\tilde{\Omega}'|\, V\, |a''\tilde{\Omega}'\rangle = w'w''(-1)^{j''-j'}\langle a' - \tilde{\Omega}'|\, V\, |a'' - \tilde{\Omega}''\rangle, \tag{2.117}$$

where V is the total coupling operator on the rhs of (2.103) or (2.114). In the rotating coordinate frame, reflection in the plane containing the molecular axis is known to play the role of inversion (see, e.g., [2.24]). The phase factor on the rhs of (2.117) is obtained using the correlation

$$|a\tilde{\Omega}\rangle \xrightarrow[R\to\infty]{} |a\tilde{\Omega}\rangle_{\mathrm{A}}$$

and the reflection operator of the electronic coordinates in the $Y'Z'$ plane,

$$\hat{\sigma}_{Y'Z'} = \hat{I}_{\mathrm{el}}\hat{C}_{\pi X'}, \tag{2.118}$$

where $\hat{I}_{\mathrm{el}}$ is the inversion operator of the electronic coordinates and $\hat{C}_{\pi X'}$ is the rotation operator about the X' axis through the angle π.

We introduce the functions

$$\chi^{+J}_{a'\Omega'p';a\Omega p} = \chi^{+J}_{a'\Omega';a\Omega} + (-1)^p \chi^{+J}_{a'\Omega';a-\Omega} + (-1)^{p'} \chi^{+J}_{a'-\Omega;a\Omega} + (-1)^{p'+p} \chi^{+J}_{a'-\Omega';a-\Omega}, \tag{2.119}$$

where $\Omega, \Omega' = |\tilde{\Omega}|, |\tilde{\Omega}'|$ and p, p' take the values 0 or 1. It can be readily shown that these functions satisfy the equations

$$\left[-\frac{1}{2\mu}\frac{d^2}{dR^2} + \frac{J(J+1)-\Omega'^2}{2\mu R^2} + U^{(0)}_{a'\Omega'}\right] \chi^{+J}_{a'\Omega'p';a\Omega p} = -\sum_{a''\Omega''p''} \langle a'\tilde{\Omega}'p'|\, V\, |a''\Omega''p''\rangle\, \chi^{+J}_{a''\Omega''p'';a\Omega p} \tag{2.120}$$

with the boundary conditions following from (2.39, 109):

$$\chi^{+J}_{a'\Omega'p';a\Omega p} \xrightarrow[R\to\infty]{} (k')^{-1/2} \left\{ 2\delta_{a'a}\delta_{\Omega'\Omega}\delta_{p'p}(1+\delta_{\Omega 0})\exp\left[-\mathrm{i}\left(kR - \frac{J\pi}{2}\right)\right] - S^J_{a'\Omega'p';a\Omega p} \exp\left[\mathrm{i}\left(k'R - \frac{J\pi}{2}\right)\right]\right\}, \tag{2.121}$$

$$\chi^{+J}_{a'\Omega'p';a\Omega p} \xrightarrow[R\to 0]{} 0. \tag{2.122}$$

Here

$$\langle a'\Omega'p'|\, V\, |a''\Omega''p''\rangle = \tfrac{1}{2}[\langle a'\Omega'|\, V\, |a''\Omega''\rangle + (-1)^{p'}\langle a'-\Omega'|\, V\, |a''\Omega''\rangle + (-1)^{p''}\langle a'\Omega'|\, V\, |a''-\Omega''\rangle + (-1)^{p'+p''}\langle a'-\Omega'|\, V\, |a''-\Omega''\rangle] \tag{2.123}$$

and

$$S^J_{a'\Omega'p';a\Omega p} = S^J_{a'\Omega';a\Omega} + (-1)^p S^J_{a'\Omega';a-\Omega} + (-1)^{p'} S^J_{a'-\Omega';a\Omega} + (-1)^{p'+p} S^J_{a'-\Omega';a-\Omega}. \tag{2.124}$$

It is seen by taking into account (2.43, 117) that

$$S^J_{a'\Omega'p';a\Omega p} = 0 \quad \text{if} \quad ww'(-1)^{p-p'}(-1)^{j-j'} = -1 \tag{2.125}$$

and that the set of (2.120) consists of two noninteracting blocks such that the equality $w'w''(-1)^{p''-p'}(-1)^{j''-j'} = 1$ is fulfilled in each block.

Consider now the connection between (2.85) and (2.114). First of all note that, as follows from (2.40) with the use of the asymptotic expressions for the Clebsch-Gordan coefficients in the limit of large J, and the relation (2.87),

$$S^J_{a'v';av} = \sum_{\tilde{\Omega}\tilde{\Omega}'} D^{j'*}_{\tilde{\Omega}'v'}(0, \pi/2, \pi)\, S^J_{a'\tilde{\Omega}';a\tilde{\Omega}} D^j_{\tilde{\Omega}v}(0, \pi/2, \pi). \tag{2.126}$$

On the other hand, if we take $V_{AB}^{(0)} = 0$, $V_{AB}^{(1)} = V_{AB}$, then $|a\tilde{\Omega}\rangle = |a\tilde{\Omega}\rangle_A$, the radial nonadiabatic coupling vanishes identically and (2.114) transforms to

$$\begin{aligned}\left(-\frac{1}{2\mu}\frac{d^2}{dR^2} + \frac{(J+1/2)}{2\mu R^2} + \varepsilon_{a'} - \mathscr{E}\right) \chi^{+J}_{a'\tilde{\Omega}';a\tilde{\Omega}} \\ = -\sum_{a''\tilde{\Omega}''} \Bigg(\langle a'\tilde{\Omega}'|\, V_{AB}\, |a''\tilde{\Omega}'\rangle\, \delta_{\tilde{\Omega}'\tilde{\Omega}''} - \frac{J+1/2}{\mu R^2}\, \delta_{a'a''} \\ \times \{\delta_{\tilde{\Omega}'';\tilde{\Omega}'+1} [(j' + \tilde{\Omega}' + 1)(j' - \tilde{\Omega}')]^{1/2} \\ + \delta_{\tilde{\Omega}'';\tilde{\Omega}'-1} [(j' + \tilde{\Omega}')(j' - \tilde{\Omega}' + 1)]^{1/2}\}\Bigg) \chi^{+J}_{a''\tilde{\Omega}'';a\tilde{\Omega}},\end{aligned} \tag{2.127}$$

where

$$\langle a'\tilde{\Omega}'|\, V_{AB}\, |a''\tilde{\Omega}'\rangle = \sum_{\varkappa} (2j' + 1)^{-1/2} \begin{bmatrix} j & \varkappa & j' \\ \tilde{\Omega}' & 0 & \tilde{\Omega}' \end{bmatrix} \langle a' \|\, W_\varkappa \,\| a''\rangle. \tag{2.128}$$

Equation (2.128) is obtained using (2.78) since, in the rotating coordinate frame $X'Y'Z'$, the polar angle of $\boldsymbol{R}$ is zero.

Direct calculation confirms that the total-interaction matrices $\langle a'v'|\, V\, |a''v''\rangle$ and $\langle a'\tilde{\Omega}'|\, V\, |a''\tilde{\Omega}''\rangle$ appearing in (2.85) and (2.114), respectively, are connected with each other in the same manner as are the S matrix elements, namely

$$\begin{aligned}\langle a'v'|\, V\, |a''v''\rangle = \sum_{\tilde{\Omega}'\tilde{\Omega}''} D^{j'*}_{\tilde{\Omega}'v'}(0, \pi/2, \pi)\, \langle a'\tilde{\Omega}'|\, V\, |a''\tilde{\Omega}''\rangle \\ \times D^{j''}_{\tilde{\Omega}''v''}(0, \pi/2, \pi).\end{aligned} \tag{2.129}$$

This may be interpreted in the following way. Let the rotating coordinate frame $X'Y'Z'$ be such that the X' axis is normal to the collision plane defined by the vectors $\hat{\boldsymbol{k}}$ and $\hat{\boldsymbol{k}}'$ (Fig. 2.3). By definition, the quantum numbers $\tilde{\Omega}$ specify the projection of $\boldsymbol{j}$ on the Z' axis of the rotating coordinate frame. Then, as follows from (2.126, 129), the quantum numbers v specify the projection of $\boldsymbol{j}$ on the Z_1' axis of the rotating coordinate frame $X_1'Y_1'Z_1'$ obtained from the frame $X'Y'Z'$ by rotation through the Euler

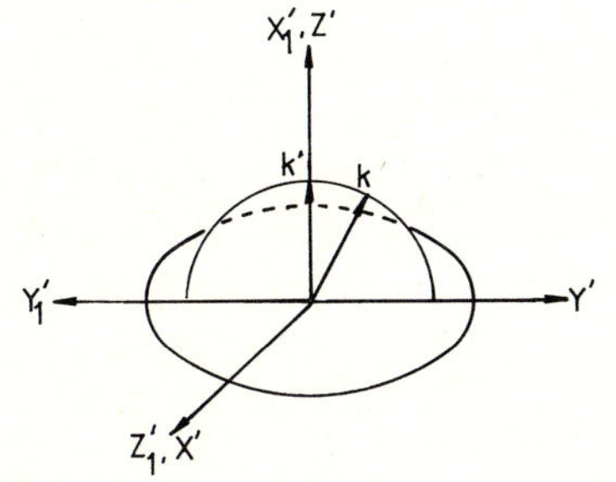

Fig. 2.3. Body-fixed coordinate system X_1', Y_1', Z_1'

angles $\alpha = 0$, $\beta = \pi/2$, $\gamma = \pi$. From the semiclassical point of view this means, as mentioned earlier, that the matrix elements $S^J_{a'v';av}$ describe scattering in the rotating coordinate frame with the quantization axis normal to the collision plane, along the conserved vector $\boldsymbol{J}$.

2.3 Collisions of Two Many-Electron Atoms

2.3.1 Scattering Matrix and Scattering Equations

A direct generalization of the above model of the collision of an atom with a structureless particle is provided by a model of inelastic collision of two many-electron atoms, resulting in transitions at large internuclear distances (e.g. transitions between fine-structure components [2.29–31]). This model assumes that the electrons belonging to atom A (coordinates referred to nucleus A denoted by $\boldsymbol{r}_1^{\mathrm{A}}, \ldots, \boldsymbol{r}_{N_{\mathrm{A}}}^{\mathrm{A}}$) and to atom B (coordinates referred to nucleus B denoted by $\boldsymbol{r}_1^{\mathrm{B}}, \ldots, \boldsymbol{r}_{N_{\mathrm{B}}}^{\mathrm{B}}$) are distinguishable. Such an assumption is justified if the main contribution to the cross section of the process being considered is due to large impact parameters, permitting one to neglect the exchange of electrons between atoms.

If the additional momenta of the electrons traveling with the nucleus (Sect. 2.1) are neglected, the wave function of this model system in the asymptotic region at $R \to \infty$ can be represented in a form completely similar to (2.3):

$$\Psi_{R\to\infty} \to \exp(\mathrm{i}\boldsymbol{k}\boldsymbol{R})\,|cjm\rangle + \sum_{c'j'm'} f_{c'j'm';cjm}(\hat{\boldsymbol{k}}', \hat{\boldsymbol{k}}) \times R^{-1}\exp(\mathrm{i}k'R)\,|c'j'm'\rangle. \tag{2.130}$$

Here the functions $|cjm\rangle$ are expressed via the functions $|a_1m_1\rangle_{\mathrm{A}}$ and $|a_2m_2\rangle_{\mathrm{B}}$ of the free atoms A and B (electronic coordinates referred to nuclei A and B, respectively) as follows:

$$|cjm\rangle = \sum_{m_1m_2}\begin{bmatrix} j_1 & j_2 & j \\ m_1 & m_2 & m \end{bmatrix} |a_1m_1\rangle_{\mathrm{A}}\,|a_2m_2\rangle_{\mathrm{B}}. \tag{2.131}$$

Accordingly, c stands for the set of quantum numbers a_1, a_2. When using the R-helicity representation, the expansion of the total wave function Ψ over functions with fixed J and M is similar to (2.38), but the functions $|a'\tilde{\Omega}'\rangle$ are replaced by molecular wave functions $|a_1' a_2' j'\tilde{\Omega}'\rangle$. The latter are eigenfunctions of the model electronic Hamiltonian for any fixed value of R, and when $R \to \infty$ they convert to linear combinations of the products of atomic wave functions having the form (2.131).

The expressions derived for the S matrix elements and scattering amplitudes in the preceding sections are obviously valid for the model discussed, provided the quantum numbers a are replaced by the quantum numbers c. The scattering equations in the atomic and molecular basis sets are of the same form as those in Sect. 2.2. In the case of the molecular basis,

$$\boldsymbol{j} = \boldsymbol{j}_{\mathrm{A}} + \boldsymbol{j}_{\mathrm{B}}. \tag{2.132}$$

In $\boldsymbol{j}_{\mathrm{A}}$ the electronic coordinates are referred to nucleus A, in $\boldsymbol{j}_{\mathrm{B}}$, to nucleus B, and the differentiation of the molecular wave functions with respect to R is performed at constant $\boldsymbol{r}_i^{\mathrm{A}}$, $\boldsymbol{r}_i^{\mathrm{B}}$.

Certain difficulties arise in the attempt to account for the indistinguishability of electrons in terms of the above scheme. This can be seen, for instance, from the following. When the scattering problem is treated in the atomic basis, a natural generalization of the expansion of the total wave function over the states (2.8) with a definite angular momentum will be an expansion of the form

$$\begin{aligned} \Psi = 2\pi \mathrm{i} k^{-1/2} R^{-1} \sum_{JM} \sum_{c'j'm'} \sum_{l'l} \langle \hat{\boldsymbol{R}}, l'j'm' | \, Y \, | \hat{\boldsymbol{k}}, ljm \rangle \\ \times \mathrm{i}^l \chi^{+J}_{c'j'l';cjl} \hat{A} \, | c'j'm' \rangle, \end{aligned} \tag{2.133}$$

where $\hat{A}$ is the antisymmetrization operator over the electron coordinates and is defined in Sect. 3.2 by (3.23). In this case, the matrix elements $\langle \hat{A} \langle c'j'm' || \, V_{\mathrm{R}} \, | \hat{A} \, | cjm \rangle \rangle$ and $\langle \hat{A} \langle c'j'm' || \, V_{\mathrm{R}}^2 \, | \hat{A} \, | cjm \rangle \rangle$ will enter into the set of equations for the radial wave functions $\chi^{+J}_{c'j'l';cjl}$.

In the above model with distinguishable electrons on different atoms, the matrix elements of this type tend to zero since differentiation is performed at constant $\boldsymbol{r}_i^{\mathrm{A}}$ and $\boldsymbol{r}_i^{\mathrm{B}}$. However, when the electrons belonging to different atoms are indistinguishable, the condition that the electron coordinates be constant must be symmetric over all electrons. Consequently, all electron coordinates must be referred to the same origin. It will be seen that, for any choice of this point, the matrix elements of the differentiation operators either decrease as $1/R$ when $R \to \infty$, or tend to constant values. In the molecular basis, the same spurious behavior is exhibited by the nonadiabatic-coupling matrix elements. The problem of the incorrect behavior (in terms of general scattering theory) of the kinematic-coupling matrix elements, when the expansion involves atomic wave functions moving together with the nuclei, or molecular wave functions, has been intensively studied.

These studies were devoted mostly to charge exchange processes, where the role of symmetry with respect to electron permutations is played by the transfer of the electron from one nucleus to another. It is particularly clear in this case that the difficulties in using expansions in molecular or atomic wave functions are caused by not taking into account the momentum transfer which accompanies electron transfer.

A straightforward way of removing this difficulty is to modify the basis set in such a way that the spurious asymptotic couplings disappear. For the problem of charge exchange, this can be readily achieved by multiplying the atomic or molecular wave functions by so-called translational factors, which take into account the electron momentum associated with the motion of the nuclei [2.32–40]. Unfortunately, the requirement of the correct asymptotic behavior of the coupling matrix elements is insufficient and still leaves much freedom in the choice of the translational factors. Certainly, if the basis set used is complete, all possible different forms of the translational factors will yield the same S matrix. But this is a formal argument only, since all practical calculations use truncated (and often very limited) basis sets. The problem of the optimal choice of the translational factors was extensively discussed in the literature [2.41–47]. The general implication of these investigations is that there exists no general solution to this problem and specific physical arguments have to be used in each particular case.

Another important point pertaining to the translational-factor method is that this approach cannot be generalized to account for the indistinguishability of the electrons in collisions of two many-electron atoms [2.39]. An alternative approach (we shall call it the scattering-coordinate method) to the electron momentum transfer problem has been developed [2.48–52]. It can be used both for charge-exchange processes and for the collisions of many-electron atoms. In the case of charge exchange, this approach is equivalent to the translational-factor method with a certain (but not unique) choice of the translational factors [2.52]. The scattering-coordinate method uses a function which is only in part determined by the asymptotic conditions and therefore, from the standpoint of uniqueness, this method is not superior to the translational-factor method. However, it provides scattering equations similar to the familiar equations in the preceding section. These equations are especially suitable for the problems discussed in the present book since they permit one to decide, for the slow quasi-classical collisions, whether or not the electron momentum transfer is important for the process in question. Consider, following [2.50], the scattering-coordinate method for the case when no charge exchange or ionization occurs.

The above discussion shows that, from a formal standpoint, the main difficulty in accounting for the indistinguishability of electrons lies with the definition of the scattering coordinate which is symmetric in all electrons and specifies the relative position of the colliding atoms. When the ionization and charge-exchange channels can be neglected, such a scattering coordinate ϱ is defined by the nonlinear implicit equation

$$\varrho = \frac{M_A}{M_A + N_A} \boldsymbol{R}_A - \frac{M_B}{M_B + N_B} \boldsymbol{R}_B + \frac{1}{2\bar{\mu}} \sum_{i=1}^{N_A+N_B} \bar{r}_i [\bar{f}(\bar{r}_i, \varrho) - \bar{\lambda}]. \tag{2.134}$$

Here M_A and M_B are the masses of nuclei A and B, $\boldsymbol{R}_A$ and $\boldsymbol{R}_B$ are the radius vectors of nuclei A and B, $\bar{r}_i$ is the radius vector of the ith electron (all radius vectors are referred to the center of mass of the nuclei and electrons),

$$\bar{\mu} = [(M_A + N_A)(M_B + N_B)]/(M_A + M_B + N_A + N_B), \tag{2.135}$$

$$\bar{\lambda} = (M_A + N_A - M_B - N_B)/(M_A + M_B + N_A + N_B) \tag{2.136}$$

and, in other respects, the arbitrary function $\bar{f}(\bar{r}_i, \varrho)$ satisfies the boundary conditions

$$\lim_{\varrho \to \infty} \bar{f}(\bar{r}_i, \varrho) = \begin{cases} 1 & \text{if } \bar{r}_i \to \varrho\bar{\mu}/(M_A + N_A) \quad \text{(electron close to nucleus A)} \\ -1 & \text{if } \bar{r}_i \to -\varrho\bar{\mu}/(M_B + N_B) \quad \text{(electron close to nucleus B)} \end{cases}. \tag{2.137}$$

After the collision, the regions of configuration space corresponding to various distributions of the electrons among the atoms can be considered to be nonoverlapping. It can be shown, using the boundary conditions (2.137) and taking into account the fact that electrons in the bound state are localized close to the atoms, that in any such region ϱ equals the distance between the centers of mass of the free atoms. Consequently, there is every reason to choose ϱ as the vector describing the relative position of the atoms in the process of scattering.

As is seen from (2.134), ϱ differs from $\boldsymbol{R} = \boldsymbol{R}_A - \boldsymbol{R}_B$ by a value of the order of m_{el}/M_{nuc}. Therefore the total Hamiltonian $\mathscr{H}$ in the variables ϱ, $\bar{r}_i$ can be represented with an accuracy up to terms linear in m_{el}/M_{nuc}, in the form

$$\mathscr{H} \simeq \mathscr{H}^{(0)} - \frac{i}{2\mu}(\boldsymbol{C}\nabla_R + \nabla_R\boldsymbol{C}) + \frac{1}{\mu}\delta V(\boldsymbol{R}, \boldsymbol{r}_i) - \frac{1}{2(M_A + M_B)}\left(\sum_{i=1}^{N_A+N_B} \nabla_i\right)^2. \tag{2.138}$$

Here

$$\mathscr{H}^{(0)} = -\frac{1}{2\mu}\nabla_R^2 + H(R), \tag{2.139}$$

where $H(R)$ is the usual electronic Hamiltonian defined by (3.13, 14) and the vectors $\boldsymbol{r}_i$ are referred to the center of mass of the nuclei. Vector $\boldsymbol{C}$, describing the electron momentum transferred from atom to atom, is

$$\boldsymbol{C} = -\frac{i}{2}\sum_{i=1}^{N_A+N_B} [\nabla_i^2; \boldsymbol{r}_i f(\boldsymbol{r}_i, \boldsymbol{R})], \tag{2.140}$$

where $[a; b] = ab - ba$, and the new functions $f(\boldsymbol{r}_i, \boldsymbol{R})$ satisfy the boundary conditions

$$\lim_{R\to\infty} f(\boldsymbol{r}_i, \boldsymbol{R}) = \begin{cases} \dfrac{M_{\mathrm{B}}}{M_{\mathrm{A}} + M_{\mathrm{B}}} & \text{if } \boldsymbol{r}_i \to \dfrac{M_{\mathrm{B}}}{M_{\mathrm{A}} + M_{\mathrm{B}}}\boldsymbol{R}, \\ -\dfrac{M_{\mathrm{A}}}{M_{\mathrm{A}} + M_{\mathrm{B}}} & \text{if } \boldsymbol{r}_i \to -\dfrac{M_{\mathrm{A}}}{M_{\mathrm{A}} + M_{\mathrm{B}}}\boldsymbol{R} \end{cases} . \tag{2.141}$$

The potential $\delta V(\boldsymbol{r}_i, \boldsymbol{R})$ depends on the functions $f(\boldsymbol{r}_i, \boldsymbol{R})$. The last term in (2.138) is due to mass polarization. Thus, the effect of momentum transfer in the scattering-coordinate method leads to the appearance of additional terms of the order $m_{\mathrm{el}}/M_{\mathrm{nuc}}$ in the Hamiltonian. It can be readily checked that the radial scattering equations using the Hamiltonian (2.138) and the expansion (2.133) of the total wave function exhibit the correct behavior for $R \to \infty$, due to the terms depending on $f(\boldsymbol{r}_i, \boldsymbol{R})$ in (2.138) and provided the isotopic shift in the free atoms A and B is taken into account. With a rigorous solution of the problem in the variables $\boldsymbol{\varrho}, \bar{\boldsymbol{r}}_i$, and accounting for the complete set of electronic states, the ultimate results are certainly independent of $\bar{f}(\bar{\boldsymbol{r}}_i, \boldsymbol{\varrho})$. Since such a rigorous solution is in fact impossible, one must estimate the extent of the uncertainty caused by these functions, which are, to a considerably extent, arbitrary, on calculations taking the limited number of electronic states into account. This can be suitably done by using a molecular basis.

To define this basis, we introduce

$$H^{\mathrm{mod}} = H + \frac{1}{\mu}\,\delta V(\boldsymbol{r}_i, \boldsymbol{R}) + \frac{1}{\mu}\,\delta H_{\mathrm{mp}}, \tag{2.142}$$

where the additional term $\delta H_{\mathrm{mp}}/\mu$ ensures the correct mass polarization in the free atoms A and B. The expression for δH_{mp} containing the functions $f(\boldsymbol{r}_i, \boldsymbol{R})$ and symmetric in all electrons is given in [2.50]. The molecular basis functions $|cj\tilde{\Omega}\rangle_{\mathrm{mod}}$ are eigenfunctions of H^{mod} and depend on R as a parameter. To the zeroth approximation in $m_{\mathrm{el}}/M_{\mathrm{nuc}}$, these functions coincide with the usual molecular functions $|cj\tilde{\Omega}\rangle$ which behave for $R \to \infty$ as

$$\underset{R\to\infty}{|cj\tilde{\Omega}\rangle} = \hat{A} \sum_{m_1 m_2} \begin{bmatrix} j_1 & j_2 & j \\ m_1 & m_2 & \tilde{\Omega} \end{bmatrix} |a_1 m_1\rangle_{\mathrm{A}}\, |a_2 m_2\rangle_{\mathrm{B}}. \tag{2.143}$$

A detailed discussion of the properties of these functions will be given in the next chapter.

Let us represent the total wave function Ψ in a form similar to (2.38):

$$\Psi = 2\pi i k^{-1/2} \frac{1}{R} \sum_{JM} \sum_{c'j'\tilde{\Omega}'} \mathrm{i}^J \left(\frac{2J+1}{4\pi}\right) D^{J*}_{M\tilde{\Omega}'}(\varphi, \theta, 0)$$

$$\times\, D^J_{M-\tilde{\Omega}}(\alpha, \beta, 0)\, \chi^{+J}_{c'j'\tilde{\Omega}';cj\tilde{\Omega}}(R)\, |cj\tilde{\Omega}\rangle_{\mathrm{mod}}. \tag{2.144}$$

Then we obtain the following set of equations for the radial functions $\chi^{+J}_{c'j'\tilde{\Omega}';cj\tilde{\Omega}}$ taking into account that, in a rotating coordinate frame, $f(\boldsymbol{r}_i, \boldsymbol{R})$ depends only on R:

$$\left(-\frac{1}{2\mu}\frac{d^2}{dR^2}+\frac{J(J+1)-\tilde{\Omega}'^2}{2\mu R^2}+U^{\text{mod}}_{c'j'\tilde{\Omega}'}-\mathscr{E}\right)\chi^{+J}_{c'j'\tilde{\Omega}';cj\tilde{\Omega}}$$

$$=-\sum_{c''j''\tilde{\Omega}''}(\langle c'j'\tilde{\Omega}'|\,V^{\text{mod}}_{\mathbf{R}}\,|c''j''\tilde{\Omega}''\rangle$$

$$+\langle c'j'\tilde{\Omega}'|\,V^{\text{mod}}_{\text{Cor}}\,|c''j''\tilde{\Omega}''\rangle)\,\chi^{+J}_{c''j''\tilde{\Omega}'';cj\tilde{\Omega}}. \qquad (2.145)$$

Here $U^{\text{mod}}_{c'j'\tilde{\Omega}'}$ are the eigenvalues of the modified electronic Hamiltonian,

$$\langle c'j'\tilde{\Omega}'|\,V^{\text{mod}}_{\mathbf{R}}\,|c''j''\tilde{\Omega}''\rangle=\delta_{\tilde{\Omega}'\tilde{\Omega}''}$$

$$\times\left[-\frac{1}{\mu}\langle c'j'\tilde{\Omega}'|\frac{\partial}{\partial R}+\mathrm{i}C_{Z'}\,|c''j''\tilde{\Omega}''\rangle\frac{d}{dR}\right.$$

$$+\frac{1}{2\mu}\langle c'j'\tilde{\Omega}'|-\frac{\partial^2}{\partial R^2}-2\mathrm{i}\,C_{Z'}\frac{\partial}{\partial R}-\mathrm{i}\frac{\partial C_{Z'}}{\partial R}\,|c''j''\tilde{\Omega}''\rangle$$

$$+\frac{1}{2\mu}\langle c'j'\tilde{\Omega}'|\,R^{-2}(j^2_{X'}+j^2_{Y'})$$

$$+R^{-1}(C_{Y'}j_{X'}+j_{X'}C_{Y'}-C_{X'}j_{Y'}-C_{Y'}j_{X'})\,|c''j''\tilde{\Omega}''\rangle$$

$$\left.-\frac{1}{2\mu}\langle c'j'\tilde{\Omega}'|\frac{\mu}{M_{\text{A}}+M_{\text{B}}}\left(\sum_{i=1}^{N_{\text{A}}+N_{\text{B}}}\boldsymbol{V}_i\right)^2+2\delta H_{\text{mp}}\,|c''j''\tilde{\Omega}''\rangle\right], \qquad (2.146)$$

$$\langle c'j'\tilde{\Omega}'|\,V^{\text{mod}}_{\text{Cor}}\,|c''j''\tilde{\Omega}''\rangle=-\frac{1}{\mu R^2}[\lambda^+(J)\,\delta_{\tilde{\Omega}'';\tilde{\Omega}'+1}$$

$$\times\langle c'j'\tilde{\Omega}'|\,j_{X'}+RC_{Y'}\,|c''j''\tilde{\Omega}''\rangle+\lambda^-(J)\,\delta_{\tilde{\Omega}'';\tilde{\Omega}'-1}$$

$$\times\langle c'j'\tilde{\Omega}'|\,j_{X'}+RC_{Y'}\,|c''j''\tilde{\Omega}''\rangle] \qquad (2.147)$$

and the factors $\lambda^{\pm}(J)$ are given by (2.102). Within the accuracy adopted here, the molecular wave functions $|cj\tilde{\Omega}\rangle$ of zero order in $m_{\text{el}}/M_{\text{nuc}}$ must be used in the matrix elements of the nonadiabatic coupling already containing the small factor $1/\mu$. Differentiation with respect to R in the matrix elements is done at constant $\boldsymbol{r}_i$ referred to the nuclear center of mass. Direct calculation using boundary conditions (2.141, 143) readily shows that the behaviour of the nonadiabatic coupling matrix elements (2.146, 147) is correct for $R \to \infty$ due to the terms dependent on $f(\boldsymbol{r}_i, R)$.

Usually the contribution of these terms to the probabilities of inelastic processes in slow quasi-classical collisions is small. The quasi-classical condition allows one to neglect all the terms on the right side of (2.146), except the first one, on the basis of estimates completely similar to those given by (2.111, 112). In evaluating the contribution of $C_{Z'}$ to the radial coupling and of $C_{Y'}$ to the Coriolis coupling, the following relation is obtained from (2.140) if relativistic interactions are neglected:

$$\langle c'j'\tilde{\Omega}'|\, \boldsymbol{C}\, |c''j''\tilde{\Omega}''\rangle = \mathrm{i}(U_{c'j'\tilde{\Omega}'} - U_{c''j''\tilde{\Omega}''}) \times \langle c'j'\tilde{\Omega}'|\sum_i \boldsymbol{r}_i f(\boldsymbol{r}_i, R)\,|c''j''\tilde{\Omega}''\rangle. \tag{2.148}$$

At the same time, it is known that in slow collisions the main contribution to the inelastic probability is from regions where the various terms $U_{cj\tilde{\Omega}}$ approach one another or cross. For terms with the same symmetry which approach one another, the estimate is

$$\left|\langle c'j'\tilde{\Omega}'|\frac{\partial}{\partial R}|c''j''\tilde{\Omega}''\rangle\right| \sim |U_{c'j'\tilde{\Omega}'} - U_{c''j''\tilde{\Omega}'}|^{-1}, \tag{2.149}$$

and in the crossing region of terms with different $\tilde{\Omega}$,

$$|\langle c'j'\tilde{\Omega}'|\, j_{X'}\, |c''j''\tilde{\Omega}''\rangle| \sim 1. \tag{2.150}$$

Therefore in those regions where $|U_{c'j'\tilde{\Omega}'} - U_{c''j''\tilde{\Omega}''}|$ can be much smaller than unity (in atomic units), the contribution from the $f(\boldsymbol{r}_i, R)$-dependent terms to the nonadiabatic coupling can be neglected. Thus the uncertainty of the inelastic probabilities caused by the effect of momentum transfer is not significant provided these probabilities are not too small. In such cases, the role of the $f(\boldsymbol{r}_i, R)$-dependent terms in (2.145) simply ensures the mathematical correctness of the problem. This is particularly important for numerical calculations. The ultimate result will depend only slightly on the form of $f(\boldsymbol{r}_i, R)$. It follows from intuitive considerations that $f(\boldsymbol{r}_i, R)$ will attain its asymptotic R-independent form at the distance at which the exchange interaction of the atoms can be neglected, i.e., when the electrons belonging to different atoms can be considered to be distinguishable. Equations (2.145) will then reduce to those for the scattering model with distinguishable electrons discussed at the beginning of this section. In the limit as $R \to 0$, the function f will tend to zero since, in this limit, the two-atom system can be considered as one atom moving in space.

It follows from the estimates (2.149) and (2.150) that the terms in (2.145) which depend on $f(\boldsymbol{r}_i, R)$, and therefore on the electron momentum transfer, can be important in two cases. The first case is that of direct nonadiabatic transitions between states for which $|U_{c'j'\tilde{\Omega}'} - U_{cj\tilde{\Omega}}| \sim 1$ at all accessible R. The probabilities of these transitions are of interest only for very high collision energies when the impact velocity is comparable to the velocities of the atomic electrons. However, such high-energy collisions are not treated in this book. The second case where by the $f(\mathbf{r}_i, R)$-dependent terms may be important is for resonant transitions induced solely by nonadiabatic coupling to closed channels (i.e., via intermediate electronic states). An example of such a process is the spin depolarization of an alkali atom in the ground state $^2S_{1/2}$ in collisions with a noble gas atom [2.53]. Since the function $f(\mathbf{r}_i, R)$ is to a considerable extent arbitrary, estimates of the cross section of this process obtained by the methods considered in this book can only be accurate to an order of magnitude.

On the whole, it follows from the above discussion that, for slow quasi-classical collisions of two many-electron atoms, one can use all the general expressions obtained in the preceding sections for the S matrix and the scattering amplitude in the R-helicity representation with an appropriate choice of indices. However, in calculating the S matrix elements, one has to use modified radial equations (2.145) in which the terms dependent on $f(\mathbf{r}_i, R)$ play, as a rule, an auxiliary part, ensuring the correct behavior of the matrix elements only for $R \to \infty$.

2.3.2 Collisions of Identical Atoms

In considering the collisions of identical atoms, their indistinguishability must be taken into account. This manifests itself in the symmetry of the total wave function on permutation of the nuclei: for nuclei with a half-integer spin I the function will change its sign, and for nuclei with an integer spin it will remain unchanged. The construction of a function with the correct total symmetry is simplified taking into account the fact that, for a system of identical atoms, the electronic function also possesses a certain symmetry: it either remains unchanged (even function g) or changes its sign (odd function u) on inversion of the electron coordinates in the center of mass of the system. Collisions induce no transitions between states of different electronic parity so that, in an appropriate representation, the S matrix will be diagonal in the quantum number $w (w = 1$ for g states and -1 for u states), i.e., it can be represented in the form ${}^wS^J_{c'j'l';cjl}$, where $c'j'l'$ and cjl are quantum numbers of the initial and final states of a diatom. Another specification of the system state is realized by the quantum numbers a_1 (first atom) and a_2 (second atom), the states $a_1 a_2$ and $a_2 a_1$ being degenerate. The S matrices in the two representations are related as

$$S^J_{a'_1 a'_2 j' l'; a_1 a_2 jl} \equiv {}^{\mathrm{d}}S^J_{c'j'l';cjl} = \tfrac{1}{2}({}^gS^J_{c'j'l';cjl} + {}^uS^J_{c'j'l';cjl}), \tag{2.151 a}$$

$$S^J_{a'_1 a'_2 j' l'; a_2 a_1 jl} \equiv {}^{\mathrm{t}}S^J_{c'j'l';cjl} = \tfrac{1}{2} w_1 w_2 (-1)^{j_1 + j_2 - j + N} \times ({}^gS^J_{c'j'l';cjl} - {}^uS^J_{c'j'l';cjl}), \tag{2.151 b}$$

which involves the familiar definitions of the scattering matrix for a direct process ${}^{\mathrm{d}}S^{J}_{c'j'l';cjl}$ (transition $a_1 a_2 \rightarrow a'_1 a'_2$) and for a process with state transfer ${}^{\mathrm{t}}S^{J}_{c'j'l';cjl}$ (transition $a_2 a_1 \rightarrow a'_1 a'_2$). The phase factor in (2.151 b) is determined by the inversion character of the electronic wave function (2.143) (Sect. 3.2.2) and N is the number of electrons in the atom. The amplitudes of direct scattering $f^{\mathrm{d}}_{c'j'm';cjm}$ and of scattering with state transfer $f^{\mathrm{t}}_{c'j'm';cjm}$ are introduced accordingly:

$$f^{\mathrm{d}}_{c'j'm';cjm} = \tfrac{1}{2}(f^{g}_{c'j'm';cjm} + f^{u}_{c'j'm';cjm}) \tag{2.152a}$$

$$f^{\mathrm{t}}_{c'j'm';cjm} = \tfrac{1}{2} w_1 w_2 (-1)^{j_1+j_2-j+N}(f^{g}_{c'j'm';cjm} - f^{u}_{c'j'm';cjm}). \tag{2.152b}$$

Consider now a transformation of the amplitudes $f^{w}_{c'j'm';cjm}$ under a permutation of the nuclei. This procedure is equivalent to an inversion of the electron and nuclear coordinates and to a subsequent inversion of the electron coordinates. The first operation leads to the replacement of θ, φ by $(\pi - \theta)$, $(\varphi + \pi)$ and multiplication of each term in the asymptotic expression of the total wave function of type (2.3) by $w'_1 w'_2$. The second operation leads to the multiplication of the total wave function by w. Thus, under a permutation of the nuclei, $f^{w}_{c'j'm';cjm}$ transforms as follows:

$$f^{w}_{c'j'm';cjm}(\theta, \varphi) \rightarrow w w'_1 w'_2 f^{w}_{c'j'm';cjm}(\pi - \theta, \varphi + \pi). \tag{2.153}$$

The scattering amplitude correctly accounting for nuclear statistics depends on the total spin $I = I_1 + I_2$ of the nuclei, since the value of this spin determines the symmetry of the wave function under a permutation of the spin coordinates of the nuclei. At the same time, since the symmetry of the total wave function is determined by the type of nuclear statistics, the coordinate function, as well as the scattering amplitude, will depend on I. Thus, the scattering amplitude will be specified by a certain multiplicity $2I + 1({}^{2I+1}f)$.

For $I_1 = I_2 = 0$ and also $I = 0$, the amplitude will be symmetric upon a permutation of the nuclei:

$${}^{1}f^{w}_{c'j'm';cjm} = f^{w}_{c'j'm';cjm}(\theta, \varphi) + w w'_1 w'_2 f^{w}_{c'j'm';cjm}(\pi - \theta, \varphi + \pi). \tag{2.154}$$

For $I_1 = I_2 = 1/2$ two values are possible: $I = 0$ and $I = 1$. In the first case, the scattering amplitude will be symmetric (the singlet spin function is antisymmetric under a permutation of the spin coordinates) and, in the second, it will be antisymmetric (the triplet spin function is symmetric under a permutation of the spin coordinates):

$$\begin{aligned} {}^{1}f^{w}_{c'j'm';cjm} &= f^{w}_{c'j'm';cjm}(\theta, \varphi) + w w'_1 w'_2 f^{w}_{c'j'm';cjm}(\pi - \theta, \varphi + \pi), \\ {}^{3}f^{w}_{c'j'm';cjm} &= f^{w}_{c'j'm';cjm}(\theta, \varphi) - w w'_1 w'_2 f^{w}_{c'j'm';cjm}(\pi - \theta, \varphi + \pi). \end{aligned} \tag{2.155}$$

For more complex cases the scattering amplitude will be constructed in the same way.

The amplitudes ${}^{2I+1}f^{w}_{c'j'm';cjm}$ correctly describing the nuclear statistics and corresponding to a certain parity of the electron state w are used for constructing the amplitudes of direct scattering ${}^{2I+1}f^{\mathrm{d}}_{c'j'm';cjm}$ and of scattering with state transfer ${}^{2I+1}f^{\mathrm{t}}_{c'j'm';cjm}$:

$$
{}^{2I+1}f^{\mathrm{d}}_{c'j'm';cjm} = \tfrac{1}{2}({}^{2I+1}f^{g}_{c'j'm';cjm} + {}^{2I+1}f^{u}_{c'j'm';cjm}), \tag{2.156a}
$$

$$
\begin{aligned}
{}^{2I+1}f^{\mathrm{t}}_{c'j'm';cjm} &= \tfrac{1}{2} w_1 w_2 (-1)^{j_1+j_2-j+N} \\
&\quad \times ({}^{2I+1}f^{g}_{c'j'm';cjm} - {}^{2I+1}f^{u}_{c'j'm';cjm}).
\end{aligned} \tag{2.156b}
$$

For collisions involving nonpolarized nuclei the differential cross sections of the two processes are expressed via appropriate amplitudes as

$$
q^{\mathrm{d,t}} = \sum_{I} g_I \frac{k'}{k} |{}^{2I+1}f^{\mathrm{d,t}}_{c'j'm';cjm}|^2, \tag{2.157}
$$

where g_I is the statistical weight of the state with total nuclear spin I.

Strictly speaking, a collision of atomic isotopes cannot be considered as a collision of identical systems. However, the difference in the electron functions of a system of two isotopically similar and isotopically different atoms is so small that the classification of electronic states as even and odd retains its meaning to a great degree of accuracy. In this approximation the amplitudes of direct scattering and of scattering with state transfer are given by (2.152).

2.4 Integral Cross Sections for Isotropic Collisions

The expressions for the scattering amplitude and cross section $a, \bar{m} \to a'\bar{m}'$ derived in Sect. 2.1.2 refer to a specific angular momentum quantization axis chosen to coincide with the direction of the relative velocity before collision. The general expression (2.25) allows one, when needed, to treat the case of an arbitrary orientation of the quantization axis. The rather complex expressions obtained for the cross section in this case can be substantially simplified if the integral cross section for a specific transition is averaged over all orientations of the vector $\hat{\boldsymbol{k}}$ relative to the quantization axis.

This is the case of the so-called isotropic collisions – it represents, for instance, inelastic atomic collisions in a gas at thermal equilibrium. Using (2.5) for the differential cross section $q_{a'm';am}$ with the amplitude given by

(2.25), integrating $q_{a'm';am}$ over all scattering angles and averaging the integral cross section $\sigma_{a'm';am}$ over the isotropic distribution of $\hat{k}$ relative to the chosen quantization axis, we arrive at the averaged integral cross section $\bar{\sigma}_{a'm';am}$. Suppressing all quantum numbers except j and m[7], we get

$$\bar{\sigma}_{j'm';jm} = \frac{\pi}{k^2} \sum_{ln;l'n'} \left| \sum_{MJ} \begin{bmatrix} j' & l' & J \\ m' & n' & M \end{bmatrix} T^J_{j'l';jl} \begin{bmatrix} j & l & J \\ m & n & M \end{bmatrix} \right|^2 . \tag{2.158}$$

The cross sections with $m' \neq m$ correspond to population transfer either between various electronic states (for $j' \neq j$) or between Zeeman sublevels of the same state (for $j' = j$). The cross section with $m' = m$ corresponds to an elastic collision. The latter is insignificant for processes where the change of direction of the relative velocity is unimportant.

This suggests a definition of another quantity diagonal in j, m instead of the averaged elastic cross section. Such a quantity in the theory of atomic state relaxation in isotropic collisions is the total cross section for depletion of the level jm. In this way the definition (2.158) is retained for $\bar{\sigma}_{j'm';jm}$ with $\Delta j \neq 0$, any m, and $\Delta j = 0$, $\Delta m \neq 0$, and the diagonal cross section is defined by the relation

$$\bar{\sigma}_{jm;jm} = - \sum_{j'm'} \bar{\sigma}_{j'm';jm} . \tag{2.159}$$

As follows from (2.159) $\bar{\sigma}_{jm;jm}$ is negative and this corresponds to a depletion of the state j, m in collisions.

Analysis of the sum (2.158) reveals [2.54–56] that the number of parameters expressing all cross sections for transitions $jm \to j'm'$ is smaller than that obtained in using the obvious symmetry $\bar{\sigma}_{j'm';jm} = \bar{\sigma}_{jm;j'm'} = \bar{\sigma}_{j'-m';j-m}$. These parameters differ, depending on how the four Clebsch-Gordan coefficients are summed over magnetic quantum numbers.

One way, now widely used in connection with the collisional relaxation of atomic states, is to represent the sums (2.158, 159) in the form

$$\bar{\sigma}_{j'm';jm} = -(-1)^{j'-m'+j-m} \sum_{\varkappa} \begin{bmatrix} j' & j' & \varkappa \\ m' & -m' & 0 \end{bmatrix} \begin{bmatrix} j & j & \varkappa \\ m & -m & 0 \end{bmatrix} \sigma^{j';j}_{\varkappa} . \tag{2.160}$$

As follows from the triangular condition for the Clebsch-Gordan coefficients, $\varkappa$ spans the range from 0 to $\min(2j, 2j')$. The cross sections $\sigma^{j';j}_{\varkappa}$ are known as irreducible cross sections for multipole relaxation for a state j (for $j' = j$) or for multipole transfer $j \to j'$. In turn, the irreducible cross

[7] For simplicity, only transitions between the levels of one fine structure multiplet will be considered in this section, i.e., it is assumed that $\gamma = \gamma'$. The generalization for the case $\gamma \neq \gamma'$ is straightforward

sections are expressed by the scattering matrix in the form

$$\sigma_{\varkappa}^{j';j} = (\pi/k^2) \sum_{ll'} \sum_{J_\alpha J_\beta} (-1)^{j'-j+l'-l}(2J_\alpha+1)(2J_\beta+1)$$

$$\times \begin{Bmatrix} j' & \varkappa & j' \\ J_\beta & l' & J_\alpha \end{Bmatrix} \begin{Bmatrix} j & \varkappa & j \\ J_\beta & l & J_\alpha \end{Bmatrix} (S^{J_\alpha}_{aj'l';jl} S^{J^*_\beta}_{j'l';jl} - \delta_{ll'}\delta_{j'j}). \quad (2.161)$$

The transformation inverse to (2.160) has the form

$$\sigma_{\varkappa}^{j';j}\delta_{0q}\delta_{0q'} = \sum_{m',m} (-1)^{j'-m'+j-m}$$

$$\times \begin{bmatrix} j' & j' & \varkappa \\ m' & -m'+q' & q' \end{bmatrix} \begin{bmatrix} j & j & \varkappa \\ m & -m+q & q \end{bmatrix} \bar{\sigma}_{j'm';jm}. \quad (2.162)$$

For $q = q' = 0$, (2.162) expresses irreducible cross sections via the averaged population transfer cross section and, for $q \neq 0$ or $q' \neq 0$, it imposes constraints on $\bar{\sigma}_{j'm';jm}$ due to the isotropic conditions of collisions. For instance, for transitions within a state with given j there are only $2j+1$ irreducible cross sections $\sigma_{\varkappa}^{j';j}$, $\varkappa = 0 \ldots 2j$. At the same time, accounting for symmetry properties only, $\bar{\sigma}_{jm';jm} = \bar{\sigma}_{jm;jm'} = \bar{\sigma}_{j-m;j-m'}$, would result in $j(j+1)$ (for integer j) or in $(j+1/2)^2$ (for semi-integer j) different cross sections for population transfer.

To clear up the physical meaning of the cross sections $\sigma_{\varkappa}^{j';j}$, consider the change in the density matrix of the ensemble of atoms A induced by isotropic collisions with heat bath atoms B($^1S_0$). The density matrix element $\varrho_{m_1 m_2}^{j_1 j_2}$ is defined by the amplitudes $a_{m_1}^{j_1}$ of states $|j_1 m_1\rangle$ in the usual way:

$$\varrho_{m_1 m_2}^{j_1 j_2} = \langle a_{m_1}^{j_1} a_{m_2}^{j_2^*} \rangle_{\text{over ensemble}}. \quad (2.163)$$

Replacing the elements $\varrho_{m_1 m_2}^{j_1 j_2}$ by spherical components $\varrho_{\varkappa q}^{j_1 j_2}$ and using the expression

$$\varrho_{\varkappa q}^{j_1 j_2} = \sum_{m_1} (-1)^{j_2-m_1+q} \begin{bmatrix} j_1 & j_2 & \varkappa \\ m_1 & -m_1+q & q \end{bmatrix} \varrho_{m_1 m_2 - q}^{j_1 j_2} \quad (2.164)$$

enables one to make certain definite statements about the kinetics of changes in $\varrho_{\varkappa q}^{j_1 j_2}$ induced by collisions of atoms A with atoms B [2.54, 56–58], namely, for isotropic collisions, the relaxation and transfer of the components $\varrho_{\varkappa q}^{j_1 j_2}$ of different rank $\varkappa$ are not coupled and, for a given $\varkappa$, occur in the same way for all spherical components with different q.

Thus, the relaxation equations for spherical components of the density matrix are

$$(\dot{\varrho}_{\varkappa q}^{j_1' j_2'})_{\text{rel}} = \sum_{j_1 j_2} \Gamma_{\varkappa}^{j_1' j_2'; j_1 j_2} \varrho_{\varkappa q}^{j_1 j_2}. \quad (2.165)$$

The kinetic coefficients $\Gamma_{\varkappa}^{j_1' j_2'; j_1 j_2}$ are related to the generalized irreducible cross sections by

$$\Gamma_{\varkappa}^{j_1' j_2'; j_1 j_2} = N_{\mathrm{B}} \langle v \sigma_{\varkappa}^{j_1' j_2'; j_1 j_2} \rangle \tag{1.166}$$

where N_{B} is the number density of the bath atoms B. It should be noted [2.59] that the relaxation equations (2.165) can be readily obtained from the general kinetic theory developed for dilute molecular gases with internal degrees of freedom [2.60].

The cross sections $\sigma_{\varkappa}^{j_1' j_2'; j_1 j_2}$ are defined by the scattering matrix similar to (2.161), but in a more general way:

$$\sigma_{\varkappa}^{j_1' j_2'; j_1 j_2} = (\pi/k^2) \sum_{ll'} \sum_{J_\alpha J_\beta} (-1)^{j_1' - j_1 + l' - l} \begin{Bmatrix} j_1' & \varkappa & j_2' \\ J_\beta & l' & J_\alpha \end{Bmatrix} \begin{Bmatrix} j_1 & \varkappa & j_2 \\ J_\beta & l & J_\alpha \end{Bmatrix}$$
$$\times (S^{J_\alpha}_{j_1' l'; j_1 l} S^{J_\beta *}_{j_2' l'; j_2 l} - \delta_{ll'} \delta_{j_1' j_1} \delta_{j_2' j_2}). \tag{2.167}$$

For $j_1' = j_2' = j'$ and $j_1 = j_2 = j$ this expression reduces to (2.161), with $\sigma_{\varkappa}^{j'j';jj}$ identical to $\sigma_{\varkappa}^{j';j}$.

The general kinetic equation for the density matrix is usually obtained by adding the dynamic equation to the relaxation equation (2.165). For an ensemble of free atoms and neglecting all radiation processes, the dynamic equation is

$$(\dot{\varrho}_{\varkappa q}^{j_1 j_2})_{\mathrm{dyn}} = \mathrm{i} \Omega_{j_1 j_2} \varrho_{\varkappa q}^{j_1 j_2}, \tag{2.168}$$

where $\Omega_{j_1 j_2}$ are the frequencies of transitions between states j_1 and j_2. Thus, the general kinetic equations will read

$$\dot{\varrho}_{\varkappa q}^{j_1' j_2'} = \sum_{j_1 j_2} (\Gamma_{\varkappa}^{j_1' j_2'; j_1 j_2} + \mathrm{i} \delta_{j_1' j_1} \delta_{j_2' j_2} \Omega_{j_1 j_2}) \varrho_{\varkappa q}^{j_1 j_2}. \tag{2.169}$$

An important simplification is possible here if one uses the conventional arguments of perturbation theory, namely, that the components $\dot{\varrho}_{\varkappa q}^{j_1' j_2'}$ and $\varrho_{\varkappa q}^{j_1 j_2}$ of the density matrix can be decoupled provided

$$|\Omega_{j_1' j_2'} - \Omega_{j_1 j_2}| \gg | \Gamma_{\varkappa}^{j_1' j_2'; j_1 j_2}| \tag{2.170}$$

and the relaxation can be treated separately for each of these components. If all frequencies satisfy (2.170), the relaxation of the density matrix elements diagonal in j can be treated independently of the nondiagonal elements (the so-called secular approximation).

For atomic collisions in gases the condition (2.170) is fulfilled for inter- and intramultiplet transitions with Γ of the order of the gas collision frequency ($\Gamma \sim 10^9\,\mathrm{s}^{-1}$ under normal conditions). However this is not the case for transitions between hyperfine atomic states (see below).

Assuming that (2.170) is fulfilled, consider now the simplified relaxation equations for $\varrho^{jj}_{\varkappa q} \equiv \varrho^{j}_{\varkappa q}$. These read

$$(\dot{\varrho}^{j'}_{\varkappa q})_{\mathrm{rel}} = \sum_{j} \Gamma^{j';j}_{\varkappa} \varrho^{j}_{\varkappa q}, \tag{2.171}$$

where

$$\Gamma^{j';j}_{\varkappa} = N_{\mathrm{B}} \langle v \sigma^{j';j}_{\varkappa} \rangle. \tag{2.172}$$

Now, for $j_1 = j_2 = j$, the quantities $\varrho^{j}_{\varkappa q}$ defined by (2.164) are the expectation values of the qth component of the relevant $2^{\varkappa}$-state multipole [2.57, 58, 61]. Thus ϱ^{j}_{00} is proportional to the total population of the degenerate state with angular momentum j; $\varrho^{j}_{1q}(q = -1, 0, 1)$ is proportional to the spherical components of the mean magnetic moment of the atom in state j, and ϱ^{j}_{2q} to the spherical components of the electric quadrupole moment. The coefficient of proportionality is to be defined specifically depending on the nature of the physical quantity used.

Therefore, equations of the type (2.171) are often referred to as equations for the relaxation and transfer of atomic multipole moments, and the cross sections $\sigma^{j';j}_{\varkappa}$, as the cross sections for multipole relaxation and transfer.

The equations for the macroscopic components of multipoles can be obtained from (2.171). Consider first the total population N^{j} of state $|j\rangle$,

$$N^{j} = \sum_{m} \varrho^{jj}_{mm}. \tag{2.173}$$

Comparing (2.173) with (2.164), we find $\varrho^{j}_{00} = N^{j}/(2j+1)^{1/2}$. Then it immediately follows from (2.171) that the relaxation equation for populations is

$$\dot{N}^{j'} = \sum_{j} \Gamma^{j';j}_{0} \frac{(2j'+1)^{1/2}}{(2j+1)^{1/2}} N^{j}. \tag{2.174}$$

In a similar way we obtain the relaxation equation for components of the dipole moment (also called orientation) M^{j}_{q} or of the quadrupole moment Q^{j}_{q} (also called alignment),

$$\dot{M}^{j'}_{q} = \sum_{j} \Gamma^{j';j}_{1} \frac{[2j'(2j'+1)(2j'+2)]^{1/2}}{[2j(2j+1)(2j+2)]^{1/2}} M^{j}_{q}, \tag{2.175}$$

$$\dot{Q}^{j'}_{q} = -\sum_{j} \Gamma^{j';j}_{2} \frac{[(2j'-1)\,2j'(2j'+1)(2j'+2)(2j'+3)]^{1/2}}{[(2j-1)\,2j(2j+1)(2j+2)(2j+3)]^{1/2}} Q^{j}_{q}. \tag{2.176}$$

Here, for instance, the Z components ($q = 0$) of the moments are defined as

$$M^{j}_{0} = \sum_{m} m \varrho^{jj}_{mm}, \tag{2.177}$$

$$Q^{j}_{0} = \sum_{m} [m^2 - 3j(j+1)]\, \varrho^{jj}_{mm}. \tag{2.178}$$

Again, the weight factors on the rhs of (2.175, 176) are readily inferred by comparison of these equations with (2.164).

Note that just as $\Gamma_\varkappa^{j';j}$ describes the relaxation of any spherical component of the density matrix of rank $\varkappa$, it also describes, via (2.164), the relaxation of the components $\varrho_{mm'}^{jj}$ nondiagonal in m (the so-called Zeeman coherences). This means, by way of (2.162), that the averaged population transfer cross sections $\bar{\sigma}_{jm';jm}$ completely define the relaxation rates not only for the Zeeman populations, but also for the Zeeman coherences.

Consider in more detail now the transitions within the multiplet ${}^{2S+1}L_j$ assuming that the total population of all the fine structure states remains unchanged. For each level j there are $2j+1$ cross sections $\sigma_\varkappa^{j;j}$ ($\varkappa = 0, 1, 2, \dots, 2j$), one of these describing the population depletion of this state as the result of transitions to all other states. Assuming that $L \geqq S$ for this multiplet, there will be a total of $N_{LS} = (1/6)\ (2S+1)\ (2S+2)\ (6L-2S+3)$ various cross sections taking into account their symmetry, see (2.179). The unitarity condition for the S matrix in the multiplet block imposes $2S+1$ constraints and therefore $\sigma_0^{j;j}$ can be considered to depend on $\sigma_0^{j';j}$. Thus, a complete description of an isolated multiplet relaxation is given by the set of $N_{LS} - (2S+1)$ transition cross sections. Specifically, the doublet ${}^2P_j$ discussed in detail in Chap. 11 displays six independent cross sections, namely, two for population and orientation transfer between states $j = 1/2$ and $j = 3/2$, two for orientation relaxation in states $j = 1/2$ and $j = 3/2$, and two for alignment and octupole moment relaxation in state $j = 3/2$. Instead of these cross sections one can use some of their linear combinations, as first suggested in [2.62].

Consider now certain properties of the cross sections $\sigma_\varkappa^{j';j}$. It can be shown that all "diagonal" cross sections $\sigma_\varkappa^{j;j}$ ($\varkappa = 0, 1, \dots, 2j$) are negative, and the population transfer cross sections $\sigma_0^{j';j}$ ($j' \neq j$) are positive. The sign of the other cross sections $\sigma_\varkappa^{j';j}$ ($\varkappa = 1, \dots, 2j$) can be arbitrary because of the tensor character of the transferred quantity. It follows from (2.161) that the cross section for direct ($j \to j'$) and reverse ($j' \to j$) processes are related by the microscopic reversibility equations

$$k^2 \sigma_\varkappa^{j';j}(k) = (k')^2 \sigma_\varkappa^{j;j'}(k'). \tag{2.179}$$

It follows from (2.179) that the detailed balance equation for $\Gamma_\varkappa^{j';j}$ and $\Gamma_\varkappa^{j;j'}$ in the Maxwellian heat bath is

$$\Gamma_\varkappa^{j';j}/\Gamma_\varkappa^{j;j'} = \exp[(\varepsilon_j - \varepsilon_{j'})/kT]. \tag{2.180}$$

For $\varkappa = 0$ the cross sections are easily expressed by the square modulus of the S matrix elements:

$$\sigma_0^{j';j} = (2j+1)^{-1/2}(2j'+1)^{-1/2}(\pi/k^2) \sum_J (2J+1) \sum_{ll'} |S_{j'l';jl}^J|^2 > 0, \tag{2.181}$$

$$\sigma_0^{j;j} = -(2j+1)^{-1}(\pi/k^2)\sum_J (2J+1)\left[1-\sum_{ll'}|S^J_{jl';jl}|^2\right] < 0. \tag{2.182}$$

It can be seen from these relations that the cross sections $\sigma_0^{j';j}$ are connected with the total cross sections (2.30) $\sigma_{j';j}$ or $\sigma_{j;j'}$ by the relations

$$\left(\frac{2j'+1}{2j+1}\right)^{1/2}\sigma_{j';j} = \sigma_0^{j';j} = \left(\frac{2j+1}{2j'+1}\right)^{1/2}\sigma_{j;j'}\,. \tag{2.183}$$

Consider now a simplification of the general expression (2.161) under quasi-classical conditions. The angular momenta J_α, J_β, l and l' are large and, therefore, one may use the asymptotics of the 6-j symbols expressing these by the Clebsch-Gordan coefficients. Note that the sum over four indices in (2.161) involves a summation over one of these within a wide range of its variation, and that the summation over the other three indices spans a narrow range (of the order of j). Accordingly, it is reasonable to introduce the total mean momentum $\bar{J}$, the difference between J_α and J_β, and also, as before, the quantum numbers v and v':

$$\begin{aligned} \bar{J} &= (J_\alpha + J_\beta)/2,\\ \varrho &= J_\alpha - J_\beta,\\ v &= J_\alpha - l,\\ v' &= J_\alpha - l'. \end{aligned} \tag{2.184}$$

Replacing the summation over J_α, J_β, l, l' in (2.161) with that over $\bar{J}$ ϱ, v, v' and rewriting it in the form of an integral of the probability $P_\varkappa^{j';j}(\bar{J})$ over the angular momentum $\bar{J}$,

$$\sigma_\varkappa^{j';j} = (\pi/k^2)\int_0^\infty P_\varkappa^{j';j}(\bar{J})\,2\bar{J}\,d\bar{J}, \tag{2.185}$$

$$P_\varkappa^{j';j}(\bar{J}) = (2\varkappa+1)^{-1}\sum_{v,v',\varrho}\begin{bmatrix} j' & j' & \varkappa\\ v' & -v'-\varrho & -\varrho\end{bmatrix}\begin{bmatrix} j & j & \varkappa\\ v & -v-\varrho & -\varrho\end{bmatrix}$$
$$\times \exp(\mathrm{i}\pi\varrho)\,[S^{\bar{J}-\varrho/2}_{j'v';jv}\,S^{\bar{J}+\varrho/2*}_{j'v'+\varrho;jv+\varrho} - \delta_{j'j}\delta_{v'v}\exp(-\mathrm{i}\pi\varrho)]. \tag{2.186}$$

Now, if the S matrix is represented by an equation similar to (2.58), the summation over v and v' affects only the coefficients $S^A_{j'v';jv}$, and the summation over ϱ requires a calculation of the phases $2\delta^A_{j';j}(\bar{J}-\varrho/2)$. These phases are calculated via the deflection functions using the quasi-classical expression

$$2\delta^A_{j';j}(\bar{J}\pm\varrho/2) = 2\delta^A_{j';j}(\bar{J}) \pm \eta^A_{j';j}(\bar{J})\frac{\varrho}{2}\,. \tag{2.187}$$

It can be seen from (2.186) that the probability $P_\varkappa^{j';j}(\bar{J})$ contains two types of oscillatory terms. One is related to the interference of waves specified by various phases $\delta^{A}_{j';j}(J)$. The difference in the phases $\delta^{A}_{j';j}(\bar{J}) - \delta^{A'}_{j';j}(\bar{J})$ is usually large and the corresponding interference term oscillates rapidly as a function $\bar{J}$. The second type of oscillatory terms arises as the result of the phase difference $2[\delta^{A}_{j';j}(\bar{J} + \varrho/2) - \delta^{A}_{j';j}(\bar{J} - \varrho/2)] \simeq \eta^{A}_{j';j}\varrho$ which is not large.

If the rapidly oscillating contribution to $P^{j';j}_{\varkappa}(\bar{J})$ is of no interest, the representation (2.58) provides the following averaged estimation of the products of the S matrix elements in (2.186):

$$\langle S^{\bar{J}-\varrho/2}_{j'v';jv} S^{\bar{J}+\varrho/2*}_{j'v'+\varrho;jv+\varrho} \rangle_{\text{over rapid oscillations}}$$
$$= \sum_{A} \exp(-\mathrm{i}\eta^{A}_{j'j}\varrho)\, S^{A}_{j'v';jv}\, S^{A*}_{j'v'+\varrho;jv+\varrho}. \tag{2.188}$$

The rhs of (2.188) is a slowly varying function of $\bar{J}$ and, when substituted in (2.186), yields the average transition probability $\bar{P}^{j';j}_{\varkappa}$. This is sometimes sufficient for an approximate calculation of the cross section as will be elaborated in Chap. 11 for representative collision processes.

In treating transitions between fine structure components within one multiplet, it is sometimes possible to introduce a dynamic simplification connected with the smallness of the Massey parameter with respect to transitions between states with different j (highly nonadiabatic intramultiplet mixing). In this case the electron spin can be taken as fixed throughout the collision, i.e., the collisional changes in spin functions can be disregarded. Then accounting for the electron spin reduces to a simple kinematic problem: the dynamic collision problem is solved in the basis of purely orbital functions, and then the spin (S) and orbital (L) angular momenta are coupled to form j using the common rules of angular momentum addition. For this case we specify the state $|jm\rangle$ as $|LSjm\rangle$ and use the general relation for the change of representation from $LSjlJM$ to $LlKSJM$, where $\boldsymbol{K} = \boldsymbol{l} + \boldsymbol{L}$. Now, taking into account that the S operator does not act on the spin functions, we get [2.63]

$$S^{J}_{j'l';jl} \equiv S^{J}_{LSj'l';LSjl} = (-1)^{j'-j-l'+l}[(2j+1)(2j'+1)]^{1/2}$$
$$\times \sum_{K} (2K+1) \begin{Bmatrix} S & L & j' \\ l' & J & K \end{Bmatrix} \begin{Bmatrix} S & L & j \\ l & J & K \end{Bmatrix} S^{K}_{Ll';Ll}, \tag{2.189}$$

where $S^{K}_{Ll';Ll}$ is the scattering matrix of an atom in state $|L\rangle$ in the representation of the total momentum K. The sum over K on the rhs of (2.189) has only a narrow range $|J - S| \leqq K \leqq |J + S|$. In the quasi-classical limit it is convenient to introduce a summation index $\varrho = J - K$ and to pass from the 6-j symbols to the Clebsch-Gordan coefficients

$$S^{J}_{LSj'v';LSjv} = \sum_{\varrho} \begin{bmatrix} S & L & j' \\ \varrho & v'-\varrho & v' \end{bmatrix} \begin{bmatrix} S & L & j \\ \varrho & v-\varrho & v \end{bmatrix} S^{J-\varrho}_{L,v'-\varrho;L,v-\varrho} \exp(\mathrm{i}\pi\varrho). \tag{2.190}$$

Here, just as in (2.186), the change in the angular momentum $K = J - \varrho$, when summing over ϱ, must be taken into account only for the terms containing large phases δ^{A}_{LL} in the S matrix representation (2.58).

A substitution of (2.189) into (2.161) enables one to express the cross section $\sigma^{j';j}_{\varkappa}$ for the multiplet ${}^{2S+1}L_j$ by the depolarization cross sections of $\sigma^{L;L}_{\lambda}$ for a spinless atomic state $|L\rangle$. The appropriate expression is [2.64]

$$\sigma^{j';j}_{\varkappa} = \sum_{\lambda} (2\lambda + 1)(-1)^{j'-j}(2j+1)(2j'+1) \times \begin{bmatrix} \lambda & S & S & \varkappa \\ L & L & j' & j \\ L & j' & L & j \end{bmatrix} \sigma^{L;L}_{\lambda}, \quad \text{where} \begin{bmatrix} a_1 & a_2 & a_3 & a_4 \\ b_1 & b_2 & b_3 & b_4 \\ c_1 & c_2 & c_3 & c_4 \end{bmatrix} \tag{2.191}$$

is a 12-j symbol of the second kind [2.65, 66] expressed by a sum of 6-j symbol products [2.66],

$$\begin{bmatrix} a_1 & a_2 & a_3 & a_4 \\ b_1 & b_2 & b_3 & b_4 \\ c_1 & c_2 & c_3 & c_4 \end{bmatrix} = \sum_{\varkappa} (2\varkappa + 1)(-1)^{b_1-b_2-b_3-b_4} \begin{Bmatrix} c_1 & c_2 & \varkappa \\ a_3 & a_1 & b_1 \end{Bmatrix} \times \begin{Bmatrix} c_3 & c_4 & \varkappa \\ a_3 & a_1 & b_2 \end{Bmatrix} \begin{Bmatrix} c_1 & c_2 & \varkappa \\ a_4 & a_2 & b_3 \end{Bmatrix} \begin{Bmatrix} c_3 & c_4 & \varkappa \\ a_4 & a_2 & b_4 \end{Bmatrix}. \tag{2.192}$$

Thus in the fixed electron spin approximation, N_{SL} cross sections for intramultiplet transfer and relaxation are expressed by $2L$ cross sections for the depolarization of an atomic state with orbital momentum L.

Now consider briefly the problem of calculating the irreducible kinetic coefficients when the states discussed are characterized by the quantum numbers of the hyperfine, rather than by the fine structure state. The condition (2.170) will now be rewritten in the form

$$|\Omega_{F_1' F_2'} - \Omega_{F_1 F_2}| \gg |\Gamma^{F_1' F_2'; F_1 F_2}_{\varkappa}|, \tag{2.193}$$

where F_1', F_2', F_1, F_2 are quantum numbers of the total angular momentum $\boldsymbol{F} = \boldsymbol{j} + \boldsymbol{I}$ ($\boldsymbol{I}$ is the nuclear spin vector) and the $\Omega_{F_1F_2}$ are the frequencies of hyperfine transitions. The condition (2.193) is not expected to be fulfilled and general relaxation equations of type (2.165) have to be used. If, for simplicity, we confine ourselves to transitions within one fine structure state $|j\rangle$ (2.169) will read

$$\dot{\varrho}^{F_1' F_2'}_{\varkappa' Q} = \sum_{F_1 F_2} (\Gamma^{F_1' F_2'; F_1 F_2}_{\varkappa'} + \mathrm{i}\delta_{F_1' F_1}\delta_{F_2' F_2}\Omega_{F_1 F_2})\, \varrho^{F_1 F_2}_{\varkappa' Q}. \tag{2.194}$$

Here the quantum number j is omitted and $\varrho^{F_1 F_2}_{\varkappa' Q}$ are now the spherical components of rank $\varkappa'$ of the hyperfine density matrix; $\varrho^{F_1 F_1}_{\varkappa' Q}$ describe populations and coherences of the Zeeman substates belonging to the hyperfine state $|F_1\rangle$, and $\varrho^{F_1 F_2}_{\varkappa' Q}$ are various types of hyperfine coherences.

With respect to the collision time τ_c, the hyperfine transitions are always extremely nonadiabatic: $\Omega_{F_1F_2} \ll 1/\tau_c$. Hence, the fixed nuclear spin approximation [2.56, 64] can be used to express the cross sections $\sigma_{\varkappa'}^{F_1'F_2';F_1F_2}$ via $\sigma_{\varkappa}^{j;j}$. The corresponding equation generalizing (2.191) is

$$\sigma_{\varkappa'}^{F_1'F_2';F_1F_2} = \sum_{\varkappa} (2\varkappa+1)(-1)^{F_1-F_1'} \times [(2F_1'+1)(2F_1+1)(2F_2'+1)(2F_2+1)]^{1/2} \times \begin{bmatrix} \varkappa & I & I & \varkappa' \\ j & j & F_1' & F_1 \\ j & F_2' & j & F_2 \end{bmatrix} \sigma_{\varkappa}^{j;j}. \tag{2.195}$$

For cross sections with $F_1 = F_2$ and $F_1' = F_2'$, this equation coincides with (2.191) with proper change of momenta.

Note in concluding this section that other types of angular momentum coupling can be used to express the linearly independent cross sections $\bar{\sigma}_{j'm';jm}$. If, for example, the product of two 6-j symbols in (2.161) is replaced by the sum of products of three 6-j symbols, we get

$$\sigma_{\varkappa}^{j';j} = \sum_{\mu} (2\mu+1) \begin{Bmatrix} j' & j' & \varkappa \\ j & j & \mu \end{Bmatrix} s_{\mu}^{j';j} \tag{2.196}$$

with

$$s_{\mu}^{j';j} = \frac{\pi}{k^2} \sum_{ll'} \sum_{J_\alpha J_\beta} (-1)^{j'-j+J_\alpha-J_\beta}(2J_\alpha+1)(2J_\beta+1) \times \begin{Bmatrix} j' & j & \mu \\ l' & l & J_\alpha \end{Bmatrix} \begin{Bmatrix} j' & j & \mu \\ l' & l & J_\beta \end{Bmatrix} (S_{j'l';jl}^{J_\alpha} S_{j'l';jl}^{J_\beta *} - \delta_{j'j}\delta_{l'l}). \tag{2.197}$$

It then follows, upon substituting (2.196) in (2.160) (see, e.g., [2.67, 68]), that

$$\bar{\sigma}_{j'm';jm} = \sum_{\mu} \begin{bmatrix} j' & j & \mu \\ m' & -m & m'-m \end{bmatrix}^2 s_{\mu}^{j';j}. \tag{2.198}$$

3. Diatomic Electronic States

The discussion in the preceding chapter shows that the calculation of slow atomic collision cross sections can be divided into two virtually independent stages. The first involves the determination of the adiabatic electronic wave functions and terms of a diatom. The second deals with the solution of scattering equations for the nuclei.

The accurate calculation of electronic functions and terms is often a cumbersome problem, especially if many-electron atoms are considered. However, valuable information concerning the qualitative properties of adiabatic wave functions and terms and nonadiabatic couplings can be obtained using rigorous and approximate symmetry properties of the electronic Hamiltonian of a diatom. Such a qualitative discussion will be given in this chapter. Approximate quantitative approaches to the diatomic electronic structure calculations will be considered in the next chapter.

First of all, to facilitate understanding of the treatment below, the general properties of the wave functions of a free atom will be briefly discussed. Detailed descriptions of atomic structure can be found in [3.1–6].

3.1 Quantum Numbers and Wave Functions of a Free Atom

The Hamiltonian H of a free atom (or ion) can be represented as the sum of a nonrelativistic Hamiltonian H_0 and a Hamiltonian H_{rel} involving all relativistic interactions. The spin-orbit coupling is the most important of the relativistic interactions, and only this coupling will be taken into account below. For not too heavy atoms the energy levels are basically determined by the Hamiltonian H_0, whereas H_{rel} splits the H_0 levels giving the so-called fine structure of atomic energy levels. Accordingly, it is natural to first define the wave functions and energy levels in the nonrelativistic approximation, and then to account for H_{rel} using perturbation theory.

For an atom (or ion) with nuclear charge Z and N electrons, H_0 is independent of spin variables and is of the form

$$H_0 = -\frac{1}{2}\sum_{i=1}^{N} \nabla_i^2 + \frac{1}{2}\sum_{\substack{i,k=1\\ i\neq k}}^{N} \frac{1}{r_{ik}} - \sum_{i=1}^{N} \frac{Z}{r_i}, \tag{3.1}$$

where r_{ik} is the distance between electrons i and k, and r_i is the distance between electron i and the nucleus. The Hamiltonian H_0 is invariant under spatial rotations, inversion and electrons permutations. Since these operations commute, the eigenfunctions of H_0 can be classified according to the irreducible representations of the rotation, inversion and permutation groups to which they belong. The irreducible representations of the spatial rotation group are known to be specified by the quantum number L of the orbital angular momentum, and those of the inversion group by the parity w. If a certain quantization axis Z is chosen, the wave functions of a given L are also specified by the quantum number m_L of the orbital angular momentum projection on this axis.

Since the total wave function of a system with spin must be antisymmetric relative to permutations of spatial and spin electronic coordinates, not all the possible permutation symmetry types of the coordinate wave functions are allowed. As shown, for instance, in [3.7], a coordinate wave function can only belong to representations specified by Young diagrams with no more than two columns. The number of boxes in each column defines the total spin of the system S: if the first column contains n_1, and the second n_2 boxes, then $S = (n_1 - n_2)/2$. An important property of a system consisting of particles with spin 1/2 is that the total spin specifies not only the irreducible representation of the permutation group to which the coordinate wave function can belong, but also the representation of the spin rotation group to which the appropriate spin function belongs. The antisymmetric eigenfunctions $|\bar{\gamma} S m_S L m_L\rangle$ ($\bar{\gamma}$ is the set of all quantum numbers except those shown explicitly) of the Hamiltonian H_0 involving the spin variables σ_i can be expressed in the form [3.6]:

$$|\bar{\gamma} S m_S L m_L\rangle = \sum_i \Psi^{SL}_{im_L}(\boldsymbol{r}_1, \ldots, \boldsymbol{r}_N)\, f^S_{im_S}(\sigma_1, \ldots, \sigma_N). \tag{3.2}$$

Here the coordinate wave functions $\Psi^{SL}_{im_L}$ belong to line i of the above irreducible representation of the permutation group and transform properly under spatial rotations and inversion. The spin functions $f^S_{im_S}$ belong to line i of the permutation group representation associated with that to which $\Psi^{SL}_{im_L}$ belong. The Young diagrams of these associated representations contain no more than two rows. The quantum number m_S specifies the projection of the total spin on Z axis.

The expression (3.2) involves all that can be obtained on the basis of general symmetry considerations alone. For further specification of the wave function and, in particular, for elucidation of the physical meaning of the quantum numbers $\bar{\gamma}$, approximations of a dynamical nature must be invoked. For a free atom the central field approximation appears to be good enough. It assumes that the motion of every electron in a certain average central-symmetric potential field $V(r)$ created by the nucleus and the other electrons is independent of the other electrons. Notice that in the central field approximation the symmetry of the atomic Hamiltonian is much higher than that of H_0: the Hamiltonian is invariant not only under rotations of the electron system as a whole, but also under rotations of each individual electron. Consequently, each electron can be characterized by a certain value l of the orbital angular momentum quantum number.

Since in the central field approximation the total Hamiltonian of the atom is represented as a sum of one-electron Hamiltonians, every energy level ε is

$$\varepsilon = \sum_{i=1}^{N} \varepsilon_i^{(1)}. \tag{3.3}$$

Here $\varepsilon_i^{(1)}$ is the one-electron energy level occupied by the ith electron. By virtue of the spherical symmetry of $V(r)$ these levels are obtained from the Schrödinger one-electron radial equation

$$-\frac{1}{2}\frac{1}{r^2}\frac{d}{dr}\left(r^2\frac{d\varphi_{nl}}{dr}\right) + \left[\frac{l(l+1)}{r^2} + V(r)\right]\varphi_{nl} = \varepsilon_{nl}^{(1)}\varphi_{nl} \tag{3.4}$$

and are determined by two quantum numbers, the principal, n, and the orbital, l. The total one-electron wave function (atomic orbital AO) corresponding to a certain projection m of the electron angular momentum on the Z axis takes the form (in spherical coordinates r, θ, φ)

$$\varphi_{nl}^{m}(\boldsymbol{r}) = \varphi_{nl}(r)\, Y_l^m(\theta, \varphi). \tag{3.5}$$

Generally speaking, the $\varepsilon_{nl}^{(1)}$ corresponding to the same values of n but to different values of l are different. However, this is not so for the important case of the Coulomb potential. In this case, for a given value of n, all states with allowed values of l, $0 \leqq l \leqq n-1$, are degenerate. This degeneracy is the consequence of the invariance of the one-particle Hamiltonian with a Coulomb potential under rotations in four-dimensional (rather than in the usual three-dimensional) space, see, e.g., [3.8].

An important property of the Schrödinger equation with a Coulomb potential is that it allows a separation of variables not only in spherical coordinates, but also in parabolic coordinates,

$$\nu = r + z, \qquad \mu = r - z, \qquad \varphi = \arctan(y/z), \tag{3.6}$$

such that $0 \leqq \nu < \infty$, $0 \leqq \mu < \infty$, $0 \leqq \varphi \leqq 2\pi$. For the given principal quantum number n, the Coulomb AO

$$\varphi_{nn_1n_2}^{m}(\boldsymbol{r}) = C f_{n_1,|m|}^{n}(\nu)\, f_{n_2,|m|}^{n}(\mu)\exp(im\varphi), \tag{3.7}$$

where C is a normalization constant and the parabolic quantum numbers

n_1, n_2 and the azimuthal quantum number m must obey the relation

$$n_1 + n_2 + |m| + 1 = n. \tag{3.8}$$

Explicit expressions for $f^n_{n_1,|m|}(\nu)$, $f^n_{n_2,|m|}(\mu)$ may be found in [3.1, 7]. The Coulomb AOs in parabolic coordinates are especially suitable for consideration of the influence of the uniform electric field on the hydrogen-like atom and arise naturally in the problem of a one-electron diatom for large R (Sect. 3.3).

In the many-electron atom $V(r)$ differs from the Coulomb potential due to screening effects (Sect. 4.1), but it may be expected that in two cases the AOs will not differ significantly from the Coulomb ones, namely, for the inner-shell AOs localized near the nuclei in atoms with large Z, and for the highly excited Rydberg AOs with dimensions much larger than those of the atomic core. Qualitatively, the extent of the deviation of one-electron states from hydrogen-like ones is determined by the ratio of the splitting of terms with a certain n and different l to the distance from the nearest terms with other n values. The order of magnitude of this ratio for different n of various atoms is seen in Fig. 3.1.

It follows from (3.3) that a certain distribution of electrons over levels with different n and l corresponds to every value of ε. Such a distribution is referred to as a configuration and is denoted by $(n_1 l_1)^{N_1} \ldots (n_k l_k)^{N_k}$, where N_i is the number of electrons in AOs with quantum numbers n_i and l_i. According to the Pauli principle the value N_i cannot be higher than $2(2l_i + 1)$. The energy level corresponding to a given configuration possesses a high degree of degeneracy since, even taking into account the restric-

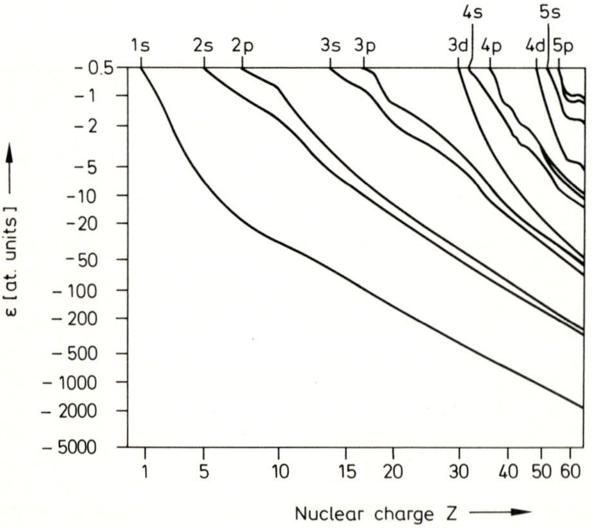

Fig. 3.1. One-electron atomic energy levels versus nuclear charge (according to [3.9])

tions inherent in the Pauli principle, there are many possible sets of quantum numbers m_{l_i} and spin projections m_{s_i} of individual electrons.

The correlation of electron motion, not accounted for in the central field approximation, results in the partial removal of this degeneracy. The most widely used scheme for constructing proper wave functions of zero-order approximation accounting for correlation is based on the fact that the interaction between electrons belonging to one shell (occupying AOs with the same n and l) is stronger than that between electrons of different shells. Therefore it is assumed at the first stage that the electrons belonging to different shells are distinguishable and the wave functions of shells $|(n_i l_i)^{N_i} S_i m_{S_i} L_i m_{L_i}\rangle$ are constructed from the antisymmetric products of AOs. These wave functions correspond to certain values of the total spin S_i of shell i, the total orbital angular momentum L_i of shell i, and their spin projections m_{S_i} and m_{L_i}. The parity w_i is determined unambiguously by the values N_i and l_i and is given by $w_i = (-1)^{N_i l_i}$.

It can be readily seen that when $l = 0$ or 1 (s and p shells) the specification of N_i and the quantum numbers S_i, m_{S_i}, L_i, m_{L_i} completely defines the wave function of the shell. If $l \geqq 2$, several different wave functions with the same S_i, m_{S_i}, L_i, m_{L_i} can be obtained for the particular N_i. Additional quantum numbers, for instance the seniority quantum number for $l = 2$ (the d shell) [3.4, 5], are then introduced for classification of the states obtained. Events requiring such quantum numbers are rather infrequent for problems connected with atomic collisions. Therefore it will always be assumed in the following that the wave functions of each shell i are completely controlled by the specification of N_i and S_i, m_{S_i}, L_i, m_{L_i}.

It might be convenient to use a representation of shell wave functions which explicitly single out the coordinates of one, for instance the first, electron. This representation is of the form

$$|(nl)^N S m_S L m_L\rangle = \sum_{S'L'} G^{SL}_{S'L'} (-1)^{N-1} \sum_{m_{S'} m_s} \sum_{m_{L'} m_l} \begin{bmatrix} S' & 1/2 & S \\ m_{S'} & m_s & m_S \end{bmatrix} \begin{bmatrix} L' & l & L \\ m_{L'} & m_l & m_L \end{bmatrix} \times |(nl)^{N-1} S' m_{S'} L' m_{L'}\rangle\, \varphi^m_{nl}(\boldsymbol{r}_1)\, \chi^{1/2}_{m_s}(\sigma_1), \tag{3.9}$$

where $G^{SL}_{S'L'}$ stands for the coefficients of fractional parentage, tabulated for instance in [3.5]; $|(nl)^{N-1} S' m_{S'} L' m_{L'}\rangle$ are antisymmetric wave functions of the shell $(nl)^{N-1}$, $\varphi^m_{nl}(\boldsymbol{r}_1)$ is the AO of the first electron and $\chi^{1/2}_{m_s}$ its spin function. The summation in (3.9) is over all values S', L' possible for the shell $(nl)^{N-1}$.

Since the AOs corresponding to different shells are orthogonal, the correct wave function of the zero-order approximation $|\bar{\gamma} S m_S L m_L\rangle$, for two shells for instance, will be

$$|\bar{\gamma} S m_S L m_L\rangle = \hat{A} \sum_{m_{S_1} m_{S_2}} \sum_{m_{L_1} m_{L_2}} \begin{bmatrix} S_1 & S_2 & S \\ m_{S_1} & m_{S_2} & m_S \end{bmatrix} \begin{bmatrix} L_1 & L_2 & L \\ m_{L_1} & m_{L_2} & m_L \end{bmatrix}$$
$$\times |(n_1 l_1)^{N_1} S_1 m_{S_1} L_1 m_{L_1}\rangle \, |(n_2 l_2)^{N_2} S_2 m_{S_2} L_2 m_{L_2}\rangle. \quad (3.10)$$

Here $\hat{A}$ is the antisymmetrization operator defined by the expression

$$\hat{A} = (N!\, N_1!\, N_2!)^{-1/2} \sum_p \delta_p \hat{P}, \quad (3.11)$$

where $\hat{P}$ stands for electron permutation operators, $\delta_p = +1$ for even and -1 for odd permutations, and the summation is over all the permutations of N electrons. It can be seen from (3.10) that, when the deviations from the central field approximation are small enough, the additional quantum numbers $\bar{\gamma}$ are represented by the configuration, as well as by the quantum numbers S_i and L_i of the shells. The expression (3.10) can be readily generalized for any number of shells.

The simplest method of accounting for the correlation of electrons is the use of perturbation theory for the electron interactions. The resulting splittings between terms of the same configuration, but with different S and L, are represented by the Slater parameters F^k and G^k [3.2, 4, 5]. More accurate calculations take account of the coupling between terms with the same S, L, w, but belonging to different configurations (configuration interaction). The most interesting implication here seems to be that at an appropriate choice of $V(r)$ even the wave functions of the central field approximation adequately represent the distribution of the electron density of atoms in the region where the density is high [3.1, 10–12].

Accounting for the spin-orbit (s.o.) coupling lowers the symmetry of the atom. The operator for this coupling $V_{\text{s.o.}}$ depends both on the spatial and on the spin coordinates of electrons and is invariant only under simultaneous spatial and spin rotations. For this reason, only the absolute value of the total atomic angular momentum and its projection on the Z axis, which correspond to the quantum numbers j and m respectively, are rigorously conserved. For not too heavy atoms the spin-orbit coupling is usually much smaller than the splitting between terms with different L and S. The LS-coupling approximation, assuming the conservation of the absolute values of the spin and orbital angular momenta precessing around the total angular momentum of the atom, appears to be valid here. The wave functions with definite j and m corresponding to the LS-coupling approximation are obtained from the wave functions $|S m_S L m_L\rangle$ using the conventional scheme of angular momenta addition,

$$|\bar{\gamma} S L j m\rangle = \sum_{m_S m_L} \begin{bmatrix} S & L & j \\ m_S & m_L & m \end{bmatrix} |\bar{\gamma} S m_S L m_L\rangle. \quad (3.12)$$

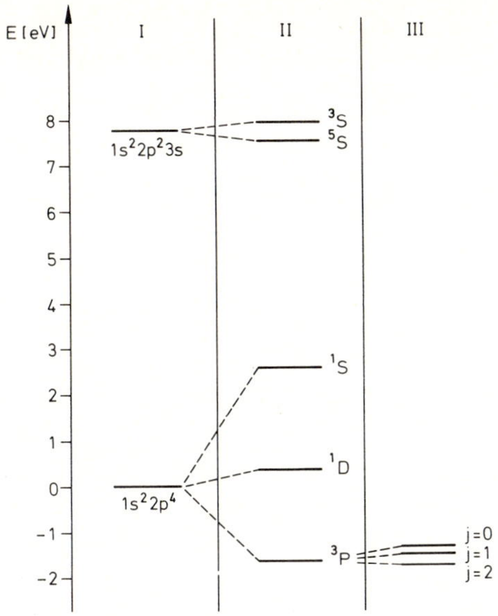

Fig. 3.2. Diagram of the lowest energy levels of the O atom. (I) refers to central field approximation (configurations), (II) to electrostatic interaction switched on (terms), (III) to spin-orbit coupling switched on (fine structure components). The fine structure splitting is given not in the scale of the figure

The mean values of $V_{s.o.}$ over the functions (3.12) give the contribution from spin-orbit coupling to atomic levels in the LS-coupling approximation.

When the spin-orbit splitting of terms with definite L and S is comparable with the energy interval between terms with different L and S, the LS-coupling approximation is invalid. However, it is often convenient to treat the secular problem arising in this case using the functions (3.12) with different L and S. The angular momentum coupling schemes yielding the best wave functions in the zero-order approximation when the spin-orbit coupling is larger than the splitting between terms with different L and S have been discussed in detail in [3.5].

Thus, if the deviations from the central field and the LS-coupling approximations are not too great, the above scheme for constructing the zero-order wave functions gives physical meaning to all the additional quantum numbers γ besides j and m that specify the state of the atom. Such a scheme based on the sequential introduction of couplings decreasing the interaction symmetry is illustrated in Fig. 3.2 for the oxygen atom.

3.2 Quantum Numbers and Wave Functions of Diatoms

3.2.1 General Classification of Adiabatic Diatomic States

The adiabatic electronic wave functions and the terms $U(R)$ of a diatom are eigenfunctions and eigenvalues of the Hamiltonian

$$H(R) = H_0(R) + V_{rel}(R). \tag{3.13}$$

The nonrelativistic Hamiltonian $H_0(R)$ of two atoms (A and B) with nuclear charges Z_A and Z_B and with a total number of electrons $N = N_A + N_B = Z_A + Z_B$ (we consider here neutral systems) is

$$H_0(R) = \sum_{i=1}^{N} \left(-\frac{1}{2} \nabla_i^2 - \frac{Z_A}{r_{iA}} - \frac{Z_B}{r_{iB}} \right) + \sum_{i>j} \frac{1}{r_{ij}} + \frac{Z_A Z_B}{R}, \tag{3.14}$$

where r_{iA}, r_{iB} stand for the distance from electron i to A and B, respectively, and r_{ij} for the separation between electrons i and j. As in the atomic case, the relativistic interaction V_{rel} is mainly determined by the spin-orbit coupling and only the latter will be accounted for hereafter.

When the relation between the spin-orbit coupling and the electrostatic interaction between the atoms is arbitrary, the symmetry group of the electronic Hamiltonian $H(R)$ is represented by the direct product $\mathcal{F} = \mathcal{P} \times \mathcal{J}$, where $\mathcal{P}$ is the electron permutation group, and $\mathcal{J}$ the geometrical symmetry group of a diatom. For a heteronuclear diatom $\mathcal{J} = C_{\infty v}$ and involves rotations around the diatomic axis and reflections in any plane containing this axis. For a homonuclear diatom $\mathcal{J} = D_{\infty h}$ and involves, in addition to the above operations, the inversion commuting with these operations. All operations of the group $\mathcal{F}$ act simultaneously on both the spatial and the spin variables of electrons.

Accordingly, the adiabatic classification of electronic molecular terms with an arbitrary relation between the spin-orbit coupling and the electrostatic interaction of the atoms is specified by the following quantum numbers:

i) Ω – the projection absolute value of the total electronic (orbital and spin) angular momentum along the diatomic axis;
ii) $\sigma(\sigma = +, -)$ – the character of reflection of the total wave function with $\Omega = 0$ in the plane containing the diatomic axis;
iii) $w(w = g, u)$ – the parity of the total wave function on inversion through the center of a diatom with like nuclei;
iv) δ – all other quantum numbers.

The first three quantum numbers specify the irreducible representation Q of the group $\mathcal{J}$. The molecular terms are denoted as $U_{\delta Q}(R)$; $Q = \Omega$ when $\Omega > 0$ and the nuclei are different, $Q = \Omega_w$ when $\Omega > 0$ and the nuclei are identical, $Q = 0^\sigma$ when $\Omega = 0$ and the nuclei are different, and $Q = 0^\sigma_w$ when $\Omega = 0$ and the nuclei are identical. The molecular terms with $\Omega > 0$ are doubly degenerate. Therefore, the complete specification of the respective wave functions would require one more quantum number. We shall specify the quantum number of the total angular momentum projection along the molecular axis, denoting it by $\tilde{\Omega}$, and will write the molecular wave functions as $|\delta Q \tilde{\Omega}\rangle$ for $\Omega > 0$, and as $|\delta Q\rangle$ for $\Omega = 0$.

Approximations of a dynamic nature increasing the symmetry of H must be introduced to specify the quantum numbers δ, just as in the case of an atom. In particular, when the spin-orbit coupling is weak compared to the electrostatic interaction of the atoms (Hund coupling case a) the group $\mathscr{F} = \mathscr{P} \times \mathscr{J}' \times \mathscr{R}_S$ will be the symmetry group of the zero-order Hamiltonian $H_0(R)$. Here, as above, $\mathscr{P}$ is the electron permutation group, $\mathscr{J}'$ is the geometrical symmetry group of a diatom (the operations from the latter group act only on the spatial variables of the electrons) and $\mathscr{R}_S$ is the rotation group in spin space. The appropriate adiabatic classification of molecular terms is specified by the following quantum numbers:

i) Λ ($\Lambda \geqq 0$) – the projection absolute value of the electronic orbital angular momentum along the diatomic axis;
ii) σ' ($\sigma' = +, -$) – the character of reflection of the spatial part of the electronic wave function with $\Lambda = 0$ in the plane containing the diatomic axis;
iii) w' ($w' = g, u$) – the parity of the spatial part of the electronic wave function on inversion through the center of a diatom with like nuclei;
iv) S – the total spin of the electronic system;
v) $\bar{\delta}$ – all other quantum numbers ($\bar{\delta} \neq \delta$).

The first three quantum numbers specify the irreducible representation Q' of the group $\mathscr{J}'$ and the quantum number S specifies the irreducible representation of group $\mathscr{R}_S$. The molecular terms in this approximation neglecting the spin-orbit coupling are denoted by $\bar{\delta}^{2S+1}Q'$; $Q' = \Pi, \Delta \ldots$, when $\Lambda = 1, 2, \ldots$ and the nuclei are different, $Q' = \Pi_{w'}, \Delta_{w'}$, when $\Lambda = 1, 2, \ldots$ and the nuclei are identical, $Q' = \Sigma^{\sigma'}$, when $\Lambda = 0$ and the nuclei are different and $Q' = \Sigma^{\sigma'}_{w'}$, when $\Lambda = 0$ and the nuclei are identical. In the Hund coupling case *a*, each term is $(2S + 1)$-fold degenerate over the total spin projections. Moreover, the additional double degeneracy for $\Lambda > 0$ is due to the possibility of two different signs of the orbital angular momentum projection along the diatomic axis. Consequently, the complete specification of the wave functions would also require prefixed quantum numbers m_S and $\tilde{\Lambda}$ (for $\Lambda > 0$) of the spin and orbital angular momentum projections, respectively, on the diatomic axis. The molecular wave functions obtained in this way for the Hund coupling case a will be denoted by $|\bar{\delta}Q'\tilde{\Lambda}Sm_S\rangle$ for $\Lambda > 0$, and by $|\bar{\delta}Q'Sm_S\rangle$ for $\Lambda = 0$.

Since the general classification can be applied to the Hund case a functions, there is a connection between the quantum numbers of the two classifications. The connection of $\tilde{\Omega}$ with m_S and $\tilde{\Lambda}$ is given by the obvious relation $\tilde{\Omega} = m_S + \tilde{\Lambda}$, and thus $\Omega = |\tilde{\Omega}| = |m_S + \tilde{\Lambda}|$. It is known that in the nonrelativistic case, the inversion does not involve the spin variables [3.3, 6] For this reason $w' = w$. To evaluate σ from the given quantum numbers for the Hund case a, it must be taken into account that the reflection operation in the plane containing the diatomic axis (the Z axis of the coordinate frame in which the molecular wave functions are determined) is $\hat{\sigma}_v = \hat{I} \times \hat{C}_{\pi Y}$, where $\hat{I}$ is the inversion operator, and $\hat{C}_{\pi Y}$ the operator for rotations through the angle π around the axis normal to this plane (below, this axis is taken to be the Y axis). Since inversion has no effect on the spin variables, $\hat{\sigma}_v$ can be represented as $\hat{\sigma}_v = \hat{\sigma}_{v'} \hat{C}^S_{\pi Y}$ where $\hat{\sigma}_{v'}$ acts only on the spatial and $\hat{C}^S_{\pi Y}$ only on the spin variables. The quantum number $\tilde{\Omega}$ can be zero either for $\tilde{\Lambda} = 0$, $m_S = 0$, or for $\tilde{\Lambda} + m_S = 0$. In the former case the function $|\bar{\delta}\Sigma^{\sigma'} S0\rangle$ already transforms correctly under the operator σ_v and

$$\hat{\sigma}_v |\bar{\delta}\Sigma^{\sigma'} S0\rangle = \hat{\sigma}_{v'} \hat{C}^S_{\pi Y} |\bar{\delta}\Sigma^{\sigma'} S0\rangle = \sigma'(-1)^S |\bar{\delta}\Sigma^{\sigma'} S0\rangle, \tag{3.15}$$

that is

$$\sigma = \sigma'(-1)^S. \tag{3.16}$$

Here use was made of

$$\hat{C}_{\pi Y} |jm\rangle = (-1)^{j-m} |j - m\rangle. \tag{3.17}$$

When $\tilde{\Lambda} + m_S = 0$, the functions transforming correctly under the action of the operator $\hat{\sigma}_v$ are

$$|\bar{\delta}\Lambda S p 0^\sigma\rangle = (|\bar{\delta}\Lambda\tilde{\Lambda} S m_S\rangle + (-1)^p |\bar{\delta}\Lambda - \tilde{\Lambda} S - m_S\rangle)/\sqrt{2}, \tag{3.18}$$

where p can assume the values 0 and 1. To evaluate σ for these functions it must be taken into account that reflection in the plane containing the axis changes the direction of rotation around this axis and, therefore,

$$\hat{\sigma}_{v'} |\bar{\delta}\Lambda\tilde{\Lambda} S m_S\rangle = \sigma_{\bar{\delta}\Lambda} |\bar{\delta}\Lambda - \tilde{\Lambda} S m_S\rangle, \tag{3.19}$$

where the factor $\sigma_{\bar{\delta}\Lambda} = \pm 1$ depends on the form of the wave function (for particular cases, see below). Accounting for (3.19) we get

$$\hat{\sigma}_v |\bar{\delta}\Lambda S p 0^\sigma\rangle = (-1)^p \sigma_{\bar{\delta}\Lambda} (-1)^{S-m_S} |\bar{\delta}\Lambda S p 0^\sigma\rangle, \tag{3.20}$$

i.e.,

$$\sigma = (-1)^p \sigma_{\bar{\delta}\Lambda} (-1)^{S-m_S}. \tag{3.21}$$

Further specification of the quantum numbers $\bar{\delta}$ is attained in different ways, unlike the atomic case, depending on whether the interaction between the atoms is smaller than or comparable to the splitting between atomic terms. We now consider the former case realized at a sufficiently large atomic separation.

3.2.2 Wave Functions of a Diatom at Large Internuclear Separations

At large atomic separations, the spin-orbit coupling can be comparable with or even stronger than the electrostatic interaction between the atoms.

The latter is referred to as the Hund coupling case c. However, consider first the situation of a spin-orbit coupling so weak that, even at large separations, it makes sense to speak of the Hund case a. The small perturbation disturbing the spherical symmetry of the intra-atomic potential energy of the electrons in every atom permits the specification of the additional quantum numbers $\bar{\delta}$ for large R. These are determined simply by the quantum numbers of the free-atom states that convert to molecular states under the action of small perturbations due to the interaction between atoms. Since we consider here the Hund case a, the spin-orbit coupling must be neglected for atoms as well. Then the problem of finding the quantum numbers $\bar{\delta}$ reduces to constructing wave functions which transform correctly under the action of the group operations from the wave functions

$$|\bar{\gamma}_1 S_1 m_{S_1} L_1 m_{L_1}\rangle_A = |\gamma_1 m_{S_1} m_{L_1}\rangle_A \quad \text{of atom A} \quad \text{and}$$
$$|\bar{\gamma}_2 S_2 m_{S_2} L_2 m_{L_2}\rangle_B = |\gamma_2 m_{S_2} m_{L_2}\rangle_B \quad \text{of atom B}$$

(throughout the rest of this chapter, the atomic angular momenta are quantized along the diatomic axis). For brevity the system of such functions will be referred to as basis a. This problem was discussed back in 1928 by *Wigner* and *Witmer* [3.13] (see also [3.14, 15]).

The wave functions (as above, the sets of quantum numbers $\bar{\gamma}_1 S_1 L_1$ and $\bar{\gamma}_2 L_2 S_2$ will be denoted by γ_1 and γ_2 for short) are now introduced:

$$|\gamma_1 \gamma_2; L\tilde{\Lambda} S m_S\rangle = \hat{A} \sum_{m_{L_1} m_{L_2}} \sum_{m_{S_1} m_{S_2}} \begin{bmatrix} L_1 & L_2 & L \\ m_{L_1} & m_{L_2} & \tilde{\Lambda} \end{bmatrix} \begin{bmatrix} S_1 & S_2 & S \\ m_{S_1} & m_{S_2} & m_S \end{bmatrix}$$
$$\times |\gamma_1 m_{S_1} m_{L_1}\rangle_A |\gamma_2 m_{S_2} m_{L_2}\rangle_B . \tag{3.22}$$

Here

$$\hat{A} = [(N_A + N_B)!\, N_A!\, N_B!]^{-1/2} \sum_p \delta_p \hat{P} \tag{3.23}$$

is the antisymmetrization operator, $\hat{P}$ stands for the electron permutation operators, the factors δ_p are $+1$ for even, and -1 for odd permutations, and the summation is over all permutations of the $(N_A + N_B)$ electrons. The normalization of $\hat{A}$ is chosen such that, at R sufficiently large that the overlap of the wave functions of atoms A and B can be neglected, the functions (3.22) are normalized to unity. The functions (3.22) are antisymmetric with respect to all electrons and correspond to definite values of the total spin S, its projection m_S on the diatomic axis, and the projection $\tilde{\Lambda}$ of the total orbital angular momentum on this axis. For definiteness let $L_1 \geqq L_2$. It follows then from (3.22) that, for any possible value of S, there are two functions in the form of (3.22) with $\Lambda = L_1 + L_2$, four functions with $\Lambda = L_1 + L_2 - 1, \ldots, 2\ (2L_2 + 1)$ functions with $\Lambda = L_1 - L_2$, $2\ (2L_2 + 1)$ with $\Lambda = L_1 - L_2 - 1 \ldots$ and $2L_2 + 1$ with $\Lambda = 0$.

Obviously, the functions (3.22) are the desired correct linear combinations for a heteronuclear diatom. It remains only to determine their character σ' under reflection in the plane containing the diatom axis. When $\Lambda = 0$, this is readily done by taking into account that $\hat{\sigma}_{v'}$ commutes with $\hat{A}$, and using (3.17) and the symmetry properties of the Clebsch-Gordan coefficients. We then obtain

$$\sigma' = w_1 w_2 (-1)^L. \tag{3.24}$$

The treatment is somewhat more complex for homonuclear diatoms, since inversion is involved in the symmetry operations. Furthermore, when the separated atoms are in different states, the set of atomic wave functions and the number of possible molecular terms are twice those for the heteronuclear diatoms since, by exchanging the atomic states, we obtain the same energy values at infinite separation.

For a homonuclear molecule, $\mathscr{J}' = C_{\infty v} \mathscr{J} = D_{\infty h}$ where the group $\mathscr{J}$ consists of the inversion $\hat{I}$ which commutes with all the symmetry operations from the group $C_{\infty v}$, and an identity element. Thus the functions which transform correctly under the action of group $D_{\infty h}$ symmetry operations must be in the form of linear combinations of the functions (3.22). The construction of correct linear combinations is elementary. More complicated is the determination of their parity w. Therefore consider in more detail the action of $\hat{I}$ on (3.22). Using the representation (3.2) for the atomic wave functions and taking into account the fact that $\hat{I}$ commutes with $\hat{A}$, we get

$$\begin{aligned} \hat{I}\,|\gamma_1\gamma_2; L\tilde{\Lambda} S m_S\rangle = \hat{A} \sum_{m_{L_1} m_{L_2}} \sum_{m_{S_1} m_{S_2}} &\begin{bmatrix} L_1 & L_2 & L \\ m_{L_1} & m_{L_2} & \tilde{\Lambda} \end{bmatrix} \begin{bmatrix} S_1 & S_2 & S \\ m_{S_1} & m_{S_2} & m_S \end{bmatrix} \\ &\times \sum_{i_1 i_2} [\hat{I}\, \Psi^{\gamma_1}_{i_1 m_{L_1} \mathrm{A}}(\boldsymbol{r}_1, \ldots, \boldsymbol{r}_{N_\mathrm{A}})\, \Psi^{\gamma_2}_{i_2 m_{L_2} \mathrm{B}}(\boldsymbol{r}_{N_\mathrm{A}+1}, \ldots, \boldsymbol{r}_{N_\mathrm{A}+N_\mathrm{B}})] \\ &\times f^{S_1}_{i_1 m_{S_1}} f^{S_2}_{i_2 m_{S_2}} \end{aligned} \tag{3.25}$$

and obviously

$$\begin{aligned} &\hat{I}\, \Psi^{\gamma_1}_{i_1 m_{L_1} \mathrm{A}}(\boldsymbol{r}_1, \ldots, \boldsymbol{r}_{N_\mathrm{A}})\, \Psi^{\gamma_2}_{i_2 m_{L_2} \mathrm{B}}(\boldsymbol{r}_{N_\mathrm{A}+1}, \ldots, \boldsymbol{r}_{N_\mathrm{A}+N_\mathrm{B}}) \\ &\quad = w_1 w_2\, \Psi^{\gamma_1}_{i_1 m_{L_1} \mathrm{B}}(\boldsymbol{r}_1, \ldots, \boldsymbol{r}_{N_\mathrm{A}})\, \Psi^{\gamma_2}_{i_2 m_{L_2} \mathrm{A}}(\boldsymbol{r}_{N_\mathrm{A}+1}, \ldots, \boldsymbol{r}_{N_\mathrm{A}+N_\mathrm{B}}). \end{aligned} \tag{3.26}$$

Taking into account the fact that $\hat{A}$ involves the permutation of all the electrons between atoms A and B, and is even for even $N_\mathrm{A} = N_\mathrm{B} = N$ and odd for odd N, we obtain after substituting (3.26) into (3.25) and using the symmetry properties of the Clebsch-Gordan coefficients,

$$\hat{I}\,|\gamma_1\gamma_2;\,L\tilde{\Lambda}Sm_S\rangle = (-1)^{N+S_1+S_2-S+L_1+L_2-L}w_1w_2\,|\gamma_2\gamma_1;\,L\tilde{\Lambda}Sm_S\rangle. \tag{3.27}$$

It follows from (3.27) that, for identical atoms in different states the proper behavior on inversion is displayed by the linear combinations

$$|\gamma_1\gamma_2;\,pL\tilde{\Lambda}Sm_S\rangle = [|\gamma_1\gamma_2,\,L\tilde{\Lambda}Sm_S\rangle + (-1)^p\,|\gamma_2\gamma_1;\,L\tilde{\Lambda}Sm_S\rangle]/\sqrt{2}, \tag{3.28}$$

where p can assume the values 0 and 1. The inversion character for these functions is

$$w = (-1)^p w_1 w_2 (-1)^{N+S_1+S_2-S+L_1+L_2-L}. \tag{3.29}$$

When the atomic states are the same, the functions are given by (3.22), and

$$w = (-1)^{S+L}. \tag{3.30}$$

Thus, at large atomic separations for the Hund coupling case a, the additional quantum numbers $\bar{\delta}$ involve quantum numbers of the initial unperturbed atomic states treated without accounting for the spin-orbit coupling, as well as the quantum number L of the total orbital angular momentum. The connection with the general classification is found according to the rules described above. The expression

$$\sigma_{\bar{\delta}\Lambda} = w_1 w_2 (-1)^{L-\Lambda} \tag{3.31}$$

is obtained for the factor $\sigma_{\bar{\delta}\Lambda}$ involved in (3.19–21).

When the spin-orbit coupling is stronger than the electrostatic interaction between the atoms (Hund coupling case c), the general classification must be used from the beginning. In this case, for large R, the general classification is detailed by introducing additional quantum numbers specifying the fine structure states of the free atoms to which the molecular term converts. Here the treatment is completely similar to that for the Hund case a. For different atoms the problem of constructing functions which correctly transform under the action of group operations from the functions $|\gamma_1 j_1 m_1\rangle_A$ and $|\gamma_2 j_2 m_2\rangle_B$ of atoms A and B (the system of such functions will be referred to hereafter as basis c) is solved using the expression

$$|\gamma_1 j_1 \gamma_2 j_2;\,j\tilde{\Omega}\rangle = \hat{A}\sum_{m_1 m_2}\begin{bmatrix} j_1 & j_2 & j \\ m_1 & m_2 & \tilde{\Omega}\end{bmatrix}|\gamma_1 j_1 m_1\rangle_A\,|\gamma_2 j_2 m_2\rangle_B. \tag{3.32}$$

Here, for $\Omega = 0$, the character of reflection in the plane containing the diatomic axis is expressed as

$$\sigma = w_1 w_2 (-1)^j, \tag{3.33}$$

similar to (3.24). The same functions transform correctly under the action of $\mathscr{F}$-group operations when identical atoms are in the same state. The character of inversion is here

$$w = (-1)^j, \tag{3.34}$$

similar to (3.30). When identical atoms are in different states, the functions transforming correctly under inversion are

$$|\gamma_1 j_1 \gamma_2 j_2; pj\Omega\rangle = (|\gamma_1 j_1 \gamma_2 j_2; j\tilde{\Omega}\rangle + (-1)^p |\gamma_2 j_2 \gamma_1 j_1; j\tilde{\Omega}\rangle)/\sqrt{2}, \tag{3.35}$$

whereby

$$w = (-1)^p w_1 w_2 (-1)^{N + j_1 + j_2 - j}. \tag{3.36}$$

When the LS-coupling approximation is adequate for the atomic states considered, the functions of basis c can obviously be expressed in terms of those of basis a. In the simple case of two different atoms with $\Omega > 0$, this expression is

$$|\gamma_1 j_1 \gamma_2 j_2; j\tilde{\Omega}\rangle = \sum_{LS} \sum_{m_S \tilde{\Lambda}} [(2S+1)(2L+1)(2j_1+1)(2j_2+1)]^{1/2} \times \begin{Bmatrix} S_1 & S_2 & S \\ L_1 & L_2 & L \\ j_1 & j_2 & j \end{Bmatrix} \begin{bmatrix} S & L & j \\ m_S & \tilde{\Lambda} & \tilde{\Omega} \end{bmatrix} |\gamma_1 \gamma_2; L\tilde{\Lambda} S m_S\rangle \tag{3.37}$$

where $\{\ldots\}$ is the $9-j$ symbol. If $\Omega = 0$ and taking into account the symmetry properties of the Clebsch-Gordan coefficients, this can be rewritten as

$$|\gamma_1 j_1 \gamma_2 j_2; j0^\sigma\rangle = \sum_{LS} [(2S+1)(2L+1)(2j_1+1)(2j_2+1)]^{1/2} \times \begin{Bmatrix} S_1 & S_2 & S \\ L_1 & L_2 & L \\ j_1 & j_2 & j \end{Bmatrix} \left(\begin{bmatrix} S & L & j \\ 0 & 0 & 0 \end{bmatrix} |\gamma_1 \gamma_2; L0S0\rangle + (2)^{1/2} \sum_{m_S > 0} \begin{bmatrix} S & L & j \\ m_S & \tilde{\Lambda} & 0 \end{bmatrix} |\gamma_1 \gamma_2; L\tilde{\Lambda} S p_{SLj} 0^\sigma\rangle \right). \tag{3.38}$$

Here the functions $|\gamma_1 \gamma_2; L\tilde{\Lambda} S p_{SLj} 0^\sigma\rangle$ are connected with (3.22) by the relation (3.18), with p_{SLj} such that $(-1)^{p_{SLj}} = (-1)^{S+L-j}$.

Note that the right side of this equation contains functions with the same values of $\sigma = w_1 w_2 (-1)^j$ since $\begin{bmatrix} S & L & j \\ 0 & 0 & 0 \end{bmatrix} = 0$ for odd $S + L - j$. For identical atoms with the same values $\gamma_1 j_1$ and $\gamma_2 j_2$, the connection between the basis c and a functions is given by the same equations (3.37, 38). When

$j_1 \neq j_2$ and the quantum numbers γ_1 and γ_2 coincide, we have for $\Omega > 0$

$$|\gamma_1 j_1 \gamma_1 j_2; pj\tilde{\Omega}\rangle = \sum_{LS} \sum_{m_S \tilde{\Lambda}} [(2S+1)(2L+1)(2j_1+1)(2j_2+1)]^{1/2} \times \begin{Bmatrix} S_1 & S_2 & S \\ L_1 & L_2 & L \\ j_1 & j_2 & j \end{Bmatrix} \begin{bmatrix} S & L & j \\ m_S & \tilde{\Lambda} & \tilde{\Omega} \end{bmatrix} \frac{1}{\sqrt{2}} [1 + (-1)^{p+2S_1+L+S+j_1+j_2+j}] \times |\gamma_1 \gamma_1; L\tilde{\Lambda} S m_S\rangle \quad (3.39)$$

and a similar expression for $\Omega = 0$. Finally, if all the quantum numbers of the identical A and B atomic states are different, we get for $\Omega > 0$

$$|\gamma_1 j_1 \gamma_2 j_2; pj\tilde{\Omega}\rangle = \sum_{LS} \sum_{m_S \tilde{\Lambda}} [(2S+1)(2L+1)(2j_1+1)(2j_2+1)]^{1/2} \times \begin{Bmatrix} S_1 & S_2 & S \\ L_1 & L_2 & L \\ j_1 & j_2 & j \end{Bmatrix} \begin{bmatrix} S & L & j \\ m_S & \tilde{\Lambda} & \tilde{\Omega} \end{bmatrix} 2^{1/2} \left\{ \frac{1}{\sqrt{2}} [|\gamma_1 \gamma_2; L\tilde{\Lambda} S m_S\rangle + (-1)^{p+S_1+L_1+S_2+L_2+j_1+j_2+S+L+j} |\gamma_2 \gamma_1; L\tilde{\Lambda} S m_S\rangle] \right\}. \quad (3.40)$$

The functions of type (3.28) belonging to basis a are given on the rhs in curly brackets. A similar expression is obtained for $\Omega = 0$. Specific calculations of the transformation matrix between sets c and a for two identical atoms in the state ^2P (of halogen atoms) can be found in [3.16].

So much attention has been paid to the construction of a and c sets from the wave functions of free atoms for two reasons. First, as was shown in Sect. 2.3, to solve the problem of atomic scattering in the molecular basis, one has to know the coefficients of transformations from atomic to molecular functions at large atomic separations. Second, the calculation of atomic interactions for large R amounts (as will be shown in Sect. 4.2) to the diagonalization of the matrix of a certain effective Hamiltonian in the basis a or c. In the theory of chemical bonding this approach is referred to as the Heitler-London approximation.

When the atomic separation is smaller and the interaction between the atoms becomes comparable with the splitting of atomic terms, the coupling of various atomic terms, including those for different configurations, must be taken into account. The molecular wave function is then represented as a linear combination involving functions of the type (3.22) with different quantum numbers and comparable weights. In such a situation the additional quantum numbers δ specifying those atomic terms into which the given atomic term converts for $R \to \infty$ provide, generally speaking, no information on the type of molecular wave functions and the position of molecular terms. Obtaining such information for small R would require another approximation that would enable the introduction of additional quantum numbers for small internuclear separations. In the theory of chemical bonding this is known as the molecular-orbital approximation.

3.2.3 Molecular-Orbital Approximation

The molecular-orbital approximation for diatoms is similar to the central field approximation for atoms, described in Sect. 3.1. The electrons in a diatom are assumed to move independently in a certain effective potential field $V(\mathbf{r}, R)$ created by the nuclei and by the other electrons. The potential $V(\mathbf{r}, R)$ depends on R as a parameter and has the symmetry of the diatom considered, i.e. $C_{\infty v}$ for a heteronuclear and $D_{\infty h}$ for a homonuclear diatom.

The nonrelativistic Hamiltonian of the diatom in the molecular-orbital approximation is

$$H_{\mathrm{MO}}(R) = \sum_{i=1}^{N} \left[-\frac{1}{2} \nabla_i^2 + V(\mathbf{r}_i, R) \right] = \sum_{i=1}^{N} H_i(R). \tag{3.41}$$

Just as in the case of an atom, the symmetry of the Hamiltonian $H_{\mathrm{MO}}(R)$ is much higher than that of the Hamiltonian $H_0(R)$. The Hamiltonian $H_{\mathrm{MO}}(R)$ is invariant not only under the group $\mathscr{J}'$ of transformations acting on the system of electrons as a whole, but also under the group of transformations acting on every individual electron.

Since $H_{\mathrm{MO}}(R)$ is represented as the sum of Hamiltonians of noninteracting electrons, the molecular terms in the molecular-orbital approximation will be

$$U(R) = \sum_{i} U_i^{(1)}(R). \tag{3.42}$$

Here the summation is over all occupied electronic states and the one-electron energy levels $U_i^{(1)}(R)$ are obtained from the eigenvalue equations

$$[-\tfrac{1}{2}\nabla^2 + V(\mathbf{r}, R)]\, \varphi_i(\mathbf{r}, R) = U_i^{(1)}(R)\, \varphi_i(\mathbf{r}, R). \tag{3.43}$$

The one-electron wave functions $\varphi_i(\mathbf{r}, R)$ are usually referred to as molecular orbitals (MO).

The potential $V(\mathbf{r}, R)$ is invariant under the group of transformations $\mathscr{J}'$ and therefore the classification of one-electron molecular levels is similar to that of many-electron levels and is given by the quantum numbers:

i) λ – projection absolute value of the orbital angular momentum of the electron on the internuclear axis,

ii) w ($w = g, u$) – MO parity for inversion through the center of a diatom with like nuclei,

iii) η – additional quantum number.

The quantum number σ (the character of MO reflection in a plane containing the diatom axis when $\lambda = 0$) is not given here, since it will be seen that only $\sigma = +1$ is possible in the one-electron case. The one-electron energy levels are denoted by $U_{\eta q}^{(1)}$ where q is the irreducible representation of the group $\mathscr{J}'$. For a heteronuclear diatom $q = \sigma, \pi, \delta, \ldots$ and for a homonuclear diatom $q = \sigma_w, \pi_w, \delta_w, \ldots$ when $\lambda = 0, 1, 2, \ldots$, respectively. Since the states with $\lambda > 0$ are doubly degenerate, the quantum numbers $\tilde{\lambda} = \pm\lambda$ will also be introduced. The MO will be denoted by $\varphi_{\eta q \tilde{\lambda}}(\boldsymbol{r}, R)$ for $\lambda > 0$ and by $\varphi_{\eta \sigma}(\boldsymbol{r}, R)$ for $\lambda = 0$.

As is seen from (3.42) a certain distribution over one-electron levels with different η and q corresponds to every energy level in the molecular-orbital approximation. As in the case of an atom, such a distribution is called a configuration and is denoted by $(\eta_1 q_1)^{N_1}(\eta_2 q_2)^{N_2} \ldots (\eta_k q_k)^{N_k}$, where N_i stands for the number of electrons on the level with quantum numbers η_i, q_i. From the Pauli principle, N_i cannot be higher than 2 for nondegenerate levels σ and 4 for doubly degenerate levels with $\lambda > 0$.

As a rule, the energy level corresponding to the chosen configuration still possesses a rather high degree of degeneracy. This degeneracy is removed in part by taking into account the correlation of electrons unaccounted for in the molecular-orbital approximation. When it is a good zero-order approximation, the resultant splitting is considerably smaller than the spacing between levels corresponding to different configurations. This permits one to find the relative order of molecular terms when the one-electron energy levels are known and to ascribe to these terms certain additional quantum numbers specifying configuration, as well as the states of the electron groups in several unoccupied shells if such shells are present. The latter is possible since, as a rule, the interaction between the electrons in any one shell is stronger than that between the electrons of different shells.

It follows from the above that in order to obtain a full classification of molecular terms in the molecular-orbital approximation, one has to find the possible types of molecular terms arising from the given configuration taking into account the interaction between electrons and to construct the appropriate correct linear combinations of antisymmetrized MO products. This problem was solved at the dawn of quantum mechanics [3.17], see also [3.18]. The results obtained will be given here in brief in a notation convenient for the present purpose.

The construction of correct linear combinations can be done in this case in a way similar to that for the atom if the Clebsch-Gordan coefficients

Table 3.1. Irreducible representations arising from the direct product of two irreducible representations for the $C_{\infty v}$ group

Q_1	$\Sigma^\pm$	Σ^+	$\Sigma^\pm$	Λ_1	Λ_1
Q_2	$\Sigma^\pm$	Σ^-	$\Lambda > 0$	$\Lambda_2 < \Lambda_1$	$\Lambda_2 = \Lambda_1$
Q	Σ^+	Σ^-	$\Lambda > 0$	$\Lambda_1 + \Lambda_2$ $\Lambda_1 - \Lambda_2$	$2\Lambda_1; \Sigma^+; \Sigma^-$

are introduced for the group $\mathscr{J}'$. These coefficients are defined, just as for the rotation group, by the relation

$$|Q_1 Q_2 Q \tilde{\Lambda}\rangle = \sum_{\tilde{\Lambda}_1 \tilde{\Lambda}_2} \begin{bmatrix} Q_1 & Q_2 & Q \\ \tilde{\Lambda}_1 & \tilde{\Lambda}_2 & \tilde{\Lambda} \end{bmatrix} |Q_1 \tilde{\Lambda}_1\rangle\, |Q_2 \tilde{\Lambda}_2\rangle, \tag{3.44}$$

where the notation is evident. Table 3.1 lists possible Q's for different Q_1's and Q_2's in the case of the group $C_{\infty v}$. The Clebsch-Gordan coefficients differ in value from the obvious values of 1 and 0 only when $Q_1 = Q_2 = \Lambda_1 > 0$. For this case

$$\begin{aligned}
\begin{bmatrix} \Lambda_1 & \Lambda_1 & 2\Lambda_1 \\ \Lambda_1 & \Lambda_1 & 2\Lambda_1 \end{bmatrix} &= \begin{bmatrix} \Lambda_1 & \Lambda_1 & 2\Lambda_1 \\ -\Lambda_1 & -\Lambda_1 & -2\Lambda_1 \end{bmatrix} = 1, \\
\begin{bmatrix} \Lambda_1 & \Lambda_1 & \Sigma^+ \\ \Lambda_1 & -\Lambda_1 & 0 \end{bmatrix} &= \begin{bmatrix} \Lambda_1 & \Lambda_1 & \Sigma^- \\ \Lambda_1 & -\Lambda_1 & 0 \end{bmatrix} = \frac{1}{\sqrt{2}}, \\
\begin{bmatrix} \Lambda_1 & \Lambda_1 & \Sigma^+ \\ -\Lambda_1 & \Lambda_1 & 0 \end{bmatrix} &= -\begin{bmatrix} \Lambda_1 & \Lambda_1 & \Sigma^- \\ -\Lambda_1 & \Lambda_1 & 0 \end{bmatrix} = \frac{1}{\sqrt{2}},
\end{aligned} \tag{3.45}$$

and the other coefficients are zero. For the group $D_{\infty h}$ everything that concerns the quantum numbers $\Lambda_1, \Lambda_2, \sigma_1, \sigma_2$ is the same as for the group $C_{\infty v}$. However it must also be taken into account that the direct products of two even or two odd representations decompose to even representations, and the direct products of even and odd representations decompose into odd ones. It follows that in constructing correct linear combinations of antisymmetrized MO products corresponding to the given configuration, all the results for the heteronuclear system can be used for the homonuclear system as well. In the latter case, the parity ascribed to the wave function is

$$w = (w_1)^{N_1}(w_2)^{N_2}\ldots(w_k)^{N_k}. \tag{3.46}$$

For simplicity, a heteronuclear diatom will be discussed here.

Just as for an atom, it is convenient to construct the proper linear combinations of antisymmetrized MO products in two stages. The first will involve the construction of antisymmetrized wave functions for shells corresponding to certain Q_i, Λ_i, S_i, m_{S_i}. These will be denoted by

Table 3.2. Quantum numbers of terms which arise from configurations $(\eta q)^N$

Shell	N	Q	S
$(\eta\sigma)^N$	1	Σ^+	1/2
	2	Σ^+	0
$(\eta\lambda)^N$	1	Λ	1/2
	2	Σ^-	1
	2	Σ^+	0
	2	2Λ	0
	3	Λ	1/2
	4	Σ^+	0

$|(\eta_i q_i)^{N_i} Q_i \tilde{\Lambda}_i S_i m_{S_i}\rangle$. The total wave functions are constructed in the second stage from the antisymmetrized products of the wave functions of the shells with the help of the Clebsch-Gordan coefficients.

The quantum numbers of all possible shell types allowed by the Pauli principle are listed in Table 3.2. The form of the $(\eta\sigma)^N$ shell wave functions is obvious. For shells $(\eta\lambda)^N$ it is convenient to introduce, just as for atoms, the coefficients of fractional parentage $G^{SQ}_{S'Q'}$. With the help of these coefficients the $(\eta\lambda)^N$ shell wave function can be written in the form

$$|(\eta\lambda)^N Q\tilde{\Lambda}Sm_S\rangle = \sum_{S'Q'} G^{SQ}_{S'Q'}(-1)^{N-1} \sum_{m_{S'}m_s} \sum_{\tilde{\Lambda}'\tilde{\lambda}} \begin{bmatrix} S' & 1/2 & S \\ m_{S'} & m_s & m_S \end{bmatrix}$$
$$\times \begin{bmatrix} Q' & \lambda & Q \\ \tilde{\Lambda}' & \tilde{\lambda} & \tilde{\Lambda} \end{bmatrix} |(\eta\lambda)^{N-1} Q'\tilde{\Lambda}'S'm_{S'}\rangle \, \varphi_{\eta\lambda\tilde{\lambda}}(\mathbf{r}_1, R)\, \chi^{1/2}_{m_s}(\sigma_1), \qquad (3.47)$$

where $|(\eta\lambda)^{N-1} Q'\tilde{\Lambda}'S'm_{S'}\rangle$ are the antisymmetric wave functions of the shell $(\eta\lambda)^{N-1}$, and the summation is over all Q', S' possible for this shell. The coefficients of fractional parentage $G^{SQ}_{S'Q'}$ for the only nontrivial case $(\eta\lambda)^3$ are listed in Table 3.3 [3.19, 20].

The relative positions of terms for different configurations are obtained within the molecular-orbital approximation from the known one-electron energy levels. Certain information about the order of terms belonging to configuration can often be obtained using a rule similar to Hund's rule for atoms: the maximum S value and afterwards the maximum Λ value for a given S correspond to the lowest term. Sometimes this rule is also used to find the relative positions of other terms belonging to this configuration.

Table 3.3. Coefficients of fractional parentage for the configuration $(\eta\lambda)^3$

$(\eta\lambda)^3$	$(\eta\lambda)^2$		
	$^1\Sigma^+$	$^1(2\lambda)$	$^3\Sigma^-$
	$1/\sqrt{6}$	$-1/\sqrt{3}$	$1/\sqrt{2}$

However, it should be emphasized that, unlike the atomic case, Hund's rule is not obeyed for a diatom, sometimes not even for the lowest term, and its use for higher terms might very often lead to wrong results (for details see [3.21]).

The quantum numbers η stand for additional quantum numbers in the one-electron problem. In limiting cases these additional quantum numbers have, just as for many-electron systems, a quite definite physical meaning enabling qualitative determinations of the relative positions of one-electron levels. Such limiting cases correspond to small and large R.

As R decreases, the field in which the electron moves increasingly resembles the field of an atom with nuclear charge $Z = Z_A + Z_B$ (i.e., a united atom). Consequently the quantum numbers η can be correlated at small R with the AO quantum numbers in the united atom. A weak field with symmetry $C_{\infty v}$ causes shift and splitting of the one-electron levels of the united atom with quantum numbers n and l to $l + 1$ levels specified by the quantum numbers λ. Then the appropriate wave functions of the zero-order approximation are obviously the AOs $\varphi_{nl}^{\bar{\lambda}}(\boldsymbol{r})$. If the perturbation field has the symmetry $D_{\infty h}$, then $w = (-1)^l$. Thus for sufficiently small R, the quantum numbers n, l of the united atom one-electron levels represent the additional quantum numbers η, the molecular one-electron levels being denoted as nlq. The method of calculating molecular terms using the wave functions of the united atom as the zero-order approximation is called the united-atom approximation (one-center approximation). This has been described in detail, e.g., in [3.22, 23].

For small R the deviation of $V(\boldsymbol{r}, R)$ from a spherically symmetric potential is small and its expansion in Legendre polynomials will rapidly converge. Consequently, the main contribution to splitting of the united-atom term is from the third term of this expansion (the second term vanishes for a homonuclear system) proportional to $P_2(\cos\theta)$. Therefore for small R

$$U_{nl\lambda}^{(1)}(R) = \varepsilon_{nl}^{(1)} + \Delta\varepsilon_{nl}^{(1)}(R) + Q_{nl}(R)\,\lambda^2 . \qquad (3.48)$$

At small internuclear separations along the Z axis, the lower energy is possessed by those states with an electron density closer to the Z axis, i.e., by those states with lower λ, and thus $Q_{nl} > 0$. An expression of the type (3.48) for one-electron terms close to the united-atom limit was first proposed by *Mulliken* [3.17]. Note that for heavy atoms, the pattern of one-electron terms near the united-atom limit can be considerably modified by the spin-orbit coupling [3.24].

Under the natural assumption (see, e.g. [3.25]) that $V(\boldsymbol{r}, R) = V(r_A, r_B)$, where r_A and r_B are the distances from the electron to nuclei A and B, respectively, it is easy to show that, for small R, $\Delta\,\varepsilon_{nl}^{(1)}$ and $Q_{nl} \sim R^2$. A rigorous proof of such power behavior of the molecular terms close to the united-atom limit for a one-electron diatom is discussed in [3.26], and for a many-electron diatom in [3.23, 27].

For large R the quantum numbers η can be correlated with the quantum numbers of free one-electron atoms. It should be emphasized, however, that passing to the limit of very large R can be, generally speaking, only formal. The point is that, as R increases, the molecular-orbital approximation becomes worse. This is due to the large contribution of the ionic configurations to the approximate wave function [3.18]. The ionic configurations correspond to charge transfer from one atom to another, i.e. for large R, to positive and negative ions. For large R the energy of such states with charge transfer is, as a rule, higher than that of the neutral atoms to which the molecule must dissociate. Therefore the additional quantum numbers introduced in the molecular-orbital approximation cannot be used for large R. However, the R values can almost always be chosen such that, on the one hand, the molecular-orbital approximation is still valid and, on the other, the shift and splitting of the one-electron molecular levels due to interaction between the atoms are considerably smaller than the splittings of the one-electron energy levels of free atoms. In such cases, the role of the additional quantum numbers can evidently be played by $n_1 l_1$ or $n_2 l_2$ of the one-electron levels of atoms A or B close, for the chosen R, to the molecular levels. It can be formally accepted that, for one-electron levels, such R correspond to the separated-atom limit. This limit will be adopted hereafter exactly in this sense.

The molecular one-electron terms arise from the free atom terms as a result of the small axially symmetric perturbation caused by the atomic interaction. In the case of heteronuclear diatoms, the AOs localized on the atom with the one-electron energy level close to the molecular level of interest are used as the zero-order approximation for the MOs. With homonuclear diatoms, the MOs of the zero-order approximation are constructed as linear combinations of equivalent AOs localized on each of the atoms, i.e. it is assumed that

$$\varphi^{\tilde{\lambda}}_{nl\lambda p}(\boldsymbol{r}) = C\,[\varphi^{\tilde{\lambda}}_{nl\mathrm{A}}(\boldsymbol{r}) + (-1)^p\,\varphi^{\tilde{\lambda}}_{nl\mathrm{B}}(\boldsymbol{r})]\,, \tag{3.49}$$

where $p = 0, 1$, $\boldsymbol{r}$ in the orbitals $\varphi^{\tilde{\lambda}}_{nl\mathrm{A}}(\boldsymbol{r})$ and $\varphi^{\tilde{\lambda}}_{nl\mathrm{B}}(\boldsymbol{r})$ is referred to the nuclei A and B, respectively, and the normalization constant

$$C = \{2\,[1 + (-1)^p\, S^{nl\lambda}_{\mathrm{AB}}]\}^{-1/2}\,, \tag{3.50}$$

where

$$S^{nl\lambda}_{\mathrm{AB}} = \int \varphi^{\tilde{\lambda}*}_{nl\mathrm{A}}(\boldsymbol{r})\,\varphi^{\tilde{\lambda}}_{nl\mathrm{B}}(\boldsymbol{r})\,d^3\boldsymbol{r} \tag{3.51}$$

is the overlap integral. It can be readily seen that this is independent of the sign of $\tilde{\lambda}$. Since under inversion in the center of the diatom, $\varphi_{nlA}^{\tilde{\lambda}} \to (-1)^l \varphi_{nlB}^{\tilde{\lambda}}(\boldsymbol{r})$ and $\varphi_{nlB}^{\tilde{\lambda}}(\boldsymbol{r}) \to (-1)^l \varphi_{nlA}^{\tilde{\lambda}}$,

$$w = (-1)^{p+l} \tag{3.52}$$

for the functions (3.49).

For a heteronuclear diatom, just as in the united-atom limit, every one-electron term of the separated atoms with given n and l yields $l+1$ molecular terms with $\lambda = 0, 1, \ldots, l$. In this case, the terms are usually denoted by $q(n\,l)_{\mathrm{A}}$ or $q(n\,l)_{\mathrm{B}}$ to distinguish them from the united-atom limit. With a homonuclear diatom, the one-electron atomic term yields $l+1$ even, and $l+1$ odd molecular terms. These are denoted by $q\,n\,l$. Note that the method of calculating molecular terms implying that, in the zero-order approximation, the MOs are approximated according to the above rules by AOs localized on different nuclei, is known as the MO LCAO method (molecular orbitals – linear combinations of atomic orbitals).

The relative positions of the one-electron molecular terms arising from atomic terms at large R obey the following rules: for a heteronuclear diatom the terms with larger λ lie higher; for a homonuclear diatom, amongst the terms of the same parity those with larger λ also lie higher; for $\lambda = 0, 2$ the even terms lie lower than the odd ones, and vice versa for $\lambda = 1$.

These rules are readily interpreted in terms of perturbation theory. It appears that, for identical atoms, the lower energy is possessed by those MOs with their electron densities concentrated between the atoms. This corresponds to the intuitive notions about the nature of chemical bonding. In quantum chemistry (see, e.g. [3.28]) such MOs are referred to as bonding MOs. The MOs possessing nodal surfaces in the region between the nuclei correspond to higher energy and are often called antibonding MOs. Perturbation theory also implies that the splitting between even and odd terms is approximately proportional to $S_{\mathrm{AB}}^{nl\lambda}$ and thus decreases exponentially with distance. Consequently, for sufficiently large R the splitting of terms with different λ's, showing a power law dependence on R (multipole interaction), exceeds that between even and odd terms with the same λ.

3.3 Adiabatic States, Diabatic States, and Correlation Diagrams

3.3.1 The Noncrossing Rule and Adiabatic Correlation Diagrams

It is evident from the discussion in Sect. 3.2 that additional quantum numbers are introduced in a different way for different R. A typical situa-

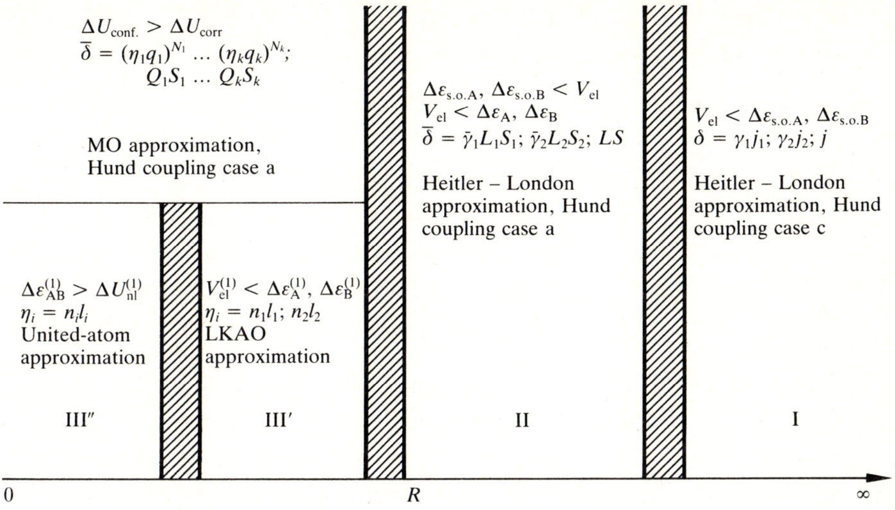

Fig. 3.3. Different sets of additional quantum numbers of a diatom and approximate methods of electronic structure calculations. V_{el} is the interaction between atoms, ΔU_{conf} is the splitting of diatomic terms belonging to different configurations, ΔU_{corr} is the splitting of diatomic terms belonging to the same configuration, $\Delta U_{nl}^{(1)}$ is the splitting of one-electron diatomic terms originating from the same term of the united atom, $V_{el}^{(1)}$ is the shift of the one-electron energy level of separated atoms due to interatomic interaction, $\Delta\varepsilon_{s.o.A}$, $\Delta\varepsilon_{s.o.B}$ are fine structure splittings in free atoms, $\Delta\varepsilon_A^{(1)}$, $\Delta\varepsilon_B^{(1)}$ are splittings of terms belonging to the same configuration in free atoms, $\Delta\varepsilon_{AB}^{(1)}$ is splitting of one-electron terms in the united atom. The regions where it is impossible to introduce unambiguously additional quantum numbers are shaded

tion is presented in Fig. 3.3. There the problem arises of correlating essentially different sets of quantum numbers. This is dealt with by invoking the noncrossing rule [3.29, 30] (see also [3.31–34]). This rule reads: the eigenvalues of the Hamiltonian $H(R)$, taken to be functions of R, do not cross if they belong to the same irreducible representations of the Hamiltonian symmetry group. The eigenvalues corresponding to different irreducible representations of this symmetry group may cross. The proof of the noncrossing rule is based on the following arguments [3.7]. Let the Hamiltonian $H(R)$ have closely lying eigenvalues $U_1(R_0)$ and $U_2(R_0)$ corresponding to the eigenfunctions $|1\rangle_0$ and $|2\rangle_0$ at a certain separation R_0 of the nuclei. Treating the Hamiltonian for R values close to R_0 as a perturbed $H(R_0)$,

$$H(R) = H(R_0) + \left.\frac{\partial H}{\partial R}\right|_{R=R_0} \delta R = \mathrm{H}(R_0) + V, \tag{3.53}$$

it can be readily seen that the difference between the energy eigenvalues at point R is

$$\Delta U_{12}(R) = (\{[U_1(R_0) + V_{11}] - [U_2(R_0) + V_{22}]\}^2 + 4|V_{12}|^2)^{1/2}, \tag{3.54}$$

where

$$V_{ik} = {}_0\langle i|V|k\rangle_0, \quad i, k = 1, 2. \tag{3.55}$$

In order that the terms cross, both terms in the radical of (3.54) must go to zero simultaneously. If $|1\rangle_0$ and $|2\rangle_0$ belong to different irreducible representations of the H(R) symmetry group, $V_{12} \equiv 0$ and (3.54) can be converted to zero by the proper choice of δR. But when $|1\rangle_0$ and $|2\rangle_0$ belong to the same irreducible representations of the Hamiltonian symmetry group, $V_{12} \not\equiv 0$ and generally speaking both terms in the radical of (3.54) cannot go to zero simultaneously, since they are functions of only one parameter δR.

We mention here the following points [3.31, 35–37]. The noncrossing rule uses a complete symmetry group of the Hamiltonian. Generally speaking, besides the symmetry groups discussed in the preceding section, this complete symmetry group might also involve a dynamic symmetry due to the specific form of the potential. The best known example of such a dynamic symmetry is that of the hydrogen atom (Sect. 3.1). The dynamic symmetry was found for one-electron diatoms (of the H_2^+-ion type) and it is connected with the possibility of separating the variables of this system in prolate spheroidal coordinates (see below). For many-electron systems such a dynamic symmetry has not been found so far. Therefore it can be expected that in using the noncrossing rule for a many-electron case it is sufficient to account for the diatomic symmetry groups discussed in Sect. 3.2.

The noncrossing rule provides a recipe for finding the correlation between terms, the relative position of which is known for two internuclear separation values R_1 and R_2. Such a correlation is usually represented by a correlation diagram (Fig. 3.4). Even a correlation diagram based on the most general formulation of the noncrossing rule unambiguously predicts

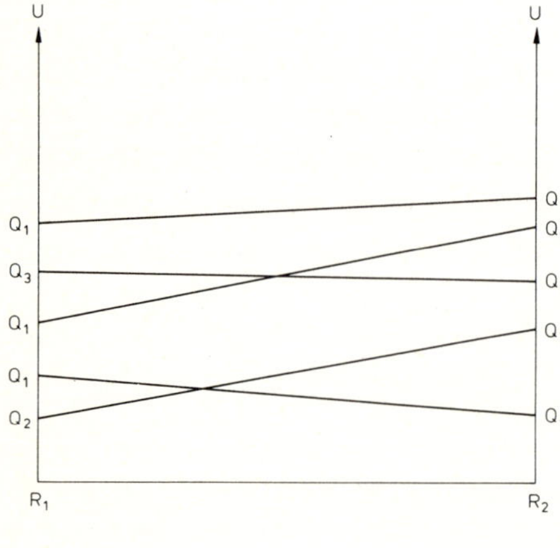

Fig. 3.4. Use of the noncrossing rule

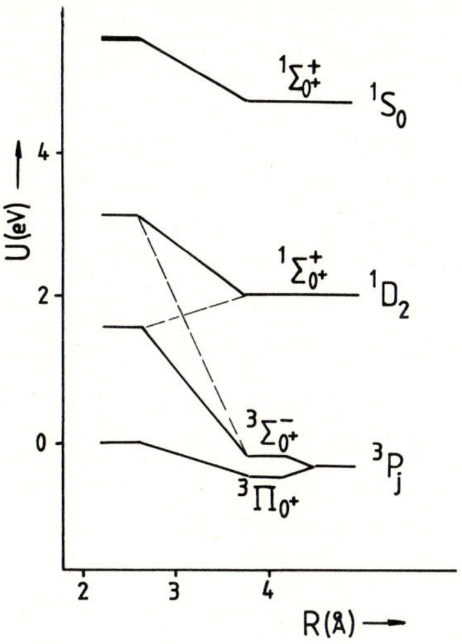

Fig. 3.5. Correlation diagram for the XeO system constructed using the noncrossing rule only. The dashed lines refer to crossing diabatic terms which do not take into account the spin-orbit coupling

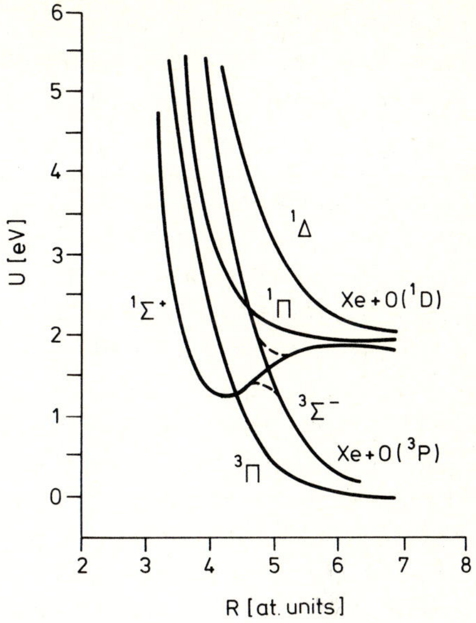

Fig. 3.6. Terms of the XeO system [3.38] neglecting spin-orbit coupling. The pseudocrossing induced by this coupling is shown by the dashed lines (not to scale)

the crossing of different symmetry terms. However the notions about the real behavior of terms obtained in this way often appear to be too crude. This is illustrated in Figs. 3.5, 6. Figure 3.5 is a correlation diagram for several low 0^+ states in the system XeO plotted from the known positions of terms for the separated atoms ($R_1 = \infty$) and from the estimated positions of terms at $R_2 \simeq 2.5$ Å. The diagram was plotted taking into account solely the requirements of the noncrossing rule in its general formulation without invoking any additional information. Figure 3.6 presents the real behavior of the terms of this system plotted with no account for the spin-orbit coupling [3.38]. The latter, of the order of 100 cm^{-1} in the case considered, results in a shift and splitting of terms (almost indiscernible on the scale of the figure) and in the appearance of a small splitting between two terms ($^1\Sigma_{0^+}^+$ and $^3\Sigma_{0^+}^-$) at the crossing point. This splitting is shown in Fig. 3.6 by the dashed line (not to scale). Such a behavior is usually referred to as an avoided crossing. Comparison of Figs. 3.5, 6 shows that a correlation diagram more correct from the standpoint of collision theory would be obtained if, contrary to the noncrossing rule, the two upper terms of the same symmetry were allowed to cross (dashed lines in Fig. 3.5). The reason for the incorrect behavior of terms constructed with direct use of the noncrossing rule in this simple example is evident. The strong electrostatic interaction between the atoms and the spin-orbit coupling, which substantially decrease the Hamiltonian symmetry, but is much weaker at the distances considered, have been assumed to be equally important.

The above example makes clear the general means for improving the correlation diagrams. Assume that the molecular terms for R_1 and R_2 are known. First the total Hamiltonian H must be partitioned into a term H_d which possesses a higher symmetry than that of H and another term V_d, in such a way that, over the entire range from R_1 to R_2 the perturbation V_d can be considered smaller than H_d. Since the symmetry of H_d is higher than that of H, the irreducible representations of the H_d symmetry group appear to be, generally speaking, reducible representations of the H symmetry group. Thus two different irreducible representations of the H_d symmetry group can contain the same irreducible representations of the H symmetry group. Therefore in the general case the H_d eigenvalues possessing the same symmetry relative to the H symmetry group may cross. This crossing will become an avoided crossing if the perturbation V_d, which is nondiagonal in the quantum numbers specifying the H_d eigenvalues, is taken into account. The eigenvalues of H_d belonging to different symmetries may also be almost parallel over a sufficiently wide range of R. The perturbation V_d in this case causes the divergence of such eigenvalues.

As a rule, when the calculation of terms is approximate, the effective symmetry of the system increases, resulting in the conversion of avoided crossings into real crossings. However, if the zero-order approximation Hamiltonian is defined with no proper account for the symmetry of the system considered, the situation might switch: the crossings of terms obtained by the rigorous solution of the problem might convert into avoided crossings. A typical example of such an approximate calculation is, as noted in [3.31, 36], the conventional variational calculation of a one-electron diatom with no account for the abovementioned dynamic symmetry of this system.

The eigenvalues and eigenfunctions of H_d possess, over a limited range of R the same properties as do the so-called diabatic molecular terms and wave functions. The notion of diabatic molecular states naturally arises when the dynamics of nonadiabatic transitions is considered. This notion is widely used and the problems connected with it will be discussed here in more detail.

3.3.2 Diabatic States and Diabatic Correlation Diagrams

The name "diabatic states" seems to have been first introduced by *Hellman* and *Syrkin* [3.39], though such states had already been used in the pioneering works of *Landau* [3.40] and *Zener* [3.41].

Usually the following qualitative definition of diabatic states is accepted. The diabatic states are approximate eigenstates of the electronic Hamiltonian such that, on the one hand, they account for the electronic interatomic interaction to a considerable extent but, on the other hand, the radial nonadiabatic coupling is weak here and may be neglected. This definition is useful if the approximate electronic Hamiltonian H_d with its diabatic eigenstates does not appreciably differ from H, and therefore the diabatic states differ substantially from the adiabatic ones only over the regions where the latter exhibit avoided crossings. The transitions between the diabatic states defined in such a way are caused by

$$V = H - H_d + V_{Cor} = V_d + V_{Cor}. \tag{3.56}$$

In particular, the diabatic states with the same values of Ω are coupled solely by the velocity-independent interaction V_d. Therefore (Sect. 7.3) the transition probability at high velocities is small here, as opposed to the adiabatic terms for which the transition probability is small at low velocities. This situation is illustrated schematically in Fig. 3.7. At low velocities the system in state I propagates along the term I–II′ (arrow a), i.e., adiabatically. At high velocities the system propagates along the term I–I′ (arrow b), i.e., diabatically.

The definition of diabatic states given here assumes that the Coriolis coupling is weaker than the electronic interaction between atoms. The situation when the Coriolis coupling cannot be considered to be weak will be discussed at the end of this section.

Lichten was the first to use this intuitive notion of diabatic states systematically in discussing the mechanisms of inelastic atomic collisions [3.9, 42–46]. His papers stimulated many other applications of this notion

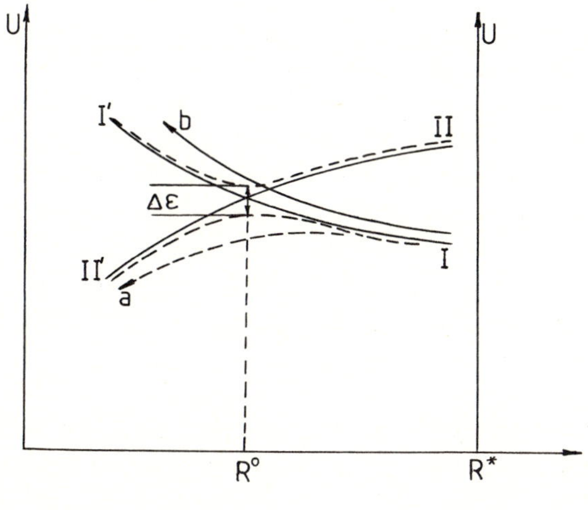

Fig. 3.7. Definition of the adiabatic and diabatic states. (*Full lines*) I – I′ and II – II′: eigenvalues of H_d, (*dashed lines*) I – II′ and II – I′: eigenvalues of $H_d + V$

to certain processes (see, e.g. [3.47–53]). At the same time, the problem of the rigorous definition of diabatic states has been repeatedly discussed [3.54–62]. This problem is not simple. Indeed, the most natural way to introduce the complete diabatic basis set is to impose the condition that the radial coupling operator $\partial/\partial R$ [or, if the electron momentum transfer is taken into account, $\partial/\partial R + \mathrm{i}\, C_{Z'}$, see (2.146)] be diagonal in this basis [3.55, 57, 62].

For simplicity, consider the collision of an atom with a structureless particle as discussed in Chap. 2. In this case the diabatic basis functions $|a\,\Omega'\rangle_\mathrm{d}$ defined in the formal way mentioned above satisfy the conditions

$$_\mathrm{d}\langle a'\,\tilde{\Omega}'|\,\frac{\partial}{\partial R}\,|a\,\tilde{\Omega}\rangle_\mathrm{d} = 0 \tag{3.57}$$

for all $a\,\Omega$; $a'\,\Omega'$. But, as noted in [3.56], if $|a\,\tilde{\Omega}\rangle_\mathrm{d}$ form a complete basis set, (3.57) is equivalent to

$$\frac{\partial}{\partial R}\,|a\,\tilde{\Omega}\rangle_\mathrm{d} = 0\,. \tag{3.58}$$

Remembering that the $|a\,\tilde{\Omega}\rangle_\mathrm{d}$ must convert to atomic wave functions in the rotating coordinate frame when $R \to \infty$, it follows from (3.58) that the $|a\,\tilde{\Omega}\rangle_\mathrm{d}$ coincide with these functions everywhere and, therefore, do not account for any of the distortion effects caused by the interatomic interaction. In the general case, when $\partial/\partial R + \mathrm{i}\, C_{Z'}$ is diagonalized in the complete basis set, the $|c\,j\,\tilde{\Omega}\rangle_\mathrm{d}$ obtained only depend on R due to electron momentum transfer effects, and do not account for distortion [3.55].

These results show that the general definition of diabatic states as states diagonalizing the nonadiabatic radial coupling everywhere may only be useful at very high collision energies when the distortion effects are unimportant. When the collision energy is of the order of the interatomic interaction, only partial diabatic representations for small numbers of states strongly coupled by nonadiabatic interactions over a limited range of R appear to be useful [3.54, 55].

If the results of the quantum-chemical calculations of adiabatic molecular terms and the radial nonadiabatic coupling matrix elements are available, the partial diabatic representation is obtained directly by diagonalization of the radial nonadiabatic coupling in the truncated basis of the strongly coupled adiabatic states. In the common situation when only two states must be taken into account, this procedure is straightforward [3.54, 57, 62]. The corresponding expressions and a discussion of the different types of two-state models of nonadiabatic couplings are given in Chap. 7.

Very often, however, information about the adiabatic molecular terms and, especially, about the nonadiabatic coupling matrix elements is poor.

In such situations, the direct method of constructing partial diabatic representations described above is not applicable and a qualitative approach is needed. A suitable qualitative approach for constructing partial diabatic representations is based on the use of approximate symmetries in the electronic Hamiltonian over different ranges of R, as discussed in Sect. 3.2. Indeed, the eigenfunctions of H_d defined in Sect. 3.3.1 depend on R and thus partially account for distortion effects. At the same time, the radial nonadiabatic couplings between the eigenstates of H_d corresponding to crossing eigenvalues vanish since, according to the noncrossing rule, these states must belong to different irreducible representations of the H_d symmetry group. But since the symmetry of H_d is higher than that of H, these crossing terms may be of the same symmetry with respect to the H symmetry group and, therefore, they are coupled by the velocity-independent interaction $V_d = H - H_d$.

Since the approximate symmetries of the electronic Hamiltonian correspond to additional quantum numbers possessing a definite physical meaning, it is possible to introduce H_d unambiguously only in the nondashed regions in Fig. 3.3. In region I,

$$H_d = H_{HL} + V_{s.o.} \tag{3.59}$$

where H_{HL} is the effective nonrelativistic Hamiltonian for the interaction of two atoms at large distances (see Sect. 4.2) and $V_{s.o.}$ is the spin-orbit coupling in free atoms. The Hamiltonian H_{HL} is diagonal in the quantum numbers $\bar{\gamma}_1 S_1 L_1; \bar{\gamma}_2 S_2 L_2$ of the free atoms. In region II,

$$H_d = H_{HL}. \tag{3.60}$$

In region III,

$$H_d = H_{MO}, \tag{3.61}$$

where H_{MO} is defined by (3.41). Such a definition of diabatic terms (sums of the one-electron energies of the MOs occupied in the given configuration) and of diabatic wave functions (determinants from the MOs belonging to one configuration) in region III coincides with the definition proposed by *Lichten* [3.43, 44]. The introduction of a one-electron diabatic Hamiltonian $H_d^{(1)}$ in region III will be discussed in the next section.

It has been assumed so far that the Coriolis coupling V_{Cor} is weaker than all the other interactions. However, in regions I and II and in the intermediate region in Fig. 3.3, the very slowly decreasing (as R^{-2}) V_{Cor} can become of the same order as $V_{s.o.}$ and the electrostatic interaction of the atoms V_{el}. The Coriolis coupling can also become of the order of the one-electron term splitting in the region III″ near the limit of the united atom. In these cases V_{Cor} must be taken into account when constructing the partial diabatic basis. A problem similar to that of the three competing interactions is known from the theory of stable diatomic molecules where different zero-order electron-nuclear wave functions diagonalizing the stronger interactions correspond to different Hund coupling cases [3.63, 64]. An extension of this notion, which will hereafter be called the dynamical Hund coupling cases, has been proposed in one of the first papers on collisional intramultiplet mixing [3.65]. Later on, this notion, though sometimes with a different terminology, was widely used in considering the collisional processes of atoms in degenerate states governed by large interatomic distances [3.55, 66–75]. A similar approach has been framed in the theory of electron-molecule collisions [3.76, 77].

Consider first the characteristics of the Hund coupling cases for stable diatoms (fixed atomic separation)[1]. Later, these considerations will be modified to enable a description of various dynamical coupling cases for a system of two colliding atoms to be made.

For stable diatoms, the various Hund cases are classified according to the relative strengths of $V_{el}(R)$ (the coupling of the electronic orbital angular momentum $\boldsymbol{L}$ with the molecular axis $\boldsymbol{n}$), $V_{s.o.}$ (coupling of $\boldsymbol{L}$ with $\boldsymbol{S}$), and V_{Cor} (coupling of $\boldsymbol{L}$ and $\boldsymbol{S}$ with the angular momentum $\boldsymbol{l}$ of the nuclei). These cases are listed in Table 3.4 which essentially coincides with the Mulliken Table [3.78] which contains cases a, b, c, d, e. Case e′ is added here for completeness so that the table covers all possible types of couplings. Since only the semiclassical approximation for nuclear motion will be discussed below, the table contains conly the electronic quantum numbers Λ, Ω, S and j (for simplicity only one of the colliding atoms is assumed to have nonzero L and S); m_S is the projection of $\boldsymbol{S}$ on $\boldsymbol{n}$, ν_S is the projection of $\boldsymbol{S}$ on the vector of the nuclear angular momentum $\boldsymbol{l}$ and ν_L, ν, are similar projections of $\boldsymbol{L}$ and $\boldsymbol{j}$, respectively. In the semiclassical approxima-

Table 3.4 Hund coupling cases

Hund coupling case	V_{el}-coupling of $\boldsymbol{L}$ with $\boldsymbol{n}$	$V_{s.o.}$-coupling of $\boldsymbol{L}$ with $\boldsymbol{S}$	V_{Cor}-coupling of $\boldsymbol{L}$, $\boldsymbol{S}$ with $\boldsymbol{l}$	Electronic quantum numbers
a	strong	intermediate	weak	Λ, S, m_S
b	strong	weak	intermediate	Λ, S, ν_S
c	intermediate	strong	weak	Ω
d	intermediate	weak	strong	L, S, ν_L, ν_S
e	weak	intermediate	strong	j, ν
e′	weak	strong	intermediate	j, ν

[1] More precisely, a certain vibrational state of the molecule, rather than a fixed separation between the atoms, would have to be invoked. The difference is immaterial for the following discussion.

tion, the direction of the vector $\boldsymbol{l}$ coincides with that of the total angular momentum of the diatom and thus defines the axis of the space-fixed coordinate frame (Sect. 6.1 and Fig. 6.1). Note that the case e′ not discussed by Mulliken coincides in its physical meaning with e. This "degeneracy" of the two cases is caused by the impossibility of a direct (rather than via $\boldsymbol{L}$) coupling of the spin with the molecular axis.

Let the separation between the atoms be R. We take as the length scale the distance R_0 for which the electrostatic interaction becomes comparable with the spin-orbit interaction, i.e. $V_{\mathrm{el}} \simeq V_{\mathrm{s.o.}}$. Then we form two dimensionless parameters $\beta = R/R_0$ (the dimensionless separation between two atoms) and $\zeta = v/\omega R_0$ (a dimensionless velocity), where v is the linear velocity of the nuclei in the rotating molecule, and ω is the transition frequency between fine structure components. Various regions of the plane ζ, β correspond to different relations between the three types of couplings (V_{el}, $V_{\mathrm{s.o.}}$, V_{Cor}). Such a representation of the various Hund coupling cases for diatoms is presented in Fig. 3.8. The regions far from boundaries correspond to pure Hund coupling cases. Close to the boundaries either the coupling type breaks down, or one type of coupling is converted into another. The most often-encountered types of coupling breakdown pertain to the intermediate a–b case. Another example is the initial stage of the $\boldsymbol{L}-\boldsymbol{n}$ decoupling which manifests itself in Λ-doubling [3.63].

Assume now that the separation of atoms is not fixed, i.e., the relative velocity is not normal to $\boldsymbol{n}$. Just as for the simplest parametrization of the relative atomic motion trajectory, we introduce the impact parameter b. Then the linear coordinate z can be defined in terms of b by the relation $z = (R^2 - b^2)^{1/2}$. For a fixed R, the adiabatic term is specified by the values of $V_{\mathrm{el}}(R)$ and $V_{\mathrm{s.o.}}(R)$ and also by the Coriolis coupling $V_{\mathrm{Cor}}(R, b) = bv/R^2$. Using the dimensionless parameters β, ζ defined above and introducing a

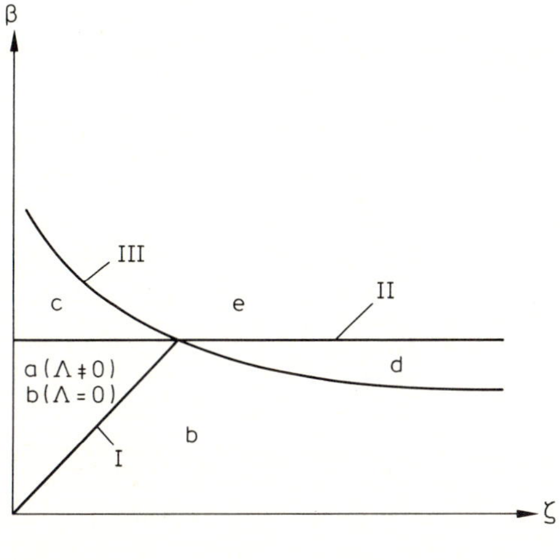

Fig. 3.8. Schematic representation of various Hund coupling cases for a diatom with a fixed internuclear separation. Lines I, II, III are the boundaries between regions of pure Hund coupling cases and refer to sections of the surfaces I, II, III in Fig. 3.9 in the plane $\xi = 0$

new parameter $\xi = z/R_0$, we construct certain surfaces separating the regions of different coupling cases in the coordinate frame β, ζ, ξ. These are shown in Fig. 3.9 for $\Lambda \neq 0$, $S \neq 0$, and $\Lambda = 0$, $S \neq 0$. If $\Lambda \neq 0$, but $S = 0$, only one surface III separates the regions of the couplings b and d. The intersection of the surfaces with the plane $\xi = 0$ is shown in Fig. 3.8. However, other intersections are of greater interest for collisions.

For slow atomic collisions when the overlap of electronic functions is small, the coupling of $\boldsymbol{L}$ and $\boldsymbol{S}$ may be assumed to be independent of R. This means that at a fixed v the parameter ζ is independent of R, i.e. the collision is described by the motion of a representative point in the plane $\zeta = \text{const}$. As mentioned above, different dynamical Hund coupling cases correspond to different diabatic basis sets and, in each region of the pure Hund coupling case, the weakest interaction may be treated as a small perturbation. Along any trajectory, not necessarily rectilinear, there occur changes in the coupling types. It is essential that, since V_{el} varies much faster with R than do $V_{\text{s.o.}}$ and V_{Cor}, the transitions between different diabatic basis sets only occur close to the boundaries between the pure-coupling-case regions. If the rate of such a change is high enough, the transition amplitude between the diabatic basis functions corresponding to the Hund cases i and k is defined simply by the projection of the basis k on the basis i. The overlap matrices for sets c and a are given in Sect. 3.2.2. The overlap matrices for sets c and e are readily obtained using the relation between the scattering equations in the atomic and molecular representations (see Sect. 2.2.2). These overlap matrix elements are simply

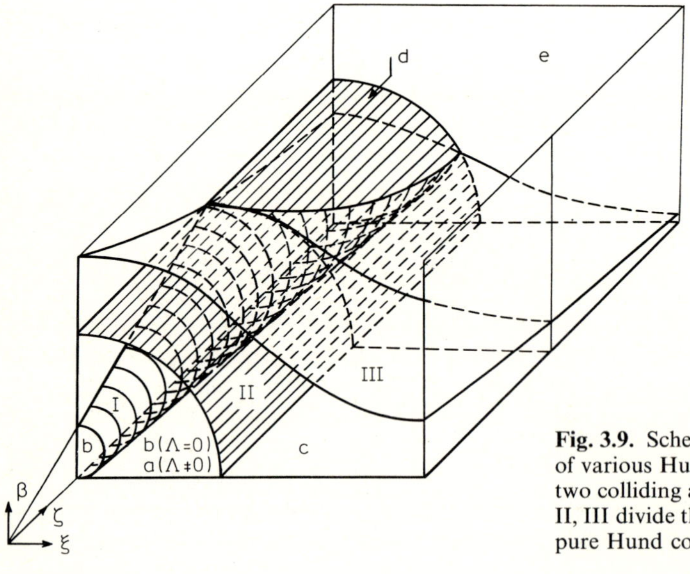

Fig. 3.9. Schematic representation of various Hund coupling cases for two colliding atoms. The surfaces I, II, III divide the regions of different pure Hund coupling cases

$D^j_{\nu\tilde{\Omega}}(0, \pi/2, 0)$. It is seen that all other overlap matrices are also expressed via the matrices for rotations through angles $\alpha = 0$, $\beta = \pi/2$, $\gamma = \pi$. If the basis switching is not too fast, one has to solve the problem of nonadiabatic coupling in a small region close to the boundary.

The use of the dynamical Hund coupling cases will be illustrated in detail in Chap. 11 where intramultiplet mixing in collisions is dealt with. The case of strong Coriolis coupling near the united-atom limit will be considered in Sect. 9.3.3.

3.3.3 One-Electron Correlation Diagrams

According to the scheme in Fig. 3.3, region III corresponding to the molecular-orbital approximation breaks up into two parts, III′ and III″. In the subregion III′ the MOs can be approximated by linear combinations of AOs localized on different atoms and, in III″, by the AOs of the united atom.

The rules discussed in Sect. 3.2.3 for the relative positions of one-electron molecular terms can be used in constructing one-electron correlation diagrams in the region between the separated and united-atom limits. Fig. 3.10 presents a correlation diagram for a homonuclear many-electron system constructed in terms of the noncrossing rule using only the symmetry properties of the total one-electron Hamiltonian. It appears, however, that for sufficiently high Z and low values of the principal quantum numbers in the separated atoms, the true adiabatic terms may behave in an essentially different way. A system of one-electron terms for a Ne_2 diatom calculated [3.25] numerically from (3.43) with $V(\boldsymbol{r}, R)$ evaluated in the statistical approximation (see Sect. 4.4.2) is presented in Fig. 3.11. The same symmetry terms with avoided crossings are seen to correlate with the states of isolated atoms both with identical n and different l (terms $2\sigma_g$ and $3\sigma_g$) and with different n (terms $3\sigma_g$ and $4\sigma_g$). Thus the situation is similar to that discussed in Sect. 3.3.1 (Fig. 3.5, 6).

Consequently the need arises for introducing the one-electron diabatic Hamiltonian $H^{(1)}_{\mathrm{d}}$. It can be seen from Fig. 3.1 that the energy splitting of inner shells with identical n and different l is small compared to that of shells with different n. Thus, it can be expected that for molecular terms correlating, for $R \to \infty$, with the atomic one-electron terms with small n, the main contribution to $V(\boldsymbol{r}, R)$ comes from the Coulomb field of the nonscreened nuclei. From these considerations, *Fano* and *Lichten* [3.42] proposed to use the Hamiltonian of a one-electron diatom with nuclear charges Z_{A} and Z_{B} as the $H^{(1)}_{\mathrm{d}}$ for the inner shell electrons. At first they only considered a symmetric system (see also [3.44]). Later on this approach was generalized for asymmetric systems as well [3.9].

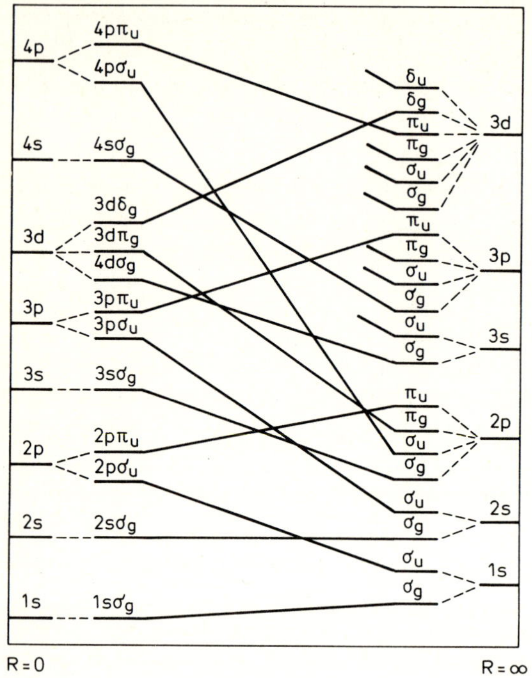

Fig. 3.10. Adiabatic correlation diagram for a homonuclear diatom, constructed using the non-crossing rule only without any dynamic considerations

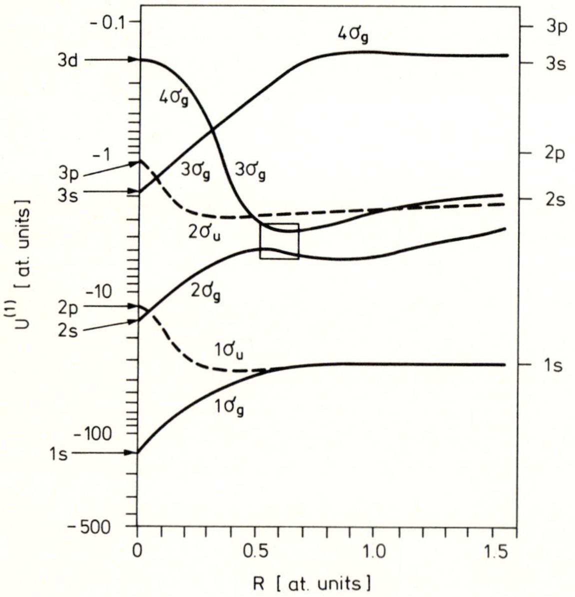

Fig. 3.11. One-electron σ terms of Ne_2 [3.79]. The full curves refer to even terms, the dashed curves to odd terms. The pseudocrossing in the square is considered in the text

The solutions of the Schrödinger equation for a one-electron diatom,

$$\left(-\frac{1}{2}\nabla^2 - \frac{Z_A}{r_A} - \frac{Z_B}{r_B}\right)\varphi(r, R) = U^{(1)}(R)\,\varphi(r, R)\,, \tag{3.62}$$

have been studied in great detail from the end of the nineteen-twenties onwards. For a discussion and references see [3.26, 80]. We shall discuss here in brief only the correlation of terms following [3.81].

The variables in (3.62) can be separated by introducing prolate spheroidal coordinates ξ, η, ϕ. These are related to the Cartesian coordinates x, y, z according to

$$\begin{aligned} x &= (R/2)\,[(\xi^2-1)(1-\eta^2)]^{1/2}\cos\phi\,,\\ y &= (R/2)\,[(\xi^2-1)(1-\eta^2)]^{1/2}\sin\phi\,,\\ z &= (R/2)\,\xi\eta\,. \end{aligned} \tag{3.63}$$

The nucleus of atom A is located at point A, with coordinates $x_A = 0$, $y_A = 0$, $z_A = -R/2$ and that of atom B is at point B, with coordinates $x_B = 0$, $y_B = 0$, $z_B = R/2$. The coordinate ξ can vary from 1 to ∞, η from -1 to 1, and ϕ from 0 to 2π. The relations inverse to (3.63) are

$$\xi = (r_A + r_B)/R\,, \quad \eta = (r_A - r_B)/R\,, \quad \phi = \arctan(y/x)\,. \tag{3.64}$$

It follows from these relations that the coordinate surfaces $\xi = \text{const}$ represent a system of confocal spheroids with foci at points A and B, and the coordinate surfaces $\eta = \text{const}$, a system of confocal parted hyperboloids of revolution with foci at the same points (Fig. 3.12a).

Putting

$$\varphi = X(\xi)\,Y(\eta)\exp(-\mathrm{i}\tilde{\lambda}\phi) \tag{3.65}$$

and substituting into (3.62), we obtain for the functions $X(\xi)$ and $Y(\eta)$

$$\frac{d}{d\xi}\left[(\xi^2-1)\frac{dX}{d\xi}\right] + \left[-\frac{\lambda^2}{\xi^2-1} + R(Z_A+Z_B)\xi - p^2\xi^2 + \mathscr{A}\right]X = 0\,, \tag{3.66}$$

$$\frac{d}{d\eta}\left[(1-\eta^2)\frac{dY}{d\eta}\right] + \left[-\frac{\lambda^2}{1-\eta^2} + R(Z_B-Z_A)\eta + p^2\eta^2 - \mathscr{A}\right]Y = 0\,, \tag{3.67}$$

where

$$p^2 = -R^2 U^{(1)}/2 \tag{3.68}$$

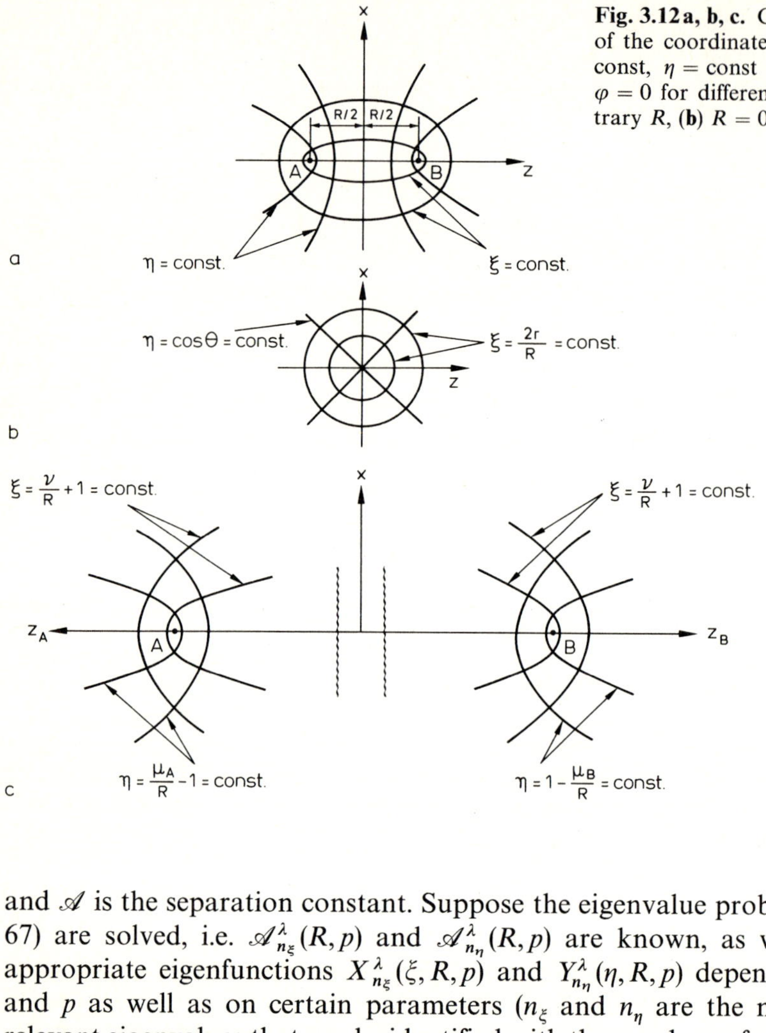

Fig. 3.12a, b, c. Cross sections of the coordinate surfaces $\xi =$ const, $\eta =$ const by the plane $\varphi = 0$ for different R. (**a**) Arbitrary R, (**b**) $R = 0$, (**c**) $R \to \infty$

and $\mathscr{A}$ is the separation constant. Suppose the eigenvalue problems (3.66, 67) are solved, i.e. $\mathscr{A}^{\lambda}_{n_\xi}(R,p)$ and $\mathscr{A}^{\lambda}_{n_\eta}(R,p)$ are known, as well as the appropriate eigenfunctions $X^{\lambda}_{n_\xi}(\xi,R,p)$ and $Y^{\lambda}_{n_\eta}(\eta,R,p)$ depending on R and p as well as on certain parameters (n_ξ and n_η are the numbers of relevant eigenvalues that can be identified with the numbers of nodes of the appropriate eigenfunctions). Then the functions in the form (3.65) will be eigenfunctions of the initial equation (3.62), if

$$\mathscr{A}^{\lambda}_{n_\xi}(R,p) = \mathscr{A}^{\lambda}_{n_\eta}(R,p)\,. \tag{3.69}$$

This equation can be solved relative to p thus leading to the energy levels $U^{(1)}_{n_\xi,\,n_\eta,\,\lambda}(R)$.

Obviously, when $R \to \infty$ the values $U^{(1)}_{n_\xi n_\eta}$ will turn into terms of hydrogen-like ions with charges Z_{A} or Z_{B}, and when $R \to 0$ into those of an ion with charge $Z_{\mathrm{A}} + Z_{\mathrm{B}}$. However, a correlation between the quantum numbers of the united and the separated atoms cannot be found here in terms of the noncrossing rule because of the abovementioned dynamical symmetry of the system considered. This additional symmetry is caused by the fact that, due to the possibility of separation of variables, the separation constant operator

$$\hat{\mathscr{A}} = -l^2 - \frac{R}{4}\left(\nabla^2 - \frac{\partial^2}{\partial z^2}\right) - R(Z_A \cos\theta_A - Z_B \cos\theta_B) - p^2 \tag{3.70}$$

commutes (along with the operator l_z of the electron angular momentum projection onto the diatomic axis) with the Hamiltonian. Here θ_A and θ_B are the angles between the z axis and the vectors $\boldsymbol{r}_A$ and $\boldsymbol{r}_B$, respectively. The eigenvalues of

$$\hat{\mathscr{A}} \text{ are } \mathscr{A}^{\lambda}_{n_\xi}(R,p) \quad\text{and}\quad \mathscr{A}^{\lambda}_{n_\eta}(R,p).$$

The possibility that terms with identical λ will cross can be shown here directly as well using the same considerations underlying the proof of the noncrossing rule. The operator $H^{(1)}_d$ in prolate spheroidal coordinates is

$$H^{(1)}_d = -\frac{2}{R^2(\xi^2-\eta^2)}\left[\frac{\partial}{\partial\xi}(\xi^2-1)\frac{\partial}{\partial\xi} + \frac{\partial}{\partial\eta}(1-\eta^2)\frac{\partial}{\partial\eta}\right.$$
$$\left. + \frac{\xi^2-\eta^2}{(\xi^2-1)(1-\eta^2)}\frac{\partial^2}{\partial\phi^2}\right] - \frac{2Z_A}{R(\xi+\eta)} - \frac{2Z_B}{R(\xi-\eta)}. \tag{3.71}$$

Differentiation of (3.71) with respect to R readily yields the expression for the operator V involved in (3.53, 55) (the values below with the superscript or subscript 0 correspond to $R = R_0$):

$$V = \left[-\frac{2}{R_0} H^{(1)}_d(R_0) - \frac{1}{R_0}\left(\frac{Z_A}{r^{(0)}_A} + \frac{Z_B}{r^{(0)}_B}\right)\right]\delta R. \tag{3.72}$$

The eigenfunctions $\varphi^{(0)}_1$ and $\varphi^{(0)}_2$ corresponding to close energy levels $U^{(1)}_{1,0}$ and $U^{(1)}_{2,0}$ are orthogonal and thus the matrix element V_{12} is

$$V_{12} = -\frac{\delta R}{R_0}\int_0^{2\pi} d\phi \int_{-1}^{1} d\eta \int_1^{\infty} d\xi \left(\frac{Z_A}{r^{(0)}_A} + \frac{Z_B}{r^{(0)}_B}\right)$$
$$\times \frac{R_0^3}{8}(\xi^2-\eta^2)\,\varphi^{(0)*}_1\varphi^{(0)}_2. \tag{3.73}$$

Substituting $\varphi^{(0)}_1$ and $\varphi^{(0)}_2$ from (3.65) and putting $\tilde{\lambda}_1 = \tilde{\lambda}_2$, we get

$$\left(\frac{1}{r^{(0)}_A}\right)_{12} = \frac{\pi R_0^2}{2}\left[\int_1^{\infty} X^{(0)}_1 \xi X^{(0)}_2 d\xi \int_{-1}^{1} Y^{(0)}_1 Y^{(0)}_2 d\eta\right.$$
$$\left. - \int_1^{\infty} X^{(0)}_1 X^{(0)}_2 d\xi \int_{-1}^{1} Y^{(0)}_1 \eta Y^{(0)}_2 d\eta\right]. \tag{3.74}$$

At the same time, it follows from (3.66, 67) that

$$(\mathscr{A}_1^{(0)} - \mathscr{A}_2^{(0)}) \int_1^\infty X_1^{(0)} X_2^{(0)} d\xi = (p_{1,0}^2 - p_{2,0}^2) \int_1^\infty X_1^{(0)} \xi^2 X_2^{(0)} d\xi ,$$
$$(\mathscr{A}_1^{(0)} - \mathscr{A}_2^{(0)}) \int_{-1}^1 Y_1^{(0)} Y_2^{(0)} d\eta = (p_{1,0}^2 - p_{2,0}^2) \int_{-1}^1 Y_1^{(0)} \eta^2 Y_2^{(0)} d\eta \tag{3.75}$$

for the functions $X_1^{(0)}$, $X_2^{(0)}$, $Y_1^{(0)}$ and $Y_2^{(0)}$.

For given values of p and R, the separation constant $\mathscr{A}$ in (3.66, 67) is an eigenvalue of the appropriate operator. Since for a one-dimensional equation the eigenvalues are not degenerate, $\mathscr{A}_1^{(0)} \neq \mathscr{A}_2^{(0)}$ (provided $\varphi_1^{(0)} \neq \varphi_2^{(0)}$ and $R \neq 0$). Thus, taking into account (3.68, 75), $(1/r_A^{(0)})_{12}$ can be written in the form

$$\left(\frac{1}{r_A^{(0)}}\right)_{12} = \frac{\pi R_0^4}{4} \frac{U_{1,0}^{(1)} - U_{2,0}^{(1)}}{\mathscr{A}_1^{(0)} - \mathscr{A}_2^{(0)}} \left(\int_1^\infty X_1^{(0)} \xi^2 X_2^{(0)} d\xi \int_{-1}^1 Y_1^{(0)} \eta\, Y_2^{(0)} d\eta - \int_1^\infty X_1^{(0)} \xi X_2^{(0)} d\xi \int_{-1}^1 Y_1^{(0)} \eta^2 Y_2^{(0)} d\eta \right). \tag{3.76}$$

A similar approach for $(1/r_B^{(0)})_{12}$ gives

$$V_{12} = \text{const.}\,(U_{1,0}^{(1)} - U_{2,0}^{(2)})\, \delta R. \tag{3.77}$$

Thus, in the approximation considered V_{12} goes to zero together with $U_{1,0}^{(1)} - U_{2,0}^{(1)}$ and consequently the crossing of terms with the same λ is possible.

The correlation between the quantum numbers of the united and the separated atoms for this case is based on the following considerations. The functions $X_{n_\xi}(\xi)$ and $X_{n_\eta}(\eta)$ represent solutions to the Sturm-Liouville problem which depends on the parameter R. The number of nodes of such functions (n_ξ and n_η) is known to remain unchanged with varying R. At the same time the prolate spheroidal coordinates turn into spherical coordinates when $R \to 0$, and into two sets of parabolic coordinates when $R \to \infty$ (see below). Thus the quantum numbers n_ξ and n_η which remain unchanged with variations in R will be expressed via the quantum numbers (numbers of nodes) of the states of the united atom treated in spherical coordinates, and of the separated atoms treated in parabolic coordinates. By virtue of the constancy of n_ξ and n_η, this ensures correlation between the quantum numbers of the united and the separated atoms.

The relation of n_ξ and n_η to the quantum numbers n and l of the united atom is readily obtainable. It follows from (3.64) that for small R

$$\xi \to 2r/R, \qquad \eta \to \cos\theta. \tag{3.78}$$

Thus the prolate spheroidal coordinates turn into spherical coordinates r, θ and ϕ and the coordinate surfaces degenerate to spheres and cones (Fig. 3.12b). Replacing η by $\cos\theta$ in (3.67) and omitting the terms going to zero when $R \to 0$, (3.67) will convert to the equation for the associated Legendre functions and

$$\mathscr{A}_{R=0} = -l(l+1), \tag{3.79}$$

$$Y^{\lambda}_{n_\eta}(\eta)_{R=0} = \text{const.}\ \mathrm{P}^{\lambda}_{l}(\cos\theta). \tag{3.80}$$

Substituting this value of $\mathscr{A}$ into (3.66), replacing ξ by $2r/R$ and retaining only terms of zero order in R, we obtain an equation for the radial wave functions of the hydrogen-like ion with charge $Z_\mathrm{A} + Z_\mathrm{B}$. Therefore

$$X^{\lambda}_{n_\xi}|_{R=0} = R_{nl}(r). \tag{3.81}$$

Since the associated Legendre polynomials $\mathrm{P}^{\lambda}_{l}(\cos\theta)$ have $l-\lambda$ zeros and the radial Coulomb functions $R_{nl}(r)$ have $n-l-1$ zeros,

$$n_\xi = n - l - 1, \tag{3.82}$$

$$n_\eta = l - \lambda. \tag{3.83}$$

For $R \to \infty$ the system of prolate spheroidal coordinates splits up into two sets of parabolic coordinates, namely, in the limit $R \to \infty$, the spheroids $\xi = \text{const.}$ convert to two paraboloids of revolution directed head-on with foci at points A and B, and the sheets of hyperboloids of revolution $\eta = \text{const.}$ corresponding to positive and negative signs of η convert to two oppositely directed paraboloids of revolution with foci at the same points (Fig. 3.12c).

It follows from (3.63) that for very large R, the variables

$$v = (\xi - 1)\,R, \tag{3.84}$$

$$\mu_\mathrm{A} = (1+\eta)\,R \tag{3.85}$$

with the additional conditions $v, \mu_\mathrm{A} \ll R$ are parabolic coordinates in the region close to the A nucleus and the z_A axis of the corresponding Cartesian coordinate frame is directed away from the B nucleus. Similarly, the variables v defined by (3.84), and

$$\mu_\mathrm{B} = (1-\eta)\,R \tag{3.86}$$

with the same conditions $v, \mu_\mathrm{B} \ll R$ are parabolic coordinates in the region close to nucleus B, and the z_B axis of the corresponding Cartesian coordinate frame is directed away from nucleus A.

Consider for simplicity a homonuclear system. In this case the Hamiltonian (3.71) is invariant under inversion through the origin of coordinates, for which $\xi \to \xi, \eta \to -\eta, \phi \to \phi + \pi$. Consequently, the function $Y^{\lambda}_{n_\eta}(\eta)$ will be either even or odd. It follows from (3.84–86) that, under inversion, $\mu_A \to \mu_B$, $\mu_B \to \mu_A$ and v remains unchanged. Thus, if atom A is in a state specified by the principal quantum number n_A, the parabolic quantum numbers n_1 and n_2 and the magnetic quantum number $\tilde{\lambda}$, the correct linear combination in the zero approximation when $R \to \infty$ will be

$$\varphi_{n_1 n_2 n_A \tilde{\lambda} \pm} = \text{const.} f^{n_A}_{n_1 \lambda}(v) \, [f^{n_A}_{n_2 \lambda}(\mu_A) \pm f^{n_A}_{n_2 \lambda}(\mu_B)] \exp(i\tilde{\lambda}\phi), \tag{3.87}$$

where $f^{n_A}_{n_1 \lambda}$, $f^{n_A}_{n_2 \lambda}$ are the Coulomb functions in parabolic coordinates (see Sect. 3.1). These functions have n_1 and n_2 nodes, respectively. When $n_1 > n_2$ the electron density for positive z_A or z_B is greater than that for negative values and vice versa.

It can be seen that after replacing ξ by v from (3.84), (3.66) converts, for $R \to \infty$, to an equation for $f^{n_A}_{n_1 \lambda}(v)$. Therefore it follows from the conservation of the number of nodes that

$$n_\xi = n_1 . \tag{3.88}$$

After replacement of η by μ_A or μ_B according to (3.85) or (3.86), (3.67) for $Y^\lambda_{n_\eta}(\eta)$ converts to an equation for the functions $f^{n_A}_{n_2 \lambda}(\mu_A)$ or $f^{n_A}_{n_2 \lambda}(\mu_B)$, respectively. Thus, taking account of the fact that both functions have n_2 nodes [moreover in the case of a minus sign an additional node appears in (3.87) for $\eta = 0$], we obtain

$$n_\eta = 2n_2 \quad \begin{cases} \text{if the term and } \lambda \text{ are even} \\ \text{if the term and } \lambda \text{ are odd}, \end{cases} \tag{3.89}$$

$$n_\eta = 2n_2 + 1 \begin{cases} \text{if the term is odd and } \lambda \text{ is even} \\ \text{if the term is even and } \lambda \text{ is odd}. \end{cases} \tag{3.90}$$

Combining (3.82, 83, 88–90) and taking into account that $n_A = n_1 + n_2 + \lambda + 1$, we obtain the following relations between the quantum numbers of the united-atom terms and the separated-atom terms which convert into the same molecular term:

$$\left.\begin{aligned} n &= n_A + n_2 \\ l &= \lambda + n_2 \end{aligned}\right\} \text{ if } l \text{ and } \lambda \text{ differ by an even number,} \tag{3.91}$$

$$\left.\begin{aligned} n &= n_A + n_2 + 1 \\ l &= \lambda + 2n_2 + 1 \end{aligned}\right\} \text{ if } l \text{ and } \lambda \text{ differ by an odd number.} \tag{3.92}$$

In constructing the correlation diagrams, one must also take into account, besides the rules (3.91, 92), the behavior of terms for large and small R. For small R the splitting of the molecular terms obeys the same rules as those for the many-electron case considered above. For large R (appreciably exceeding the effective range of the AOs), the position of one-electron terms is essentially governed by the linear Stark effect. It can be readily seen that when R is large

$$U^{(1)}_{n_1 n_2 \lambda} = -\frac{Z_A^2}{n_A^2} - \frac{Z_A}{R} - \frac{3\,n_A(2n_2 + \lambda + 1 - n_A)}{2R^2} \tag{3.93}$$

for the symmetric system considered, and thus the terms with a higher n_2 lie lower for given n_A and λ. This is consistent with the abovementioned asymmetry of the electron density distribution in the hydrogen-like ion in states with definite parabolic quantum numbers.

A correlation diagram constructed for the σ terms of a homonuclear one-electron molecule using the rules (3.91, 92) with account for Stark splitting at large R is presented in Fig. 3.13. The most important features of this diagram are: first, the crossings of terms of the same symmetry (with respect to the group $D_{\infty h}$) and, second, the increase in the principal quantum number for many terms in passing from the separated atom to the united-atom limit [this is evident even from (3.91, 92)]. This increase in the principal quantum number is usually referred to as promotion. It should be pointed out that the states lying lower at large R are more strongly promoted since they correspond to higher n_2 values.

In passing to the many-electron case, the above rules for constructing correlation diagrams remain the same in all that concerns the united-atom limit since, in this limit, the prolate spheroidal coordinates convert to spherical coordinates which allow for a separation of variables when nuclear charge screening is taken into account. Certain difficulties arise with respect to the separated-atom limit.

The point is that in view of the nuclear charge screening by the other electrons, only the quantum numbers n_A, l_A specifying the one-electron terms of the separated atoms are of importance in this limit when the variables are separated in spherical coordinates, whereas the abovementioned rules for constructing correlation diagrams operate with the parabolic quantum numbers of the separated atoms. Originally [3.9, 42, 44] this difficulty in the construction of diabatic correlation diagrams was overcome for the many-electron case in the following way. The left part of Fig. 3.13 (closer to the united-atom limit) was retained. The right part was modified in such a way that the higher-lying terms of the same symmetry converting in the initial Coulomb diagram to a term with a given n_A were correlated with the higher-lying, screening-split terms of the separated atoms with the same n_A but different l_A. The Stark effect was completely neglected. A diabatic correlation diagram constructed in this way for the homonuclear case is presented in Fig. 3.14. It shows no localized avoided crossings of terms correlating with the terms of the separated atoms with different l_A and the same n_A of the type marked by the square in Fig. 3.11.

Fig. 3.13. Correlation diagram for the σ terms of a one-electron homonuclear diatom

Fig. 3.14. Diabatic one-electron correlation diagram for the σ terms of a homonuclear many-electron diatom, constructed without accounting for the strong Stark effect [3.44]

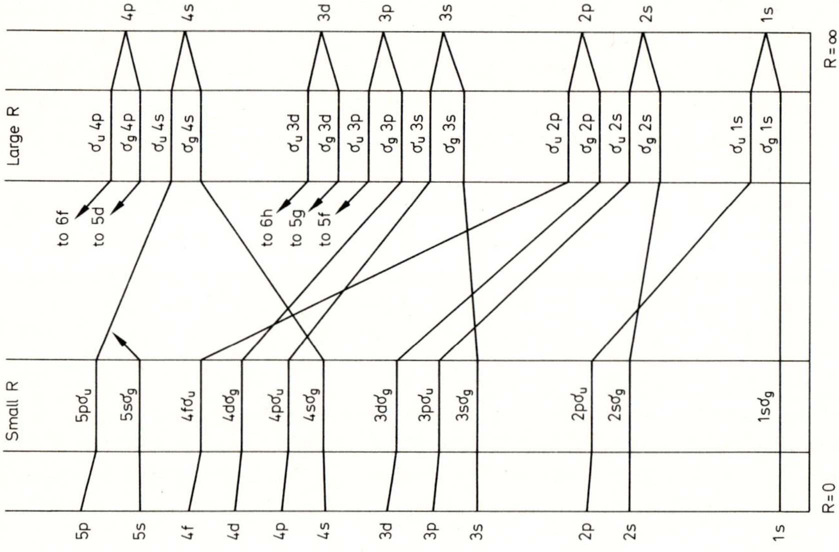

Fig. 3.14

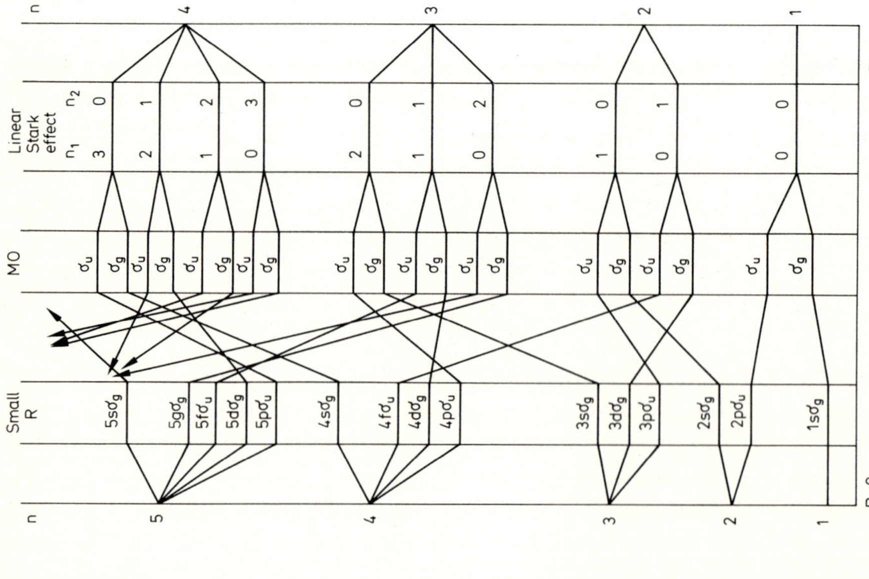

Fig. 3.13

The following qualitative considerations seem to show the inconsistency of the above approach and suggest an alternative scheme of constructing diabatic correlation diagrams [3.79, 82].

For inner atomic shells where the Coulomb field provides a satisfactory zeroth approximation, the splitting of terms with the same n_A and different l_A is much smaller than that of terms with different n_A. Estimates show that there is a distance R^* still appreciably exceeding the inner-shell radii of a many-electron atom (these are of the order $1/Z_A$) such that the Stark splitting of levels due to the interaction between atoms and calculated with no account for nuclear charge screening exceeds the splitting of atomic terms with the same n_A and different l_A. At the same time, the Stark splitting is still small compared to that of atomic terms with different n_A. In such a situation, it is natural to take as the zero-order wave functions linear combinations of AOs with different l_A and the same n_A corresponding to definite parabolic quantum numbers. The relative positions of the zero-order terms are essentially controlled by the linear Stark effect and are therefore the same as in Fig. 3.13.

For $R > R^*$ the dynamic symmetry of the Coulomb field is completely broken by screening. Therefore the diagrams correlating the terms of the separated atoms and terms modified by the Stark effect must be constructed simply according to the noncrossing rule for terms of the same symmetry. Since the splitting of g- and u terms is exponentially small for large R, this splitting can be neglected for $R > R^*$. In fact it is possible in this range to consider one atom under the action of a uniform electric field created by the other atom. This corresponds to an effective symmetry $C_{\infty v}$ in this region. A diabatic correlation diagram constructed in this way is presented in Fig. 3.15. Besides the crossings in Fig. 3.14, it also contains the above-mentioned additional crossings of terms of the same symmetry correlating with states of separated atoms with the same n_A and different l_A. These crossings are due to the fact that the lower-lying terms of separated atoms with lower l_A, which correlate (in passing from $R = \infty$ to $R = R^*$) with the lower-lying Stark sublevels with higher n_2 are more strongly promoted.

It follows from the above that the introduction of $H_d^{(1)}$ in the many-electron case is in fact possible for $R < R^*$ only and R^* is not larger than 1 Å for the inner shells. Therefore there is no need to account for the $H_d^{(1)}$ behaviour when $R \to \infty$. This permits one to go beyond the limits of the rigorous Coulomb approximation in finding $H_d^{(1)}$. The general expression for the potential $V(\mathbf{r}, R)$ allowing the separation of variables in prolate spheroidal coordinates is

$$V(\mathbf{r}, R) = \frac{4}{R^2(\xi^2 - \eta^2)} [f(\xi, R) + g(\eta, R)]. \tag{3.94}$$

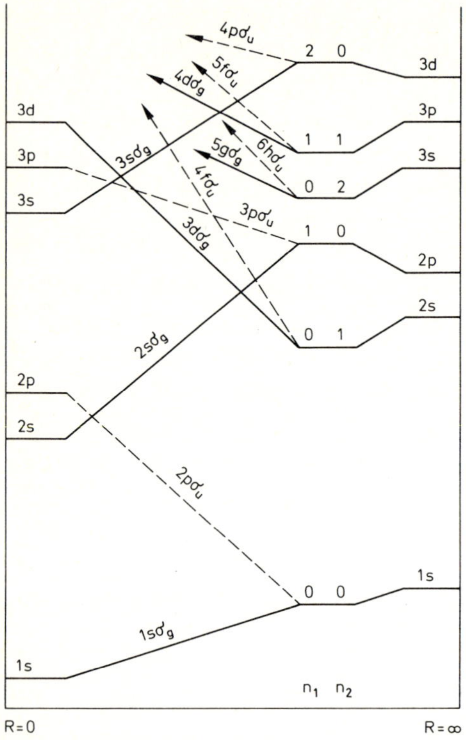

Fig. 3.15. Diabatic one-electron correlation diagram for the σ terms of a homonuclear many-electron diatom, constructed with account for the strong Stark effect

In the above case of two identical Coulomb centers each with a charge Z_A,

$$\begin{aligned} f(\xi, R) &= -2 Z_A R \xi/2, \\ g(\eta, R) &= 0. \end{aligned} \tag{3.95}$$

A natural generalization of these expressions is

$$\begin{aligned} f(\xi, R) &= -2 Z_A \frac{R\xi}{2} F\left(\frac{R\xi}{2}, R\right), \\ g(\eta, R) &= R \bar{g}(\eta, R). \end{aligned} \tag{3.96}$$

Here

$$F\left(\frac{R\xi}{2}, R\right)\Bigg|_{R\to 0} \to F(r), \tag{3.97}$$

where $F(r)$ is the screening function for the united atom (Sect. 4.1), and $\bar{g}(\eta, R)$ is an even function of η remaining finite when $R \to 0$.

The potential $V(\boldsymbol{r}, R)$ defined by (3.94, 96) evidently gives an accurate description of screening in the united-atom limit. For $R \sim R^*$, in the region close to nucleus A (B) this potential reduces to a sum of the potentials of a point charge localized at the point A (B) and of the uniform electric field created by the other nucleus. The screening effect can also be partly taken into account by choosing the potential parameters such that the point charge is not equal to Z_A (Z_B). All the above rules for constructing correlation diagrams based on the Coulomb approximation will be retained when using $V(\boldsymbol{r}, R)$ defined in this way. However, it now becomes clear that for small R these diagrams will, in fact, be much better than it would seem from the initial derivation strictly based on a Coulomb approximation. It should be emphasized in particular that for small R the screening can be almost completely (completely in the limit $R \to 0$) accounted for in $V(\boldsymbol{r}, R)$. It follows that in the region of avoided crossings the splittings of terms with the same symmetry can be much smaller than the screening-induced splittings of the united atom terms with the same l and different n. A detailed discussion of the calculation of diabatic one-electron terms based on non-Coulomb $V(\boldsymbol{r}, R)$ of the type (3.94, 96) can be found in [3.83–87].

The above discussion concerning the construction of diabatic correlation diagrams pertained to a homonuclear diatom. No fundamental changes would be involved for a heteronuclear diatom. Only finding the corre-

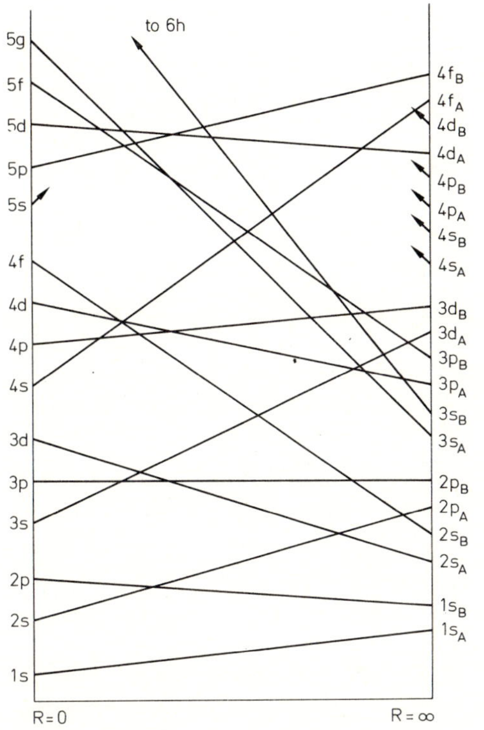

Fig. 3.16. Diabatic one-electron correlation diagram for the σ terms of a slightly asymmetric many-electron heteronuclear diatom ($Z_A > Z_B$), constructed with account for the strong Stark effect [3.79]

lation between n_η and the parabolic quantum numbers would be somewhat more difficult. This question has been discussed in [3.26, 81]. Examples of diabatic correlation diagrams for heteronuclear diatoms in the Coulomb and the many-electron cases can be found in [3.9]. It should be pointed out that the above statement about the inconsistency of neglecting the strong Stark effect refers to these diagrams in the many-electron case as well. A diabatic correlation diagram for a slightly asymmetric diatom accounting for a strong Stark effect is given in Fig. 3.16 [3.79].

3.4 Coupling Between Electronic States. Selection Rules

The general expressions for the marix elements of nonadiabatic coupling terms between accurate adiabatic electronic states are given in Sects. 2.2.2 and 2.3.1. In practice such expressions are often inconvenient. Usually the eigenfunctions of various diabatic Hamiltonians are introduced in the various regions presented in Fig. 3.3, whereas $V_d = H - H_d$ is treated as a perturbation along with the radial coupling V_R and the Coriolis coupling V_{Cor}. Here we shall consider the coupling selection rules in various regions under the assumption that the Coriolis coupling is weaker than other interactions. This corresponds to the Hund coupling cases a and c which usually occur at not very large separations. Often only the couplings between two states may be considered at such separations. This provides a direct connection between the selection rules and nonadiabatic transition probabilities. The Hund coupling cases b, d, e, and e′ usually occur at very large separations where the interactions between diabatic states is of minor importance. Here the transition probabilities are mainly controlled by the abrupt changes in the Hund coupling schemes. The corresponding, essentially many-state, dynamical problem usually cannot be treated within the framework of perturbation theory.

As a rule, the calculation of terms and wave functions accurately accounts for the spin-orbit coupling only for region I and for the region intermediate between I and II (Fig. 3.3). Therefore there is reason to speak about coupling between exact adiabatic states only with respect to these regions. The coupling operator V^{I} in these regions is actually

$$V^{\mathrm{I}} = V_R + V_{Cor}, \tag{3.98}$$

where the matrix elements of the radial coupling V_R and of the Coriolis coupling V_{Cor} are defined by (2.104, 108, 146, 147).

For smaller R (regions II, III, and the intermediate region in Fig. 3.3), the molecular terms are usually calculated with no account for spin-orbit coupling. Consequently, even with the accurate solution of the nonrelativistic Schrödinger equation with the Hamiltonian H_0, these terms are partly diabatic and $H_d = H_0$, $V_d = V_{s.o.}$. Since the eigenfunctions of H_0 depend on R, the coupling operator in these regions will be represented as

$$V^{II,III} = V_R + V_{Cor} + V_{s.o.}, \tag{3.99}$$

where the matrix elements of V_R and V_{Cor} are defined by the same expressions as those in (3.98).

Further modification of the coupling $V^{II,III}$ is based on the possibility of introducing the diabatic Hamiltonians H_{HL} in region II and H_{MO} in region III. Consequently it is often convenient to take

$$V^{II} = V_R + V_{Cor} + V_{s.o.} + (H_0 - H_{HL}) \tag{3.100}$$

as the coupling for region II and

$$V^{III} = V_R + V_{Cor} + V_{s.o.} + (H_0 - H_{MO}) \tag{3.101}$$

as the coupling for region III. For the region intermediate between III and II, it is only justifiable to speak about the coupling $V^{II,III}$.

The different nature of the couplings and wave functions in various regions also make the selection rules for possible transitions different.

a) Region I and the Intermediate Region Between I and II

These regions can play an important part in resonant and quasi-resonant excitation transfer, in transitions between fine-structure components and depolarization [3.72, 73, 88, 89]. The molecular wave functions are expressed to a good approximation here via antisymmetrized products of free atom wave functions (Sect. 3.2.2). The most general selection rules are valid here:

$$V_R\colon \Delta\Omega = 0, \quad 0^+ \leftrightarrow 0^+, \quad 0^- \leftrightarrow 0^-, \quad g \leftrightarrow g, \quad u \leftrightarrow u \tag{3.102}$$

$$V_{Cor}\colon \Delta\Omega = \pm 1, \quad g \leftrightarrow g, \quad u \leftrightarrow u\,. \tag{3.103}$$

b) Region II

In this region the wave functions are still expressed as linear combinations of antisymmetrized products of the atomic wave functions defined in Sect. 3.2.2 (3.22). In principle, the crossing of molecular terms which correlate with various atomic terms for the given atomic electronic configurations is possible. For region II it can be assumed to a good approximation (Sect. 4.5) that

$$V_{s.o.} = V_{s.o.A} + V_{s.o.B}, \tag{3.104}$$

where $V_{\text{s.o.A}}$ and $V_{\text{s.o.B}}$ are operators of the spin-orbit coupling in the free atoms A and B. Since the Heitler-London approximation does not take into account the interaction between states with different sets of quantum numbers $\bar{\gamma}_1 S_1 L_1, \bar{\gamma}_2 S_2 L_2$, we obtain the following selection rules for V^{II}:

$$V_R\colon \Delta S = 0, \quad \Delta m_S = 0, \quad \Delta\Lambda = 0, \quad \mathrm{g}\leftrightarrow\mathrm{g}, \quad \mathrm{u}\leftrightarrow\mathrm{u},$$
$$(\bar{\gamma}_1 S_1 L_1; \bar{\gamma}_2 S_2 L_2) \leftrightarrow (\bar{\gamma}_1 S_1 L_1; \bar{\gamma}_2 S_2 L_2) \tag{3.105}$$

$$V_{\text{Cor}}\colon \Delta S = 0, \quad \Delta m_S = 0, \quad \pm 1, \quad \Delta\Lambda = 0, \quad \pm 1, \quad \mathrm{g}\leftrightarrow\mathrm{g}, \quad \mathrm{u}\leftrightarrow\mathrm{u},$$
$$(\bar{\gamma}_1 S_1 L_1; \bar{\gamma}_2 S_2 L_2) \leftrightarrow (\bar{\gamma}_1 S_1 L_1; \bar{\gamma}_2 S_2 L_2) \tag{3.106}$$

$$V_{\text{s.o.}}\colon \begin{cases} \Delta S = 0, \quad \pm 1, \quad \Delta m_S = 0, \quad \pm 1, \quad \Delta\Lambda = 0, \quad \pm 1, \\ \Sigma^+ \leftrightarrow \Sigma^-, \quad \mathrm{g}\leftrightarrow\mathrm{g}, \quad \mathrm{u}\leftrightarrow\mathrm{u}, \\ \text{transitions } {}^1\Sigma^+ \leftrightarrow {}^1\Sigma^- \text{ are forbidden}, \\ (\bar{\gamma}_1 S_1 L_1; \bar{\gamma}_2 S_2 L_2) \leftrightarrow (\bar{\gamma}_1 S_1' L_1'; \bar{\gamma}_2 S_2 L_2) \\ \text{if} \quad \Delta S_1 = 0, \quad \pm 1, \quad \Delta L_1 = 0, \quad \pm 1, \\ (\bar{\gamma}_1 S_1 L_1; \bar{\gamma}_2 S_2 L_2) \leftrightarrow (\bar{\gamma}_1 S_1 L_1; \bar{\gamma}_2 S_2' L_2') \\ \text{if} \quad \Delta S_2 = 0, \quad \pm 1, \quad \Delta L_2 = 0, \quad \pm 1 \end{cases} \tag{3.107}$$

$$H_0 - H_{\text{HL}}\colon \begin{cases} \Delta S = 0, \quad \Delta m_S = 0, \quad \Delta\Lambda = 0, \quad \Sigma^+ \leftrightarrow \Sigma^+, \\ \Sigma^- \leftrightarrow \Sigma^-, \quad \mathrm{g}\leftrightarrow\mathrm{g}, \quad \mathrm{u}\leftrightarrow\mathrm{u}, \\ (\bar{\gamma}_1 S_1 L_1; \bar{\gamma}_2 S_2 L_2) \leftrightarrow (\bar{\gamma}_1' S_1' L_1'; \bar{\gamma}_2' S_2' L_2') \\ \text{if} \quad (\bar{\gamma}_1 S_1 L_1; \bar{\gamma}_2 S_2 L_2) \neq (\bar{\gamma}_1' S_1' L_1'; \bar{\gamma}_2' S_2' L_2'). \end{cases} \tag{3.108}$$

The matrix elements of V_{Cor} in the given region are expressed via the matrix elements of the atomic angular momenta calculated over the wave functions of the free atoms. Explicit expressions for the matrix elements of $V_{\text{s.o.}}$ will be given in Sect. 4.5.

c) Region Intermediate Between II and III

Generally speaking, there is no possibility here of introducing additional quantum numbers $\bar{\delta}$ having any specific physical meaning. Therefore only the most general selection rules for the Hund coupling case a will be given for coupling described by the operator $V^{\text{II,III}}$:

$$V_R\colon \begin{cases} \Delta S = 0, \quad \Delta m_S = 0, \quad \Delta\Lambda = 0, \quad \Sigma^+ \leftrightarrow \Sigma^+, \\ \Sigma^- \leftrightarrow \Sigma^-, \quad \mathrm{g}\leftrightarrow\mathrm{g}, \quad \mathrm{u}\leftrightarrow\mathrm{u} \end{cases} \tag{3.109}$$

$$V_{\mathrm{Cor}}: \begin{cases} \Delta S = 0, \quad \Delta m_S = 0, \quad \pm 1, \quad \Delta\Lambda = 0, \quad \pm 1, \\ \quad \mathrm{g} \leftrightarrow \mathrm{g}, \quad \mathrm{u} \leftrightarrow \mathrm{u} \end{cases} \tag{3.110}$$

$$V_{\mathrm{s.o.}}: \begin{cases} \Delta S = 0, \quad \pm 1, \quad \Delta m_S = 0, \quad \pm 1, \quad \Delta\Lambda = 0, \quad \pm 1, \\ \quad \Sigma^+ \leftrightarrow \Sigma^-, \quad \mathrm{g} \leftrightarrow \mathrm{g}, \quad \mathrm{u} \leftrightarrow \mathrm{u}, \\ \quad \text{transitions } {}^1\Sigma^+ \leftrightarrow {}^1\Sigma^- \text{ are forbidden.} \end{cases} \tag{3.111}$$

The evaluation of the coupling matrix elements in this region will be considered in Sect. 4.5.

d) Region III

The molecular-orbital approximation implying that the electron wave functions can be approximated by determinants containing MOs of the same configuration can be used for this region. In this case, coupling between the terms corresponding to different configurations which exhibit crossings or avoided crossings will be of interest. An avoided crossing can occur, for instance, if the configurations involve different one-electron states of the same symmetry. When using the diabatic one-electron terms defined in Sect. 3.3.3, such avoided crossings may degenerate to crossings. Then an additional term $H_{\mathrm{MO}} - H^{\mathrm{d}}_{\mathrm{MO}}$ must be added to the nonadiabatic coupling.

Since the matrix elements are calculated here over functions in the form of determinants, it is particularly important to know whether the sums of one-electron or two-electron operators are involved. From the general expressions in Sects. 2.2.2 and 2.3.1, V_{Cor} is seen to be a sum of one-electron operators. The same is true, to a good approximation, for the spin-orbit coupling (Sect. 4.5). The radial coupling operator V_{R} also behaves as a sum of one-electron operators [3.53, 90]. The operator $H_{\mathrm{MO}} - H^{\mathrm{d}}_{\mathrm{MO}}$ is by construction a sum of one-electron operators. The only operator in V^{III} which includes the sum of two-electron operators is $H_0 - H_{\mathrm{MO}}$ corresponding to electron correlation.

Consequently the contribution to transitions between states $(\eta_1 q_1)^{N_1} \ldots (\eta_i q_i)^{N_i} \ldots (\eta_j q_j)^{N_j} \ldots (\eta_k q_k)^{N_k}$ and $(\eta_1 q_1)^{N_1} \ldots (\eta_i q_i)^{N_i - 1} \ldots (\eta_{i'} q_{i'})^{N_{i'}+1} \ldots (\eta_j q_j)^{N_j - 1} \ldots (\eta_{j'} q_{j'})^{N_{j'}+1} \ldots (\eta_k q_k)^{N_k}$ is from $H_0 - H_{\mathrm{MO}}$ only, and the selection rules are

$$H_0 - H_{\mathrm{MO}}: \begin{cases} \Delta S = 0, \quad \Delta m_S = 0, \quad \Delta\Lambda = 0, \quad \mathrm{g} \leftrightarrow \mathrm{g}, \\ \quad \mathrm{u} \leftrightarrow \mathrm{u}, \quad \Sigma^+ \leftrightarrow \Sigma^+, \quad \Sigma^- \leftrightarrow \Sigma^- \\ \quad \text{provided the products of the representa-} \\ \quad \text{tions } q_i \times q_j \text{ and } q_{i'} \times q_{j'} \text{ contain the same} \\ \quad \text{irreducible representations.} \end{cases} \tag{3.112}$$

All the terms involved in V^{III} contribute to transitions between states corresponding to configurations $(\eta_1 q_1)^{N_1} \ldots (\eta_i q_i)^{N_i} \ldots (\eta_k q_k)^{N_k}$ and $(\eta_1 q_1)^{N_1} \ldots (\eta_i q_i)^{N_i - 1} \ldots (\eta_{i'} q_{i'})^{N_{i'} + 1} \ldots (\eta_k q_k)^{N_k}$. Here the selection rules are

$$V_{\text{R}}: \begin{cases} \Delta S = 0, \quad \Delta m_S = 0, \quad \Delta\Lambda = 0, \quad \text{g} \leftrightarrow \text{g}, \quad \text{u} \leftrightarrow \text{u}, \\ \Sigma^+ \leftrightarrow \Sigma^+, \quad \Sigma^- \leftrightarrow \Sigma^-, \quad q_i = q_{i'} \end{cases} \tag{3.113}$$

$$V_{\text{Cor}}: \begin{cases} \Delta S = 0, \quad \Delta m_S = 0, \quad \pm 1, \quad \Delta\Lambda = 0, \quad \pm 1, \\ \text{g} \leftrightarrow \text{g}, \quad \text{u} \leftrightarrow \text{u}, \quad \lambda_i - \lambda_{i'} = 0, \quad \pm 1 \end{cases} \tag{3.114}$$

$$V_{\text{s.o.}}: \begin{array}{l} \Delta S = 0, \quad \pm 1, \quad \Delta m_S = 0, \quad \pm 1, \quad \Delta\Lambda = 0, \quad \pm 1, \\ \text{g} \leftrightarrow \text{g}, \quad \text{u} \leftrightarrow \text{u}, \quad \lambda_i - \lambda_{i'} = 0, \quad \pm 1, \quad \Sigma^+ \leftrightarrow \Sigma^-, \\ \text{transitions } {}^1\Sigma^+ \leftrightarrow {}^1\Sigma^- \text{ and } \lambda_i = \lambda_{i'} = 0 \\ \text{are forbidden} \end{array} \tag{3.115}$$

$$H_{\text{MO}} - H_{\text{MO}}^{\text{d}}: \begin{cases} \Delta S = 0, \quad \Delta m_S = 0, \quad \Delta\Lambda = 0, \quad \text{g} \leftrightarrow \text{g}, \\ \text{u} \leftrightarrow \text{u}, \quad \Sigma^+ \leftrightarrow \Sigma^+, \quad \Sigma^- \leftrightarrow \Sigma^-, \\ q_i = q_{i'} \end{cases} \tag{3.116}$$

$$H_0 - H_{\text{MO}}: \begin{cases} \Delta S = 0, \quad \Delta m_S = , \quad \Delta\Lambda = 0, \quad \text{g} \leftrightarrow \text{g}, \\ \text{u} \leftrightarrow \text{u}, \quad \Sigma^+ \leftrightarrow \Sigma^+, \quad \Sigma^- \leftrightarrow \Sigma^- \\ \text{provided there is at least one } \varphi_{\eta_j q_j} \\ \text{amongst the MOs identical in both} \\ \text{configurations, such that the direct} \\ \text{products of the representations } q_j \times q_i \\ \text{and } q_j \times q_{i'} \text{ contain the same irreducible} \\ \text{representations.} \end{cases} \tag{3.117}$$

Only the most important couplings are accounted for in the above selection rules. Two additional types of couplings have not been discussed since they are very weak. One is the hyperfine coupling which induces transitions between the terms 0^+ and 0^- in region I and in the region intermediate between I and II. Transitions with changes in electron parity are also possible for collisions between different isotopes [3.91].

4. Approximate Calculation of the Electronic States of Diatoms

The evaluation of adiabatic electronic wave functions and of terms of a diatom is a typical problem in quantum chemistry. A rigorous ab initio approach commonly solves this problem by the conventional variational method with use of trial wave functions, often chosen for mathematical convenience, but having no sufficiently clear physical meaning.

Such a direct approach needs very cumbersome and expensive specific calculations for every system and, for the present state of computing techniques, is feasible only for systems with a rather small number of electrons. Many papers deal with the methods and results of such nonempirical calculations [4.1–10].

A rather crude estimate of the molecular terms and of their coupling is often sufficient for practical calculation of inelastic atomic collisions. Therefore, because of the complexity of ab initio quantum-chemical calculations, various approximate and semi-empirical methods of constructing molecular electronic terms and wave functions have been developed. These are based on some version of perturbation theory, on systematic use of the symmetry properties of the system studied and on known characteristics of free atoms (electron density distribution, multipole moments, polarizability). Some typical methods of this kind suitable for calculation of the electronic states and terms for various limited ranges of R will be described here. Only neutral systems for which charge transfer interactions are unimportant will be considered. (For the latter see, e.g., [4.11, 12].) To facilitate understanding of the basic material and to introduce some widely used notions, we begin with a brief discussion of the properties of the electron potential and AOs in free atoms.

4.1 Atomic Potential and Atomic Orbitals

4.1.1 Hartree-Fock Screening Function and Atomic Orbitals

The Hartree-Fock method is the one most consistent for the definition of AOs [4.13, 14]. This method has been widely used in numerical calculations, and rather detailed numerical tables of radial wave functions have been published [4.15–19].

The basic physical effect for which the Hartree-Fock method accounts is the screening of the nuclear charge (as seen by one electron) by the other electrons. In rigorous Hartree-Fock calculations this screening depends on the nature of the AOs occupied by all the electrons [4.13, 14]. Thus, the Hartree-Fock potential acting on the electron in the ith AO in the $\bar{\gamma}$ configuration can be written in the form

$$V_{i\bar{\gamma}}(r) = \frac{Z}{r} F_{i\bar{\gamma}}(r), \tag{4.1}$$

where $F_{i\gamma}$ is the screening function. Its explicit expression for variants of the Hartree-Fock method can be found, e.g., in [4.14]. The function $F_{i\gamma}(r)$ obviously satisfies the following boundary conditions:

$$F_{i\bar{\gamma}}(0) = 1, \qquad F_{i\bar{\gamma}}(r) \xrightarrow[r \to \infty]{} 1/Z. \tag{4.2}$$

If the $F_{i\gamma}$ were strongly dependent on the quantum numbers i and $\bar{\gamma}$, the central field approximation discussed in Sect. 3.1 would obviously be inadequate. However repeated calculations by the Hartree-Fock method show that the dependence of $F_{i\gamma}$ on i and $\bar{\gamma}$ is quite weak. Therefore, it can be expected that there is a certain universal screening function $F(r)$ which, on average, correctly approximates the Hartree-Fock calculation results and yields the potential $V(r)$ of the central field approximation. Since $F(r)$ does not account, by definition, for the shell effects, it would be natural to obtain it in terms of some variant of the statistical theory of atoms [4.20, 21]. The statistical Fermi-Amaldi atomic model [4.20] seems to be most convenient for this purpose. It differs from the original statistical Thomas-Fermi model in that it approximately omits the electron self-interaction, and thus $F_{\mathrm{F-A}}(r)$ behaves correctly at large distances from the nucleus.

The Fermi-Amaldi screening function $F_{\mathrm{F-A}}(r)$ is

$$F_{\mathrm{F-A}}(r) = \begin{cases} \chi(r/\mu^*) + \dfrac{r}{r_0 Z} & \text{if} \quad r \leqq r_0 \\ 1/Z \quad \text{if} \quad r > r_0, \end{cases} \tag{4.3}$$

where $\chi(x)$ ($x = r/\mu^*$, $x_0 = r_0/\mu^*$) satisfies the equation

$$\chi''(x) = \chi^{3/2}(x)/x^{1/2} \tag{4.4}$$

with boundary conditions

$$\chi(0) = 1, \qquad \chi(x_0) = 0, \qquad x_0 \chi'(x_0) = -1/Z \tag{4.5}$$

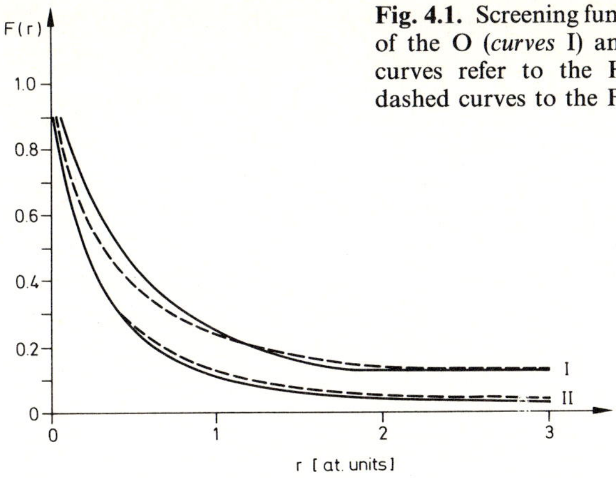

Fig. 4.1. Screening function $F(r)$ for the outer p electrons of the O (*curves* I) and Kr (*curves* II) atoms. The full curves refer to the Hartree-Fock method [4.19], the dashed curves to the Fermi-Amaldi model [4.23]

and

$$\mu^* = 0.8853\, Z^{-1/3}[Z/(Z-1)]^{2/3}. \tag{4.6}$$

The equations in the Fermi-Amaldi model which define χ, a function of the dimensionless distance x, for a neutral atom coincide with those for a singly charged positive ion in the Thomas-Fermi model. Consequently, χ and x_0 can be obtained using the tables in [4.20, 22, 23]. Figure 4.1 presents as an example the screening functions calculated for the O and Kr atoms by the Hartree-Fock-Slater method, which implies that the exchange interaction can be approximated by the local potential field [4.19] and the appropriate functions $F_{\mathrm{F-A}}(r)$. The screening functions calculated by different methods are in good agreement. This is suported also by extensive calculations of binding energies [4.22]. Note that several analytical expressions for the screening function based on various approximations of the Thomas-Fermi and the Thomas-Fermi-Dirac equations for atoms and ions have been proposed recently [4.24–30].

According to the Koopmans theorem [4.14] the one-electron energy levels $\varepsilon_{nl}^{(1)}$ found by the Hartree-Fock method are close to the ionization potentials of the relevant shells (with a minus sign). Consequently, it can be expected that the results of AO calculations will improve if the screening function is modified in such a way that the solution of (4.4) provides the experimental values of the shell ionization potentials. The one-electron scheme discussed here will remain quite consistent even if different screening functions have to be introduced for different l. The orbitals $\varphi_{nl}^{m}(\boldsymbol{r})$ obtained with the same l and different n are orthogonal due to the orthogonality of the radial wave functions, and those with different l are also orthogonal due to the orthogonality of the spherical functions. The effectiveness of such an approach to the definition of the screening function, using $F_{F-A}(r)$ as the basis, has been demonstrated in [4.23]. The function $F_{F-A}(r)$ was modified by replacing μ^* with $\alpha_{nl}\mu^*$, i.e., by varying the size of the atom, with the result that the scaling parameters α_{nl} leading to the experimental ionization potentials were only slightly different from unity and virtually independent of n for a given l.

The screening function is close to unity near the nucleus and equals the constant $1/Z$ outside the atom (for $r > r_0$). Therefore the AOs with small principal quantum numbers n localized near the nucleus, and those with large n for $r \gg r_0$ differ only slightly from hydrogen-like orbitals. Since the extent of a hydrogen-like orbital is of the order of n/Z, the deviations from the Coulomb field at small n are small if

$$(n/Z\mu^*)\,\chi'(0) < 1\,. \tag{4.7}$$

Taking into account (4.6) and also the fact that $\chi'(0)$ as a function of Z varies only slightly about the value 1.6 [4.20], (4.7) actually reads

$$n < 0.5\,Z^{2/3}\,. \tag{4.8}$$

For large n, when the effective nuclear charge can be taken as unity, the AO will be almost hydrogen-like if

$$n > x_0\mu^*\,. \tag{4.9}$$

The results of AO calculations are usually presented either as numerical tables, or by some analytical approximation. Most conventional is the approximation of radial AOs by linear combinations of Slater orbitals φ_n^{Sl} given by

$$\varphi_n^{\mathrm{Sl}} = C_n r^{n-1}\exp(-\alpha_{\mathrm{Sl}}r)\,, \tag{4.10}$$

$$C_n = (2\alpha_{\mathrm{Sl}})^{n+1/2}(2n!)^{-1/2}\,. \tag{4.11}$$

The wide use of these functions is due to the successful results obtained for $\varepsilon_{nl}^{(1)}$ even when the AO is approximated by just one Slater orbital. Simple rules for finding α_{Sl} using such an approximation were proposed by Slater as early as 1930. These were described, e.g., in [4.31] and were repeatedly revised later [4.32]. It should be emphasized that the Slater orbitals have no nodes and consequently can provide an approximation to the true Hartree-Fock AO for $l < n-1$ only in the region to the right of the last extremum (Fig. 4.2). In this sense they correspond somewhat to the AOs in the pseudopotential method (see below) [4.33].

All the above pertains to distances where the electron density is high. The calculation of the electron density at large distances from the nucleus, where it is exponentially small, is somewhat more complicated. A fundamental difficulty is that, generally speaking, the variational and the pertur-

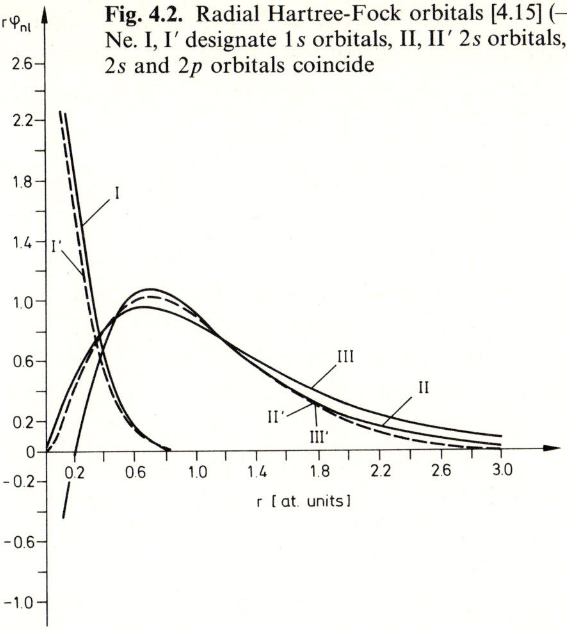

Fig. 4.2. Radial Hartree-Fock orbitals [4.15] (——) and Slater orbitals (– – –) for Ne. I, I′ designate 1*s* orbitals, II, II′ 2*s* orbitals, III, III′ the 2*p* orbital. The Slater 2*s* and 2*p* orbitals coincide

bation theory methods can provide no asymptotic solutions to differential equations. Consequently, the Hartree-Fock method as a variational method does not provide, in general, the true asymptotic behaviour of the electron density distribution far from the nucleus. Physically this is due to the fact that the $\varepsilon_{nl}^{(1)}$ values differ from the ionization potentials. Yet it is qualitatively clear and it has been rigorously shown [4.34–36] that the asymptotic behaviour of an accurate many-electron wave function of the atom, as one of the electrons, say electron 1, recedes from the nucleus is

$$|\bar{\gamma} S m_S L m_L\rangle \underset{r_1 \to \infty}{\simeq} f(r_1)\exp(-\alpha r_1), \tag{4.12}$$

where $\alpha = (2I)^{1/2}$, I is the exact lowest ionization potential and $f(r_1)$ is a function which varies more slowly than the exponential. In the Hartree-Fock approximation $\alpha = (-2\varepsilon_{\min})^{1/2}$, where $\varepsilon_{\min}$ is the energy of the highest occupied AO.

Until recently, rigorous proof existed only for a crude upper limit of $f(r_1)$ [4.34, 35, 37] in the case of a many-electron atom:

$$|f(r_1)| < C r_1^{(Z/\alpha)-1}. \tag{4.13}$$

Nevertheless it has been assumed [4.12, 38] in calculating the overlap effects that $f(r_1)$ behaves as

$$f(r_1) = C r_1^{(1/\alpha)-1} \tag{4.14}$$

as expected for the motion of a particle with energy $-I$ in the potential field behaving as $-r_1^{-1}$ for $r_1 \to \infty$. A rigorous proof of this asymptotic behavior has been obtained quite recently [4.39, 40]. Obviously, in the semiempirical method of the AO calculation described above [4.23], the highest occupied AO shows a Coulomb asymptotic behavior:

$$\varphi_{nl}(r) \xrightarrow[r \to \infty]{} C_{nl} r^{(1/\alpha)-1} \exp(-\alpha r) \tag{4.15}$$

and this automatically leads to the asymptotic function (4.14) for the total antisymmetric wave function.

In a rigorous Hartree-Fock calculation it would be most natural to improve the highest occupied AO by matching it with the asymptotic expression (4.15) at distances of order $1/\alpha$. Such a matching is, as a rule, sufficiently smooth since $-\varepsilon_{\min}$ usually differs only slightly from I. For instance, for helium $-\varepsilon_{\min} = 0.92$, $I = 0.90$, for argon $-\varepsilon_{\min} = 0.59$, $I = 0.58$ and for sodium $-\varepsilon_{\min} = 0.18$, $I = 0.19$, where the values of $\varepsilon_{\min}$ are taken from [4.15]. The deviations of $-\varepsilon_{\min}$ from I are due first to the difference between the Hartree-Fock energies of the singly-charged ion and the "frozen" atomic core, and second to the different correlation energies of the ion and atom. These two effects are in part mutually compensated, which results in the small difference between $-\varepsilon_{\min}$ and I. The matching constants C for many atoms are listed in [4.41, 42].

The following difficulty arises for lower-lying AOs. At first sight their asymptotic form seems to depend on the appropriate $\varepsilon_{nl}^{(1)}$. However this is not so [4.43, 44] because the exchange term involved in the Hartree-Fock equations couples various AOs when the Pauli principle is taken into account, and this coupling decreases only slowly with distance. Thus the asymptotic behavior of all AOs is determined by linear combinations of the form

$$\sum_i f_i(r) \exp[(-2\varepsilon_i^{(1)})^{1/2} r],$$

where the summation is over all occupied AOs. Nevertheless, the Hartree-Fock calculation results suggest that for low-lying AOs the coefficients $f_i(r)$ pertaining to higher excited AOs are very small. Therefore, taking into account the large difference in ionization potentials of different AOs, it is natural to assume that at distances of physical interest, the asymptotic form of the inner shell AOs is defined by the same expression (4.15), where the appropriate $\varepsilon_{nl}^{(1)}$ or the ionization potential are substituted. The general approach can be simplified if interest is centered on valence shell electrons. This is the case for low-energy collisions which do not appreciably disturb the inner atomic shells.

Accordingly, it has been attempted for more than forty years to avoid the explicit introduction of the coordinates of the inner (core) electrons and to describe their effect on the outer (valence) electrons by means of an appropriately chosen potential.

4.1.2 The Pseudopotential Method for Valence Electrons of Atoms

In terms of the central field approximation the inner electrons display a dual function. First, they contribute to the creation of an average potential $V(r)$ and second, they occupy, by virtue of the Pauli principle, all low-lying levels in this potential well. Only the upper levels with a much lower binding energy are accessible to valence electrons. This can be interpreted as the inner electrons adding to $V(r)$ a certain repulsive potential (known as the pseudopotential ${}^{\mathrm{p}}V$) such that the lower levels of the valence electrons in the effective potential $V_{\mathrm{eff}} = V + {}^{\mathrm{p}}V$ coincide with the unoccupied levels in the potential V. Such considerations, tracing back to *Hellman* [4.45] and *Gombas* [4.20, 46], underlie the pseudopotential method that has been intensely developed in recent years [3.47–64].

By now the pseudopotential method is sufficiently advanced for the case of one electron outside closed shells (e.g., an alkali metal atom) or for electron scattering on an inert gas atom. It is formulated most rigorously in terms of the field-theoretical formulation of the many-body problem, the part of the effective potential being played by the operator of the irreducible electron self-energy involved in the equation for the one-electron Green's function [4.52]. Such a treatment implies rigorously accounting both for the Pauli principle effect mentioned above and for the effect of correlation. It is essential that in this case, V is nonlocal and dependent on the outer electron energy. Detailed calculations using this approach were made for the Li atom [4.65] and for electron scattering on the He atom [4.66].

In practice, the pseudopotential method based on the Hartree-Fock method or on the central field approximation is used most often. In the formulation of this method [4.47, 55, 59, 64] the AO of the valence electron ${}^{\mathrm{v}}\varphi_{nl}^{m}$ is represented in a form which explicitly ensures its orthogonality to the core AOs ${}^{\mathrm{c}}\varphi_{nl}^{m}$:

$$ {}^{\mathrm{v}}\varphi_{nl}^{m} = {}^{\mathrm{p}}\varphi_{nl}^{m} - \sum_{n'} {}^{\mathrm{c}}\varphi_{nl}^{m} \langle {}^{\mathrm{c}}\varphi_{nl}^{m} | {}^{\mathrm{p}}\varphi_{nl}^{m} \rangle , \tag{4.16} $$

where the summation is over all core states with specified values of l and ${}^{\mathrm{p}}\varphi_{nl}^{m}$ are the so-called (still unknown) normalized pseudo AOs. As is consistent with the above considerations of the effect of the inner electrons, the radial pseudo AO ${}^{\mathrm{p}}\varphi_{nl}(r)$ for the lowest possible state of the valence electron with a given l must have no nodes. This smoothness condition can be expressed analytically by demanding that the functional [4.55, 56]

$$I = \frac{1}{2} \int |\nabla\, {}^{\mathrm{p}}\varphi^m_{nl}|^2 \, d^3\boldsymbol{r} - {}^{\mathrm{v}}\varepsilon^{(1)}_{nl} \langle {}^{\mathrm{p}}\varphi^m_{nl} | {}^{\mathrm{p}}\varphi^m_{nl} \rangle \tag{4.17}$$

be a minimum. Hence,

$$\langle {}^{\mathrm{c}}\varphi^m_{n'l} | {}^{\mathrm{p}}\varphi^m_{nl} \rangle \, ({}^{\mathrm{v}}\varepsilon^{(1)}_{nl} - {}^{\mathrm{c}}\varepsilon^{(1)}_{n'l}) = - \langle {}^{\mathrm{c}}\varphi^m_{n'l} | \, V \, | {}^{\mathrm{p}}\varphi^m_{nl} \rangle . \tag{4.18}$$

On the other hand, substituting (4.16) into the Schrödinger equation and taking (4.18) into account, we obtain the eigenvalue equation

$$\left[-\frac{1}{2} \frac{1}{r^2} \frac{d}{dr} \left(r^2 \frac{d}{dr} \right) + \frac{l(l+1)}{2r^2} + V(r) \right] {}^{\mathrm{p}}\varphi_{nl}$$

$$- \sum_{n'} \int_0^\infty {}^{\mathrm{c}}\varphi_{n'l}(r') \, V(r') \, {}^{\mathrm{p}}\varphi_{nl}(r') \, r'^2 \, dr' \, {}^{\mathrm{c}}\varphi_{n'l}(r) = {}^{\mathrm{v}}\varepsilon^{(1)}_{nl} \, {}^{\mathrm{p}}\varphi_{nl}(r) . \tag{4.19}$$

The spectrum of (4.19) coincides with the Hartree-Fock spectrum of the valence electron if the Hartree-Fock potential is used as $V(r)$. Thus, this equation formally solves the problem. However, to find $V(r)$ and ${}^{\mathrm{c}}\varphi_{n'l}$ the Hartree-Fock equation must be solved for all electrons. The next step in the simplification consists of taking $V(r)$ and ${}^{\mathrm{c}}\varphi_{n'l}$ to be the same for all states of the valence electron, and equal to the appropriate functions in the ion (frozen-core approximation). In this approximation the effective potential defining the energy levels of the valence electron can be calculated explicitly. This potential is nonlocal and, as can be seen from (4.19), is defined by the relation

$$V_{\mathrm{eff}} \, {}^{\mathrm{p}}\varphi_{nl}(r) = V(r) \, {}^{\mathrm{p}}\varphi_{nl}(r) - \sum_{n'} \int_0^\infty {}^{\mathrm{c}}\varphi_{n'l}(r') \, V(r') \, {}^{\mathrm{p}}\varphi_{nl}(r') \, r'^2 \, dr' \, {}^{\mathrm{c}}\varphi_{n'l}(r) . \tag{4.20}$$

If the core wave functions were to represent a complete set, the second term on the right side would be equal to the first, and the interaction would be completely compensated. In reality this is certainly not so, but for a rather heavy atom, the set of core AOs for small l is sufficient for an appreciable compensation of the interaction in a region of the order of the core size. Since $V(r) < 0$, it follows that in this region the nonlocal term in V_{eff}, which must be identified with the above pseudopotential ${}^{\mathrm{p}}V$, is of a repulsive nature on average.

A possible way of obtaining a local approximation for V_{eff} is to factor out ${}^{\mathrm{p}}\varphi_{nl}$ and replace ${}^{\mathrm{p}}\varphi_{nl}(r)$ and ${}^{\mathrm{p}}\varphi_{nl}(r')$ by C/r^l and $C/(r')^l$. This substitution seems to be justified by the smoothness of ${}^{\mathrm{p}}\varphi_{nl}(r)$ in the core region (for small r). The result is

$$V_{\mathrm{eff}} = V(r) - \sum_{n'} \left[\int_0^\infty {}^{\mathrm{c}}\varphi_{n'l}(r') \, V(r') \, r'^{l+2} \, dr' \right] \frac{{}^{\mathrm{c}}\varphi_{n'l}(r)}{r^l} . \tag{4.21}$$

V_{eff} is seen to depend essentially on l, as there are different numbers of core AOs for different l values. For instance, for the Na atom with its core configuration $(1s)^2(2s)^2(2p)^6$, the s states of the valence electron have two preceding core states ($n = 1$ and $n = 2$), the p states have one preceding core state ($n = 2$), and the d states have no preceding core state. Accordingly ${}^{\mathrm{P}}V_{l=0} > {}^{\mathrm{P}}V_{l=1}$ and ${}^{\mathrm{P}}V_{l=2} = 0$.

This is illustrated in Fig. 4.3 which presents V_{eff} calculated using (4.21) in the frozen-core approximation for the Na atom. Figure 4.4 presents the $3s$ pseudo AO and the appropriate true Hartree-Fock AO. It can be seen that outside the core these AOs are very close, as would be expected.

Equations (4.20, 21) provide theoretical grounds for the introduction of model potentials for the valence electron [4.47, 55, 64, 69–73]. These potentials V_{mod} are usually expressed in the quasi-local form reflecting the above dependence of V_{eff} on l

$$V_{\mathrm{mod}} = \sum_{l} V_l(r)\, \hat{P}_l, \tag{4.22}$$

where $\hat{P}_l$ stands for the projection operators on states with given l. The functions $V_l(r)$ are expressed in an analytical form qualitatively representing the behavior expected from (4.21) and depending on the fitting parameters. These parameters are chosen from the condition that the calculated value of the lowest level of the valence electron for given l coincides with the experimental value. After the above treatment, one expects that this set of parameters would also give good results for the excited states with the same l. As a rule, this is actually the case [4.64]. The correspondence between the model potentials and V_{eff} defined by (4.21) is illustrated in Fig. 4.5, where several model potentials are shown for the valence s electron of the Na atom.

A rigorous generalization of the pseudopotential method for the case of several valence electrons meets considerable difficulties, the most fundamental being the modification of the interaction between the valence electrons as a result of their interaction with the core [4.63, 64]. However, the most simple approach, describing the interaction of each electron with the core by a model potential of the type (4.22) and assuming that the interaction between the valence electrons is a Coulomb one, often yields quite satisfactory results [4.47, 50, 56, 64].

For an isolated atom, the pseudopotential method described above represents, in fact, even with the use of model potentials, another formulation of the Hartree-Fock method, and thus, generally speaking, does not simplify the calculation of the electron density distribution. The semi-empirical method based on the Fermi-Amaldi model described earlier is, in principle, by no means more complicated. The advantages of the pseu-

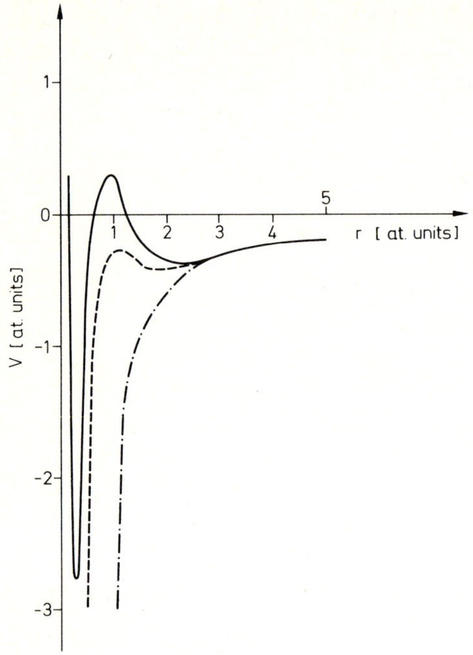

Fig. 4.3

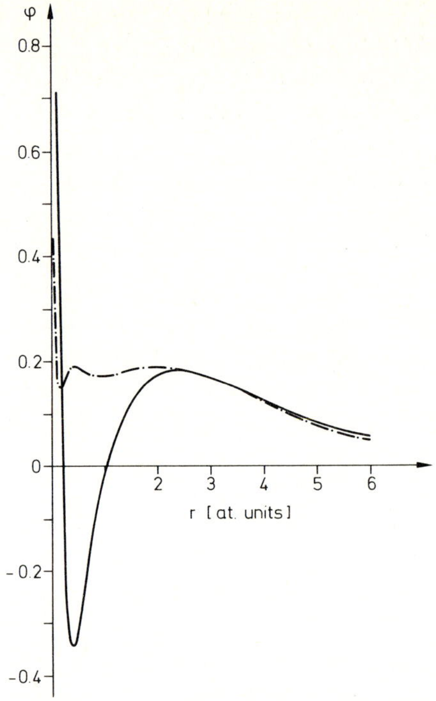

Fig. 4.4

Fig. 4.3. Effective potentials for a valence electron in Na, calculated in the frozen-core approximation [4.67]. The full line refers to V_{eff} for s states, the dashed line to V_{eff} for p states, the dot-dashed line to the Hartree-Fock potential $V(r)$

Fig. 4.4. Radial orbitals of a Na valence electron in the lowest s state. The full line refers to the Hartree-Fock $3s$ orbital [4.68], the dash-dotted line to the corresponding pseudoorbital [4.67]

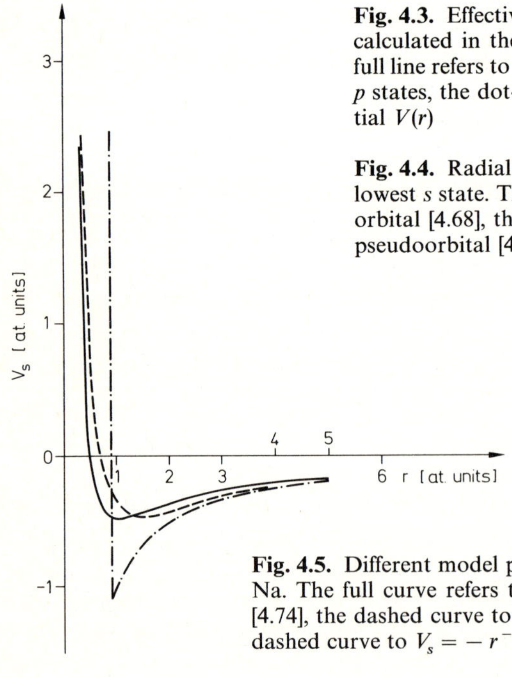

Fig. 4.5. Different model potentials $V_s(r)$ for a valence s electron of Na. The full curve refers to $V_s = -r^{-1} + 2.685\,r^{-1}\exp(-1.446\,r)$ [4.74], the dashed curve to $V_s = -r^{-1} + 0{,}5105\,r^{-2}$ [4.75], the dot-dashed curve to $V_s = -r^{-1}$ for $r > 0.91$, $V_s = \infty$ for $r \leqq 0.91$ [4.76]

dopotential method become evident on passing to two interacting atoms. When these atoms are so far removed that the cores do not appreciably overlap, calculations of the diatomic terms can account for the valence electrons only, assuming that their interaction with the cores is defined by potentials of the type (4.22). The specific use of the pseudopotential method for the calculation of the interaction between two atoms will be discussed later.

4.2 Diatomic Interactions at Large Distances and the Heitler-London Approximation

For R sufficiently large that the interaction between atoms is much weaker than the splitting between terms of isolated atoms, the Heitler-London approximation can be used in the calculation of the molecular terms (Fig. 3.3). This approximation is in fact equivalent to the perturbation treatment of the interatomic interactions. It differs from conventional perturbation theory in that, for large R, the indistinguishability of electrons impedes an unambiguous partitioning of the Hamiltonian H into $H^{(0)}$ and V, which correspond to the Hamiltonian of the isolated atoms and to a perturbation.

Several formal schemes of perturbation theory which account for electron indistinguishability have been proposed [4.3, 77–86]. If only the first order is retained, all approaches define the diatomic energy as an average value of H over the antisymmetrized product of the wave functions of the free atoms. A second-order term involves the usual dispersion interaction and the contributions arising in connection with the requirement that the first-order correction for the wave function be antisymmetric. The second-order terms differ for various approaches, which is due to the different definitions of the perturbation. In the calculation of these terms for rather complex systems great difficulties arise and therefore calculations which account for second-order exchange corrections have been carried out so far only for simple systems: for H_2 [4.87] and He_2, Ne_2 [4.88]. It can be expected, however, that the second-order exchange corrections, proportional (for large R) to a small distortion of the wave function and to an exponentially small overlap of AOs localized on different nuclei, will be small compared to the first-order exchange terms and the dispersion interaction. The results for H_2 and He_2 support this expectation [4.87–89]. Moreover, it should be noted that the finite number of terms in the perturbation series expansion cannot correctly account for the distortion of the exponentially small tails of the atomic wave functions, due to correlation of the exchanging electrons, resulting in an important modification of the exchange interaction [4.90–93].

Taking into account all the above considerations, we now look at a variant of the Heitler-London method convenient for the practical application to many-electron atoms [4.12, 94–96], which permits, in principle, the treatment of the interactions of any many-electron atom at sufficiently large distances. But, as will be seen later, it is particularly adequate for interactions between atoms whose ionization potentials are not very different. In this case, at distances where perturbation theory is still valid, it is possible to estimate relatively readily the exchange interaction, and then to retain only the leading terms in an electrostatic interaction expansion over multipoles, symmetric for both atoms.

If the ionization potential of one atom (say atom A) is much lower than that of the other, another approach discussed in the next section appears to be more convenient.

4.2.1 Effective Hamiltonian

Let the free atoms A and B be in states specified by the quantum numbers $\bar{\gamma}_1, S_1, L_1$ and $\bar{\gamma}_2, S_2, L_2$, respectively. For conciseness these will be referred to hereafter as γ_1 and γ_2. For the time being, the spin-orbit coupling will be neglected. We denote as set D the entity of atomic states degenerate in the projections of the orbital and spin electronic angular momenta, and which, for identical atoms, are also degenerate for exchange of states between atoms. We number the electrons in some manner and introduce nonantisymmetrized linear combinations of the atomic wave function products $|\gamma_1\gamma_2 L\tilde{\Lambda} S m_S D\rangle$ defined by (3.22) with operator $\hat{A}$ omitted.

The effective Hamiltonian H_{HL} is introduced, for which, at the distances of interest, the diatomic terms correlating, for $R \to \infty$, with the abovementioned free-atom terms are obtained from the secular equation

$$\mathrm{Det}\langle\gamma_1\gamma_2 L\tilde{\Lambda} S m_S D|\, H_{\mathrm{HL}} - U\, |\gamma_1'\gamma_2' L'\tilde{\Lambda}' S' m_{S'} D\rangle = 0. \tag{4.23}$$

The effective Hamiltonian defined in such a way coincides with the partially diabatic Hamiltonian H_{HL} introduced in Sect. 3.3.2 for the region II in Fig. 3.3. This Hamiltonian takes into account the effect of antisymmetrization and, generally speaking, involves the contribution from excited states. The construction of H_{HL} is similar to that of the spin-Hamiltonian in the theory of ferromagnetism with direct exchange. Here H_{HL} will be constructed according to *Herring* [4.91, 92] ([4.92] contains a detailed list of references for problems connected with the derivation of the spin-Hamiltonian).

Consider two hypothetical interacting atoms A and B such that the electrons $1 \ldots N_A$ and $N_A + 1 \ldots N_A + N_B$ belonging to one atom are indistinguishable, and the electrons belonging to different atoms are distinguishable. The Hamiltonian of this system is

$$H_0 = H_{0A} + H_{0B} + V_{AB}, \tag{4.24}$$

where H_{0A} and H_{0B} are the nonrelativistic Hamiltonians of the atoms A and B involving the coordinates of electrons $1 \ldots N_A$ and $N_A + 1 \ldots N_A + N_B$, respectively and V_{AB} is the operator for the electrostatic interaction between the atoms.

Herring [4.91] gave convincing arguments in favour of the possibility that certain linear combinations $|\gamma_1\gamma_2 L\tilde{\Lambda} S m_S P\rangle$ can be constructed from accurate wave functions possessing the following properties:

i) These functions correspond to localization of electrons $1 \ldots N_A$ at the A atom, and of electrons $N_A + 1 \ldots N_A + N_B$ at the B atom and are antisymmetric under permutations of the electrons in each atom.
ii) For $r_i \to \infty$ (r_i is the distance from the electron to "its" nucleus), $|\gamma_1\gamma_2 L\tilde{\Lambda} S m_S P\rangle \sim \exp(-\alpha_{A,B} r_i)$, where $\alpha_{A,B} = (2I_{A,B})^{1/2}$ and $I_{A,B}$ stands for the ionization potential of the appropriate atom.
iii) These functions are similar in symmetry to the functions $|\gamma_1\gamma_2 L\tilde{\Lambda} S m_S D\rangle$.
iv) $|\gamma_1\gamma_2 L\tilde{\Lambda} S m_S P\rangle \to |\gamma_1\gamma_2 L\tilde{\Lambda} S m_S D\rangle$ for $R \to \infty$.

The antisymmetrized functions $\hat{A}\,|\gamma_1\gamma_2 L\tilde{\Lambda} S m_S P\rangle$ describe the allowed physical states. The exact equation for the diatomic terms corresponding to a definite value of the total spin and converting, when $R \to \infty$, to atomic terms of the set D becomes

$$\operatorname{Det}\langle\gamma_1\gamma_2 L\tilde{\Lambda} S m_S P|\,(H_0 - U)\,\hat{A}\,|\gamma_1'\gamma_2' L'\tilde{\Lambda}' S' m_S' P\rangle = 0. \tag{4.25}$$

Actual evaluation of the diatomic terms would require one first to find approximate expressions for $|\gamma_1\gamma_2 L\tilde{\Lambda} S m_S P\rangle$ and then to calculate the Hamiltonian matrix elements using these approximate functions. Obviously, for large R, the functions $|\gamma_1\gamma_2 L\tilde{\Lambda} S m_S P\rangle$ will be slightly different from $|\gamma_1\gamma_2 L\tilde{\Lambda} S m_S D\rangle$. However, this small difference due to the correlation of the electrons of different atoms is very important, since it is responsible for the dispersion interaction of atoms and the correct asymptotic form of the exchange interaction. The dispersion interaction is due to the distortion of the electronic density in those regions where it is high. At the same time, the exchange interaction is caused by the exponentially small tails of the atomic wave functions, and thus the distortion of wave functions far from the nuclei is important in calculating this interaction. For large enough R these regions do not overlap and this leads to the very important suggestion that the exchange interaction can be calculated independently of the dispersion interaction. This suggestion has been discussed by *Margenau* [4.97]. In recent years it has often been used in calculating the interactions of many-electron atoms at moderate internuclear distances [4.94, 98–109].

The satisfactory results of these calculations support the validity of the above suggestion.

The matrix elements in (4.25) can be represented as the sum of terms involving a different number of electron permutations between atoms. Obviously, the contribution to the matrix element corresponding to the permutation of n electron pairs will be proportional to S_{AB}^{2n} where S_{AB} is the overlap integral of AOs localized on different atoms, (3.51). For the present purpose, as shown by the results of many calculations for specific systems, all terms of the order $\varepsilon_{int} S_{AB}^2$ (ε_{int} is the interaction energy) can be neglected. It is sufficient here to account only, in $\hat{A}$, for electron permutations in each atom and for the pair-transfer terms between atoms. Moreover, the nonorthogonality can be neglected in normalizing the wave functions. The problem of accounting for higher-order permutations has been discussed in [4.81, 110–114]. It seems, however, that the corresponding terms can become essential only for such values of R at which the atomic interaction becomes comparable to the splitting of the atomic terms, and thus one cannot confine oneself to accounting for states that belong to the D set only.

From all the above and using only second-order perturbation theory for the Coulomb interaction, the matrix elements in (4.25) can be written in the form

$$\langle \gamma_1 \gamma_2 L \tilde{\Lambda} S m_S D | H_{0A} + H_{0B} + V_{AB} + V_{AB}^{(2)} + V_{ex} - U | \gamma_1' \gamma_2' L' \tilde{\Lambda}' S' m_{S'} D \rangle . \tag{4.26}$$

Here $V_{AB}^{(2)}$ is the operator of the Coulomb interaction in second-order perturbation theory (referred to hereafter as the dispersion interaction and defined in the next section), and

$$\langle \gamma_1 \gamma_2 L \tilde{\Lambda} S m_S D | V_{ex} | \gamma_1' \gamma_2' L' \tilde{\Lambda}' S' m_{S'} D \rangle = - N_A N_B \times \langle \gamma_1 \gamma_2 L \tilde{\Lambda} S m_S P | (H_0 - U) P_{1;N_A+1}^{\sigma} P_{1;N_A+1}^{x} | \gamma_1' \gamma_2' L' \tilde{\Lambda}' S' m_{S'} P \rangle , \tag{4.27}$$

where $P_{1;N_A+1}^{\sigma}$ is the permutation operator of the spin coordinates, and $P_{1;N_A+1}^{x}$ permutes the spatial coordinates of electrons 1 and $N_A + 1$. Property (i) of the functions $|\gamma_1 \gamma_2 L \tilde{\Lambda} S m_S P\rangle$ was used in deriving (4.27). The methods of calculating the exchange interaction matrix elements will be discussed in Sect. 4.2.4.

It follows from (4.25–27) that

$$H_{\mathrm{HL}} = H_{0\mathrm{A}} + H_{0\mathrm{B}} + V_{\mathrm{AB}} + V_{\mathrm{AB}}^{(2)} + V_{\mathrm{ex}}. \tag{4.28}$$

It will be seen that H_{HL} is diagonal in the quantum numbers $\tilde{\Lambda}$, S and m_S and the corresponding matrix elements do not depend on m_S and on the sign of $\tilde{\Lambda}$. The term $H_{0\mathrm{A}} + H_{0\mathrm{B}}$ simply provides the same contributions $\varepsilon_{\gamma_1} + \varepsilon_{\gamma_2}$ to all diagonal matrix elements. It can be taken as zero for all diatomic terms converging to the same energy level of two free atoms.

So far, the spin-orbit coupling has been neglected and the basis a (Sect. 3.2.2) used. In region II in Fig. 3.3 this coupling is usually considered as a perturbation which induces transitions between partially diabatic states corresponding to the Hund coupling case a. Therefore the calculation of the spin-orbit coupling in this region will be considered in Sect. 4.5.

In region I of Fig. 3.3 corresponding to the Hund case c the spin-orbit coupling must be taken into account when calculating the potential curves. This is usually done with the assumption that at such large distances the spin-orbit coupling does not differ much from that in free atoms, i.e.

$$V_{\mathrm{s.o.}} = V_{\mathrm{s.o.A}} + V_{\mathrm{s.o.B}}, \tag{4.29}$$

where $V_{\mathrm{s.o.A}}$ and $V_{\mathrm{s.o.B}}$ are operators of the spin-orbit coupling in free atoms A and B. Moreover, in accordance with the above, the terms of order $V_{\mathrm{s.o.}} S_{\mathrm{AB}}^2$ must be neglected, since for the distances of interest, the spin-orbit coupling is of the same order as the interaction between atoms. This means that when the basis c wave functions defined by (3.32) with the operator $\hat{A}$ omitted are used, the quantities $\varepsilon_{\gamma_1 j_1} + \varepsilon_{\gamma_2 j_2}$ ($\varepsilon_{\gamma j}$ is the energy of a fine structure level) will appear in the diagonal elements of the secular determinant. All the other interactions, V_{AB}, $V_{\mathrm{AB}}^{(2)}$, V_{ex} are diagonal in $\tilde{\Omega}$ in this basis. If, as is often the case, the LS-coupling approximation can be used for free atoms, the matrix elements of V_{AB}, $V_{\mathrm{AB}}^{(2)}$, V_{ex} in the basis c are readily expressed via the corresponding matrix elements in basis a.

The Coulomb and dispersion interactions of atoms at large R have been thoroughly studied. All relevant problems have been extensively discussed [4.3, 12, 38, 80, 115–120]. Consequently only some of the general expressions and definitions used will be given here and the methods of evaluating the constants in these expressions will be described in brief. The exchange interaction at large distances will be considered in terms of the asymptotic method [4.12, 91, 92, 95, 96].

4.2.2 Coulomb Interaction

For large R the Coulomb interaction operator can be expanded in multipoles retaining only the first few terms of the expansion. A convenient form of such an expansion has been obtained [4.121–123]. In the case considered, when the radius vectors of electrons $1 \ldots N_{\mathrm{A}}$ are referred to the nucleus A, those of the electrons $N_A + 1 \dots N_A + N_B$ to the nucleus B, and the axes z_A and z_B of the relevant coordinate frames are parallel to $\boldsymbol{R}$, the expansion becomes

$$V_{AB} = \sum_{\varkappa_1=1}^{\infty} \sum_{\varkappa_2=1}^{\infty} R^{-(\varkappa_1+\varkappa_2+1)} \sum_{q=-\varkappa_{min}}^{\varkappa_{min}} \times \frac{(-1)^{\varkappa_2}(\varkappa_1+\varkappa_2)!}{[(\varkappa_1+q)!\,(\varkappa_1-q)!\,(\varkappa_2+q)!\,(\varkappa_2-q)!]^{1/2}} Q^{\varkappa_1}_{qA} Q^{\varkappa_2}_{-qB}. \tag{4.30}$$

Summation over $\varkappa_1, \varkappa_2$ starts from unity, rather than from 0, since the total atomic charge is zero. Here $Q^{\varkappa_1}_{qA}$, $Q^{\varkappa_2}_{qB}$ are the spherical components of the multipole moment operators of ranks $\varkappa_1$ and $\varkappa_2$ for atoms A and B, respectively. These are defined as

$$Q^{\varkappa}_q = -\sum_i r_i^{\varkappa} \left(\frac{4\pi}{2\varkappa+1}\right)^{1/2} Y^q_{\varkappa}(\theta_i, \varphi_i) = \sum_i Q^{\varkappa}_q(r_i, \theta_i, \varphi_i) \tag{4.31}$$

where r_i, θ_i, φ_i are the spherical coordinates of the ith electron in a coordinate frame whose z axis is directed along $\boldsymbol{R}$ and whose origin is at the atomic nucleus. Such a definition of the multipole moment operators coincides with that in [4.124] (the minus sign corresponds to the negative electronic charge). In practice one usually deals with dipole-dipole and quadrupole-quadrupole interactions. The appropriate operators are

$$V_{ABd} = -\frac{2}{R^3} \sum_{q=-1}^{1} \frac{1}{(1-q)!\,(1+q)!} Q^{(1)}_{qA} Q^{(1)}_{-qB}, \tag{4.32}$$

$$V_{ABq} = \frac{24}{R^5} \sum_{q=-2}^{2} \frac{1}{(2-q)!\,(2+q)!} Q^{(2)}_{qA} Q^{(2)}_{-qB}. \tag{4.33}$$

The dipole-dipole interaction contributes to first-order perturbation theory only when the set D involves pairs of isolated atomic states of different parity. Such a situation occurs, for instance, in the resonant and almost-resonant excitation transfer in the collisions of alkali metal atoms [4.100, 125]. The matrix elements of V_{ABd} over the functions of the basis sets a and c are

$$\begin{aligned}\langle \gamma_1\gamma_2 L\tilde{\Lambda} S m_S | V_{ABd} | \gamma_1'\gamma_2' L'\tilde{\Lambda}' S' m_{S'} \rangle &= \delta_{SS'}\delta_{m_S m_{S'}}(-2/R^3) \\ &\times \langle \gamma_1 \| Q^{(1)}_A \| \gamma_1' \rangle \langle \gamma_2 \| Q^{(1)}_B \| \gamma_2' \rangle \sum_{\varkappa q} [(1-q)!\,(1+q)!]^{-1} \begin{bmatrix} 1 & 1 & \varkappa \\ q & -q & 0 \end{bmatrix} \\ &\times \begin{bmatrix} L' & \varkappa & L \\ \tilde{\Lambda}' & 0 & \tilde{\Lambda} \end{bmatrix} [(2L'+1)(2\varkappa+1)]^{1/2} \begin{Bmatrix} \varkappa & L & L' \\ 1 & L_1 & L_1' \\ 1 & L_2 & L_2' \end{Bmatrix},\end{aligned} \tag{4.34}$$

$$\begin{aligned}&\langle \gamma_1 j_1 \gamma_2 j_2 j \tilde{\Omega} | V_{ABd} | \gamma_1' j_1' \gamma_2' j_2' j' \tilde{\Omega} \rangle \\ &= \delta_{\tilde{\Omega}\tilde{\Omega}'}(-2/R^3) \langle \gamma_1 j_1 \| Q^{(1)}_A \| \gamma_1' j_1' \rangle \langle \gamma_2 j_2 \| Q^{(1)}_B \| \gamma_2' j_2' \rangle\end{aligned}$$

$$\times \sum_{\varkappa q} [(1-q)!\,(1+q)!]^{-1} \begin{bmatrix} 1 & 1 & \varkappa \\ q & -q & 0 \end{bmatrix} \begin{bmatrix} j' & \varkappa & j \\ \tilde{\Omega}' & 0 & \tilde{\Omega} \end{bmatrix}$$

$$\times [(2j'+1)(2\varkappa+1)]^{1/2} \begin{Bmatrix} \varkappa & j & j' \\ 1 & j_1 & j_1' \\ 1 & j_2 & j_2' \end{Bmatrix}. \tag{4.35}$$

In the *LS*-coupling approximation the quantum numbers of the total orbital angular momentum and spin can be explicitly introduced into the reduced matrix elements

$$\langle \gamma_1 j_1 \| Q_A^{(1)} \| \gamma_1' j_1' \rangle, \langle \gamma_2 j_2 \| Q_B^{(1)} \| \gamma_2' j_2' \rangle.$$

For instance,

$$\langle \gamma_1 j_1 \| Q_A^{(1)} \| \gamma_1' j_1' \rangle = \delta_{S_1 S_1'} (-1)^{L_1' + S_1 + j_1 + 1}$$

$$\times [(2j_1+1)(2j_1'+1)]^{1/2} \begin{Bmatrix} L_1 & j_1 & S_1 \\ j_1' & L_1' & 1 \end{Bmatrix} \langle \gamma_1 \| Q_A^{(1)} \| \gamma_1' \rangle. \tag{4.36}$$

The square moduli of the reduced matrix elements $\langle \gamma j \| Q^{(1)} \| \gamma' j' \rangle$ are connected with the experimentally observed oscillator strengths of the dipole transition $f(\gamma j; \gamma' j')$ by [4.124]

$$f(\gamma j; \gamma' j') = \frac{2}{3} \frac{(\varepsilon_{\gamma' j'} - \varepsilon_{\gamma j})}{2j+1} |\langle \gamma j \| Q^{(1)} \| \gamma' j' \rangle|^2. \tag{4.37}$$

Therefore, in principle, the dipole-dipole atomic interaction can be calculated from the known emission intensities of free atoms. Detailed information concerning the experimentally measured oscillator strengths of the atomic transitions can be found in [4.126, 127]. Many papers dealing with semi-empirical and purely theoretical calculations of oscillator strengths have been published. A survey of the methods used and the results obtained can be found in [4.124, 128–133].

The contribution from the quadrupole-quadrupole interaction to that between two atoms in states with $L \neq 0$ is substantial. The matrix elements of V_{ABq} over the functions of basis a are

$$\langle \gamma_1 \gamma_2 L \tilde{\Lambda} S m_S | V_{ABq} | \gamma_1' \gamma_2' L' \tilde{\Lambda}' S' m_{S'} \rangle = \delta_{\tilde{\Lambda}\tilde{\Lambda}'} \delta_{SS'} \delta_{m_S m_{S'}}$$

$$\times (24 R^{-5}) \langle \gamma_1 \| Q_A^{(2)} \| \gamma_1' \rangle \langle \gamma_2 \| Q_B^{(2)} \| \gamma_2' \rangle$$

$$\times \sum_{\varkappa q} [(2-q)!\,(2+q)!]^{-1} \begin{bmatrix} 2 & 2 & \varkappa \\ q & -q & 0 \end{bmatrix} \begin{bmatrix} L' & \varkappa & L \\ \tilde{\Lambda}' & 0 & \tilde{\Lambda} \end{bmatrix}$$

$$\times [(2L'+1)(2\varkappa+1)]^{1/2} \begin{Bmatrix} \varkappa & L & L' \\ 2 & L_1 & L_1' \\ 2 & L_2 & L_2' \end{Bmatrix}. \tag{4.38}$$

The quadrupole moment of the atom is governed by the electrons in open shells. For instance, if there is one open shell $(nl)^N$,

$$\langle\gamma \| Q^{(2)} \| \gamma'\rangle = \langle (nl)^N SL \| Q^{(2)} \| (nl)^N SL'\rangle \, \delta_{SS'}. \tag{4.39}$$

The right side is the reduced matrix element of the operator of the quadrupole moment of N electrons in this shell over wave functions of the form (3.9). Here [4.124],

$$\langle (nl)^N SL \| Q^{(2)} \| (nl)^N SL'\rangle = \langle r^2\rangle_{nl} \left[\frac{l(l+1)(2l+1)}{(2l+3)(2l-1)}\right]^{1/2} \langle (nl)^N SL \| \hat{u}_q^{(2)} \| (ml)^N SL'\rangle, \tag{4.40}$$

$$\langle r^2\rangle_{nl} = \int_0^\infty r^2 \varphi_{nl}^2(r)\, r^2\, dr, \tag{4.41}$$

$$\hat{u}_q^{(2)} = \sum_{i=1}^{N} \hat{u}_{qi}^{(2)} \tag{4.42}$$

and the one-electron tensor operators $\hat{u}_{qi}^2$ are defined by the condition

$$\langle l \| \hat{u}_i^{(2)} \| l'\rangle = \delta_{ll'}. \tag{4.43}$$

The tables of reduced matrix elements $\langle (nl)^N SL \| \hat{u}_q^{(2)} \| (nl)^N SL'\rangle$ dependent on l and N only are given in [4.124] for $l = 1, 2$ and all values of N possible in these cases. The $\langle r^2\rangle_{nl}$ values can be found in [4.15–17, 19, 42] for a large number of atoms and different n and l, calculated for the Hartree-Fock orbitals. Rather good estimates can also be obtained using AOs in the form (4.10) with parameters α_{Sl} and n defined by the Slater rules. Then

$$\langle r^2\rangle_{nl} = \frac{(2n+2)(2n+1)}{4\alpha_{\mathrm{Sl}}^2}. \tag{4.44}$$

The matrix elements of V_{ABq} over the functions of basis c are expressed via those of basis a in a manner similar to that for the transition from (4.34) to (4.35); the relevant expression is not given here.

4.2.3 Dispersion Interaction

Let the set D contain states of isolated atoms of the same parity. In this case, the contribution from V_{ABd} to the interaction matrix in second-order perturbation theory is proportional to R^{-6}. This contribution is known as the dispersion or the van der Waals interaction. Since the set D is defined in such a way that the splittings between its energy levels (usually of the order of spin-orbit coupling) are small compared to the splitting of the free-atom terms, the differences between the energy levels of set D can be neglected in the expressions of second-order perturbation theory. The relevant unsplit energy levels will be denoted by ε_{01} for atom A and by ε_{02} for atom B.

When calculating the second-order contribution, we shall use, following [4.134], the identity

$$\frac{1}{a+b}=\frac{2}{\pi}\int_0^\infty \frac{ab}{(a^2+\omega^2)(b^2+\omega^2)}\,d\omega. \tag{4.45}$$

Then the matrix elements of $V^{(2)}_{\mathrm{ABd}}$ in basis a for two different atoms (the case of two identical atoms with a possible exchange of states is treated similarly) are represented as

$$\begin{aligned}\langle\gamma_1\gamma_2 L\tilde{\Lambda} S m_S|\, V^{(2)}_{\mathrm{ABd}}\,|\gamma_1\gamma_2 L'\,\tilde{\Lambda}' S' m_{S'}\rangle &= \delta_{SS'}\delta_{m_S m_{S'}}\delta_{\tilde{\Lambda}\tilde{\Lambda}'} \\ &\times \sum_{m_{L_1}m_{L_2}}\sum_{m'_{L_1}m'_{L_2}}\begin{bmatrix} L_1 & L_2 & L \\ m_{L_1} & m_{L_2} & \tilde{\Lambda}\end{bmatrix}\begin{bmatrix} L_1 & L_2 & L' \\ m'_{L_1} & m'_{L_2} & \tilde{\Lambda}\end{bmatrix}(-R^{-6}) \\ &\times \sum_{qq'}[(1-q)!\,(1+q)!\,(1-q')!\,(1+q')!]^{-1} \\ &\times \frac{2}{\pi}\int_0^\infty d\omega\,\langle\gamma_1 m_{L_1}|\,\alpha^{(11)}_{qq'\mathrm{A}}(\omega)\,|\gamma_1 m'_{L_1}\rangle\,\langle\gamma_2 m_{L_2}|\,\alpha^{(11)}_{-q-q'\mathrm{B}}(\omega)\,|\gamma_2 m'_{L_2}\rangle,\end{aligned} \tag{4.46}$$

where

$$\begin{aligned}\langle\gamma m_L|\,\alpha^{(11)}_{qq'}(\omega)\,|\gamma m'_L\rangle &= 2\sum_{\gamma^*\neq\gamma}\sum_{m^*_L}\frac{(\varepsilon_{\gamma^*}-\varepsilon_\gamma)}{(\varepsilon_{\gamma^*}-\varepsilon_\gamma)^2+\omega^2} \\ &\times\langle\gamma m_L|\,Q^{(1)}_q\,|\gamma^* m^*_L\rangle\,\langle\gamma^* m^*_L|\,Q^{(1)}_{q'}\,|\gamma m'_L\rangle\end{aligned} \tag{4.47}$$

is the dipole dynamic polarizability at an imaginary frequency $i\omega$. Thus, the two-center problem of calculating the dispersion interaction of two atoms reduces to the calculation of the dynamic polarizability of isolated atoms. The operator $\alpha^{(11)}_{qq'}(\omega)$ defined by (4.47) transforms under rotation as the direct product of two irreducible tensor operators of rank 1. It is convenient to introduce irreducible tensor operators

$$\alpha^{\varkappa}_{\mu}(\omega)=\sum_{\mu_1\mu_2}\begin{bmatrix} 1 & 1 & \varkappa \\ \mu_1 & \mu_2 & \mu\end{bmatrix}\alpha^{11}_{\mu_1\mu_2}(\omega). \tag{4.48}$$

Here the polarizability of the atom in a degerate state (neglecting spin-orbit coupling) is obviously defined by three reduced matrix elements $\langle\gamma\|\,\alpha^{\varkappa}(\omega)\,\|\gamma\rangle$ with $\varkappa=0,1,2$. The following expression is readily obtained for these elements:

$$\langle\gamma\| \alpha^{\varkappa}(\omega) \|\gamma\rangle = (2\varkappa + 1)^{1/2}(-1)^{L+\varkappa} \sum_{\gamma^* \neq \gamma} \frac{\varepsilon_{\gamma^*} - \varepsilon_{\gamma}}{(\varepsilon_{\gamma^*} - \varepsilon_{\gamma})^2 + \omega^2}$$

$$\times (-1)^{L^*} \begin{Bmatrix} L^* & 1 & L \\ \varkappa & L & 1 \end{Bmatrix} |\langle\gamma\| Q^1 \|\gamma^*\rangle|^2. \tag{4.49}$$

In particular

$$\langle\gamma\| \alpha^0(\omega) \|\gamma\rangle = -3\left(\frac{2L+1}{3}\right)^{1/2} \alpha_{\gamma}(\omega), \tag{4.50}$$

where

$$\alpha_{\gamma}(\omega) = \sum_{m_L} (2L+1)^{-1} \langle\gamma m_L| \alpha_{00}^{11}(\omega) |\gamma m_L\rangle \tag{4.51}$$

is the atom polarizability in an electric field directed along the z axis averaged over all degenerate states. For an atom in a state with $L = 0$, only the reduced matrix element $\langle\gamma\| \alpha^0(\omega) \|\gamma\rangle$ survives.

Using the reduced matrix elements $\langle\gamma_1\| \alpha_{\mathrm{A}}^{\varkappa}(\omega) \|\gamma_1\rangle$ and $\langle\gamma_2\| \alpha_{\mathrm{B}}^{\varkappa}(\omega) \|\gamma_2\rangle$, (4.46) can be written as

$$\langle\gamma_1\gamma_2 L\tilde{\Lambda} S m_S| V_{\mathrm{ABd}}^{(2)} |\gamma_1\gamma_2 L'\tilde{\Lambda}' S' m_{S'}\rangle = \delta_{SS'}\delta_{m_S m_{S'}}\delta_{\tilde{\Lambda}\tilde{\Lambda}'}(-R^{-6})$$

$$\times \sum_{\varkappa}\sum_{\varkappa_1}\sum_{\varkappa_2}\sum_{qq'} [(1-q)!\,(1+q)!\,(1-q')!\,(1+q')!]^{-1}$$

$$\times [(2L'+1)(2\varkappa+1)]^{1/2} \begin{bmatrix} 1 & 1 & \varkappa_1 \\ q & q' & q+q' \end{bmatrix} \begin{bmatrix} 1 & 1 & \varkappa_2 \\ -q & -q' & -(q+q') \end{bmatrix}$$

$$\times \begin{bmatrix} \varkappa_1 & \varkappa_2 & \varkappa \\ q+q' & -q-q' & 0 \end{bmatrix} \begin{bmatrix} L' & \varkappa & L \\ \tilde{\Lambda} & 0 & \tilde{\Lambda} \end{bmatrix} \begin{Bmatrix} \varkappa & L & L' \\ \varkappa_1 & L_1 & L'_1 \\ \varkappa_2 & L_2 & L'_2 \end{Bmatrix} \varepsilon(\varkappa_1, \varkappa_2), \tag{4.52}$$

where

$$\varepsilon(\varkappa_1, \varkappa_2) = \frac{2}{\pi} \int_0^{\infty} d\omega \langle\gamma_1\| \alpha_{\mathrm{A}}^{\varkappa_1}(\omega) \|\gamma_1\rangle \langle\gamma_2\| \alpha_{\mathrm{B}}^{\varkappa_2}(\omega) \|\gamma_2\rangle. \tag{4.53}$$

It can readily be checked that the terms involving $\varepsilon(1, 0)$, $\varepsilon(0, 1)$, $\varepsilon(1, 2)$ and $\varepsilon(2, 1)$ in (4.52) are equal to zero. Thus the dispersion interaction of two atoms in orbitally degenerate states is governed in the general case by five parameters: $\varepsilon(0, 0)$, $\varepsilon(1, 1)$, $\varepsilon(2, 0)$, $\varepsilon(0, 2)$, $\varepsilon(2, 2)$. For the interaction of two identical atoms in the same states the number of parameters reduces to four since $\varepsilon(2, 0) = \varepsilon(0, 2)$; two parameters, $\varepsilon(0, 0)$ and $\varepsilon(0, 2)$, remain for $L_1 \neq 0$, $L_2 = 0$, and one, $\varepsilon(0, 0)$, for $L_1 = L_2 = 0$. In the two latter cases the expressions for the matrix elements will be

$$\langle \gamma_1 \gamma_2 L_1 \tilde{\Lambda} S m_S | V_{\mathrm{ABd}}^{(2)} | \gamma_1 \gamma_2 L_1 \tilde{\Lambda} S m_S \rangle$$

$$= -R^{-6} \frac{\varepsilon(0,0)}{2(2L_1+1)^{1/2}} + R^{-6} \begin{bmatrix} L_1 & 2 & L_1 \\ \tilde{\Lambda} & 0 & \tilde{\Lambda} \end{bmatrix} \frac{\varepsilon(2,0)}{2[2(2L_1+1)]^{1/2}}, \quad (4.54)$$

$$\langle \gamma_1 \gamma_2 00 S m_S | V_{\mathrm{ABd}}^{(2)} | \gamma_1 \gamma_2 00 S m_S \rangle = -\frac{\varepsilon(0,0)}{2R^6}. \quad (4.55)$$

Again, the matrix elements over the functions of basis c can be expressed via those in basis a, but the relevant expressions are not given here.

The methods of calculating the dynamic polarizabilities of atoms which determine the constants of the dispersion interaction can be divided into two groups. One involves the direct theoretical methods based on many-body perturbation theory [4.135–137], time-dependent Hartree-Fock theory [4.138, 139], and variational methods [4.140]. The second group embraces semi-empirical methods using available experimental data on the optical and dielectric properties of the system studied [4.141–144]. If these data are sufficiently diverse, very reliable upper and lower bounds for the constants of the dispersion interaction of atoms in orbitally nondegenerate ground-states can be obtained in terms of the semi-empirical methods [4.142, 143, 145, 146]. Moreover, the semi-empirical approaches enable one to obtain the so-called combination rules which express the unknown dispersion interaction constants of different atoms via the known constants of the dispersion interaction of the same atoms [4.147, 148].

Unfortunately, the overwhelming majority of ab initio and semi-empirical calculations of dynamic polarizabilities and dispersion interaction constants have been performed for spherically symmetric atoms in ground states. There seem to be only a few papers about calculations concerned with orbitally degenerate atoms [4.135, 149, 150], yet the interactions of orbitally degenerate or excited atoms are of particular interest in the treatment of inelastic atomic collisions.

To estimate the dynamic polarizability in these cases it is convenient to use the closure approximation which replaces $\varepsilon_{\gamma*} - \varepsilon_\gamma$ in (4.47) by a certain constant average excitation energy $\Delta\varepsilon_\gamma$. In this approximation

$$\langle \gamma m_L | \alpha_\mu^\varkappa(\omega) | \gamma m'_L \rangle = \frac{\Delta\varepsilon_\gamma}{\Delta\varepsilon_\gamma^2 + \omega^2} \langle \gamma m_L | \phi_\mu^\varkappa | \gamma m'_L \rangle, \quad (4.56)$$

where

$$\phi_\mu^\varkappa = \sum_{qq'} \begin{bmatrix} 1 & 1 & \varkappa \\ q & q' & \mu \end{bmatrix} Q_q^1 Q_{q'}^1. \quad (4.57)$$

It should be noted that $\phi_\mu^{(1)} = 0$. If the static polarizability is calculated in the same approximation, then $\langle \gamma m_L | \alpha_\mu^\varkappa(\omega) | \gamma m_L' \rangle$ for $\varkappa \neq 1$ will be

$$\langle \gamma m_L | \alpha_\mu^\varkappa(\omega) | \gamma m_L' \rangle = \frac{\Delta \varepsilon_\gamma^2}{\Delta \varepsilon_\gamma^2 + \omega^2} \langle \gamma m_L | \alpha_\mu^\varkappa(0) | \gamma m_L' \rangle . \tag{4.58}$$

For $\varkappa = 1$, the passage from (4.56 to 58) cannot be obtained since, in the approximation used, $\langle \gamma m_L | \alpha_\mu^{(1)}(\omega) | \gamma m_L' \rangle \equiv 0$.

The value $\Delta \varepsilon_\gamma$ must correspond in order of magnitude to the energy of the resonant transition for the state considered. For a more definite estimate the following considerations may be taken into account [4.134]. It follows from the sum rule for the oscillator strengths that

$$\alpha_\gamma(\omega) \xrightarrow[\omega \to \infty]{} \frac{N}{\omega^2} , \tag{4.59}$$

where N is the total number of electrons in the atom.

Then, demanding that (4.56) have the same asymptotic behavior, we get

$$\Delta \varepsilon_\gamma = \left[\frac{N}{\alpha_\gamma(0)} \right]^{1/2} . \tag{4.60}$$

For a spherically symmetric atom in the ground state the closure approximation with $\Delta \varepsilon_\gamma$ defined by (4.60) is known [4.142] to yield a rigorous upper bound for the dynamic polarizability. However it follows from the available data on the interactions of noble gas and alkali metal atoms in ground states that, to obtain a satisfactory agreement with the results of more rigorous calculations, N in (4.60) must be taken to be equal to the number of outer shell electrons. We come in this way to the Slater-Kirkwood approximation [4.151]. The constants of the dispersion interaction for alkali metal and inert gas atoms in ground states are obtained in the Slater-Kirkwood approximation to an accuracy no worse than 25% [4.12]. For the atoms in ground states the average static polarizabilities needed for the evaluation of $\Delta \varepsilon_\gamma$ by (4.60) can be found, e.g. in [4.152–154].

The Slater-Kirkwood approximation is, in fact, based on the assumption that the atom polarizability is essentially governed by the valence electrons. Consequently, when using this approximation for the calculation of the dynamic polarizability and the dispersion interaction, one must account explicitly only for the coordinates of the valence electrons in the wave functions of the atom and in Q_μ^1 in (4.56) and (4.57). Such an approximation is consistent with the pseudopotential model for valence electrons described in Sect. 4.1.2.

Consider the case of an atom with a valence shell $(nl)^N$. When the matrix elements of $\phi_\mu^\varkappa$ are calculated over the functions (3.9), the terms in the form $Q_q^1(r_i, \theta_i, \varphi_i)\, Q_{q'}^1(r_j, \theta_j, \varphi_j)$ do not contribute to the matrix element when $i \neq j$, as Q_q^1 is an odd operator. Therefore it is sufficient to calculate the matrix elements of the operator

$$\tilde{\Phi}_\mu^\varkappa = \sum_{i=1}^{N} \sum_{qq'} \begin{bmatrix} 1 & 1 & \varkappa \\ q & q' & \mu \end{bmatrix} Q_q^{(1)}(r_i, \theta_i, \varphi_i)\, Q_{q'}^{(1)}(r_i, \theta_i, \varphi_i)$$

$$= -\begin{bmatrix} 1 & 1 & \varkappa \\ 0 & 0 & 0 \end{bmatrix} \sum_{i=1}^{N} r_i^{2-\varkappa} Q(r_i, \theta_i, \varphi_i). \tag{4.61}$$

It was taken into account in deriving (4.61) that the charge of the electrons is -1, and the known expression for the product of two spherical functions was used. For the reduced matrix elements in this approximation,

$$\langle (nl)^N SL \| \alpha^{(0)}(\omega) \| (nl)^N SL \rangle = -2N \langle r^2 \rangle_{nl} \left(\frac{2L+1}{3} \right)^{1/2} \frac{\Delta \varepsilon_\gamma}{\Delta \varepsilon_\gamma^2 + \omega^2}, \tag{4.62}$$

$$\langle (nl)^N SL \| \alpha^{(2)}(\omega) \| (nl)^N SL \rangle = -2(2/3)^{1/2} \langle (nl)^N SL \| Q^{(2)} \| (nl)^N SL \rangle \frac{\Delta \varepsilon_\gamma}{\Delta \varepsilon_\gamma^2 + \omega^2}, \tag{4.63}$$

where the reduced matrix elements $Q_q^{(2)}$ are defined by (4.40).

When the atom A with an open valence shell $(n_1 l_1)^{N_1}$ interacts with atom B in a spherically symmetric state with a static polarizability $\alpha_{\gamma 2}$, we obtain from (4.58, 62, 63) the coefficients $\varepsilon(0,0)$ and $\varepsilon(2,0)$ involved in (4.54):

$$\varepsilon(0,0) = \alpha_{\gamma 2} \langle r^2 \rangle_{n_1 l_1} 2(2L_1+1)^{1/2} N_1 \frac{\Delta \varepsilon_{\gamma 2}}{\Delta \varepsilon_{\gamma_1} + \Delta \varepsilon_{\gamma_2}}, \tag{4.64}$$

$$\varepsilon(2,0) = \alpha_{\gamma_2} 2\sqrt{2} \langle (n_1 l_1)^{N_1} S_1 L_1 \| Q^{(2)} \| (n_1 l_1)^{N_1} S_1 L_1 \rangle \frac{\Delta \varepsilon_{\gamma 2}}{\Delta \varepsilon_{\gamma_1} + \Delta \varepsilon_{\gamma_2}}, \tag{4.65}$$

where $\Delta\varepsilon_{\gamma_1}$, $\Delta\varepsilon_{\gamma_2}$ are the average excitation energies of atoms A and B. The expression derived from (4.54, 64, 65) for the dispersion interaction between an atom in a spherically symmetric state and an atom with an open valence shell was first obtained in [4.155]. When $\Delta\varepsilon_{\gamma_1} \ll \Delta\varepsilon_{\gamma_2}$ (atom A highly excited), then the $\varepsilon(0,0)$ and $\varepsilon(2,0)$ values are defined only by $\langle r^2 \rangle_{n_1 l_1}$ and α_{γ_2} [4.12, 156–159]. This is the situation, for instance, for the interaction of an excited alkali atom with a noble gas atom.

4.2.4 Exchange Interaction

The general expression for the matrix elements of the exchange interaction is given by (4.27). To obtain expressions convenient for calculation, we assume that only the valence shell electrons can be explicitly accounted for. This assumption is plausible, since at distances of practical interest the wave functions of the inner electrons decrease much faster than do the wave functions of the valence electrons (Sect. 4.1.1), and this is corroborated by accurate calculations of the interaction between noble gas atoms in the van der Waals minimum region [4.105]. Thus, for the exchange interaction (as in the case of the dispersion interaction and seemingly with greater accuracy) use can be made of the atom model which implies that the electrons move in a pseudopotential field. In terms of this model, H_0 in (4.25) is modified. This modification was discussed in Sect. 4.1.2 for a free atom. For two atoms, the interaction potential of an electron with a "foreign" atom is, generally speaking, unknown. It may differ from the pseudopotential in a free atom. However, in view of the small deformation of the cores, this difference is usually neglected [4.47, 56, 160, 161].

Relatively simple analytical equations can be obtained only when the specific form of the pseudopotential is not very important for the exchange interaction at large distances, as in the case of the interaction between two atoms in states with close ionization potentials I_1 and I_2 at distances

$$R > \frac{1}{I_1} + \frac{1}{I_2}. \tag{4.66}$$

When this inequality is satisfied, the main contribution to the interaction comes from the configurational space of electrons near the internuclear axis along which the electron motion is quasi-classical. Therefore the exchange interaction can be calculated by the *Herring-Gor'kov-Pitaevskii method* [4.90–93, 162–164]. This method makes use of the quasi-classical character of the motion of the exchanging electrons in that region near the internuclear axis which is far from the nuclei (i.e., in the classically inaccessible region). The above feature allows one to account accurately for the Coulomb interaction of the electrons, and to construct explicitly the function $|\gamma_1\gamma_2 L\tilde{\Lambda} S m_S P\rangle$. The matrix elements in (4.27) are calculated to represent this expression in the form of a surface integral, which is possible provided the terms of the order of $\varepsilon_{\mathrm{int}} S_{\mathrm{AB}}^2$ are neglected. A detailed description of the calculation procedure for $|\gamma_1\gamma_2 L\tilde{\Lambda} S m_S P\rangle$ and the matrix elements of V_{ex} in terms of the asymptotic model can be found in [4.12, 92, 94–96]. The rather cumbersome calculations are not discussed here, and only the final expression for the matrix elements in the case of the interaction of two atoms with open s or p valence shells will be given:

$$\langle (n_1 l_1)^{N_\mathrm{A}} S_1 L_1; (n_2 l_2)^{N_\mathrm{B}} S_2 L_2; L\tilde{\Lambda} S m_S |\, V_{\mathrm{ex}} \,| (n_1' l_1')^{N_\mathrm{A}} S_1' L_1';$$

$$(n_2' l_2')^{N_B} S_2' L_2'; L' \tilde{\Lambda}' S' m_{S'}\rangle = \delta_{SS'} \delta_{m_S m_{S'}} \delta_{\tilde{\Lambda}\tilde{\Lambda}'} N_A N_B$$

$$\times \sum_{S_1'' L_1''} \sum_{S_2'' L_2''} G_{S_1'' L_1''}^{S_1 L_1} G_{S_1'' L_1''}^{S_1' L_1'} G_{S_2'' L_2''}^{S_2 L_2} G_{S_2'' L_2''}^{S_2' L_2'}$$

$$\times (-1)^{S_2 + S_2' - 2S_2''} [(2S_1 + 1)(2S_2 + 1)(2S_1' + 1)(2S_2' + 1]^{1/2}$$

$$\times \begin{Bmatrix} S_1'' & 1/2 & S_1 \\ 1/2 & S_2'' & S_2 \\ S_1' & S_2' & S \end{Bmatrix} \langle L_1'' L_1; L_2'' L_2; L \tilde{\Lambda}| V_{ex} | L_1'' L_1'; L_2'' L_2'; L' \tilde{\Lambda}\rangle, \quad (4.67)$$

where

$$\langle L_1'' L_1; L_2'' L_2; L \tilde{\Lambda}| V_{ex} | L_1'' L_1'; L_2'' L_2'; L' \tilde{\Lambda}\rangle$$

$$= \sum_{m_{L_1''} m_{L_2''}} \sum_{m_{L_1} m_{L_2}} \sum_{m_{L_1'} m_{L_2'}} \sum_{m_1 m_2} \sum_{m_1' m_2'} \begin{bmatrix} L_1 & L_2 & L \\ m_{L_1} & m_{L_2} & \tilde{\Lambda} \end{bmatrix} \begin{bmatrix} L_1' & L_2' & L' \\ m_{L_1'} & m_{L_2'} & \tilde{\Lambda} \end{bmatrix}$$

$$\times \begin{bmatrix} L_1'' & l_1 & L_1 \\ m_{L_1''} & m_1 & m_{L_1} \end{bmatrix} \begin{bmatrix} L_1'' & l_1' & L_1' \\ m_{L_1''} & m_1' & m_{L_1'} \end{bmatrix} \begin{bmatrix} L_2'' & l_2 & L_2 \\ m_{L_2''} & m_2 & m_{L_2} \end{bmatrix} \begin{bmatrix} L_2'' & l_2' & L_2' \\ m_{L_2''} & m_2' & m_{L_2'} \end{bmatrix}$$

$$\times \langle n_1 l_1 m_1; n_2 l_2 m_2 | I_{as} | n_1' l_1' m_1'; n_2' l_2' m_2'\rangle \, \delta_{m_1 + m_2; m_1' + m_2'} . \quad (4.68)$$

The quantities $\langle n_1 l_1 m_1; n_2 l_2 m_2 | I_{as} | n_1' l_1' m_1'; n_2' l_2' m_2\rangle$ are known as two-electron exchange integrals (the subscript as means asymptotic calculation). They are expressed via the asymptotic splitting for singlet and triplet terms of a hypothetical pair of one-electron atoms whose valence electrons move in a field which coincides with that acting on the electrons in atoms A and B.

To clarify the rather complicated (4.67) we can resort to the conventional Heitler-London approximation when treating the initial expression (4.27). This approximation consists in replacing $|\gamma_1 \gamma_2 L \tilde{\Lambda} S m_S P\rangle$ by $|\gamma_1 \gamma_2 L \tilde{\Lambda} S m_S D\rangle$ and in directly calculating the interaction matrix element. Neglecting, as we have done everywhere, the terms of the order of $\varepsilon_{int} S_{AB}^2$ and using the Dirac identity

$$P_{ij}^{\sigma} = \tfrac{1}{2}(1 + 4 s_i s_j) \quad (4.69)$$

where s_i, s_j are the spin operators of the electrons i and j, we obtain

$$V_{ex}^{HL} = - \sum_{i=1}^{N_A} \sum_{j=N_A+1}^{N_A+N_B} \frac{1}{2}(1 + 4 s_i s_j) V_{ij} P_{ij}^{x}, \quad (4.70)$$

where

$$V_{ij} = V_{eff\,B}(r_i) + V_{eff\,A}(r_j) + |r_i - r_j|^{-1} + R^{-1}. \quad (4.71)$$

$V_{eff\,B}$ and $V_{eff\,A}$ are the effective potentials of the valence electrons in atoms B and A. It can readily be verified using the conventional methods of angular momentum theory that the matrix elements of (4.70) are given by the same expressions (4.68) involving, instead of the asymptotic exchange integrals, the usual integrals of the Heitler-London theory:

$$\langle n_1 l_1 m_1; n_2 l_2 m_2 | I_{\mathrm{HL}} | n_1' l_1' m_1'; n_2' l_2' m_2' \rangle$$
$$= \int d^3 r_1 \int d^3 r_2 \, \varphi^{m_1^*}_{n_1 l_1 \mathrm{A}}(r_1) \, \varphi^{m_2^*}_{n_2 l_2 \mathrm{B}}(r_2) \, V_{12} \, \varphi^{m_1'}_{n_1' l_1' \mathrm{A}}(r_2) \, \varphi^{m_2'}_{n_2' l_2' \mathrm{B}}(r_1) . \qquad (4.72)$$

It should be noted that the symmetry under rotation around the molecular axis and reflection in the planes containing the axis results in the asymptotic and Heitler-London exchange integrals having the following properties:

$$\langle n_1 l_1 m_1; n_2 l_2 m_2 | I | n_1' l_1' m_1'; n_2' l_2' m_2' \rangle = 0 \qquad (4.73)$$

for $m_1 + m_2 \neq m_1' + m_2'$,

$$\langle n_1 l_1 m_1; n_2 l_2 m_2 | I | n_1' l_1' m_1'; n_2' l_2' m_2' \rangle$$
$$= \langle n_1 l_1 - m_1; n_2 l_2 - m_2 | I | n_1' l_1' - m_1'; n_2' l_2' - m_2' \rangle . \qquad (4.74)$$

The property (4.73) is explicitly accounted for in (4.68) by the introduction of the factor $\delta_{m_1 + m_2; m_1' + m_2'}$. Since the substitution $(m_1 m_2 m_1' m_2') \to (-m_1 - m_2 - m_1' - m_2')$ is equivalent, as seen from (4.72), to complex conjugation of the exchange integral, it follows from (4.74) that the exchange integrals are real. Within the set D, the matrix of the exchange integrals is Hermitian, since only the cases $n_1' l_1' = n_1 l_1$ and $n_2' l_2' = n_2 l_2$ (different atoms or integrals without excitation transfer for identical atoms) or $n_1' l_1' = n_2 l_2$ and $n_2' l_2' = n_1 l_1$ (integrals with excitation transfer for identical atoms) are possible. Consequently, taking into account that the exchange integrals are real, we have

$$\langle n_1 l_1 m_1; n_2 l_2 m_2 | I | n_1' l_1' m_1'; n_2' l_2' m_2' \rangle$$
$$= \langle n_1' l_1' m_1'; n_2' l_2' m_2' | I | n_1 l_1 m_1; n_2 l_2 m_2 \rangle . \qquad (4.75)$$

The development of the asymptotic method of calculating exchange interactions was stimulated by the incorrect behavior of the Heitler-London exchange interaction between H atoms in the ground state for $R \to \infty$, as first noted by *Herring* [4.91]. The Heitler-London exchange integral changes its sign for $R \simeq 50$ thus making the $^3\Sigma_u^+$ state of H_2 lower than the $^1\Sigma_g^+$ state. This contradicts the theorem about the absence of nodes in the ground state wave function. However, at such large distances the exchange interaction is so weak that it becomes immaterial, at least for atomic collisions.

The conventional Heitler-London method seems to be preferable around the van der Waals minimum region and for smaller distances (when the exchange interaction becomes comparable with the dispersion and Coulomb interactions or exceeds these). But its serious drawback is that the calculation of the integrals (4.72) can only be numerical. Since the terms involved often compensate each other, the calculations must be extremely accurate, which is possible only with fast computers. Moreover, the evaluation of $V_{\text{eff A}}$ and $V_{\text{eff B}}$ for many-electron atoms is very complicated. At the same time the expressions for

$$\langle n_1 l_1 m_1; n_2 l_2 m_2 | I_{\text{as}} | n'_1 l'_1 m'_1; n'_2 l'_2 m'_2 \rangle$$

are rather simple. These expressions have been obtained for the cases encountered most often [4.163, 164] (Tables 4.1–4).

The functions I_{kn} and I'_k in these tables are

$$\begin{aligned} I_{kn} = {} & R^{(2/\alpha)+(2/\beta)-(1/\mu)-2} \exp(-\mu R)\, \tfrac{9}{2} C_1^2 C_2^2 \Gamma(1/\mu) \\ & \times (2/\mu^2)^{1/\mu} \left\{ (\mu/2\beta)^{2/\alpha} \beta^{2/\mu} \int_0^1 \exp\left[-(\alpha-\beta) Rt + \frac{1}{\beta}(t-1) \right] \right. \\ & \times (1+t)^{(2/\alpha)-(2/\beta)+(1/\mu)-n} (1-t)^{(2/\beta)-(1/\mu)+n} \\ & \times [\mu + (\beta-\alpha) t]^{-[2+(1/\mu)+k]} dt + (\mu/2\alpha)^{2/\beta} \alpha^{2/\mu} \\ & \times \int_0^1 \exp\left[(\alpha-\beta) Rt + \frac{1}{\alpha}(t-1) \right] (t+1)^{(2/\beta)-(2/\alpha)+(1/\mu)+n} \\ & \left. \times (1-t)^{(2/\alpha)-(1/\mu)-n} [\mu + (\alpha-\beta) t]^{-[2+(1/\mu)+k]} dt \right\}, \end{aligned} \tag{4.76}$$

$$\begin{aligned} I'_k = {} & R^{(2/\alpha)+(2/\beta)-(1/\mu)-2} \exp(-\mu R)\, C_1^2 C_2^2 \Gamma(1/\mu) (2\mu)^{-1/\mu} \\ & \times 9[\alpha\beta(2\mu)^2]^{-1} (\mu/2)^{(1/\alpha)+(1/\beta)} (\alpha)^{-1/\beta} (\beta)^{-1/\alpha} \\ & \times \int_0^1 \exp\left[\left(\frac{1}{2\alpha} + \frac{1}{2\beta} \right)(t-1) \right] (1+t)^{(1/\mu)+k} (1+t^2)^k \\ & \times (1-t)^{-(1/\mu)-k+(1/\alpha)+(1/\beta)} dt, \end{aligned} \tag{4.77}$$

Table 4.1. Exchange integrals $\langle n_1 s0; n_2 l_2 m_2 | I_{\text{as}} | n_1 s0; n_2 l_2 m_2 \rangle$

$\langle n_1 s0; n_2 s0 \| I_{\text{as}} \| n_1 s0; n_2 s0 \rangle$	$-\dfrac{R}{9} I_{00}$
$\langle n_1 s0; n_2 p0 \| I_{\text{as}} \| n_1 s0; n_2 p0 \rangle$	$-\dfrac{R}{3} I_{00}$
$\langle n_1 s0; n_2 p1 \| I_{\text{as}} \| n_1 s0; n_2 p1 \rangle$	$\dfrac{1}{3\mu} I_{1-1}$

Table 4.2. Exchange integrals with excitation transfer $\langle n_1 s0; n_2 l_2 m_2 | I_{as} | n_2 l_2 m_2; n_1 s0 \rangle$

$\langle n_1 s0; n_2 s0 \| I_{as} \| n_2 s0; n_1 s0 \rangle$	$-\frac{R}{9} \mu^2 I'_0$
$\langle n_1 s0; n_2 p0 \| I_{as} \| n_2 p0; n_1 s0 \rangle$	$\frac{R}{3} \mu^2 I'_0$
$\langle n_1 s0; n_2 p1 \| I_{as} \| n_2 p1; n_1 s0 \rangle$	$-\frac{\mu}{3}\left[1 + \frac{\alpha}{\beta}\left(1 + \frac{1}{\mu}\right)\right] I'_0$

where C_1, C_2 are constants in the asymptotic expressions (4.15) for the radial AOs $\varphi_{n_1 l_1}$ and $\varphi_{n_2 l_2}$, $\alpha = (2I_1)^{1/2}$, $\beta = (2I_2)^{1/2}$, $\mu = \alpha + \beta$ and $\Gamma(1/\mu)$ is the gamma function. The R-dependence of the exchange integrals with excitation transfer is singled out explicitly. This cannot be done for exchange integrals without excitation transfer when $\alpha \neq \beta$. However, the integrals entering (4.76) can readily be obtained with computers. When $|(\alpha - \beta) R| < 1$, the explicit dependence of these integrals on R can be obtained by expanding the exponent in integrands in powers of $(\alpha - \beta) R$. Obviously, if $|(\alpha - \beta) R| < 1$, then $|\alpha - \beta| \ll 1$, as the values of R consid-

Table 4.3. Exchange integrals $\langle n_1 pm_1; n_2 pm_2 | I_{as} | n_1 pm'_1; n_2 pm'_2 \rangle$

m_1	m_2 \ m'_1 / m'_2	1 / 1	1 / 0	0 / 1
1	1	$-\left[2 + \left(2 + \frac{1}{\mu}\right)\left(1 + \frac{1}{\mu}\right)\right] R^{-1} I_{20}$	0	0
1	0	0	$\frac{1}{\mu} I_{11}$	$\left(2 + \frac{1}{\mu}\right) I_{10}$
0	1	0	$\left(2 + \frac{1}{\mu}\right) I_{10}$	$\frac{1}{\mu} I_{1-1}$
1	-1	0	0	0
-1	1	0	0	0
0	0	0	0	0

ered are large enough. Therefore we can put $\alpha = \beta$ in the coefficients of the expansion in powers of $(\alpha - \beta) R$. Then the expansion term linear in $(\alpha - \beta) R$ goes to zero and consequently a good approximation will be given even by the terms of zero order in $(\alpha - \beta) R$, corresponding to the interaction of two atoms with the same ionization potentials.

The question arises of the smallest value of R for which the simple asymptotic expressions for the exchange integrals can be used. The answer can be obtained only by comparison either with the results of variational calculations for R not very large, or with experimental data. For two H atoms in the ground state, the singlet-triplet splitting calculated asymptotically shows good agreement with the results obtained by the Heitler-London method and by a rigorous variational method [4.12] down to $R \simeq 4$, i.e., to the boundary of the quasi-classical approximation's validity, defined by the inequality (4.66).

Such a direct comparison of the exchange interactions calculated by various methods cannot be made for many-electron atoms. The terms which account for all types of interactions must be considered here, and then be compared with either the results of variational calculations or with experimental data. Such calculations have been carried out for various systems [4.101, 102, 106, 107, 109]. It was found that the region of validity

1	-1	0
-1	1	0
0	0	0
0	0	0
0	0	0
$\frac{R^{-1}}{\mu}\left(1 - \frac{1}{\mu}\right) I_{20}$	$-\left[2 + \left(2 + \frac{1}{\mu}\right)\left(1 + \frac{1}{\mu}\right)\right] R^{-1} I_{20}$	$\frac{1}{\mu} I_{10}$
$-\left[2 + \left(2 + \frac{1}{\mu}\right)\left(1 + \frac{1}{\mu}\right)\right] R^{-1} I_{20}$	$\frac{R^{-1}}{\mu}\left(1 - \frac{1}{\mu}\right) I_{20}$	$\frac{1}{\mu} I_{10}$
$\frac{1}{\mu} I_{10}$	$\frac{1}{\mu} I_{10}$	$-R I_{00}$

Table 4.4. Exchange integrals with excitation transfer $\langle n_1 p m_1; n_2 p m_2 | I_{as} | n_2 p m_1'; n_1 p m_2' \rangle$

m_1	m_2	$m_1' = 1$, $m_2' = 1$	$m_1' = 1$, $m_2' = 0$	$m_1' = 0$, $m_2' = 1$
1	1	$-R^{-1}\left[4+\frac{3}{\mu}+\frac{1}{\mu^2}+\frac{(\alpha-\beta)^2}{\alpha\beta}\left(1+\frac{1}{\mu}\right)\right]I_0'$	0	0
1	0	0	I_1'	$\mu\left[1+\frac{\beta}{\alpha}\times\left(1+\frac{1}{\mu}\right)\right]I_0'$
0	1	0	$\mu\left[1+\frac{\alpha}{\beta}\times\left(1+\frac{1}{\mu}\right)\right]I_0'$	I_1'
1	-1	0	0	0
-1	1	0	0	0
0	0	0	0	0

of the asymptotic method for calculating the exchange interaction was defined by the inequality (4.66) for the many-electron atoms as well. It is essential that this inequality covers a rather wide range of R-values and that within this range the exchange interaction is stronger than the dispersion and the Coulomb interactions. The O_2 [4.107] and I_2 [4.101] terms obtained by using the asymptotic exchange integrals are presented as examples in Figs. 4.6 and 4.7 together with the variational calculation results and the terms constructed from spectroscopic data.

On the whole, the problem of calculating the exchange interaction for sufficiently large R for atoms whose ionization potentials are close can be considered as essentially solved. Things are more complicated when I_1 and I_2 substantially differ. If the inequality

$$|(\alpha-\beta)R| \gg 1 \tag{4.78}$$

1	-1	0
-1	1	0
0	0	0
0	0	0
0	0	0
$\frac{R^{-1}}{\mu}\left(1-\frac{1}{\mu}\right)I'_0$	$-R^{-1}\left[4+\frac{3}{\mu}+\frac{1}{\mu^2}+\frac{(\alpha-\beta)^2}{\alpha\beta}\left(1+\frac{1}{\mu}\right)\right]I'_0$	I'_0
$-R^{-1}\left[4+\frac{3}{\mu}+\frac{1}{\mu^2}+\frac{(\alpha-\beta)^2}{\alpha\beta}\left(1+\frac{1}{\mu}\right)\right]I'_0$	$\frac{R^{-1}}{\mu}\left(1-\frac{1}{\mu}\right)I'_0$	I'_0
I'_0	I'_0	$-\mu^2 R I'_0$

is satisfied, the exchange interaction is governed essentially by the region close to the atom with the higher ionization potential (let it be atom B), where the quasi-classical approximation, and thus the Herring-Gor'kov-Pitaevskii method, are inapplicable.

Such a situation usually occurs when atom A has one more or less loosely bound valence electron (an alkali atom or an atom in an excited Rydberg state).

Two fundamentally different approaches are possible here. When the valence electron is not too strongly excited in the sense that α and β are of the same order, use can be made of the method described above, but the calculation of the exchange integral must be modified. Either a modification of the Herring-Gor'kov-Pitaevskii method can be used, taking the integration surface to be the sphere around the unexcited atom with a

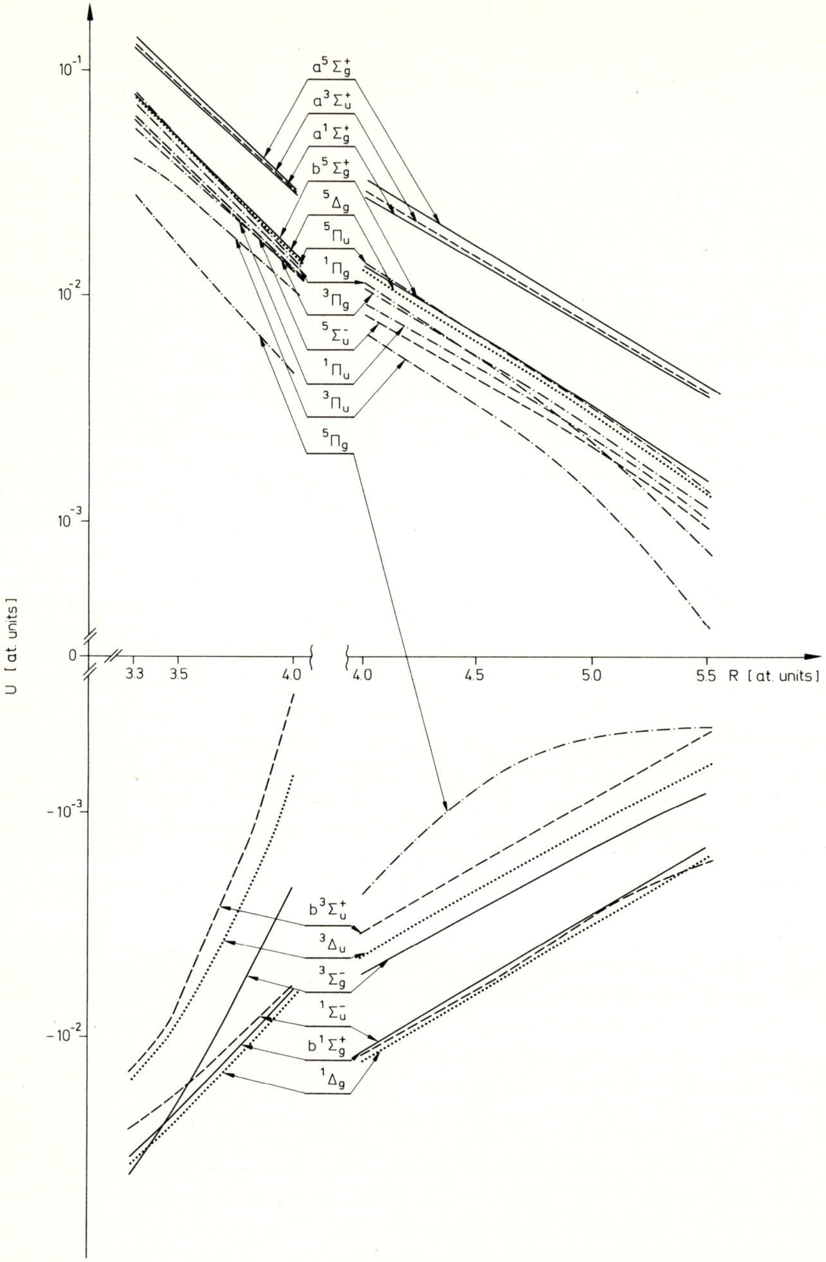

Fig. 4.6. Terms of the system O (^3p) + O (^3p). The results of asymptotic calculations [4.107] are given for $R > 4$. The curves for $R < 4$ are the results of variational calculations taking the configuration interaction into account [4.165]. (——) refers to Σ_g^- terms, (– – –) to Σ_u^- terms, (–·–·–) to Π terms, (–··–··–) to Δ terms

radius of the order of the atomic size [4.166], or the conventional Heitler-London method. At the same time, if there is only one valence electron it is convenient to invoke the notion of an effective potential for the interaction of this electron with atom B as an entity. Here the cases $\alpha \gtrsim \beta$ and $\alpha \ll \beta$ can be treated from a more or less unified standpoint. Consider the approach based on the Heitler-London approximation (the effective potential aproximation will be discussed in the next section).

Let $l_1 = 0, 1$; the cases $l_1 > 1$ are infrequent in practice and can be treated in the same way. In calculating the exchange integral (4.72) with the assumption that the inequalities (4.66, 78) are satisfied simultaneously, $\varphi^{m_1}_{n_1 l_1 \mathrm{A}}$ will be

$$\varphi^{m_1}_{n_1 l_1 \mathrm{A}} \simeq (r/R)^{|m_1|} Y^{m_1}_{l_1}(|m_1|\theta, \varphi) \exp(-\alpha r \cos\theta)\, \varphi_{n_1 l_1 \mathrm{A}}(R), \tag{4.79}$$

where θ and φ are the polar and azimuthal angles of the vector $\boldsymbol{r}$ in the

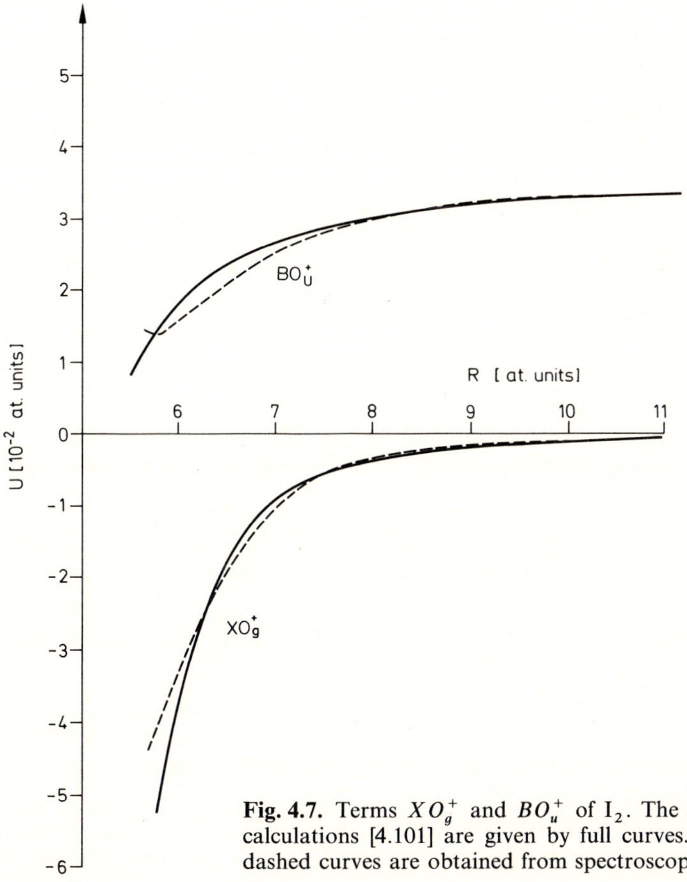

Fig. 4.7. Terms XO_g^+ and BO_u^+ of I_2. The results of asymptotic calculations [4.101] are given by full curves. The terms given by dashed curves are obtained from spectroscopic data

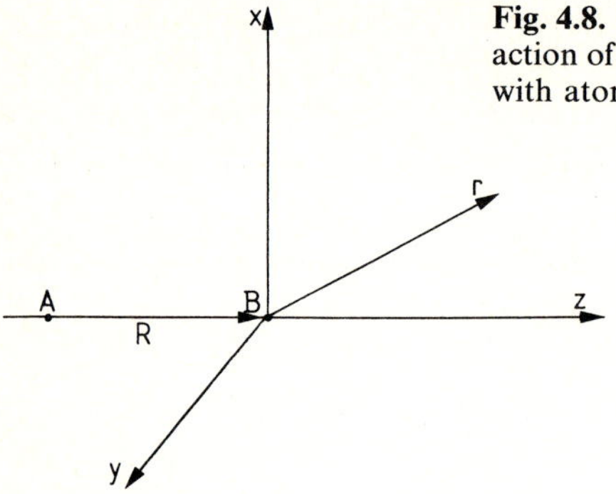

Fig. 4.8. Coordinate system used in describing the interaction of atom A (possessing one loosely bound electron) with atom B

coordinate frame $x\,y\,z$ whose origin is at the B nucleus (Fig. 4.8). Substituting (4.79) into (4.72) and taking into account that at the given distances $V_{\mathrm{eff\,A}}(\boldsymbol{r}_2) \simeq -1/R$, the exchange integral will be

$$\langle n_1 l_1 m_1; n_2 l_2 m_2 | I_{\mathrm{HL}} | n_1 l_1 m_1'; n_2 l_2 m_2' \rangle = C_{m_1 m_2; m_1' m_2'} R^{-|m_1|-|m_1'|} \varphi^2_{n_1 l_1 \mathrm{A}}(R), \tag{4.80}$$

where the constants $C_{m_1 m_2; m_1' m_2'}$ are given by the one-center integral

$$C_{m_1 m_2; m_1' m_2'} = \int d^3 \boldsymbol{r}_1 \int d^3 \boldsymbol{r}_2 \exp(-\alpha r_1 \cos\theta_1) r_1^{|m_1|} Y_{l_1}^{m_1^*}(|m_1|\theta_1, \varphi_1) \times \varphi^{m_2^*}_{n_2 l_2 \mathrm{B}}(\boldsymbol{r}_2)[V_{\mathrm{eff\,B}}(r_1) + r_1^{-1}{}_2] \exp(-\alpha r_2 \cos\theta_2) r_2^{|m_1'|} \times Y_{l_1}^{m_1'}(|m_1'|\theta_2, \varphi_2) \varphi^{m_2'}_{n_2 l_2 \mathrm{B}}(\boldsymbol{r}_1). \tag{4.81}$$

Explicit expressions for $C_{m_1 m_2; m_1' m_2'}$ have been obtained for the case when $l_1 = 0$ and B is a hydrogen atom in the ground state [4.103]. The calculation of these coefficients for the general case needs knowledge of $\varphi^{m_2}_{n_2 l_2 \mathrm{B}}(\boldsymbol{r})$ and of the relevant one-electron level $\varepsilon^{(1)}_{n_2 l_2 \mathrm{B}}$ or $V_{\mathrm{eff\,B}}(r)$.

Thus, when the inequalities (4.66, 78) are satisfied, the R-dependence of the exchange interaction is completely determined by the wave function of the loosely bound electron. Moreover it can be concluded from (4.81) that the constants $C_{m_1 m_2; m_1' m_2'}$ are of the order of unity and their signs for $l_1 = 0, 1, l_2 = 0, 1$ coincide with those of the asymptotic exchange integrals in Tables 4.1, 3. In simple cases even this information is sufficient for crude determination of the region where the transition between the Hund coupling cases c and a occurs, and for the constructing of correlation diagrams.

4.3 Pseudopotential Method for Interatomic Interactions

The pseudopotential method described in Sect. 4.1.2 for an atom with one valence electron is now widely used for the calculation of the interaction between atoms. In the general case of a diatom with several valence electrons the terms are usually calculated by the self-consistent configuration-interaction methods using computers. The only fundamental difference from the customary many-electron quantum-chemical calculations lies in the use of effective (often nonlocal) potentials of interaction between electron and atomic cores. It is usually assumed that these effective potentials calculated for free atoms do not change as the two atoms approach each other. Examples of pseudopotential calculations of diatoms with many valence electrons can be found in [4.56, 167–174].

We shall consider here the use of the pseudopotential method for the interaction between a noble gas atom B and an atom A with one loosely bound valence electron. This may be an alkali atom or any many-electron atom with one sufficiently excited electron. This system is sufficiently simple and analytical results can sometimes be obtained even for R not very large.

Two different variants of the pseudopotential method have been developed for this system. The first (it will be called the model potential method) is a direct generalization of the model potential method for an atom with one valence electron described in Sect. 4.1.2. The second (it will be called the multiple-scattering method) departs from the paper by *Fermi* [4.175]. Here the interaction potential between atom B and the electron is not introduced explicitly, and the diatomic terms are expressed directly via the characteristics of the low energy electron scattering on atom B.

4.3.1 The Model Potential Method

The calculation of the A–B interaction reduces to finding the valence electron energy levels $U^{(1)}_{\eta q}$ in the field of the core A^+B. In terms of the Hartree-Fock method valid here up to the separated-atom limit, we obtain an equation for the valence electron pseudo-orbitals ${}^{\mathrm{p}}\varphi_{\eta q\tilde{\lambda}}$, just as in Sect. 4.1.2, [see (4.19)]:

$$[-1/2\,\nabla^2 + V(\boldsymbol{r},R)]\,{}^{\mathrm{p}}\varphi_{\eta q\tilde{\lambda}}(\boldsymbol{r}) - \sum_{\eta'}\int {}^{\mathrm{c}}\varphi_{\eta' q\tilde{\lambda}}(\boldsymbol{r}')$$
$$\times\, V(\boldsymbol{r}',R)\,{}^{\mathrm{p}}\varphi_{\eta q\tilde{\lambda}}(\boldsymbol{r}')\,d^3r'\,{}^{\mathrm{c}}\varphi_{\eta' q\tilde{\lambda}}(\boldsymbol{r}) = U^{(1)}_{\eta q}(R)\,{}^{\mathrm{p}}\varphi_{\eta q\tilde{\lambda}}(\boldsymbol{r})\,, \qquad (4.82)$$

where ${}^{\mathrm{c}}\varphi_{\eta' q\tilde{\lambda}}$ is the MO of the AB core and $V(\boldsymbol{r},R)$ is the Hartree-Fock potential of the electron in the molecule AB.

If R is larger than the size of the A core and atom B (this will be assumed throughout the following), ${}^{\mathrm{c}}\varphi_{\eta' q\tilde{\lambda}}$ will be only slightly different from the AO of the A core and atom B, and $V(\boldsymbol{r},R)$ will be given simply by the sum of the Hartree-Fock potential of the outer electron of A and the static potential of atom B. Taking this into account, we get for ${}^{\mathrm{P}}\varphi_{\eta q \tilde{\lambda}}$

$$[-1/2\,\nabla^2 + V_{\mathrm{eff\,A}} + V_{\mathrm{eff\,B}}]\,{}^{\mathrm{P}}\varphi_{\eta q \tilde{\lambda}} = U^{(1)}_{\eta q}\,{}^{\mathrm{P}}\varphi_{\eta q \tilde{\lambda}}\,, \tag{4.83}$$

where $V_{\mathrm{eff\,A}}$ and $V_{\mathrm{eff\,B}}$ are given by an equation similar to (4.20).

The model quasi-local potential of the type (4.22) is commonly used as $V_{\mathrm{eff\,A}}$. The main contributions here are the long-range Coulomb attraction and the short-range repulsion by the core (Fig. 4.5). Effects such as polarization of the A core by the electron, not accounted for in the Hartree-Fock approximation, are relatively immaterial in this case.

On the other hand, the polarization effects are of great importance for the interaction of the electron with the neutral atom. Therefore $V_{\mathrm{eff\,B}}$ is always written in the form

$$V_{\mathrm{eff\,B}} = V_{\mathrm{s.r.B}} + V_{\mathrm{p.B}}\,, \tag{4.84}$$

where $V_{\mathrm{s.r.B}}$ is the short-range part of the potential which can be calculated in principle in the Hartree-Fock approximation, and $V_{\mathrm{p.B}}$ is the polarization potential. So far only the semiempirical approaches have been used in constructing $V_{\mathrm{eff\,B}}$. The difference between various approaches consists in the choice of adjustable parameters and in the form accepted for $V_{\mathrm{s.r.B}}$.

The form of $V_{\mathrm{p.B}}$ is basically the same in all approaches. It is obtained by taking into account only the dipole polarizability of B and using the adiabatic approximation with respect to the valence electron. Since the polarization of B is produced both by the electron and by the A core, then for large R and r (r is the distance from the electron to the atom B),

$$V_{\mathrm{p.B}} = -\frac{\alpha_{\mathrm{B}}}{2R^4} - \frac{\alpha_{\mathrm{B}}}{2r^4} - \frac{\alpha_{\mathrm{B}}\cos\theta}{r^2R^2}\,, \tag{4.85}$$

where α_{B} is the dipole polarizability of B and r, θ are the spherical coordinates of the electron in the coordinate frame defined in Fig. 4.8. The first two terms represent the energies due to the polarization of B by the A core and by the electron at rest, respectively. The third term, which ensures correct ($\sim R^{-6}$) asymptotic behavior of the interaction between A and B at large R is due to the superposition of these two polarization effects.

Expression (4.85) is not valid for values of r comparable with the size of B. Since theory gives only the limiting behavior of the polarization interaction between the electron and the atom for $r \to \infty$ and $r \to 0$ [4.176], the interpolation expression for $V_{\mathrm{p.B}}$ is used for practical calculations of the interactions between atoms:

$$V_{\mathrm{p.B}} = -\frac{\alpha_{\mathrm{B}}}{2}\,[R^{-4} + r^{-4}f_1(r) + \cos\theta\, r^{-2}R^{-2}f_2(r)]\,. \tag{4.86}$$

Here $f_1(r)$ and $f_2(r)$ are functions which ensure the regularity of $V_{p.B}$ for $r \to 0$ and which possess the property $f_1(r), f_2(r) \to 1$ for $r \to \infty$. Only the purely empirical truncating functions f_1 and f_2 which involve one adjustable parameter r_0 characterizing the size of B have been used uptil now [4.115, 156, 177–191].

Several different approaches have been used for $V_{s.r.B}$. In [4.115, 156, 180, 181, 185, 191] this potential has been obtained nonempirically in the Gombas statistical approximation [4.20, 46] yielding a local $V^{G}_{s.r.B}$. Since atom B has closed shells, $V^{G}_{s.r.B}$ is repulsive due to the Pauli principle. When $V_{s.r.B}$ is defined in such a way, the only adjustable parameter is r_0 entering into $V_{p.B}$. The latter is obtained here by adjusting the ground state potential well depth to the experimental data for each individual AB pair. A purely empirical local repulsive $V_{s.r.B}$ has also been used, adjusting it together with r_0 to the experimental potential well depth [4.188, 189].

A more rigorous approach has been adopted in [4.182, 184, 187, 190]. Here $V_{s.r.B}$ has the form

$$V_{s.r.B} = V^{n-loc}_{s.r.B} + V^{loc}_{s.r.B} \,. \tag{4.87}$$

The most important term $V^{n-loc}_{s.r.B}$ is nonlocal thus ensuring (in accordance with the general ideas about the pseudopotential method, see Sect. 4.1.2) the orthogonality of the wave functions of the valence electron and the occupied AO in atom B. The local term $V^{loc}_{s.r.B}$ is of minor importance. It is optimized together with r_0 in $V_{p.B}$ to reproduce experimental electron scattering phase shifts at low energies.

The optimization of $V_{eff\,B}$ to the electron scattering phase shifts has also been adopted in [4.177–179, 186] using however the local form for $V_{s.r.B}$. As noted in [4.184], this meets with the following difficulty. The *s*-scattering phase shift δ_s for noble gas atoms at an electron energy going to zero is known to tend to $n\pi$, rather than to zero (for He, $n = 1$, for Ne, $n = 2$) [4.192]. According to the Levinson theorem [4.193] (see also Sect. 5.2.4), a local potential reproducing such behavior of δ_s should support the bound states which are absent for He and Ne. Hence it is important to use the nonlocal $V_{eff\,B}$ which yields correct scattering phase shifts and does not lead to spurious bound states.

The approach using the optimization of $V_{eff\,B}$ for low energy scattering has been applied to the weakly polarizable He and Ne atoms only. It seems that for the interaction of an alkali atom in the ground or low-excited state with a heavy noble gas atom, this approach in its present form is inadequate. This is because the calculation of $V_{p.B}$ is essentially based on the adiabatic (relative to the valence electron) approximation. However in this case, the velocity of the valence electron is only slightly smaller than the velocities of the noble gas outer shell electrons and therefore nonadiabatic corrections to $V_{p.B}$ might appear to be substantial. These would lead to a decrease in the polarization attraction, since the electron density of B cannot immediately adjust itself to the motion of the valence electron. This qualitative consideration is supported by the general theorem [4.194] implying that the energy of the ground state calculated in the adiabatic approximation gives a low estimate of the energy. Specific nonadiabatic corrections have been introduced in calculating the Rydberg energy levels of valence electrons [4.66, 195, 196]. It was found that even for the Rydberg states of He nonadiabatic corrections can contribute to the energy, thus compensating an appreciable (up to 20%) fraction of the adiabatic polarization interaction. It can be expected that in the case of interest, the weakening of the polarization interaction due to the valence electron motion will be greater. For a potential of the form (4.86) this corresponds to an increase in r_0 with increasing binding energy (and thus velocity) of the valence electron, the increase being greater, the lower is the ionization potential of the noble gas atom.

It seems that when r_0 is optimized independently for each AB diatom [4.156, 180, 197], the nonadiabatic effect is empirically taken into account. For example, for fixed $B =$ Ar, Kr, Xe, the value of r_0 decreases in passing from Li to Cs (in parallel with the valence electron binding energy), to a higher extent for Xe, than for Kr and Ar. Such a behavior agrees with the expected manifestation of the nonadiabatic effects.

To obtain the total AB interaction energy at small distances the interaction due to the overlap of the electron densities of the A core and B atom must be accounted for. In the frozen-core approximation, in terms of the pseudopotential method, this is simply the Hartree-Fock interaction energy $U^{HF}_{A^+B}$ between A^+ and B. Thus,

$$U_{\eta q} = U^{(1)}_{\eta q} + U^{HF}_{A^+B}. \tag{4.88}$$

The approximate methods of calculating $U^{HF}_{A^+B}$ will be discussed in Sect. 4.4.1.

Provided $V_{\text{eff A}}$ and $V_{\text{eff B}}$ are chosen, the determination of $U^{(1)}_{\eta q}$ for ground and weakly excited states is a routine problem of quantum chemistry. It is usually solved by the variational method. Specific calculations of this sort will be considered in Sect. 11.1. However, the use of the variational method meets with fundamental difficulties when highly excited states are considered. This is due to the greater number of the strongly interacting states. Consider now an approach overcoming these difficulties which, moreover, permits one to obtain analytical results going beyond perturbation theory even for low-lying states of a valence electron in the case of a weakly polarizable noble gas atom.

4.3.2 Multiple Scattering Method

As mentioned above, in this method the interaction between A and B is expressed directly through the characteristics of the low-energy e–B scattering, thus eliminating explicit consideration of $V_{\mathrm{s.r.B}}$, which is not well known. A general formulation of the problem has been given in [4.198–202] in terms of multiple scattering theory (see, e.g. [4.203]).

To consider this formulation, we introduce the Green's function of atom A's valence electron at an energy $U_{\eta q}^{(1)}$,

$$G_{\mathrm{A}}(U_{\eta q}^{(1)}) = [U_{\eta q}^{(1)} + 1/2\,\nabla^2 - V_{\mathrm{eff\,A}}]^{-1}, \tag{4.89}$$

and the operators $t_{\mathrm{B}}^{\alpha}(\varepsilon)$, ($\alpha = p, +$), which obey the equations

$$t_{\mathrm{B}}^{\alpha}(\varepsilon) = V_{\mathrm{eff\,B}} + V_{\mathrm{eff\,B}}\, G_0^{\alpha}(\varepsilon)\, t_{\mathrm{B}}(\varepsilon), \tag{4.90}$$

where

$$G_0^{+}(\varepsilon) = [\varepsilon + \mathrm{i}\gamma + (1/2)\,\nabla^2]), \tag{4.91 a}$$

$$G_0^{p}(\varepsilon) = \mathscr{P}[\varepsilon + (1/2)\,\nabla^2]^{-1}. \tag{4.91 b}$$

Here γ is an infinitesimal positive quantity, $\mathscr{P}$ denotes the principal value, and ε may be arbitrary. In the coordinate representation, the free motion Green's functions (4.91) are

$$\langle \boldsymbol{r}|G_0^{+}(\varepsilon)|\boldsymbol{r}'\rangle = \begin{cases} -\dfrac{\exp[\mathrm{i}(2\varepsilon)^{1/2}|\boldsymbol{r}-\boldsymbol{r}'|]}{2\pi|\boldsymbol{r}-\boldsymbol{r}'|} & \text{for } \varepsilon \geqslant 0 \\[2ex] -\dfrac{\exp[-(-2\varepsilon)^{1/2}|\boldsymbol{r}-\boldsymbol{r}'|]}{2\pi|\boldsymbol{r}-\boldsymbol{r}'|} & \text{for } \varepsilon < 0, \end{cases} \tag{4.92}$$

$$\langle \boldsymbol{r}|G_0^{p}(\varepsilon)|\boldsymbol{r}'\rangle = \begin{cases} -\dfrac{\cos[(2\varepsilon)^{1/2}|\boldsymbol{r}-\boldsymbol{r}'|]}{2\pi|\boldsymbol{r}-\boldsymbol{r}'|} & \text{for } \varepsilon \geqslant 0 \\[2ex] -\dfrac{\exp[-(-2\varepsilon)^{1/2}|\boldsymbol{r}-\boldsymbol{r}'|]}{2\pi|\boldsymbol{r}-\boldsymbol{r}'|} & \text{for } \varepsilon < 0. \end{cases} \tag{4.93}$$

Equations (4.90) are called the Lippman-Schwinger equations [4.203, 204]; the operators $t_{\mathrm{B}}^{+}(\varepsilon)$ and $t_{\mathrm{B}}^{p}(\varepsilon)$ are related to the scattering characteristics in the potential $V_{\mathrm{eff\,B}}$, namely, if R is large enough so that only the second term in (4.86) must be taken into account, then for $\varepsilon > 0$,

$$\langle \boldsymbol{k}|t_{\mathrm{B}}^{+}(\varepsilon)|\boldsymbol{k}'\rangle|_{k^2=k'^2=2\varepsilon} = -\frac{1}{4\pi^2}$$

$$\times \sum_{l=0}^{\infty}(2l+1)(k\cot\mathrm{an}\,\delta_l - \mathrm{i}k)^{-1}\,\mathrm{P}_l(\cos\theta) = -\frac{f_{\mathrm{B}}(\theta)}{4\pi^2}, \tag{4.94}$$

where δ_l are the e–B elastic scattering phase shifts, $f_{\rm B}(\theta)$ ($\cos\theta = \hat{\boldsymbol{k}} \cdot \hat{\boldsymbol{k}}'$) is the scattering amplitude (Chap. 5) and the plane waves are normalized as

$$|\boldsymbol{k}\rangle = (2\pi)^{-3/2}\, {\rm e}^{{\rm i}\boldsymbol{k}\cdot\boldsymbol{r}}. \tag{4.95}$$

The matrix elements of $t^p_{\rm B}(\varepsilon)$, which is called the reaction matrix, are

$$\langle \boldsymbol{k}|t^p_{\rm B}(\varepsilon)|\boldsymbol{k}'\rangle|_{k^2=k'^2=2\varepsilon} = -\frac{1}{4\pi^2 k}\sum_{l=0}(2l+1)\tan\delta_l\,{\rm P}_l(\cos\theta). \tag{4.96}$$

Using (4.89, 90) and some operator algebra, (4.83) can be rewritten

$$\tau_{{\rm B}\eta q} = t^\alpha_{\rm B}(\varepsilon)\,[G_{\rm A}(U^{(1)}_{\eta q}) - G_0(\varepsilon)]\,\tau_{{\rm B}\eta q}, \tag{4.97}$$

where

$$\tau_{{\rm B}\eta q} = V_{\rm eff\,B}\,{}^{\rm P}\varphi_{\eta q\bar{\lambda}}. \tag{4.98}$$

The energy levels $U^{(1)}_{\eta q}$ are obtained from the usual condition that (4.97) possess nontrivial solutions.

Equation (4.97) is simpler to solve than the usual Schrödinger equation (4.83) only when the matrix elements of $t^\alpha_{\rm B}(\varepsilon)$, which are of main importance in solving (4.97), can be obtained directly from the on-energy shell matrix elements (4.94) and (4.96). Fortunately this occurs under conditions when the usual variational method meets difficulties, namely, for a highly excited atom A and large R.

If A is excited to a level such that the inequality

$$r_{\rm B}/\bar{\lambda} \ll 1 \tag{4.99}$$

($r_{\rm B}$ is the range of the potential $V_{\rm eff\,B}$ and $\bar{\lambda}$ is the electron wavelength) holds, then the zero-range potential model may be used for atom B. A detailed discussion of this model and its possible application to atomic physics can be found in [4.205, 206]. The zero-range potential model will be considered here in order to elucidate the main peculiarities of the multiple scattering method.

The zero-range potential model contains only one parameter – the scattering length $L_{\rm B}$, and in the momentum representation the matrix elements of $t^+_{0\rm B}$ and t^p_{0B} do not depend on $\boldsymbol{k}$, $\boldsymbol{k}'$. These are, respectively [4.207],

$$\begin{aligned} \langle \boldsymbol{k}|t^+_{0\rm B}(\varepsilon)|\boldsymbol{k}'\rangle &= \frac{1}{4\pi^2}[L_{\rm B}^{-1} + {\rm i}(2\varepsilon)^{1/2}]^{-1} = T^+_{\rm B}(\varepsilon), \\ \langle \boldsymbol{k}|t^p_{0\rm B}(\varepsilon)|\boldsymbol{k}'\rangle &= \frac{1}{4\pi^2}[L_{\rm B}^{-1} + {\rm i}\,\theta(-\varepsilon)(2\varepsilon)^{1/2}]^{-1} = T^p_{\rm B}(\varepsilon), \end{aligned} \tag{4.100}$$

where

$$\theta(x) = \begin{cases} 1 \text{ for } x \geqslant 0 & (4.101) \\ 0 \text{ for } x < 0. & (4.102) \end{cases}$$

In the coordinate representation with origin at atom B (Fig. 4.8), these operators are

$$\langle \boldsymbol{r} | t^{+}_{0\mathrm{B}}(\varepsilon) | \boldsymbol{r}' \rangle = 8\pi^3 T^{+}_{\mathrm{B}}(\varepsilon)\, \delta(\boldsymbol{r})\, \delta(\boldsymbol{r}'), \tag{4.103}$$

$$\langle \boldsymbol{r} | t^{p}_{0\mathrm{B}}(\varepsilon) | \boldsymbol{r}' \rangle = 8\pi^3 T^{p}_{\mathrm{B}}(\varepsilon)\, \delta(\boldsymbol{r})\, \delta(\boldsymbol{r}'). \tag{4.104}$$

Since, in the zero-range potential model,

$$V_{\mathrm{eff\,B}} \sim \delta(\boldsymbol{r}), \tag{4.105}$$

it is seen from (4.98) that, as long as ${}^{\mathrm{P}}\varphi_{\eta\sigma}$ only has no node at the diatom axis, all $\tau_{\mathrm{B}\eta q}$ except $\tau_{\mathrm{B}\eta\sigma}$ are zero. This means that only the σ-states differ from the free A-atom states for the zero-range potential model. For $\tau_{\mathrm{B}\eta\sigma}$, we can write

$$\tau_{\mathrm{B}\eta\sigma} = \tau_{\mathrm{B}\eta\sigma}(0)\, \delta(\boldsymbol{r}). \tag{4.106}$$

It should be noted also that $\langle \boldsymbol{r} | G_{\mathrm{A}}(U^{(1)}_{\eta q}) | \boldsymbol{r}' \rangle$ in the coordinate frame of Fig. 4.8 is connected with the representation $\langle \tilde{\boldsymbol{r}} | G_{\mathrm{A}}(U^{(1)}_{\eta q}) | \tilde{\boldsymbol{r}}' \rangle_{\mathrm{A}}$ in the coordinate frame whose origin is at atom A,

$$\langle \boldsymbol{r} | G_{\mathrm{A}}(U^{(1)}_{\eta q}) | \boldsymbol{r}' \rangle = \langle \boldsymbol{r} + \boldsymbol{R} | G_{\mathrm{A}}(U^{(1)}_{\eta q}) | \boldsymbol{r}' + \boldsymbol{R} \rangle_{\mathrm{A}}. \tag{4.107}$$

Substituting (4.103, 104, 106) into (4.97) and taking (4.107) into account, we get

$$\tau_{\mathrm{B}\eta\sigma}(0) = 8\pi^3 T_{\mathrm{B}}(\varepsilon) \lim_{\boldsymbol{r}' \to 0} [\langle \boldsymbol{R} | G_{\mathrm{A}}(U^{(1)}_{\eta\sigma}) | \boldsymbol{R} + \boldsymbol{r}' \rangle_{\mathrm{A}} - \langle 0 | G^{\alpha}_{0}(\varepsilon) | \boldsymbol{r}' \rangle]\, \tau_{\mathrm{B}\eta\sigma}(0), \tag{4.108}$$

and therefore the energy levels must be determined from the transcendental equation

$$1 - 8\pi^3 T^{\alpha}_{\mathrm{B}}(\varepsilon) \lim_{\boldsymbol{r}' \to 0} [\langle \boldsymbol{R} | G_{\mathrm{A}}(U^{(1)}_{\eta\sigma}) | \boldsymbol{R} + \boldsymbol{r}' \rangle_{\mathrm{A}} - \langle 0 | G^{\alpha}_{0}(\varepsilon) | \boldsymbol{r}' \rangle] = 0. \tag{4.109}$$

This equation was first considered in [4.208] for a purely Coulomb potential $V_{\mathrm{eff\,A}}$, the explicit expression for $\langle \boldsymbol{r} | G_{\mathrm{A}}(U^{(1)}_{\eta\sigma}) | \boldsymbol{r}' \rangle_{\mathrm{A}}$ being known [4.209]. In this case the exact numerical solution of (4.109) is possible (an example may be found in [4.210]), and the behavior of the diatomic term pattern possesses some peculiar features which are due to the Coulomb degeneracy of atom A's energy levels [4.198, 210]. This will not be considered here, since even for highly excited alkali atoms the splitting between s and p levels belonging to the same principal quantum number n is of the same order of magnitude as the spacing between levels with neighbouring n.

For a non-Coulombic case, but with R large enough so that

$$V_{\mathrm{eff\,A}}(R) = -1/R\,, \tag{4.110}$$

(4.109) may be simplified, provided two additional conditions are fulfilled. First, the energy shift of the level n of atom A due to perturbation by atom B is small, in the sense that

$$|U_{\eta\sigma}^{(1)} - \varepsilon_{nl}| \ll \Delta\varepsilon\,, \tag{4.111}$$

where $\Delta\varepsilon$ is the distance from the level $n\,l$ to the next nearest level. This enables identification of η with $n\,l$. Second, in the vicinity of atom B, the electron motion in the potential $V_{\mathrm{eff\,A}}$ is quasi-classical. For the potential (4.110), this is so if

$$\alpha_{nl} R \gg 1\,, \tag{4.112}$$

where $\alpha_{nl} = (-2\,\varepsilon_{nl})^{1/2}$. Under these conditions the Green's function $\langle \boldsymbol{R} + \boldsymbol{r}\,|G_{\mathrm{A}}(U_{nl\sigma}^{(1)})|\,\boldsymbol{R} + \boldsymbol{r}'\rangle$ in the region where $\boldsymbol{r}$ and $\boldsymbol{r}'$ vary such that

$$|\boldsymbol{r} - \boldsymbol{r}'| \ll \alpha_{nl}^2 R^2 \tag{4.113}$$

will be [4.199, 201]

$$\langle \boldsymbol{R} + \boldsymbol{r}\,|G_{\mathrm{A}}(U_{nl\sigma}^{(1)})|\,\boldsymbol{R} + \boldsymbol{r}'\rangle_{\mathrm{A}} = \sum_m \frac{\varphi_{nl\mathrm{A}}^m(\boldsymbol{R}+\boldsymbol{r})\,\varphi_{nl\mathrm{A}}^{m*}(\boldsymbol{R}+\boldsymbol{r}')}{U_{nl\sigma}^{(1)} - \varepsilon_{nl}}$$
$$+ \langle \boldsymbol{R} + \boldsymbol{r}\,|G_0^p(U_{nl\sigma}^{(1)} + 1/R)|\,\boldsymbol{R} + \boldsymbol{r}'\rangle. \tag{4.114}$$

The inequality (4.113) guarantees that the variation of the electron wavelength in the Coulomb potential in a region of size $\sim |\boldsymbol{r} - \boldsymbol{r}'|$ near atom B may be neglected.

Substituting (4.114) into (4.109) we find, in the limit $\boldsymbol{r}, \boldsymbol{r}' \to 0$, that for α both $+$ and p, (4.109) reduces to

$$1 - 8\,\pi^3\,T_{\mathrm{B}}^p(U_{nl\sigma}^{(1)} + 1/R)\,\frac{|\varphi_{nl\mathrm{A}}^0(\boldsymbol{R})|^2}{U_{nl\sigma}^{(1)} - \varepsilon_{nl}} = 0\,. \tag{4.115}$$

Note that this equation does not contain an arbitrary energy ε, see (4.109). If $L_{\mathrm{B}} < 0$ or $1/L_{\mathrm{B}} > (2\,|U_{nl\sigma}^{(1)} + 1/R|)^{1/2}$, then taking into account (4.111) we get

$$U_{nl\sigma}^{(1)} = \varepsilon_{nl} + 8\,\pi^3\,T_{\mathrm{B}}^p(\varepsilon_{nl} + 1/R)\,|\,\varphi_{nl\mathrm{A}}^0(\boldsymbol{R})|^2\,. \tag{4.116}$$

This result for the classically accessible region ($\varepsilon_{nl} + 1/R > 0$) coincides with the Fermi result [4.175], and for the classically forbidden region ($\varepsilon_{nl} + 1/R < 0$) with the expression for the energy shift first obtained in [4.211]. It should be emphasized that the result (4.116), though based on the assumption (4.111), cannot be obtained using conventional perturbation theory. The reson for this is that perturbation theory can be applied if (4.111) is fulfilled and the distortion of the wave function is small everywhere. However in the zero-range potential model the wave function is strongly distorted in the region where the disturbing atom B is located.

It can be shown using the quasi-classical expression for $\varphi^0_{nl\mathrm{A}}(\boldsymbol{R})$ that even in the classically allowed region, the inequality (4.111) is fulfilled if $\alpha_{nl} L_\mathrm{B} \ll 1$. The adequacy of the pole approximation (4.114) which leads to (4.116) has also been supported by numerical solution of the exact equation (4.109) for the purely Coulomb $V_{\mathrm{eff\,A}}$ and the perturbing noble gas atoms [4.210].

The zero-range potential model is valid when $V_{\mathrm{eff\,B}}$ truncates at $r \simeq \lambda$. Therefore, (4.116) can reproduce neither the dispersion intraction, proportional to R^{-6}, for very large R, nor the n-dependence of the shift of the highly excited atomic levels in an inert gas. The first shortcoming of this expression is usually removed by adding the dispersion term $C_{nl\sigma}/R^6$ to (4.116) "by hand" [4.12, 211]. The first attempts to explain the n-dependence of the energy shift are reported in [4.212, 213].

A method which enables such shortcomings of the zero-range potential model to be removed is provided by the general multiple-scattering equation (4.97). When the pole approximation (4.114) is valid and the negative ion B^- has no energy levels near ε_{nl}, it follows immediately from (4.97) that

$$U^{(1)}_{nl\lambda} = \varepsilon_{nl} + \langle nl\lambda | t^p_\mathrm{B}(\varepsilon_{nl} + 1/R) | nl\lambda \rangle , \tag{4.117}$$

which is a direct generalization of (4.116).

Consider first the classically allowed region where the electron motion is quasi-classical [the inequality (4.112) holds] and let the inequality (4.99) also be fulfilled. It is exactly these conditions which hold for the problem of calculating the pressure shifts of highly excited atomic levels [4.175, 212, 213]. Here we need only account for the second term in the polarization potential (4.86), since in the classically allowed region the probability of the electron appearing in the vicinity of B is great. Here, therefore, $t^p_\mathrm{B}(\varepsilon_{nl} + 1/R)$ in (4.117) is the reaction matrix for scattering of the electron by the free atom B, allowing for both the short-range repulsion and the long-range polarization attraction.

Consider now the σ-terms which exhibit a considerable shift under these conditions and note that the integral in (4.117) is controlled by $\boldsymbol{r}$ and $\boldsymbol{r}'$ lying in a region of size $\sim r_\mathrm{B}$ near B. However, the inequalities (4.99) and (4.112) guarantee the fulfillment of (4.113). Therefore the quasi-classical wave functions $\varphi^0_{nl\mathrm{A}}(\boldsymbol{R} + \boldsymbol{r})$ with $l \ll n$ can be approximated in (4.117) by the standing plane waves

$$\varphi^0_{nl\mathrm{A}} \simeq F(R) \cos[(\boldsymbol{k}_n(R)\boldsymbol{r} + \psi_{nl}(R)] , \tag{4.118}$$

where

$$F(R) = \left(\frac{2l+1}{4\pi}\right)^{1/2} \left(\frac{2}{\pi}\frac{\partial \varepsilon_{nl}}{\partial n}\right)^{1/2} [k_{nl}(R)]^{-1/2}, \tag{4.119}$$

$$\boldsymbol{k}_{nl}(R) = [2(\varepsilon_{nl} + 1/R)]^{1/2} \hat{\boldsymbol{R}} \tag{4.120}$$

and $\psi_{nl}(R)$ is a large phase. Substituting (4.118) into (4.117) we obtain the expression for $U^{(1)}_{nl\sigma}$ directly through the on-energy shell reaction matrix elements (4.96) [4.199, 202, 214]

$$U^{(1)}_{nl\sigma} = \varepsilon_{nl} - \pi \sum_{l'=0}^{\infty} (2l'+1)\tan(\delta_{l'}/k_{nl}) \times [1 + (-1)^{l'} \cos 2\Psi_{nl}] F^2(R), \tag{4.121}$$

where $\delta_{l'}$ are phase shifts of the e–B elastic scattering at an energy $k^2_{nl}/2$. Since the inequality (4.99) holds, the effective-range expansion may be used. For the electron-atom potential which behaves as $-\alpha_B/2r^4$ for large r this expansion reads [4.215]

$$\tan(\delta_0/k_{nl}) = -L_B - \frac{\pi}{3}\alpha_B k_{nl} - \frac{4L_B\alpha_B k^2_{nl}}{3}\log k_{nl} + a_0 k^2_{nl} + O(k^s), \tag{4.122}$$

$$\tan(\delta_1/k_{nl}) = \frac{\pi}{15}\alpha_B k_{nl} + a_1 k^2_{nl} + O(k^s), \tag{4.123}$$

$$\tan\delta_{l'}/k_{nl} = \frac{\alpha_B k_{nl}}{(2l'+3)(2l'+1)(2l'-1)} + O(k^s), \quad l' \geqslant 2, \tag{4.124}$$

where $s > 2$. Substituting (4.122–124) into (4.121) and taking into account that $\sum_{l'=1}^{\infty} (2l'+3)^{-1}(2l'-1)^{-1} = 1/3$, we get for the nonoscillatory part $\langle \Delta\varepsilon_{nl} \rangle$ of the level shift,

$$\langle \Delta\varepsilon_{nl} \rangle = \pi F^2(R)\left[L_B + \frac{4L_B\alpha_B k^2_{nl}}{3}\log k_{nl} - (a_0 + a_1)k^2_{nl} + O(k^2)\right]. \tag{3.125}$$

Note that the term linear in k_{nl} disappears from the final expression. The first term of this expansion coincides, as expected, with the nonoscillatory part of the level shift in the zero-range potential model. Other terms describe the state dependence of the level shift [4.214].

In the classically forbidden region where $\varepsilon_{nl} + R^{-1} < 0$ two problems arise. First, the complete V_{pB} given by (4.86), which is not spherically symmetric, must be considered here. Second, although the energy shift is again expressed through the on-energy shell reaction matrix elements, these must be taken at negative energies. However, when the potential possesses a long-range tail $\sim r^{-4}$, the procedure of extending the calculation of the scattering amplitude to negative energies is not straightforward.

The following approach to these problems, similar to that used in deriving the effective range expansion (4.122–124) [4.215], has been adopted in [4.200, 201, 216]. Assume that the radius r_{B} can be introduced such that the following inequalities simultaneously hold:

$$r_{\mathrm{B}} \ll R, \tag{4.126 a}$$

$$r_{\mathrm{B}} \gg r_0, \tag{4.126 b}$$

$$\alpha_{nl} r_{\mathrm{B}} < 1, \tag{4.126 c}$$

$$\alpha_{\mathrm{B}}/r_{\mathrm{B}}^2 \ll 1, \tag{4.126 d}$$

where r_0 is the distance at which (4.86) reaches its asymptotic form (4.85). The inequalities (4.126 a) and (4.126 b) allow $V_{\mathrm{eff\,B}}$ to be expressed in the form

$$V_{\mathrm{eff\,B}} = V_{\mathrm{0B}} + V'_{\mathrm{pB}} \tag{4.127}$$

where V_{0B} is zero for $r > r_{\mathrm{B}}$ and coincides with the spherically symmetric e–B potential for $r \leqslant r_{\mathrm{B}}$, and

$$V'_{\mathrm{pB}} = \begin{cases} 0 \text{ for } r < r_{\mathrm{B}} \\ -\dfrac{\alpha_{\mathrm{B}}}{2}\left(\dfrac{1}{r^4} + \dfrac{2\cos\theta}{r^2 R^2} + \dfrac{1}{R^4}\right) \quad \text{for } r \geqslant r_{\mathrm{B}}. \end{cases} \tag{4.128}$$

The inequality (4.126 c) allows one to use the zero-range potential model for V_{0B} and neglect the energy dependence of $T_{\mathrm{B}}^{p}(\varepsilon)$, and (4.126 d) makes it possible to account for V'_{pB} in first order perturbation theory. Noting that V'_{pB} is zero in that range of $\boldsymbol{r}$-values where the wave function is substantially perturbed by V_{0B}, we have from (4.117) for the σ-terms,

$$U_{nl\sigma}^{(1)} = \varepsilon_{nl} + 2\pi L_{\mathrm{0B}} |\varphi_{nl\mathrm{A}}^{0}(\boldsymbol{R})|^2 + \langle nl0|V'_{\mathrm{pB}}|nl0\rangle, \tag{4.129}$$

where L_{0B} is the scattering length for the potential V_{0B}. This expression contains two unknown parameters, L_{0B} and r_{B}. However it has been found [4.200, 201] that if $\varphi_{nl\mathrm{A}}^2$ is represented in the form

$$\varphi_{nl\mathrm{A}}^2 = \sum_{j=0}^{N} c_j \varrho^j \exp(-2\alpha_{nl}\varrho), \quad \varrho = |\boldsymbol{R} + \boldsymbol{r}| \tag{4.130}$$

which is exact for the hydrogen atom, then the final result is

$$U_{nl\sigma}^{(1)} = \varepsilon_{nl} + 2\left(L_{\mathrm{0B}} - \frac{\alpha_{\mathrm{B}}}{r_{\mathrm{B}}}\right)|\varphi_{nl\mathrm{A}}^{0}(\boldsymbol{R})|^2 + U_{nl\sigma}^{(\mathrm{d})}(R), \tag{4.131}$$

where $U^{(d)}_{nl\sigma}$ does not contain unknown parameters. On the other hand, $L_{OB} - \alpha_B/r_B$ can be identified with the scattering length, since calculation of the total scattering length using the same approach gives $L_B \simeq L_{OB} - \alpha_B/r_B$ [4.215]. The expressions for $U^{(d)}_{nl}$ for some specific cases can be found in [4.201, 216], and the results of calculations using (4.131) will be discussed in Chap. 11. Here we note only that $U^{(d)}_{nl\sigma}$ exhibits the desirable asymptotic behavior

$$U^{(d)}_{nl\sigma}|_{R\to\infty} \to C_{nl\sigma}/R^6 . \quad (4.132)$$

The coefficient $C_{nl\sigma}$ coincides with that given by (4.54, 64, 65) where $\Delta\varepsilon_{\gamma_1}$ is neglected compared to $\Delta\varepsilon_{\gamma_2}$.

The case when B is in the classically forbidden region is of particular interest for not too small $|\varepsilon_{nl}|$, e.g. for the first excited states of alkali atoms when $|\varepsilon_{nl}| \simeq 0.1$. The corresponding $\alpha_{nl} \simeq 0.5$, and it is seen that a value for r_B simultaneously satisfying the inequalities (4.126 c, d) exists only if $\alpha_B < 4$, i.e. only for He and marginally for Ne. Therefore, (4.131) may be considered to be rigorously founded for moderately excited atoms A only in the case of these noble gas atoms. Note that the latter are just those for which nonadiabatic effects in e–B scattering mentioned in the previous section are expected to be unimportant. For heavy noble gas atoms, (4.131) can be used only if $|\varepsilon_{nl}| \ll 0.1$, where the classically forbidden region seems to be of little interest.

The general equation (4.117), though adequate at any ε_{nl}, provided (4.111, 112) are fulfilled, seems to be of no great use for $|\varepsilon_{nl}| \simeq 0.1$ in the case of heavy noble gas atoms for two reasons. Since the inequalities (4.126 c, d) cannot be satisfied, there is no simple method of continuing the reaction matrix to such negative energies even if the energy independent e–B potential exists. Also, the existence of an energy-independent $V_{\mathrm{eff\,B}}$ is doubtful here, as noted in the previous section. Therefore, the energy dependence of the reaction matrix might be quite different from that for the energy-independent potential reproducing the elastic scattering phase shifts at low positive energies.

4.4 Short-Range Atomic Interactions

Another R range where a relatively simple and general approach to the calculation of molecular terms is possible is that involving an appreciable overlap of the inner electron shells of the colliding atoms. In the potential-curve plots, this range lies to the left of the ground electronic state equilibrium distance and corresponds to a strong repulsion between the atoms. In Fig. 3.3 it covers part of region III, starting from $R = 0$ and involving the left boundary of region III′. These distances do not in fact exceed 1.5 Å.

The basic simplification which gives satisfactory quantitative results for molecular terms involves using the MO approximation. This implies that the electrons move independently in a certain axially symmetric field $V(\boldsymbol{r}, R)$ (Sect. 3.2.3). Strictly speaking, this field and the appropriate MO must be calculated in terms of the Hartree-Fock method. A detailed discussion of the applications of this method to the calculation of the electronic structure of molecules can be found, e.g. in [4.1–3, 7, 9, 10, 217, 218]. However, for many-electron molecules, the Hartree-Fock calculations are much more cumbersome than for atoms. This is due to the two-center nature of the problem for fixed R and the necessity of finding the dependence of the wave functions and the terms on R. Specific difficulties arise here. First, good convergence would need the expansion of the MOs over AOs localized on different atoms for large R, and the use of a one-center approximation for small R [4.219]. Second, since the Hartree-Fock one-electron Hamiltonian is constructed using only occupied MOs corresponding to low-lying electron levels, this Hamiltonian would change abruptly in passing through the crossing point of terms, and this might result in jumps of the one-electron terms and of the total energy [4.220]. To overcome this difficulty in the vicinity of one-electron term crossing points, one would have to account in some way for the interaction of configurations, i.e. to pass, in fact, beyond the scope of the Hartree-Fock approximation.

The above difficulties make the use of the Thomas-Fermi statistical model or one of its modifications, attractive particularly for qualitative and semi-quantitative treatment of atomic collisions at high energies (10^2–10^4 eV). The general scheme of such treatment is similar to that described in Sect. 4.1 for atoms. First of all, the Thomas-Fermi molecular equation (or a modification thereof) is solved, giving the electron potential in the molecule $V(\boldsymbol{r}, R)$, and then the Schrödinger equation involving this potential is also solved. Such an approach was discussed even in the nineteen thirties [4.221, 222] (see also [4.20]). The interest in such calculations resurged recently in connection with the theory of atomic collisions at high energies [4.223–229].

The Thomas-Fermi equation (or a modification) gives in addition to $V(\boldsymbol{r}, R)$, the electron density distribution in the diatom $\varrho(\boldsymbol{r}, R)$. This distribution can be substituted into the electron energy functional $H\{\varrho(\boldsymbol{r}, R)\}$ for the statistical model used, and the electron energy of the ground molecular state $U(R)$, as well as the potential energy of the ground state interaction

$$\Delta U(R) = \frac{Z_A Z_B}{R} + U(R) - U(\infty) \tag{4.133}$$

will then be obtained.

Teller [4.230] found that the function $\Delta U(R)$ calculated in terms of the statistical Thomas-Fermi and Thomas-Fermi-Dirac models can have no minimum. Therefore, for the interaction of atoms with open valence shells where chemical bonding is possible, the function $\Delta U(R)$ calculated in terms of the statistical theory would only have meaning for small R where the repulsion due to the overlap of the inner shells substantially exceeds the splitting of the molecular terms caused by the valence electrons. For atoms and ions with closed shells, the applicability of the statistical models is somewhat wider. The statistically calculated $\Delta U(R)$ can be used, for instance, to find the "average" trajectory of relative motion when calculating the probability of inelastic collisions at high energies in terms of the semiclassical approximation. For interactions of alkali atoms with noble gas atoms (which will be discussed in Chap. 11), $\Delta U(R)$ provides a satisfactory approximation to the interaction energy of an alkali core with a noble gas atom. Most papers on the application of statistical models to diatoms deal just with the calculation of $\Delta U(R)$. For a general discussion of the problems involved, see [4.3, 20, 231–235].

Strictly speaking $V(\boldsymbol{r}, R)$ and $\Delta U(R)$ must be found simultaneously from the Thomas-Fermi equation or from its modification for the system used, with subsequent calculation of $H\{\varrho(\boldsymbol{r}, R)\}$. However only a few such calculations have been reported [4.227–229, 236–238] probably due to the difficulties in solving the nonlinear Thomas-Fermi equation for a two-center system. As a rule, $\Delta U(R)$ has been calculated using first-order statistical perturbation theory [4.20]. The zeroth-approximation electron density $\varrho_0(\boldsymbol{r}, R)$ substitued into $H\{\varrho(\boldsymbol{r}, R)\}$ is usually written in the form

$$\varrho_0(r, R) = \varrho_A(\boldsymbol{r}) + \varrho_B(\boldsymbol{r}), \tag{4.134}$$

where $\varrho_A(\boldsymbol{r})$, $\varrho_B(\boldsymbol{r})$ stand for the electron densities in the free atoms A and B. The validity of such an approach and the results obtained will be discussed in brief in Sect. 4.4.1.

The zeroth approximation for $\varrho(\boldsymbol{r}, R)$ defined by (4.134) is, generally speaking, insufficient for the calculation of $U_{\eta q}^{(1)}(R)$, particularly in following the transition between the limits of the united and separated atoms. An improved method ensuring the correct correlation of $V(\boldsymbol{r}, R)$ for the united and separated atoms will be discussed in Sect. 4.4.2.

4.4.1 The Energy of Atomic Interaction at Small Distances

At sufficiently small distances ($R < 1$ Å), the main contribution to $\Delta U(R)$ is from the changes in the kinetic energy of the electrons and in the Coulomb interaction of electrons with each other and with the nuclei as the atoms approach. Consequently, the simple Thomas-Fermi model would be sufficient to estimate $\Delta U(R)$ over this range of distances. In this model,

$$H^{\mathrm{TF}}\{\varrho(\boldsymbol{r},R)\} = 3/5\,\lambda \int \varrho^{5/3}\,d^3r - \left(\frac{Z_{\mathrm{A}}}{r_{\mathrm{A}}}+\frac{Z_{\mathrm{B}}}{r_{\mathrm{B}}}\right)\varrho\,d^3r + \iint \frac{\varrho(\boldsymbol{r}',R)\,\varrho(\boldsymbol{r},R)}{|\boldsymbol{r}'-\boldsymbol{r}'|}\,d^3r\,d^3r'\,, \tag{4.135}$$

where $\lambda = (3\pi)^2/2$. The first term in (4.135) refers to the kinetic energy of the electrons and the second and third to the Coulomb interaction of the electrons with the nuclei and with each other. Substituting $\varrho^{\mathrm{TF}}(\boldsymbol{r},R)$, which represents the solution of the Thomas-Fermi equation, into (4.135) we obtain the minimum value of this functional denoted by $U^{\mathrm{TF}}(R)$.

In the Thomas-Fermi model, the first order statistical perturbation theory result is obtained by substituting into (4.135)

$$\varrho_0^{\mathrm{TF}}(\boldsymbol{r},R) = \varrho_{\mathrm{A}}^{\mathrm{TF}}(\boldsymbol{r}) + \varrho_{\mathrm{B}}^{\mathrm{TF}}(\boldsymbol{r})\,, \tag{4.136}$$

where $\varrho_{\mathrm{A}}^{\mathrm{TF}}$, $\varrho_{\mathrm{B}}^{\mathrm{TF}}$ are the Thomas-Fermi electron densities in the free atoms A and B.

It appears that for the Thomas-Fermi model the accuracy of statistical first order perturbation theory can be rigorously estimated. This has been done by *Firsov* [4.239] who also introduced, besides $H^{\mathrm{TF}}\{\varrho(\boldsymbol{r},R)\}$, the functional

$$\tilde{H}\{f(\boldsymbol{r},R)\} = \frac{1}{8\pi}\int f\nabla^2 f\,d^3r - \frac{2}{5\lambda^{3/2}}\int\left(\frac{Z_{\mathrm{A}}}{r_{\mathrm{A}}}+\frac{Z_{\mathrm{B}}}{r_{\mathrm{B}}}-f\right)^{5/2} d^3\boldsymbol{r}\,, \tag{4.137}$$

which can assume any high negative value depending on $f(\boldsymbol{r},R)$. The $f^{\mathrm{TF}}(\boldsymbol{r},R)$ giving the maximum value of this functional was found to be related to $\varrho^{\mathrm{TF}}(\boldsymbol{r},R)$ as

$$f^{\mathrm{TF}} = \frac{Z_{\mathrm{A}}}{r_{\mathrm{A}}}+\frac{Z_{\mathrm{B}}}{r_{\mathrm{B}}} - \lambda(\varrho^{\mathrm{TF}})^{2/3}\,. \tag{4.138}$$

The maximum value of $\tilde{H}\{f(\boldsymbol{r},R)\}$ coincides with $U^{\mathrm{TF}}(R)$. Thus, choosing certain approximate ϱ_0^{TF} and f_0^{TF} one can estimate $U^{\mathrm{TF}}(R)$ from above and below. The difference between the corresponding approximate $U_0^{\mathrm{TF}}(R)$ and $\tilde{U}_0^{\mathrm{TF}}(R)$ is a measure of the accuracy of the approximation used.

To estimate the accuracy of statistical first-order perturbation theory, it is natural to use

$$f_0^{\mathrm{TF}}(\boldsymbol{r},R) = f_{\mathrm{A}}^{\mathrm{TF}}(\boldsymbol{r}) + f_{\mathrm{B}}^{\mathrm{TF}}(\boldsymbol{r})\,, \tag{4.139}$$

where f_A^{TF} and f_B^{TF} are connected with ϱ_A^{TF}, ϱ_B^{TF} by relations similar to (4.138):

$$f_{A,B}^{TF} = \frac{Z_{A,B}}{r_{A,B}} - \lambda(\varrho_{A,B}^{TF})^{2/3}. \tag{4.140}$$

Naturally, with such ϱ_0 and f_0,

$$U_0(\infty) = \tilde{U}_0(\infty) = U^{TF}(\infty) \tag{4.141}$$

and therefore $\Delta U_0^{TF}(R)$ and $\Delta \tilde{U}_0^{TF}(R)$ give the estimates for $\Delta U^{TF}(R)$ from above and below.

It has been found [4.239] that $\Delta U_0^{TF}(R)$ and $\Delta \tilde{U}_0^{TF}(R)$ differ by no more than 5% over the whole range of R variation. It should be noted that the attempts to obtain a similarly rigorous estimate of the accuracy of statistical first-order perturbation theory for the Thomas-Fermi-Dirac model failed [4.240, 241].

Though the results obtained in [4.239] showed a quite satisfactory accuracy for statistical first-order perturbation theory of the Thomas-Fermi model (it was even higher than that of the model as such), the integrals obtained were rather complicated. It was concluded [4.242] from the limiting cases $R \to 0$, $Z_A Z_B^{-1}, Z_A^{-1} \to 0$, and from the integrals calcu-

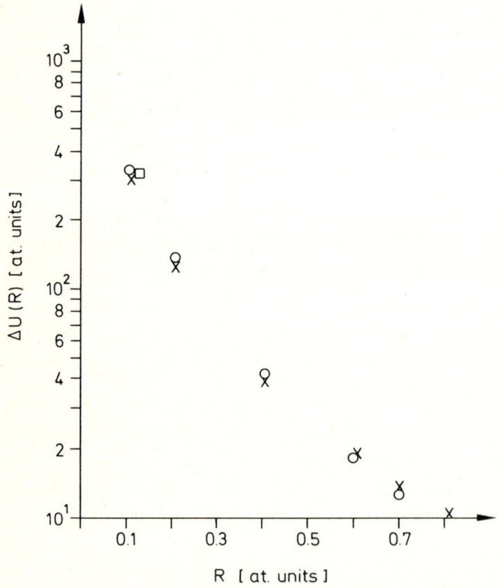

Fig. 4.9. Short-range interaction of N atoms, calculated by different methods. (o) Hartree-Fock method [4.243], (×) (4.142), (□) exact Thomas-Fermi result [4.237]

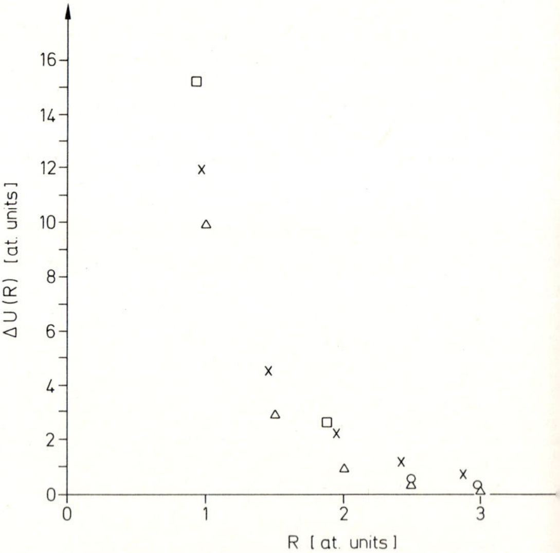

Fig. 4.10. Short-range interaction of Ne atoms, calculated by different methods, (o) Hartree-Fock method [4.244], (△) combined statistical method [4.245], (×) (4.142), (□) exact Thomas-Fermi results [4.237]

lated for several intermediate values of R, Z_A, and Z_B, that for $R < 1$ Å the function $\Delta U^{TF}(R)$ is (with an accuracy up to 20%),

$$U^{TF}(R) = \frac{Z_A Z_B}{R} \chi \left\{ [(Z_A)^{1/2} + (Z_B)^{1/2}]^{2/3} \frac{R}{0.8853} \right\}, \tag{4.142}$$

where $\chi(x)$ is the known Thomas-Fermi function (Sect. 4.1.1).

The expression (4.142) is very attractive in its universality and simplicity and gives satisfactory results for very small $R < 0.5$ Å. This can be seen from Figs. 4.9–12 which present ΔU^{TF} together with the results of more accurate calculations. When $0.5 < R < 1$ Å, (4.142) yields overestimated results, and for $R > 1$ Å it becomes quite invalid. This is due to the fact that the Thomas-Fermi model predicts a too slow (nonexponential) decrease in the atomic electron density with increasing r. Moreover, when $R > 1$ Å the exchange correction comes into play. This takes into account the fact that the Pauli principle forbids the presence of electrons with the same spin directions at one point in space and thus decreases the energy of interrepulsion of the electrons.

To account for these two circumstances, a combined calculation method was proposed, rather a long time ago [4.248–250]. It uses the electron energy functional of the statistical Thomas-Fermi-Dirac model,

$$H^{TFD}\{\varrho(\boldsymbol{r}, R)\} = H^{TF}\{\varrho(\boldsymbol{r}, R)\} - \frac{3}{4}\left(\frac{3}{\pi}\right)^{1/3} \int \varrho^{4/3} d^3\boldsymbol{r}. \tag{4.143}$$

The second term on the right side is the exchange correction. The calculation of $\Delta U(R)$ is again carried out in terms of statistical first-order perturbation theory. However to improve the results for $R > 1$ Å, the Hartree-Fock electron densities ϱ_A^{HF} and ϱ_B^{HF} are used instead of those calculated by the statistical method. The integrals can be solved by numerical calculations only. However these calculations are much simpler than those in the framework of the Hartree-Fock method, and with the use of modern computers they present no fundamental difficulties.

Such a combined method has no rigorous theoretical grounds, yet the results concerning the repulsive interaction of atoms and ions with closed shells [4.231, 233, 245, 250–259] show a rather good agreement with those obtained by the Hartree-Fock method [4.244, 247, 260], as well as with the experimental results [4.234]. The appropriate results for N_2, Ne_2, Ar_2 and $Li^+ + Ar$ obtained by the combined and the Hartree-Fock methods, as well as in terms of the Thomas-Fermi model, are presented in Figs. 4.9–12.

It has been suggested [4.233] that the good results given by the combined method were probably due to a compensation of errors caused by use of the Thomas-Fermi-Dirac electron energy functional and the approximation (4.134) for the electron density. Additional quantum and

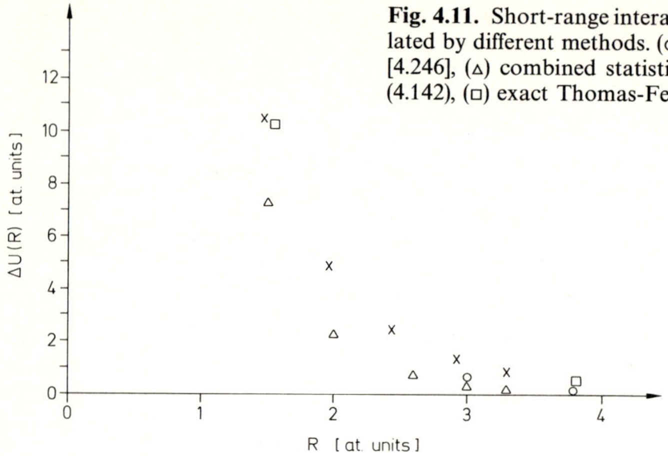

Fig. 4.11. Short-range interaction of Ar atoms, calculated by different methods. (o) Hartree-Fock method [4.246], (△) combined statistical method [4.245], (×) (4.142), (□) exact Thomas-Fermi result [4.237]

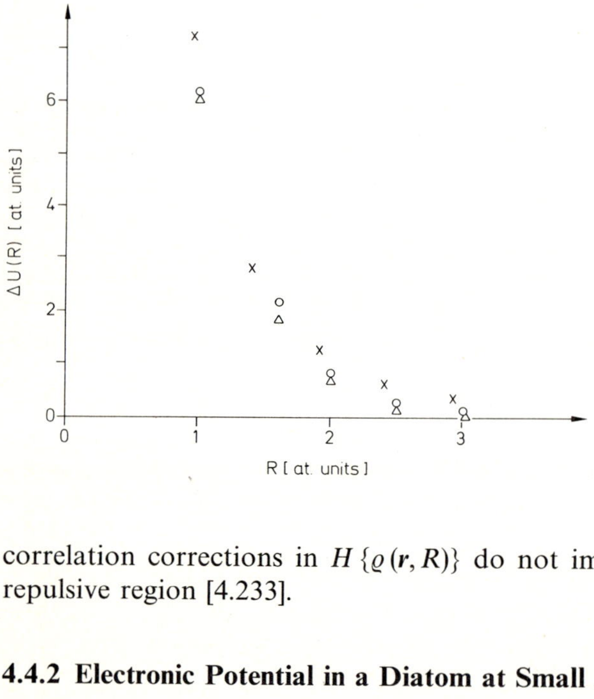

Fig. 4.12. Short-range interaction of Ar with Li^+, calculated by different methods. (o) refers to the Hartree-Fock method [4.247], (△) to the combined statistical method [4.233], (×) to (4.142)

correlation corrections in $H\{\varrho(\boldsymbol{r}, R)\}$ do not improve the results in the repulsive region [4.233].

4.4.2 Electronic Potential in a Diatom at Small *R*

As stated above, the electron density distribution in a diatom and the appropriate representation of $V(\boldsymbol{r}, R)$ as a superposition of the free atom potentials V_A and V_B are, generally speaking, insufficient for the evaluation of $U^{(1)}_{\eta q}(R)$. However, it may be expected that $V(\boldsymbol{r}, R)$ can be accurately approximated by the superposition of two spherically symmetric potentials $V_A(\boldsymbol{r}, R)$ and $V_B(\boldsymbol{r}, R)$, the parameters of which depend on R. The general approach based on this idea has been developed in [4.228, 229] in terms of the Thomas-Fermi-Dirac-Weizsäcker statistical model. The results obtained, however, do not qualitatively differ from those obtained using a much simpler variable-screening model. This model, first developed by *Hund* [4.221] for homonuclear diatoms, was recently generalized for heteronuclear diatoms [4.223, 224]. It is based on the following reasoning.

Let us search for the solution to the Thomas-Fermi equation

$$\nabla^2 V(\boldsymbol{r}, R) = -\frac{8\,(2)^{1/2}}{3\,\pi}\,[-V(\boldsymbol{r}, R)]^{3/2} \tag{4.144}$$

in the form

$$V(\boldsymbol{r}, R) = \tilde{V}_{\mathrm{A}}(\boldsymbol{r}, R) + \tilde{V}_{\mathrm{B}}(\boldsymbol{r}, R) + W(\boldsymbol{r}, R), \tag{4.145}$$

where $\tilde{V}_{\mathrm{A}}$ and $\tilde{V}_{\mathrm{B}}$ are certain spherically symmetric potentials of atoms A and B dependent on R as a parameter. The potentials $\tilde{V}_{\mathrm{A}}$ and $\tilde{V}_{B}$ must be chosen such that their sum gives the best approximation to $V(\boldsymbol{r}, R)$. Then, assuming that this is the case and that $W(\boldsymbol{r}, R)$ is small compared to $\tilde{V}_i\,(i = \mathrm{A, B})$, (4.144) can be rewritten, taking into account solely the terms linear in W, as

$$\delta + \left[\nabla^2 W - \frac{4\,(2)^{1/2}}{\pi}\,(-\tilde{V}_{\mathrm{A}} - \tilde{V}_{\mathrm{B}})^{1/2}\,W\right] = 0\,, \tag{4.146}$$

where

$$\delta = \nabla^2\,\tilde{V}_{\mathrm{A}} + \nabla^2\,\tilde{V}_{\mathrm{B}} + \frac{8\,(2)^{1/2}}{3\,\pi}\,(-\tilde{V}_{\mathrm{A}} - \tilde{V}_{\mathrm{B}})^{3/2}\,. \tag{4.147}$$

If $\delta = 0$, then $W = 0$ is the exact solution to (4.146), and $V(\boldsymbol{r}, R)$ can be represented as the sum of two spherically symmetric potentials of the atoms A and B. This is so in the united-atom limit, if $\tilde{V}_{\mathrm{A}} + \tilde{V}_{\mathrm{B}}$ is the solution to the Thomas-Fermi equation for the united atom. In the limit of separated atoms, $\tilde{V}_{\mathrm{A}}$ and $\tilde{V}_{\mathrm{B}}$ must be the solutions to the Thomas-Fermi equations for the free atoms A and B. In order to obtain the minimum possible $|\delta|$ for the general case and the correct behavior of $V(\boldsymbol{r}, R)$ in the above limiting cases, it is convenient to represent V_i in the form

$$\tilde{V}_i(\boldsymbol{r}, R) = -\frac{Z_i}{r_i}\,F\,[r_i, \tilde{Z}_i(x_i)]\,, \tag{4.148}$$

where $F\,(\boldsymbol{r}, Z)$ is the screening function of the Thomas-Fermi model for an atom with charge Z (Sect. 4.1.1),

$$x_i = 2\,r_i/R\,, \tag{4.149}$$

and $Z_i(x_i)$ is a certain, generally speaking arbitrary, function satisfying the boundary conditions

$$\tilde{Z}_i(x_i) = \begin{cases} Z_i & \text{for} \quad x_i \to 0 \\ Z_A + Z_B & \text{for} \quad x_i \to \infty . \end{cases} \tag{4.150}$$

This function must be chosen such that the value $|\delta|$ is a minimum. It has been proposed [4.221, 223, 224] that $\tilde{Z}_i(x_i)$ be taken in the form

$$\tilde{Z}_i(x_i) = \frac{Z_i \eta^2 + (Z_A + Z_B) x_i}{\eta^2 + x_i^2}, \tag{4.151}$$

where η is an adjustable parameter which does not considerably differ from 2, since, crudely speaking, the line $x_i = 2$ separates the regions of configuration space where the electron is essentially affected by one nucleus only, or by the "united" nucleus. A detailed study [4.221] for the F_2 and N_2 diatoms showed that the minimum $|\delta|$ was obtained for $\eta^2 = 3$. The correction W is no higher here than 1% of $\tilde{V}_A + \tilde{V}_B$. *Eichler* and *Wille* [4.223, 224] also concluded that the optimum value of η^2 is close to 3. The variations in η^2 from 2 to 4 have only a small effect on the results obtained.

So far we have spoken of the Thomas-Fermi electron potential in a diatom. However this potential possesses the essential disadvantage of falling faster than $1/r$ for $r \to \infty$, since the electron self-action is not excluded in the Thomas-Fermi model. For a free atom it can be eliminated approximately by using, for instance, the Fermi-Amaldi model, or some other analytical semiempirical potential (Sect. 4.1.1). It has been suggested for a diatom [4.223, 224] that the self-action can be allowed for by replacing the Thomas-Fermi screening functions in (4.148) by those of a model accounting for this effect. Then $\tilde{Z}_i(x_i)$, as well as any other parameter $\tilde{\beta}_i$ on which the screening functions might depend, will be defined by the equation

$$\beta_i = \frac{\beta_i \eta^2 + \beta_{A+B} x_i}{\eta^2 + x_i^2}, \tag{4.152}$$

where β_{A+B} is the parameter β for the united atom. A certain justification for such an interpolation is that for finite R and not too large x_i, the potential $V(\mathbf{r}, R)$ does not differ much from the Thomas-Fermi potential, and thus the above considerations are valid, whereas for large x_i when deviations from the Thomas-Fermi model are essential, $V(\mathbf{r}, R)$ is close to the united atom potential.

Many calculations of the one-electron terms, using

$$V(\mathbf{r}, R) = -\frac{Z_A}{r_A} F[r_A, \tilde{Z}_A(x_A)] - \frac{Z_B}{r_B} F[r_B, \tilde{Z}_B(x_B)] \tag{4.153}$$

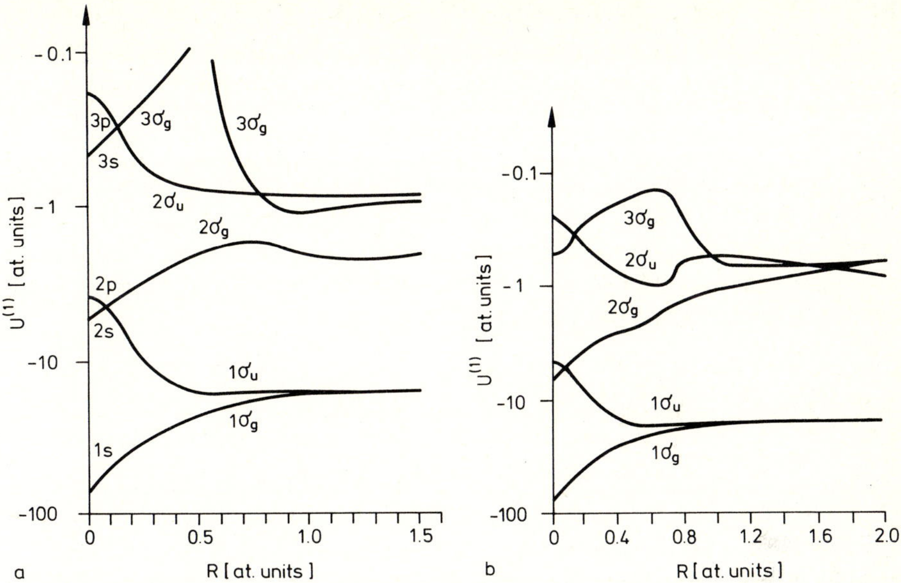

Fig. 4.13a, b. One-electron σ terms of N_2. (**a**) Terms calculated using the potential (4.153) [4.224] and the screening function from [4.28]. (**b**) Hartree-Fock results [4.219]

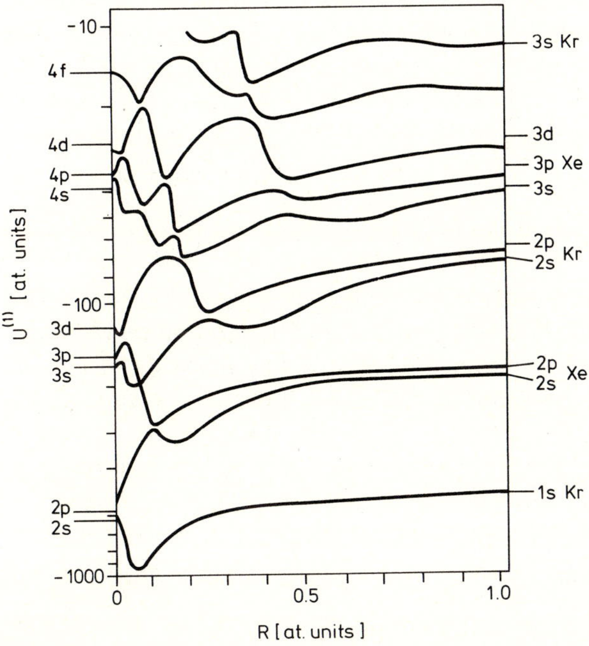

Fig. 4.14. One-electron σ terms of the system Xe-Kr, calculated using the potential (4.153) (according to [4.226])

have been reported [4.224–226]. Fig. 4.13 presents the results for one-electron terms of the N_2 molecule together with the Hartree-Fock method results [4.219]. The term patterns are seen to show rather good agreement. Moreover, the method described here is free from difficulties of the Hartree-Fock method connected with term crossing (see the beginning of this section) and thus it yields "smooth" terms. Another advantage of the method is that it enables one to obtain, without substantial complication of the calculations, an at least qualitatively correct pattern of one-electron terms for heavy systems, such as Xe–Kr, for which the Hartree-Fock calculations have not been reported (Fig. 4.14).

4.5 Coupling Between Electronic States

The interactions which must be taken into account in calculating couplings between electronic states in different regions in Fig. 3.3 have been defined in Sect. 3.4 together with the appropriate selection rules. The actual calculation of the interaction matrix elements needs an explicit form of the coupling operator, as well as knowledge of the electronic wave functions. The calculation of the spin-orbit coupling (Sect. 4.5.1) and the qualitative features of the radial nonadiabatic coupling near the avoided crossing (Sect. 4.5.2) will be considered here. The matrix elements for Coriolis coupling can be calculated for regions I and II in Fig. 3.3 using the familiar methods of angular momentum quantum theory. In region III these matrix elements can be expressed using a similar technique for groups $C_{\infty v}$ and $D_{\infty h}$ [4.261, 262] through the matrix elements of $(l_q + s_q)$ (components of the total angular momentum of one electron) between molecular orbitals. The coupling in region II induced by $H_0 - H_{HL}$ can be evaluated by a straightforward generalization of the approach outlined in Sect. 4.2.

4.5.1 Spin-Orbit Coupling

So far there exists no derivation from first principles of the spin-orbit coupling in the multi-center many-electron case. The most general expression for $V_{s.o.}$ used in practice is obtained by postulating the additivity of the Breit-Pauli two-particle Hamiltonians [4.7, 263, 264]. For a diatom this expression is

$$V_{s.o.} = (c^{-2}/2) \sum_{i=1}^{N} Z_A r_{iA}^{-3} [\boldsymbol{r}_{iA} \times \boldsymbol{p}_i] \boldsymbol{s}_i + (c^{-2}/2) \sum_{i=1}^{N} Z_B r_{iB}^{-3} [\boldsymbol{r}_{iB} \times \boldsymbol{p}_i] \boldsymbol{s}_i - (c^{-2}/2) \sum_{i \neq j} r_{ij}^{-3} [\boldsymbol{r}_{ij} \times \boldsymbol{p}_i] (\boldsymbol{s}_i + 2\boldsymbol{s}_j), \tag{4.154}$$

where c is the velocity of light and $\boldsymbol{p}_i$ is the ith electron momentum. Each term in (4.154) has a simple physical meaning. The first two are the sums of one-electron spin-orbit couplings in the Coulomb fields of nuclei A and B. Two parts can be distinguished in the third term representing the so-called spin-other orbit interaction. One, with the factor $\boldsymbol{s}_i$, is the spin-orbit coupling of electron i in the Coulomb field of electron j and the other, with the factor $\boldsymbol{s}_j$, represents the interaction of electron j's spin-magnetic moment with the orbital current of electron i.

Rigorous calculations of fine-structure splittings in stable diatomic molecules, using the $V_{\mathrm{s.o.}}$ given by (4.154) [4.263], show good agreement with spectroscopic data. These calculations, see also [4.265–267], reveal that the fine-structure splittings or the nondiagonal spin-orbit coupling matrix elements are controlled mainly by the quality of the wave functions used. As far as the form of $V_{\mathrm{s.o.}}$ and the calculation of its matrix elements are concerned, two considerable simplifications can safely be adopted here [4.263, 268].

The first simplification consists of taking into account the spin-other orbit coupling (which is the sum of two-electron operators) by introducing the screening of the nuclear charge. A similar approximation is successfully used in the theory of atomic fine structure [4.269, 270]. The spin-orbit coupling is determined by the electron coordinates near the nuclei A or B. At the same time the electron potential in these regions is controlled mostly by the spherically symmetric contribution from the relevant atom. Therefore the effective spin-orbit coupling operator

$$V_{\mathrm{s.o.}}^{\mathrm{eff}} = \sum_{i=1}^{N} f_{\mathrm{A}}(r_{i\mathrm{A}}, R)[\boldsymbol{r}_{i\mathrm{A}} \times \boldsymbol{p}_i]\boldsymbol{s}_i + \sum_{i=1}^{N} f_{\mathrm{B}}(r_{iB}, R)[\boldsymbol{r}_{i\mathrm{B}} \times \boldsymbol{p}_i]\boldsymbol{s}_i \tag{4.155}$$

can be considered here instead of the $V_{\mathrm{s.o.}}$ given by (4.154). By analogy with the spin-orbit coupling of one electron moving in a central field [4.271], we can write

$$f_{\mathrm{A,B}}(r_{i\mathrm{A,B}}, R) = (2c^2 r_{i\mathrm{A,B}})^{-1}\, \partial \tilde{V}_{\mathrm{A,B}}(r_{i\mathrm{A,B}} R)/\partial r_{i\mathrm{A,B}} \,. \tag{4.156}$$

These equations can be considered to be the definition of the effective electronic potentials which determine the contributions from atoms A and B to the spin-orbit coupling. Generally speaking, these contributions are different from $\tilde{V}_{\mathrm{A}}$ and $\tilde{V}_{\mathrm{B}}$ in (4.145), but it seems that this difference is not considerable, at least qualitatively.

The functions f_{A} and f_{B} must obviously satisfy the boundary relations

$$\begin{aligned} &f_{\mathrm{A,B}}(r_{i\mathrm{A,B}}, R) \xrightarrow[R\to\infty]{} f_{\mathrm{A,B}}^{(0)}(r_{i\mathrm{A,B}}), \\ &f_{\mathrm{A}}(r_{i\mathrm{A}}, R) + f_{\mathrm{B}}(r_{i\mathrm{B}}, R) \xrightarrow[R\to 0]{} f_{\mathrm{AB}}^{(0)}(r_i), \end{aligned} \tag{4.157}$$

where $f_A^{(0)}, f_B^{(0)}, f_{AB}^{(0)}$ are the spin-orbit coupling coefficients in the free atoms A and B and in the united atom. It appears that the R-dependence of f_A and f_B is weak and $f_A^{(0)}$ and $f_B^{(0)}$ can be used in all regions of Fig. 3.3 except III″ and the region intermediate between III′ and III″. In these two regions one can adopt an interpolation procedure similar to that considered in Sect. 4.4.2.

The second simplification consists in neglecting all two-center integrals, which is possible when calculating the matrix elements of $V_{\text{s.o.}}$ or $V_{\text{s.o.}}^{\text{eff}}$ in all regions of Fig. 3.3 except the two mentioned above.

The reliability of these two simplifications is illustrated in Fig. 4.15. The matrix elements $\langle {}^3\Pi | V_{\text{s.o.}} | {}^1\Sigma^+ \rangle$ are shown as functions of R for the systems Kr–O and Xe–O (see Fig. 3.6), calculated (a) without any simplifications [4.266], and (b) using $f_A^{(0)}$ and $f_B^{(0)}$ in (4.155) and neglecting the two-center integrals [4.272]. Configuration interaction wave functions of approximately the same quality were used in the above papers. The rather strong R-dependence of the coupling matrix elements is mainly due to an increasing admixture of ionic configurations with decreasing R, which leads to the involvement of strong spin-orbit couplings in Kr^+ and Xe^+.

Calculations of the spin-orbit coupling (4.155) matrix elements are most simple for those regions of R where the Heitler-London or the molecular-orbital approximations for the wave functions are adequate (regions II and III in Fig. 3.3).

For region III use is made of the fact that $V_{\text{s.o.}}^{\text{eff}}$ has the form

$$V_{\text{s.o.}}^{\text{eff}} = \sum_i \boldsymbol{o}_i \boldsymbol{s}_i , \tag{4.158}$$

where $\boldsymbol{o}_i$ is the pseudovector operator depending on the spatial variables of electron i. This enables the matrix elements of $V_{\text{s.o.}}^{\text{eff}}$ between wave functions constructed from shell functions of type (3.47) to be expressed through the one-electron matrix elements $\langle \eta\, q\, \tilde{\lambda} | o_m | \eta'\, q'\, \tilde{\lambda}' \rangle$ of the mth component of the pseudovector $\boldsymbol{o}$, the coefficients of fractional parentage and the analogs of $6-j$ and $9-j$ symbols for the groups $C_{\infty v}$ and $D_{\infty h}$ [4.261, 262]. Explicit expressions for the $V_{\text{s.o.}}^{\text{eff}}$ matrix elements in some frequently encountered situations are given in [4.262].

In region III′ of Fig. 3.3 where the MO-LKAO approximation is valid, $f_A^{(0)}$ and $f_B^{(0)}$ can be used, the two-center spin-orbit coupling integrals neglected and the matrix elements $\langle \eta\, q\, \tilde{\lambda} | o_m | \eta'\, q'\, \tilde{\lambda}' \rangle$ expressed through the one-electron atomic spin-orbit coupling constants $\zeta_{n_1 l_1 A}$ and $\zeta_{n_2 l_2 B}$ [4.273] defined as

$$\zeta_{nl} = \int_0^\infty f^{(0)}(r)\, \varphi_{nl}^2(r)\, r^2\, dr \tag{4.159}$$

In region III″ these matrix elements can be expressed through the spin-orbit coupling constants $\zeta_{nl\,AB}$ of the united atom [4.274].

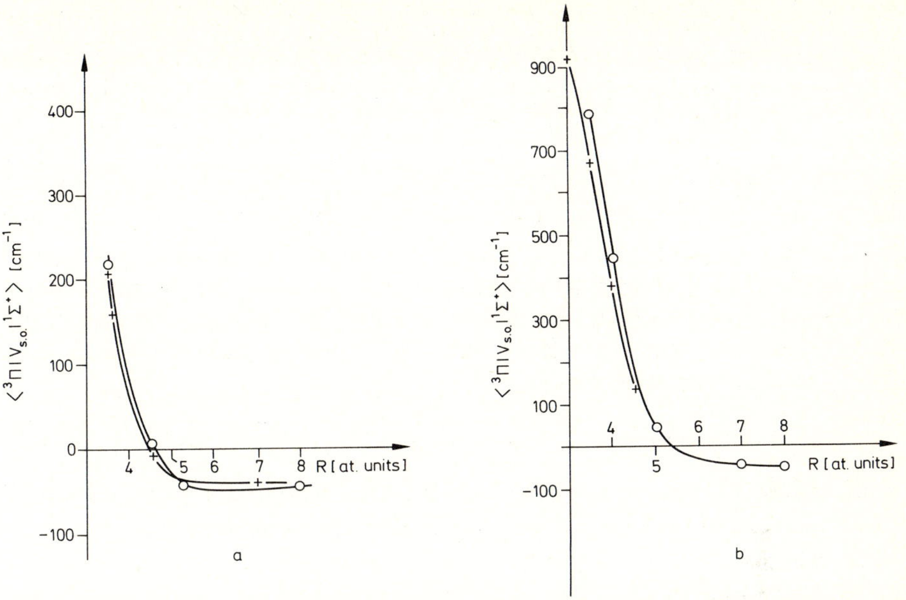

Fig. 4.15 a, b. Spin-orbit coupling matrix elements for the systems KrO (**a**) and XeO (**b**). (+) refers to exact calculations using (4.154) [4.266], (o) to simplified calculations (see text) [4.272]

In region II of Fig. 3.3 both simplifications hold and $f_A^{(0)}$ and $f_B^{(0)}$ can be substituted into (4.155). Thus, the calculation of the spin-orbit coupling in this region reduces to the calculation of matrix elements of the sum of the spin-orbit interactions in the free atoms $V_{s.o.A} + V_{s.o.B}$, between the non-antisymmetrized wave functions given by (3.22) where $\hat{A}$ is omitted.

This method of handling spin-orbit coupling at values of R for which the Heitler-London approximation is adequate was first suggested in [4.275, 276] and since then has been extensively used [4.95, 98, 101, 102, 156, 244, 277–279].

Consider, for example, the case of two atoms with open shells $(n_1 l_1)^{N_A}$ and $(n_2 l_2)^{N_B}$. Here the matrix elements of $V_{s.o.}^{eff}$ are expressed through the atomic spin-orbit coupling constants:

$$\langle (n_1 l_1)^{N_A} S_1 L_1; (n_2 l_2)^{N_B} S_2 L_2; L \tilde{\Lambda} S m_S | V_{s.o.}^{eff} | (n_1 l_1)^{N_A} S_1' L_1';$$

$$(n_2 l_2)^{N_B} S_2' L_2'; L' \tilde{\Lambda}' S' m_S' \rangle = \sum_{\tilde{\Omega} j j_1 j_2} \begin{bmatrix} S & L & j \\ m_S & \tilde{\Lambda} & \tilde{\Omega} \end{bmatrix} \begin{bmatrix} S' & L' & j \\ m_S' & \tilde{\Lambda}' & \tilde{\Omega} \end{bmatrix}$$

$$\times (2j_1 + 1)(2j_2 + 1)\,[(2S + 1)\,(2L + 1)\,(2S' + 1)\,(2L' + 1)]^{1/2}$$

$$\times \begin{Bmatrix} S_1 & L_1 & j_1 \\ S_2 & L_2 & j_2 \\ S & L & j \end{Bmatrix} \begin{Bmatrix} S_1' & L_1' & j_1 \\ S_2' & L_2' & j_2 \\ S' & L' & j \end{Bmatrix} \Big\{ (-1)^{L_1 + S_1' + j_1} [l_1 (l_1 + 1)(2 l_1 + 1)]^{1/2}$$

$$\times \begin{Bmatrix} S_1 & L_1 & j_1 \\ L_1' & S_1' & 1 \end{Bmatrix} \zeta_{n_1 l_1} \langle (n_1 l_1)^{N_A} S_1 L_1 \| V^{11} \| (n_1 l_1)^{N_A} S_1' L_1' \rangle \delta_{S_2 S_2'} \delta_{L_2 L_2'}$$

$$+ (-1)^{L_2 + S_2' + j_2} [l_2 (l_2 + 1)(2 l_2 + 1)]^{1/2} \begin{Bmatrix} S_2 & L_2 & j_2 \\ L_2' & S_2' & 1 \end{Bmatrix} \zeta_{n_2 l_2} \langle (n_2 l_2)^{N_B}$$

$$\times S_2 L_2 \| V^{11} \| (n_2 l_2)^{N_B} S_2' L_2' \rangle \, \delta_{S_1 S_1'} \delta_{L_1 L_1'} \Big\} . \tag{4.160}$$

The reduced matrix elements $\langle (n l)^N S L \| V^{11} \| (n l)^N S' L' \rangle$ for the configurations $(n p)^N$ and $(n d)^N$ are tabulated in [4.124]. Note that when only those matrix elements which are diagonal, say, in S_1, L_1 are considered, then $V_{\mathrm{s.o.A}}$ can be written in the form

$$V_{\mathrm{s.o.A}} = \mathscr{A}_{S_1 L_1} \boldsymbol{S}_{\mathrm{A}} \boldsymbol{L}_{\mathrm{A}} , \tag{4.161}$$

where $\boldsymbol{S}_{\mathrm{A}}$ and $\boldsymbol{L}_{\mathrm{A}}$ are the operators of the total spin and angular momenta of atom A. For instance, in the case of one open shell $(n_1 l_1)^{N_A}$,

$$\begin{aligned} \mathscr{A}_{S_1 L_1} = \zeta_{n_1 l_1} & [l_1 (l_1 + 1)(2 l_1 + 1)]^{1/2} \\ & \times [S_1 (S_1 + 1)(2 S_1 + 1)(L_1 + 1) L_1 (2 L_1 + 1)]^{-1/2} \\ & \times \langle (n_1 l_1)^{N_A} S_1 L_1 \| V^{11} \| (n_1 l_1)^{N_A} S_1 L_1 \rangle . \end{aligned} \tag{4.162}$$

4.5.2. Radial Coupling in the Avoided Crossing Region

Consider two adiabatic states 1 and 2 which exhibit an avoided crossing. Part of the radial nonadiabatic coupling between these states, important under quasi-classical conditions, is [see the discussions before (2.114) and (2.146)]

$$g(R) = \langle 1 | (\partial / \partial R) + \mathrm{i} C_{Z'} | 2 \rangle , \tag{4.163}$$

where the differentiation is performed for constant electronic coordinates (referred to the center of mass of the nuclei), and $C_{Z'}$ is given by (2.140). It was argued in Sect. 2.3 that the ambiguous term $\mathrm{i} C_{Z'}$ can usually be omitted in the avoided crossing region. Therefore it will not be considered here.

When the adiabatic wave functions $|1\rangle$ and $|2\rangle$ are exact, their differentiation can be avoided in calculating $g(R)$ by use of a relation similar to (2.106). If the electronic Hamiltonian is represented in the form

$$H = T_{\mathrm{e}} + V , \tag{4.164}$$

where T_{e} is the electronic kinetic energy, and V is the total interaction, then

$$g(R) = \langle 1 | \partial V / \partial R | 2 \rangle / (U_2 - U_1) . \tag{4.165}$$

The part of V which depends explicitly on R is due to the electron-nucleus interaction and is represented as the sum of one-electron operators, see (3.14) and (4.154). Therefore, as follows from (4.165), from the viewpoint of the matrix elements calculation and the selection rules, $\partial/\partial R$ will be similar to the sum of one-electron operators, as already accepted in Sect. 3.4. When the spin-orbit coupling is not included in V, which is usually the case for all regions in Fig. 3.3, except region I, an expression equivalent to (4.165) is obtained for $g(R)$ [4.280]:

$$g(R) = R^{-1}\left[(1/2)(U_1 - U_2)\left\langle 1\left|\sum_{i=1}^{N} r_i^2\right|2\right\rangle + (U_1 - U_2)^{-1}\langle 1|T_e|2\rangle\right]. \tag{4.166}$$

The use of (4.165) instead of the original (4.163) is computationally advantageous, since $\partial V/\partial R$ is usually explicitly known, but the numerical differentiation of a general configuration interaction wave function is a formidable problem. However the approach using (4.165) meets the objection that for approximate wave functions, which are almost always the only ones available, (4.165) is not equivalent to (4.163) [4.1]. Indeed, let $|1\rangle_{ap}$ and $|2\rangle_{ap}$ be the eigenfunctions of H in the subspace spanned by some K orthogonal functions $\varphi_i(R)$. The corresponding eigenvalues are U_1^{ap} and U_2^{ap}. Then [4.281–283],

$$\begin{aligned} g_{ap}(R) &= {}_{ap}\langle 1|\partial/\partial R|2\rangle_{ap} = {}_{ap}\langle 1|\partial V/\partial R|2\rangle_{ap}(U_2^{ap} - U_1^{ap})^{-1} \\ &+ \left\{(U_2^{ap} - U_1^{ap})^{-1}\sum_{i=1,j=1}^{K}[{}_{ap}\langle 1|\varphi_i\rangle(\langle\partial\varphi_i/\partial R|H|\varphi_j\rangle\right. \\ &+ \langle\varphi_i|H|\partial\varphi_j/\partial R\rangle)\langle\varphi_j|2\rangle_{ap}] \\ &+ \left.\sum_{i=1,j=1}^{K}{}_{ap}\langle 1|\varphi_i\rangle\langle\varphi_i|\partial/\partial R|\varphi_j\rangle\langle\varphi_j|2\rangle_{ap}\right\}. \end{aligned} \tag{4.167}$$

The first term on the right side of (4.167) comes directly from (4.165) using approximate wave functions. Generally speaking, the term in braces which identically vanishes for the complete basis set $\{\varphi_i\}$ can make a considerable contribution in the case of a finite basis set. In this connection, methods implying either direct differentiation of the approximate wave functions or representations of the type (4.167) have been developed for the calculation of $g_{ap}(R)$ [4.282, 284–286]. An alternative approach to this problem would be to find the conditions which the approximate wave functions must satisfy in order to minimize the effect of the term in the braces in (4.167) which is most difficult to calculate. In particular, it has been argued in [4.287] that the effect of this term is small if the Hellman-Feynman theorem [4.7, 288] is fulfilled.

An important conclusion which can be drawn from the available results of these complicated approaches is that the term in braces in (4.167) provides no considerable contribution, at least in the avoided crossing region [4.281, 284, 289]. Therefore the qualitative features of the radial coupling in the avoided crossing region obtained from (4.165) or (4.166) seem to remain unchanged for approximate wave functions as well.

It can be assumed from the arguments used in proving the noncrossing rule (Sect. 3.3.1) that in the avoided crossing region, exact adiabatic wave functions $|1\rangle$ and $|2\rangle$ can be adequately represented in the form [compare with (7.7)]

$$\begin{aligned} |1\rangle &= \varphi_1(R)\cos\vartheta(R) + \varphi_2(R)\sin\vartheta(R), \\ |2\rangle &= -\varphi_1(R)\sin\vartheta(R) + \varphi_2(R)\cos\vartheta(R). \end{aligned} \tag{4.168}$$

Here

$$\tan 2\vartheta(R) = 2H_{12}/\Delta H, \tag{4.169}$$

where $\Delta H = H_{11} - H_{22}$, $H_{aa'} = \langle\varphi_a|H|\varphi_{a'}\rangle$, $(a, a' = 1, 2)$ and φ_1 and φ_2 are such that the "zero order" terms $H_{11}(R)$ and $H_{22}(R)$ cross at some $R = R_0$ in the avoided crossing region. The representation (4.168) assumes that H_{12} is real, which is usually the case. Now

$$g(R) = g_1(R) + g_2(R), \tag{4.170}$$

where, when using (4.169), $g_1(R) = -\vartheta'(R)$ (the prime denotes differentiation with respect to R), is readily expressed in the form

$$g_1(R) = (H_{12}\Delta H' - \Delta H\, H'_{12})(\Delta U)^{-2}, \tag{4.171}$$

where

$$\Delta U = U_1 - U_2 = [(\Delta H)^2 + 4H_{12}^2]^{1/2}. \tag{4.172}$$

It has been shown in [4.290] for a purely electrostatic V that, according to (4.166) and the virial theorem, $g_2(R) = \langle\varphi_1|\partial/\partial R|\varphi_2\rangle$ can be approximately expressed in the form

$$g_2(R) = R^{-1}\Delta H\langle\varphi_1|V|\varphi_2\rangle(\Delta U)^{-2} \tag{4.173}$$

over the range of R variation around R_0, where $\langle\varphi_1|T_e|\varphi_2\rangle$ and $\Delta H'$ do not change sign.

The functions φ_1 and φ_2 can be called quasi-diabatic because H_{11} and H_{22} cross, but generally speaking $g_2(R) \neq 0$ and can be appreciable. Therefore, these functions differ from the diabatic wave functions $|1d\rangle$ and $|2d\rangle$ defined in Sect. 7.1, where we demand that $\langle 1d|\partial/\partial R|2d\rangle = 0$. The matrix $H_{aa'}$ also differs from the matrix $V_{aa'}$ defined by (7.11).

It is seen from (4.171) that if ΔH and H_{12} are slowly varying functions of R near R_0, then $g_1(R)$ is a bell-shaped function which attains its maximum value near R_0. At the same time, if (4.173) holds, $g_2(R)$ passes through zero at R_0. Therefore in some region near R_0, g_2 can be neglected compared to g_1, and φ_1 and φ_2 can be considered to be exactly diabatic. At larger distances from R_0 corresponding to the wings of $g_1(R)$, $g_2(R)$ can compete with $g_1(R)$. Since $g_2(R)$ changes sign at R_0, the total $g(R)$ can pass through zero due to this competition. Such a behavior of $g(R)$ is illustrated in Fig. 4.16 which shows the radial coupling between two pseudocrossing $^2\Sigma$ terms which induces the charge transfer reaction

$$N^{3+}[(2s)^2, {}^1S] + H(1s, {}^2S) \rightarrow N^{2+}[(2s)^2 3s, {}^2S] + H^+ . \quad (4.174)$$

This radial coupling was calculated in [4.281] using the configuration interaction wave functions and (4.167) with the term in braces omitted.

In practice the problem often arises of how to evaluate $g(R)$ in the avoided crossing region if $U_1(R)$ and $U_2(R)$ are known (e.g., from experiment), but information concerning the wave functions is lacking. Generally speaking, the solution of this problem is not unique, due to the possible effect of $g_2(R)$. It can be inferred, however, from the reasoning above that this problem may be considered by assuming first that $g_2(R)$ can be neglected. The second assumption would be that ΔH and H_{12} are slowly varying functions of R near R_0, and that because of this, only the linear terms can be retained in the expansion of ΔH and H_{12} in powers of $(R - R_0)$:

$$\Delta H = \Delta H'(R_0)(R - R_0), \quad (4.175\,a)$$

$$H_{12} = H_{12}(R_0) + H'_{12}(R_0)(R - R_0) \quad (4.175\,b)$$

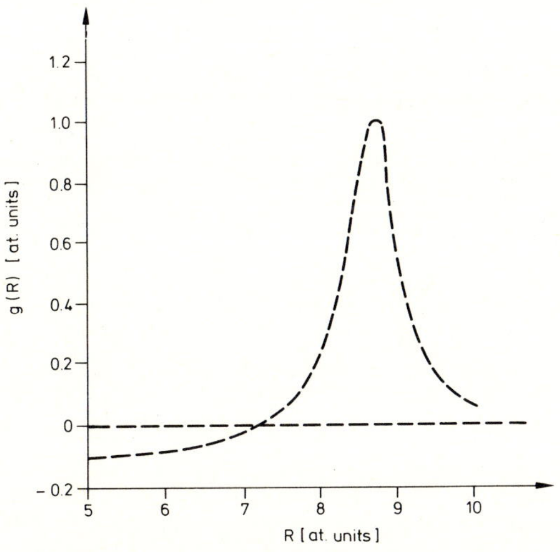

Fig. 4.16. Radial coupling between two pseudocrossing $^2\Sigma$ terms of the system NH^{3+} [4.281]

Noting that $\vartheta(R_0) = \pi/4$ and assuming that $H_{12}(R_0) > 0$, this gives

$$\Delta U = u\{[2\beta(R - R_0) + \gamma]^2 + 1\}^{1/2}, \tag{4.176}$$

$$g(R) = \beta\{[2\beta(R - R_0) + \gamma]^2 + 1\}^{-1}, \tag{4.177}$$

$$\vartheta(R) = (\pi/4) + (1/2)\arctan\gamma - (1/2)\arctan[2\beta(R - R_0) + \gamma], \tag{4.178}$$

where

$$u = 2|H_{12}(R_0)\Delta H'(R_0)|\{[\Delta H'(R_0)]^2 + 4[H'_{12}(R_0)]^2]^{-1/2}, \tag{4.179}$$

$$\beta = \{[\Delta H'(R_0)]^2 + 4[H'_{12}(R_0)]^2\}/4H_{12}(R_0)\Delta H'(R_0), \tag{4.180}$$

$$\gamma = 2H'_{12}(R_0)/\Delta H'(R_0). \tag{4.181}$$

For definiteness let $\Delta U(R) > 0$ in the avoided crossing region. Then obviously, β as given by (4.180) is also positive in this region. The adiabatic splitting ΔU attains its minimum value at some R_{max} near R_0. It follows from (4.176) and (4.177) that $\gamma(R)$ attains its maximum value at R_{max} and

$$R_0 = R_{max} + (\gamma/2\beta), \tag{4.182}$$

$$\beta = [\Delta U''(R_{max})/4\Delta U(R_{max})]^{1/2}. \tag{4.183}$$

Thus, $g(R)$ given by (4.177) is uniquely determined by three quantities: R_{max}, $\Delta U(R_{max})$, and $\Delta U''(R_{max})$, assumed to be known. The parameter γ which is needed for a precise definition of the diabatic states in the avoided crossing region, see (4.178), is usually small compared to unity. This is due to the fact that for the avoided crossing region to be well localized, $|H'_{12}(R_0)| \sim H_{12}(R_0)/\Delta R$ must be much smaller than $|\Delta H'(R_0)| \sim \Delta U(\infty)/\Delta R$ [ΔR is the characteristic length of the order of an atomic unit and $\Delta U(\infty)$ is the adiabatic splitting at large distances]. The iterative procedure starting from (4.176–178), which yields the diabatic interaction matrix $H_{aa'}$ in the finite range of R variation around R_0 when only U_1 and U_2 are known, has been considered in [4.291, 292].

We note finally that the simple approximate expression (4.177) for nonadiabatic coupling underlies the widely used model of nonadiabatic transitions which will be considered in detail in Chap. 8.

5. Elastic Scattering

This chapter deals with a process occurring over one elastic channel. One can confine oneself to elastic scattering only if the initial and final states of colliding atoms are adiabatically isolated from all other states. Also, the degeneracies of the colliding partners A and B, if any, must be immaterial for a scattering process. Two closed-shell atoms, both in ^1S states, colliding at low energy, best satisfy the above requirement. When the spins S_A and S_B are not zero, but both atoms are in states with $L_A = 0$ and $L_B = 0$, then again the elastic approximation for each molecular $^{2S+1}\Sigma$ state is very satisfactory. To a good approximation the S matrix is diagonal (in the appropriate representation) in S quantum numbers. The probability of transitions between different spin states, i.e. the spin depolarization for a given molecular state, is very low and can be calculated in terms of perturbation theory using elastic scattering as a zero-order approximation. The same arguments pertain to two-channel scattering of symmetric systems, when the scattering matrix can be made diagonal with respect to the parity w (Sect. 2.3.2).

Thus, in considering elastic scattering, we take one adiabatic potential $U(R)$ and suppress all indices indicating the states of colliding partners. The discussion will be based on expressions derived in Chap. 2 and will be aimed at illustrating the general quasi-classical approach using the most simple, though typical, cases. Scattering in the Coulomb field will be omitted, as will the Regge representation of the scattering amplitude and one-channel scattering by the complex potential.

For more extensive discussion of elastic scattering see the reviews [5.1–11] and books [5.12–16].

5.1 Quasi-classical Scattering Amplitude

For elastic scattering, the relative angular momentum l is conserved and coincides with the total angular momentum J. The S matrix is diagonal in l

$$S = \exp(2i\delta_l), \tag{5.1}$$

where $\delta_l = \delta_l(k)$ is the scattering phase shift. This phase shift is found from the asymptotic behaviour of the radial wave function for elastic scattering. Equation (2.39) describes the asymptotic behavior of the standing wave

$$\chi^{+J}(R) \equiv \chi^l(R)|_{R \to \infty} \simeq \sin[kR - l\pi/2 + \delta_l(k)]. \tag{5.2}$$

The function $\chi^l(R)$ satisfies the wave equation which follows from (2.114),

$$\frac{d^2\chi^l}{dR^2} + \mathscr{K}_l^2\chi^l = 0, \quad \text{with} \tag{5.3}$$

$$\mathscr{K}_l^2 = k^2[1 - U(R)/E - (l + 1/2)^2/2\mu R^2 E]. \tag{5.4}$$

Note that the total energy $\mathscr{E}$ in (2.114) is replaced here by the relative energy E. Also, the replacement of $l(l+1)$ in the exact wave equation by $(l + 1/2)^2$ (made in Chap. 2 on the grounds of the small relative difference between these values for large l) is additionaly substantiated for elastic scattering by the so-called Langer correction (see, e.g. [5.17]), which ensures the proper behavior of the radial quasi-classical wave function for small R, in the classically inaccessible region.

With (5.1), the general expression (2.28) for scattering amplitudes becomes

$$f(\theta) = (1/2ik) \sum_{l=0}^{\infty} (2l + 1)\, \mathrm{P}_l(\cos\theta)\, [\exp(2i\delta_l) - 1], \tag{5.5}$$

where the $\mathrm{P}_l(\cos\theta)$ are the Legendre polynomials.

Two equations, for small angle scattering and scattering through classical angles, follow from (5.5). These are simplified forms of (2.50) and (2.56):

$$f(\theta) = (\mathrm{i}/k) \int_0^\infty \mathrm{J}_0(l\theta)\, [1 - \exp(2\mathrm{i}\delta_l)]\, l\,dl \quad \text{with } \theta \ll 1, \tag{5.6}$$

$$f(\theta) = -\mathrm{i}(2\pi k^2 \sin\theta)^{-1/2} \int_0^\infty [\exp(\mathrm{i}l\theta + \mathrm{i}\theta/2 - \mathrm{i}\pi/4) + \exp(-\mathrm{i}l\theta - \mathrm{i}\theta/2 + \mathrm{i}\pi/4)]\exp(2\mathrm{i}\delta_l)\, l^{1/2}dl \quad \text{with } \theta \gg 1/l^*. \tag{5.7}$$

The meaning of l^* is explained in Sect. 2.1.2.

The following uniform approximation for scattering through all angles is valid (provided, as assumed here, $l^* \gg 1$):

$$f(\theta) = (\mathrm{i}/k)\, (\theta/\sin\theta)^{1/2} \int_0^\infty \mathrm{J}_0(\theta l)\, [1 - \exp(2\mathrm{i}\delta_l)]\, l\,dl. \tag{5.8}$$

Now, under conditions of large and smoothly varying scattering phase shifts, the integral in the rhs of (5.7) can be calculated by the stationary phase approximation.

In the case of an isolated stationary phase point at $l = l^\chi$ the contribution $f^\chi(\theta)$ to $f(\theta)$ is (Sect. 2.1.2)

$$\begin{aligned} f^\chi(\theta) &= -\mathrm{i}[{}^{\mathrm{cl}}q^\chi(\theta)]^{1/2}\exp[\mathrm{i}A^\chi(\theta) + \mathrm{i}\alpha^\chi], \\ \alpha^\chi &= (w^\chi + v^\chi)\,\pi/4, \end{aligned} \tag{5.9}$$

with

$${}^{\mathrm{cl}}q^\chi(\theta) = k^{-2}(\sin\theta)^{-1}l\,|\partial\eta_l/\partial l|^{-1}|_{l=l^\chi(\theta)}, \tag{5.10}$$

$$A^\chi(\theta) = [2\delta_l - w^\chi(l + 1/2)\,\theta]\,|_{l=l^\chi(\theta)}. \tag{5.11}$$

Here, according to (2.59), $l^\chi(\theta)$ can be found from the equations, (M: integer)

$$\begin{aligned} 2\pi M \pm \theta &= \eta_l, \\ \eta_l &= 2\partial\delta_l/\partial l, \end{aligned} \tag{5.12}$$

where η_l is the deflection function, and, as explained in connection with (2.61), w^χ and v^χ are: $w^\chi = \pm 1$, the sign corresponding to the sign before θ in the lhs of (5.12), and $v^\chi = \pm 1$ according to the sign of $\partial\eta_l/\partial l$.

With $l^\chi(\theta)$ satisfying (5.11) and (5.12), we get

$$dA^\chi/d\theta = -w^\chi[l^\chi(\theta) + 1/2]. \tag{5.13}$$

Collecting all the contributions together, we arrive at a primitive semiclassical approximation (Sect. 2.1.2),

$${}^{\mathrm{ps}}f(\theta) = \sum_\chi f^\chi(\theta). \tag{5.14}$$

Now, if two (or more) points $l^{\chi'}$ and $l^{\chi''}$ are close in the sense that the condition $|A^{\chi'} - A^{\chi''}| \gg 1$ is not fulfilled, different kinds of uniform approximations can be used [5.2, 10]. For two close stationary points $l^{\chi'}$ and $l^{\chi''}$, a cubic approximation for δ_l (corresponding to a quadratic approximation for η_l) in a region including both $l^{\chi'}$ and $l^{\chi''}$ is often quite satisfactory. Note that in this case $w^{\chi'} = w^{\chi''}$, but $v^{\chi'} = -v^{\chi''}$. The joint contribution $f^{\chi'\chi''}(\theta)$ of two coalescent stationary phase points $l^{\chi'}$ and $l^{\chi''}$ to the scattering amplitude $f(\theta)$, allowing for the change of the integrand factor $l^{1/2}$ in that region, is (the so-called uniform Airy approximation)

$$\begin{aligned} f^{\chi'\chi''}(\theta) = &-\mathrm{i}\{[({}^{\mathrm{cl}}q^{\chi'})^{1/2} + ({}^{\mathrm{cl}}q^{\chi''})^{1/2}]\,\pi^{1/2}z^{1/4}\mathrm{Ai}(-z) \\ &-\mathrm{i}[({}^{\mathrm{cl}}q^{\chi'})^{1/2} - ({}^{\mathrm{cl}}q^{\chi''})^{1/2}]\,\pi^{1/2}z^{-1/4}d\,\mathrm{Ai}(-z)\,dz\} \\ &\times \exp[\mathrm{i}(A^{\chi'} + A^{\chi''})/2 + \mathrm{i}(w^{\chi'} + w^{\chi''})\,\pi/8], \\ z = {} & [3(A^{\chi'} - A^{\chi''})/4]^{2/3}. \end{aligned} \tag{5.15}$$

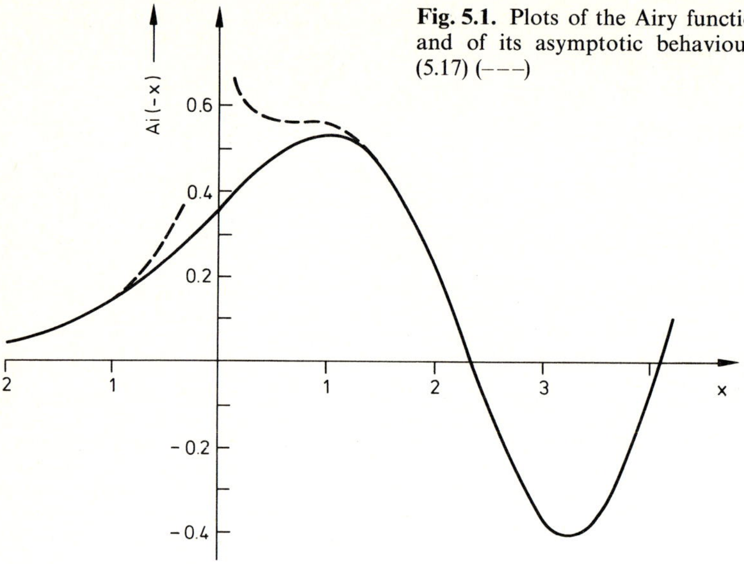

Fig. 5.1. Plots of the Airy function $\mathrm{Ai}(-x)$ (—) and of its asymptotic behaviour according to (5.17) (- - -)

The Airy function $\mathrm{Ai}(x)$ is defined by the integral

$$\mathrm{Ai}(x) = \frac{1}{2\pi} \int_{-\infty}^{\infty} \cos(ux + u^3/3)\, du. \tag{5.16}$$

Asymptotically, for $|x| \gg 1$, it behaves as

$$\mathrm{Ai}(x) = \begin{cases} 2^{-1}\pi^{-1/2}x^{-1/4}\exp(-2x^{3/2}/3), & x > 0 \\ \pi^{-1/2}|x|^{-1/4}\sin(2|x|^{3/2}/3 + \pi/4), & x < 0. \end{cases} \tag{5.17}$$

A plot of this function is shown in Fig. 5.1.

Note that when using (5.15) for $f^{\chi'\chi''}(\theta)$, the information used is the same as that needed to separately calculate $f^{\chi'}$ and $f^{\chi''}$.

When $A^{\chi'} - A^{\chi''}$ becomes large, $f^{\chi'\chi''}(\theta)$ defined by (5.15) goes over to ${}^{\mathrm{ps}}f(\theta)$, i.e. to the sum of $f^{\chi'}(\theta)$ and $f^{\chi''}(\theta)$.

5.2 Quasi-classical Scattering Matrix

Under quasi-classical conditions the scattering phase shift δ_l can be represented as a formal series in powers of the Planck constant $\hbar$.

$$\delta_l = \delta_l^{\mathrm{JWKB}} + \delta_l^{(1)} + \delta_l^{(2)}. \tag{5.18}$$

The leading term δ_l^{JWKB} is given by the JWKB (Jeffreys-Wentzel-Kramers-Brillouin) approximation and is proportional to $\hbar^{-1}$ [5.17]. The next terms

$\delta_l^{(1)} \sim \hbar$ and $\delta_l^{(2)} \sim \hbar^2$ [5.18]. The discussion in this and other sections will be based on the JWKB approximation only.

5.2.1 JWKB Scattering Phase Shifts

The JWKB approximation replaces δ_l by δ_l^{JWKB} which will be denoted hereafter by $\delta(b)$, making use of the relation between l and the impact parameter b, $kb = (l + 1/2)$. The notion of the impact parameter b is familiar in classical scattering, which will be reviewed briefly in Sect. 5.3. The scattering phase shift $\delta(b)$ is calculated by (5.2) as the difference between the full JWKB phase and the phase for free motion $kR - \pi l/2$. In this way [5.17],

$$\delta(b) = \lim_{R\to\infty}\left[\int_{R^t}^{R} \mathscr{K}(b, R)\, dR + \pi/4 - kR + \pi l/2\right]$$

$$= \lim_{R\to\infty}\left[\int_{R^t}^{R} \mathscr{K}(b, R)\, dR - kR + \pi kb/2\right], \tag{5.19}$$

where

$$\mathscr{K}(b, R) = k[1 - U(R)/E - b^2/R^2]^{1/2}$$

and R^t is the turning point where the radial velocity is zero, viz. $\mathscr{K}(b, R^t) = 0$. The JWKB treatment assumes that for $R > R^t$ the conditions of quasi-classical motion are fulfilled everywhere. Note that the additional phase $\pi/4$ in (5.19) which disappears when l is replaced by b, is due to the quantum correction for motion close to R^t [5.17] calculated by using the linear approximation for the effective potential (Sect. 5.2.3). When the potential $U(R)$ for a given b is small compared to the kinetic energy, the high-energy approximation follows from (5.19), expanding $\mathscr{K}(b, R)$ in powers of $U(R)$ and retaining the first term only:

$$\delta(b) = -\int_{R^t}^{\infty} U(R)\, (\mu/2E)^{1/2}(1 - b^2/R^2)^{-1/2}\, dR$$

$$= -\int_{0}^{\infty} U[(b^2 + z^2)^{1/2}]\, (\mu/2E)^{1/2}\, dz. \tag{5.20}$$

For higher-order corrections see [5.19]. According to (5.19) the JWKB deflection function η_l^{JWKB} is

$$\eta_l^{\mathrm{JWKB}} = 2\partial\delta_l^{\mathrm{JWKB}}/\partial l = 2\partial\delta(b)/k\partial b$$

$$= \pi - 2\int_{R^t}^{\infty} [1 - U(R)/E - b^2/R^2]^{-1/2}\, b\, dR/R^2. \tag{5.21}$$

The rhs of (5.21) is recognized to be the classical deflection angle for a trajectory with impact parameter b. This is discussed in more detail in

Sect. 5.3 and is denoted there by $\eta(b)$. Thus, at the JWKB level there exists an important relation between the scattering phase shift (a quantity characterizing quantum scattering) and the deflection angle (a quantity describing classical scattering):

$$\eta(b) = 2\partial\delta(b)/k\partial b. \tag{5.22}$$

This quantum-classical correspondence provides an interpretation of the scattering phase shifts in terms of the classical action $\mathscr{S}$ calculated along the given trajectory from an initial point $\boldsymbol{R} = \boldsymbol{R}_\mathrm{i}$ to the final point $\boldsymbol{R} = \boldsymbol{R}_\mathrm{f}$ after collision. By definition,

$$\mathscr{S}_\mathrm{if} = \int_\mathrm{i}^\mathrm{f} \boldsymbol{P}(\boldsymbol{R})\, d\boldsymbol{R}, \tag{5.23}$$

where $\boldsymbol{P}(\boldsymbol{R})$ is the linear momentum at a point $\boldsymbol{R}$ of the trajectory. Now, representing the scalar product $\boldsymbol{P}\,d\boldsymbol{R}$ in terms of radial and tangential components, we have

$$\mathscr{S}_\mathrm{if} = \int_\mathrm{i}^\mathrm{f} \mathscr{K}(R)\, dR + kb \int_\mathrm{i}^\mathrm{f} d\psi, \tag{5.24}$$

where ψ is the angular coordinate of $\boldsymbol{R}$ and the radial integral calculation must take into account the change in sign of the root at the turning point R^t. In the limit of high R_i, R_f, (5.24) yields

$$\mathscr{S}_\mathrm{if} = k(R_\mathrm{i} + R_\mathrm{f}) - k\pi b + 2\delta(b) + kb[\pi - \eta(b)]. \tag{5.25}$$

Here, the first three terms give the asymptotic value of the radial integral in (5.24) taking (5.19) into account and the last term contains the total turning angle $\pi - \eta(b)$ of the vector $\boldsymbol{R}$. Since the first term is the action $\mathscr{S}_\mathrm{if}^0$ for free motion, (5.25) relates the scattering-induced action increment $\Delta\mathscr{S} = \mathscr{S}_\mathrm{if} - \mathscr{S}_\mathrm{if}^0$ to the phase shift,

$$\Delta\mathscr{S}(b) = 2\delta(b) - kb\eta(b). \tag{5.26}$$

Note that $\Delta\mathscr{S}(b)$ bears no subscript since in the limit of large R_i, R_f, it is independent of these values. Comparing (5.26) with (5.11) and using the relation (5.12) between η and θ, we get

$$A^\varkappa(\theta) = \Delta\mathscr{S}(b)|_{b=b^\varkappa(\theta)}. \tag{5.27}$$

As an example, in the case of a repulsive potential, when there exists only one stationary phase point $b^{(1)}$, $A^{(1)}$ can be explicitly calculated to give [5.20]

$$A^{(1)} = 2\delta(b^{(1)}) - kb^{(1)}\theta = \int_{R^t}^{\infty} \frac{(2\mu)^{1/2}(dU/dR)\,R\,dR}{[E - U - E(b^{(1)}/R)^2]^{1/2}}, \qquad b^{(1)} = b^{(1)}(\theta). \tag{5.28}$$

In particular, for the scattering of rigid spheres of collision radius R_0, the integral in (5.28) is proportional to the change of the radial component of linear momentum at $R = R_0$ and $A^{(1)} = -2kR_0 \sin\theta/2$.

5.2.2 Violation of Quasi-classical Conditions in Localized Regions. Connection Problem

The recovery of phase shift $\delta(b)$ making use of the deflection function $\eta(b)$ is possible only under quasi-classical conditions, i.e. to the accuracy with which the wave function χ can be represented in the form

$$\chi(R) = \mathscr{K}^{-1/2}\left[b^- \exp\left(-\mathrm{i}\int_{R_0}^{R} \mathscr{K}\, dR\right) + b^+ \exp\left(\mathrm{i}\int_{R_0}^{R} \mathscr{K}\, dR\right)\right], \quad (5.29)$$

where R_0 is the distance chosen in some way or other, b^+ and b^- are constant coefficients depending on R_0, and $\mathscr{K}(R)$ is taken to be real over the whole range R to R_0. For simplification the subscript l will be omitted here and hereafter. Neither tunneling nor overbarrier reflection are taken into account in the approximation (5.29).

In the limit $R \to \infty$, when $\mathscr{K}(R) \to k$, the function $\chi(R)$ tends to

$$\chi(R) \sim \exp(-\mathrm{i}kR + \mathrm{i}\pi l/2) - S\exp(\mathrm{i}kR - \mathrm{i}\pi l/2) \quad (5.30)$$

containing the scattering matrix.

In many quasi-classical scattering problems for which the general criterion $kR^* \gg 1$ is valid, the JWKB approximation is violated in localized regions of relatively small length ΔR (referred to hereafter as L-regions). In this case the complete solution of the problem, $\chi(R)$, can be constructed by matching the solutions in the quasi-classical regions with exact solutions in the L-regions. The main idea of this approach is that due to the small length of the L-region, the function $\mathscr{K}(R)$ in (5.3) is approximated by a simple reference function $\tilde{\mathscr{K}}(R)$ for which the exact analytical solution $\tilde{\chi}(R)$ of (5.3) exists (the so-called method of comparison equation) [5.21]. At a certain distance from this region the function $\tilde{\chi}(R)$ tends to asymptotes which are different to the left and to the right of the L-region:

$$\begin{aligned}\tilde{\chi}(R) &= \tilde{\mathscr{K}}^{-1/2}\left[b_<^- \exp\left(-\mathrm{i}\int_{R^<}^{R} \tilde{\mathscr{K}}\, dR\right) + b_<^+ \exp\left(\mathrm{i}\int_{R^<}^{R} \tilde{\mathscr{K}}\, dR\right)\right] \\ &\quad \text{at } R < R^<, \\ \tilde{\chi}(R) &= \tilde{\mathscr{K}}^{-1/2}\left[b_>^- \exp\left(-\mathrm{i}\int_{R^>}^{R} \tilde{\mathscr{K}}\, dR\right) + b_>^+ \exp\left(\mathrm{i}\int_{R^>}^{R} \tilde{\mathscr{K}}\, dR\right)\right] \\ &\quad \text{at } R > R^>.\end{aligned} \quad (5.31)$$

Here $R^<$ and $R^>$ are reasonably chosen boundaries of the L-region. If the R-dependence of the potential U and consequently of the function $\mathscr{K}(R)$ is sufficiently smooth, the function $\tilde{\chi}(R)$ reaches its quasi-classical asymptotic behavior (5.31) before the appearance of a noticeable deviation of $\tilde{\mathscr{K}}(R)$ from $\mathscr{K}(R)$. In this case the representation of (5.31) coincides with that of χ in the form (5.29) for an appropriate choice of the constants in (5.29).

Equation (5.31) shows that a transition over the L-region results in a transformation of the amplitude $b^{\alpha'}_<$ ($\alpha' = +, -$) to $b^{\alpha''}_>$, when passing from the side of smaller R to that of larger R (the so-called connection problem). This is represented by a certain connection matrix

$$b^{\alpha'}_> = \sum_{\alpha''=+,-} \mathscr{L}^{\alpha'\alpha''} b^{\alpha''}_< , \tag{5.32}$$

the form of which depends both on the choice of $\tilde{\mathscr{K}}(R)$ and on the distances $R^<$ and $R^>$. When these are properly chosen, the phases of the matrix elements $\mathscr{L}^{\alpha'\alpha''}$ will be relatively small; it is just these phases which provide corrections to the quasi-classical phases of traveling waves for $R > R^>$ and $R < R^<$. If there are m separated L-regions located near $R_1, R_2, \ldots, R_{m-1}, R_m$, and if the connection matrix $\mathscr{L}^{\alpha'\alpha''}(R_k)$ is known for each, the general solution of the scattering equation (5.3) in any quasi-classical region can be constructed by matching (Fig. 5.2).

To this end we introduce the functions

$$\varphi^{\alpha'}(R, R') = \exp\left[i\alpha' \int_{R'}^{R} \mathscr{K}(R)\, dR \right] \tag{5.33}$$

and apply repeatedly the connection equation (5.32). Representing $\chi(R)$ for $R > R_m$ in the form

$$\chi(R) = \chi^-(R) + \chi^+(R), \tag{5.34}$$

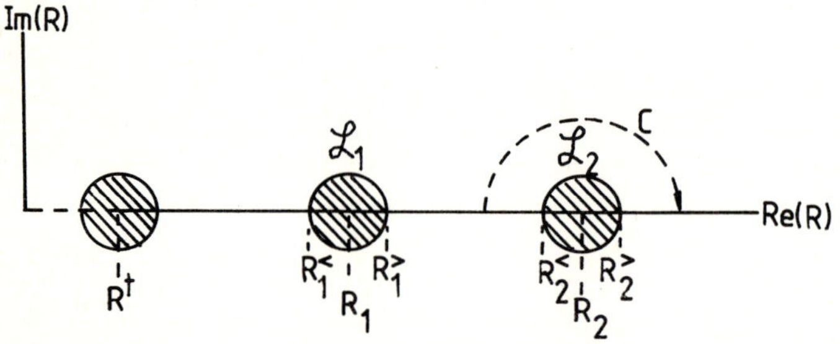

Fig. 5.2. The $\mathscr{L}$-regions in the complex R-plane (\\\\) for the turning point R^t on the left. The classically inaccessible part of the real axis is dashed. Line C is a countour circumventing the $\mathscr{L}(R_2)$ region

we obviously obtain

$$\chi^{\alpha'}(R) = [\mathscr{K}(R)]^{-1/2} \sum_{\alpha'',\ldots} \varphi^{\alpha'}(R, R_m^>)\, \mathscr{L}^{\alpha'\alpha''}(R_m)\, \varphi^{\alpha''}(R_m^<, R_{m-1}^>)$$
$$\times \ldots \mathscr{L}^{\alpha''\ldots\alpha''\ldots'}(R_1)\, \varphi^{\alpha''\ldots'}(R_1^<, R^t)\, b^{\alpha''\ldots'}, \tag{5.35}$$

where $b^{\alpha''\ldots'}$ ($b^{\alpha''\ldots'} = b^+, b^-$) are the coefficients in (5.31) representing $\chi(R)$ in the first quasi-classical region to the right of the left turning point R^t.

Equations (5.34, 35) are accurate to within the accuracy of the quasi-classical condition for every interval $R_k - R_{k-1}$ separating the L-regions. A large accumulation of phases in every quasi-classical region

$$\Delta(R_{k-1}^>, R_k^<) = \int_{R_{k-1}^>}^{R_k^<} \mathscr{K}(R)\, dR \gg 1 \tag{5.36}$$

is a necessary condition for this approximation.

If for a certain region the phase $\Delta(R_{k-1}^>, R_k^<)$ is small, the two adjoining L-regions can be united into one and another comparison equation must be considered.

From (5.34, 35) we find the scattering matrix

$$S = -\lim_{R\to\infty} [\exp(-2ikR + i\pi l)\, \chi^+(R)/\chi^-(R)]$$
$$= -\lim_{R\to\infty} \exp(-2ikR + i\pi l)$$
$$\times \frac{\sum_{\alpha',\ldots} \varphi^+(R, R_m^>)\, \mathscr{L}^{+\alpha'}(R_m) \ldots \varphi^{\alpha''\ldots'}(R_1^<, R^t)\, b^{\alpha''\ldots'}}{\sum_{\alpha',\ldots} \varphi^-(R, R_m^>)\, \mathscr{L}^{-\alpha'}(R_m) \ldots \varphi^{\alpha''\ldots'}(R_1^<, R^t)\, b^{\alpha''\ldots'}}\,. \tag{5.37}$$

It is seen from this equation that if the quasi-classical conditions are violated in localized regions, the determination of the S matrix amounts to a calculation of the $\mathscr{L}$-matrices and to finding the relation between the amplitudes b^+ and b^- in the first quasi-classical region.

Turn now to the latter problem. The most direct way of finding an $\mathscr{L}$-matrix is to attempt to solve a comparison equation in terms of known functions, or, as a last resort, numerically.

In certain cases the explicit solution of a comparison equation can be avoided by using the method of phase integrals [5.22, 23]. The idea of the method is to extend the quasi-classical asymptotes of the functions $\tilde{\chi}^+$ and $\tilde{\chi}^-$ to the complex plane R along a contour C connecting two JWKB regions and circumventing the L-region (Fig. 5.2). In order to accomplish this, the absolute value of the complex phase integral

$$\mathscr{S}(R) = \int_{R_n}^{R} \tilde{\mathscr{K}}(R)\, dR, \tag{5.38}$$

taken from the suitably defined center R_n of the L-region to the point R lying on the contour, is required to be large.

When R is complex, expressions of the type

$$\tilde{\chi}(R) = b^- \tilde{\mathscr{K}}^{-1/2} \exp[-i\tilde{\mathscr{S}}(R)] + b^+ \mathscr{K}^{-1/2} \exp[i\tilde{\mathscr{S}}(R)] \tag{5.39}$$

with constant coefficients $b^{\pm}$ represent the solution of the comparison equation with a relative accuracy $(dU/dR)/(d\tilde{\mathscr{S}}/dR)\,U$. Generally speaking, one of the terms in (5.39) will exponentially dominate another and therefore no meaning is to be ascribed to the former when it is smaller than the already neglected first quasi-classical correction to the larger term. In connection with this, it is expedient to find the lines along which both exponents are of the same order (lines of real phase called the anti-Stokes lines L_{AS}), and the lines along which the ratio of exponents take extreme values (the Stokes lines L_S). These are defined by the relations

$$\mathrm{Im}\{\tilde{\mathscr{S}}(R)\} = 0 \ (L_{AS} \text{ lines}), \qquad \mathrm{Re}\{\tilde{\mathscr{S}}(R)\} = 0 \ (L_S \text{ lines}). \tag{5.40}$$

When passing from one L_{AS} line to another along a contour lying solely in the quasi-classical region, the coefficients change, since in a certain region between these lines the representation (5.39) is no longer accurate, as the larger correction terms have already been neglected compared to the small exponential. This effect, known as the Stokes phenomenon, is often formulated as a statement that in moving along the quasi-classical contour and crossing the Stokes lines the coefficient in front of the small exponential abruptly changes.

In simple cases, such changes can be found by resorting to general properties of the exact solutions $\tilde{\chi}$. This approach will be exemplified now by a simple isolated turning point.

5.2.3 Isolated Turning Point

The solution of this problem provides a relation between b^+ and b^-, and also allows one to evaluate the coefficient for an exponentially decaying wave function in the classically forbidden quasi-classical tunneling region.

Close to the isolated turning point R^t a linear approximation is used for $\tilde{\mathscr{K}}^2(R)$:

$$\tilde{\mathscr{K}}^2 = g(R - R^t) \tag{5.41}$$

so that the appropriate comparison equation with a variable $x = g^{1/3}(R - R^t)$ (it is assumed for definiteness that $g > 0$ which corresponds to the right turning point) is

$$d^2\tilde{\chi}/dx^2 - x\tilde{\chi} = 0. \tag{5.42}$$

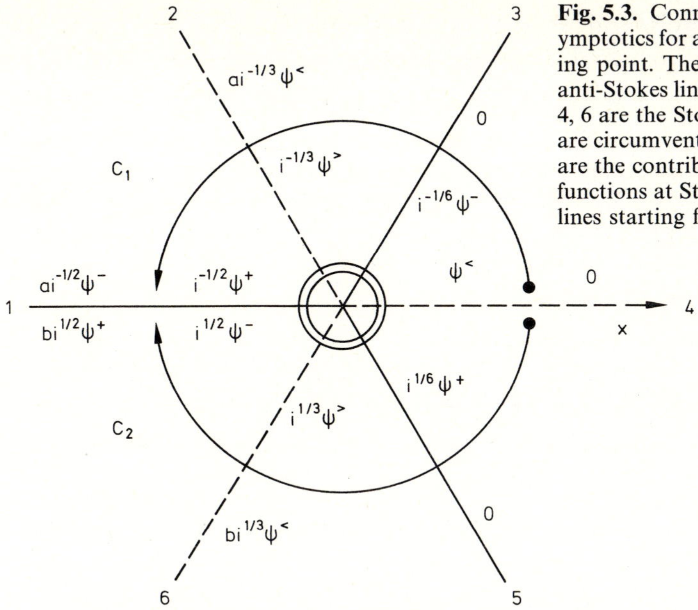

Fig. 5.3. Connection of JWKB asymptotics for an isolated right turning point. The full lines 1, 3, 5 are anti-Stokes lines, the dashed lines 2, 4, 6 are the Stokes lines, C_1 and C_2 are circumventing contours. Shown are the contributions to the JWKB functions at Stokes and anti-Stokes lines starting from line 4

The general solution of this equation is expressed via the Airy function $\mathrm{Ai}(x)$ [5.17] already mentioned, [see (5.16, 17)].

The asymptotic behavior of $\mathrm{Ai}(x)$ for $x < -1$ makes it immediately possible to evaluate the ratio of the coefficients b^+ and b^-. However to illustrate the phase integral method, consider the possibility of relating b^+ and b^- without using the explicit solution of (5.42) [5.22–25]. Consider the function

$$\Psi(z) = z^{-1/4} \exp[(2/3) z^{3/2}]. \tag{5.43}$$

On the two-sheet Riemann surface [1] with a cut along the negative part of the x axis, this function is analytical. Define, then,

$$\Psi(z) = \begin{cases} \psi_1(z) = z^{-1/4} \exp[(2/3) z^{3/2}] & \text{on the upper sheet,} \\ \psi_2(z) = z^{-1/4} \exp[(-2/3) z^{3/2}] & \text{on the lower sheet.} \end{cases} \tag{5.44}$$

The AS and S lines, as found from (5.40) are shown in Fig. 5.3.

When $x < 0$ both ψ_1 and ψ_2 oscillate and when $x > 0$ one (ψ_1) increases and the other (ψ_2) decreases. Denote also by $\psi^<(\varrho)$, $\psi^>(\varrho)$, $\psi^+(\varrho)$

[1] Strictly speaking, the function $\Psi(z)$ is one-valued on the four Riemann sheets. However, of importance here is the behaviour of exponents, and two sheets are sufficient for discussion.

and $\psi^-(\varrho)$ the functions which behave along the ray $z = \varrho \exp(i\varphi)$, $\varphi = \text{const.}$ as

$$\begin{aligned} \psi^< &= \varrho^{-1/4} \exp[-(2/3)\,\varrho^{3/2}], \\ \psi^> &= \varrho^{-1/4} \exp[(2/3)\,\varrho^{3/2}], \\ \psi^+ &= \varrho^{-1/4} \exp[(2\mathrm{i}/3)\,\varrho^{3/2}], \\ \psi^- &= \varrho^{-1/4} \exp[-(2\mathrm{i}/3)\,\varrho^{3/2}]. \end{aligned} \tag{5.45}$$

The exact solution $\tilde{\chi}(z)$ of (5.42) can be represented in the quasi-classical region, i.e. for $|z| \gg 1$, as a linear combination of ψ_1 and ψ_2,

$$\tilde{\chi}(z) = c_1\psi_1(z) + c_2\psi_2(z). \tag{5.46}$$

On the S-line both terms ψ_1 and ψ_2, and on the AS-line only one, $\psi^>$, contribute to $\tilde{\chi}(z)$. In other regions the smallest, ψ_k ($k = 1, 2$), has a meaning if it is larger than the leading term neglected in the JWKB expansion.

The function $\tilde{\chi}(z)$ as represented by (5.46) is not analytic on one sheet. Yet, as follows from the theory of differential equations, it must be analytic. This seeming contradiction is resolved by making allowance for the sudden change of the coefficient in front of the small exponential when crossing the S-line.

Figure 5.3 shows the varying contributions to the function from two components $c_1\psi_1$ and $c_2\psi_2$ along the quasi-classical contours originating at line 4, crossing lines 3 and 2 (contour C_1) and lines 5, 6 (contour C_2), and circumventing the turning point $z = 0$. In the region within the double circle the quasi-classical approach is in general inapplicable.

The change of component $c_1\psi_1$ is shown on the inner side of the contours, and that of $c_2\psi_2$ on the outer one. For motion around contour C_1 we obtain the function above the real axis, and for that around contour C_2 we have another function below the real axis. Move along contour C_1 assuming that along the real axis the component $c_1\psi_1$ is equal to $\psi^<$, and the component $c_2\psi_2$ is zero, as expected for the classically forbidden region. The function $c_1\psi_1$, equal to $\psi^<$ on line 4, changes smoothly, and for motion along C_1 it becomes $\mathrm{i}^{-1/2}\psi^+$ on line 1. The function $c_2\psi_2$, zero on line 4, suffers a jump on the Stokes line 2, so that after arrival at line 1 the function has the form of a linear combination with an unknown coefficient a. Another combination with coefficient b is obtained for $\tilde{\chi}$ moving along contour C_2. Since $\tilde{\chi}$ must be one-valued, we immediately obtain $a = \mathrm{i}$, $b = -\mathrm{i}$, whence directly follows the asymptotic expression for $\tilde{\chi}$ ($x \ll -1$). In this way, the asymptotic expressions for the function

$\tilde{\chi}(x)$ are found for both sides of the turning point, without explicit solution of (5.42):

$$\begin{aligned} \tilde{\chi}(x) &= 2\,|x|^{-1/4} \sin[(2/3)\,|x|^{3/2} + \pi/4], \qquad |x| \gg 1,\ x < 0, \\ \tilde{\chi}(x) &= x^{-1/4} \exp(-2x^{3/2}/3), \qquad x \gg 1,\ x > 0. \end{aligned} \tag{5.47}$$

These coincide apart from the normalization factor with the asymptotics of the Airy function given by (5.17).

Equation (5.46) provides the relation between the coefficients b^+ and b^- for the traveling waves χ^+ and χ^- to the left of the right turning point,

$$b^+ = ib^-. \tag{5.48}$$

Similarly, for the left turning point,

$$b^+ = -ib^-. \tag{5.49}$$

In both cases, after reflection from the potential wall, the wave loses a phase factor of $\pi/2$.

Note that the factors $(2/3)\,x^{3/2}$ in the sine and exponential functions in (5.47) are the quasi-classical phases $\int_0^x \tilde{\mathscr{K}}(x)\,dx$ for $\tilde{\mathscr{K}}(x)$ given by (5.41). The additional phase $\pi/4$ is an important correction due to the breakdown of the JWKB approximation close to the turning point. It is also seen that its value is controlled by the linear approximation of the potential near $x = 0$. This linear approximation must be valid in some region with $|x| \gg 1$, thus making possible either a quasi-classical circumvention of the point $x = 0$, or use of the Airy function as a solution of the comparison equation. Both approaches provide the asymptotics of $\tilde{\chi}(x)$ for $x > 0$ and $x < 0$, as well as giving quasi-classical solutions on the AS-lines 3 and 5. These will be used later for calculation of the nonadiabatic transition probability (Chap. 6).

5.2.4 Two Close Turning Points

The solution to this problem describes the quantum effects for motion of particles near the top of the potential barrier. To account for two close zeros of the function $\mathscr{K}^2(R)$, a parabolic approximation is adopted for $\mathscr{K}^2$,

$$\tilde{\mathscr{K}}^2(R) = g(R^{t<} - R)(R - R^{t>}). \tag{5.50}$$

The comparison equation for $\tilde{\chi}$ in dimensionless variables takes the form

$$d^2\tilde{\chi}/dx^2 + (\varepsilon - x^2/2) = 0, \tag{5.51}$$

where the parameter ε is the particle energy referred to the barrier maximum U^m and expressed in units of the vibrational quantum ω^m in a well associated with an inverted barrier, $\varepsilon = (E - U^m)/\omega^m$. The solution $\tilde{\chi}$ is expressed by parabolic cylinder functions [5.25–27]; the positions of the Stokes and anti-Stokes lines for $\varepsilon > 0$ and $\varepsilon < 0$ are given in Fig. 5.4.

For both cases the matrix $\mathscr{L}$ can be represented in a unified way (the numbering of lines and columns are in the order +, −).

$$\mathscr{L} = \begin{pmatrix} (1 + e^{-2\pi\varepsilon})^{1/2} e^{-i\phi} & -ie^{-\pi\varepsilon} \\ ie^{-\pi\varepsilon} & (1 + e^{-2\pi\varepsilon})^{1/2} e^{i\phi} \end{pmatrix}, \tag{5.52}$$

where

$$\phi(\varepsilon) = \arg\Gamma(1/2 + i\varepsilon) - \varepsilon \ln|\varepsilon| + \varepsilon. \tag{5.53}$$

The odd function $\phi(\varepsilon)$ increases from zero for $\varepsilon = 0$, passes through a maximum $\phi_{max} \approx 0.15$ when $\varepsilon \approx 0.2$, and then decreases as $\phi(\varepsilon) \approx 1/24\varepsilon$ when $\varepsilon > 1$. $R^<$ and $R^>$ [see (5.31)] are identified with the classical turning points $R^{t<}$ and $R^{t>}$ for $E < U^m$, and with the coordinate of the barrier maximum R^m for $E > U^m$.

Instead of the matrix $\mathscr{L}$ connecting the amplitudes of the positive ($\tilde{\chi}^+$) and negative ($\tilde{\chi}^-$) fluxes on the left side of the L-region to the amplitudes of relevant fluxes on the right side, it is convenient to introduce a matrix $\tilde{\mathscr{L}}$ connecting the amplitudes of fluxes incident on the two sides to those of the outgoing fluxes. This matrix will be

$$\tilde{\mathscr{L}} = \begin{pmatrix} r & t \\ t & r \end{pmatrix} = e^{-i\phi(\varepsilon)} \begin{pmatrix} -i\mathscr{R}^{1/2} & \mathscr{T}^{1/2} \\ \mathscr{T}^{1/2} & -i\mathscr{R}^{1/2} \end{pmatrix}, \tag{5.54}$$

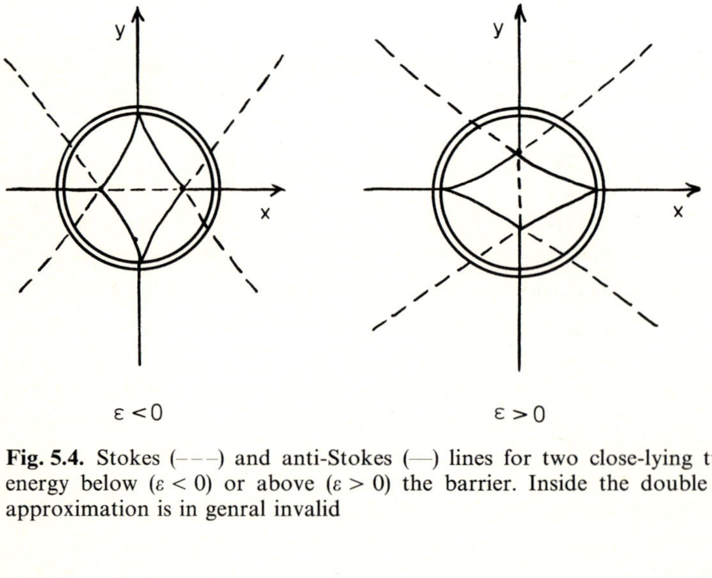

Fig. 5.4. Stokes (- - -) and anti-Stokes (—) lines for two close-lying turning points with energy below ($\varepsilon < 0$) or above ($\varepsilon > 0$) the barrier. Inside the double circles the JWKB approximation is in genral invalid

where $\mathscr{R}$ and $\mathscr{T}$ are the real reflection and transmission coefficients

$$\begin{aligned} \mathscr{R} &= [1 + \exp(2\pi\varepsilon)]^{-1}, \\ \mathscr{T} &= [1 + \exp(-2\pi\varepsilon)]^{-1}. \end{aligned} \tag{5.55}$$

The matrices $\mathscr{L}$ and $\tilde{\mathscr{L}}$ for a parabolic barrier can be generalized for barriers of any form, provided the latter become different from the parabolic barrier in a region where the tunneling factor is small. This generalization is attained by redefinition of ε:

$$\begin{aligned} \varepsilon &= -(\pi)^{-1} \left| \int_{R^<}^{R^>} \mathscr{K}\, dR \right| \quad \text{for } E < U^{\mathrm{m}}, \\ \varepsilon &= (\mathrm{i}\pi)^{-1} \int_{R^+}^{R^-} \mathscr{K}\, dR \quad \text{for } E > U^{\mathrm{m}}. \end{aligned} \tag{5.56}$$

Here $R^<$ and $R^>$ are the real roots of $\mathscr{K}(R)$ for the tunneling and R^- and R^+ are the complex roots of $\mathscr{K}(R)$ for the overbarrier case [5.13].

If the parabolic approximation is violated before the solution $\tilde{\chi}$ of (5.51) has reached the quasi-classical asymptote, then the generalization of $\mathscr{L}$ and $\tilde{\mathscr{L}}$ for barriers of an arbitrary form must account for the highest corrections to the quasi-classical approximation. This question will not be discussed here (for details see [5.22, 28–32]) as, in fact, the necessity for such a refinement does not often arise. We shall mention in particular only the paper [5.33] on penetration of the Eckart barrier calculated with higher JWKB corrections.

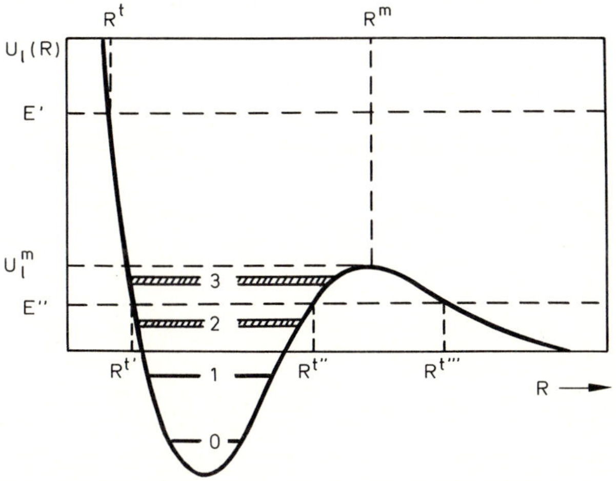

Fig. 5.5. The three-turning-point case exemplified by the isolated-turning-point ($R^{t'}$) and two-turning-point ($R^{t''}, R^{t'''}$) problems. Shown are the energy levels of two stationary (0, 1) and two quasistationary (2, 3) states

To illustrate the application of the two-turning-point problem consider the most frequently encountered case: scattering by a potential with a well and a barrier characterized by one turning point $R^{t'}$ (or R^{t}) and two turning points $R^{t''}$ and $R^{t'''}$ that can eventually coalesce (Fig. 5.5). This type of scattering is known as shape resonance scattering, different from the Feshbach resonance occurring in multi-channel scattering (Chap. 6).

The one-plus-two-turning-point problem has been thoroughly investigated [5.27, 34–37], mainly in connection with resonance scattering from the centrifugal barrier for simple interatomic potentials of the Lennard-Jones type.

The following conventions are used below:

$$R^{>} = \begin{cases} R^{t'''} & \text{for } E < U_l^{m}, \\ R^{m} & \text{for } E > U_l^{m}, \end{cases} \tag{5.57}$$

where U_l^{m} is the maximum of the effective potential $U_l(R) = U(R) + (l + 1/2)^2/2\mu R^2$. Let also

$$R^{<} = \begin{cases} R^{t''} & \text{for } E < U^{m}, \\ R^{m} & \text{for } E > U^{m}, \end{cases}$$
$$R_0 = \begin{cases} R^{t'} & \text{for } E < U^{m}, \\ R^{t} & \text{for } E > U^{m}. \end{cases} \tag{5.58}$$

Then, according to (5.37),

$$S = -\exp(2\mathrm{i}\delta_l^{>}) \frac{\mathscr{L}^{++}\exp(\mathrm{i}\Delta_l)\, b^{+} + \mathscr{L}^{+-}\exp(-\mathrm{i}\Delta_l)\, b^{-}}{\mathscr{L}^{-+}\exp(\mathrm{i}\Delta_l)\, b^{+} + \mathscr{L}^{--}\exp(-\mathrm{i}\Delta_l)\, b^{-}}. \tag{5.59}$$

Here

$$\delta_l^{>} = \lim_{R \to \infty} \left(\int_{R^{>}}^{R} \mathscr{K}_l dR' - kR + \pi l/2 \right), \tag{5.60}$$

which is a partial contribution to the scattering phase shift from the region $R > R^{>}$ and thus contains no quasi-classical correction phase $\pi/4$ [see (5.19)]. Δ_l is defined as

$$\Delta_l = \int_{R_0}^{R^{<}} \mathscr{K}_l dR. \tag{5.61}$$

The matrix elements $\mathscr{L}^{\alpha'\alpha''}$ are given by (5.52) and also depend on l via ε which is now $\varepsilon_l = (E - U_l^{m})/\omega_l^{m}$. Substituting $\mathscr{L}^{\alpha'\alpha''}$ from (5.52) in (5.59) and using (5.49), we get

$$S_l = \exp(2\mathrm{i}\delta_l) = \exp[2\mathrm{i}\delta_l^{>} - \mathrm{i}\phi(\varepsilon_l) + \mathrm{i}\pi/2]$$
$$\times \frac{(1 + \mathscr{R}^{1/2}) + \mathrm{i}(1 - \mathscr{R}^{1/2})\tan[\Delta_l - \phi(\varepsilon_l)/2]}{(1 + \mathscr{R}^{1/2}) - \mathrm{i}(1 - \mathscr{R}^{1/2})\tan[\Delta_l - \phi(\varepsilon_l)/2]}. \tag{5.62}$$

It follows from (5.62) that δ_l can be written as

$$\delta_l = \delta_l^> + \delta_l^< + \pi/4, \quad \text{where} \tag{5.63}$$

$$\delta_l^< = \arctan\left[\frac{1-\mathscr{R}^{1/2}}{1+\mathscr{R}^{1/2}}\tan(\Delta_l - \phi/2)\right] - \phi/2, \tag{5.64}$$

$$\mathscr{R} = \mathscr{R}(\varepsilon_l), \qquad \phi = \phi(\varepsilon_l)$$

and the multi-valued arctangent function is evaluated for a branch corresponding to a continuous transition from the high energy limit down to the energy E. This continuous transition is, in fact, possible only due to the correction phase $\phi(\varepsilon_l)$.

In the high energy limit representing energies well above the barrier (e.g., $E = E'$ in Fig. 5.5), $\varepsilon_l \to \infty$ and $\mathscr{R} \to 0$, $\phi \to 0$, so that

$$\delta_l = \delta_l^> + \Delta_l + \pi/4, \tag{5.65}$$

i.e. δ_l is given by the JWKB scattering phase shift.

For energies well below the barrier (e.g., $E = E''$ in Fig. 5.5), $|\varepsilon_l| > 1$, $\mathscr{R} \to 1$ and $\phi \to 0$. Then

$$\delta_l = \delta_l^> + \pi/4 + \delta_l^{\text{res}}. \tag{5.66}$$

Here the resonance phase δ_l^{res} can be expressed via the real and imaginary parts of the complex energy $\mathscr{E}_{nl} = E_{nl} - \mathrm{i}\Gamma_{nl}/2$ for quasi-stationary states localized to the left of the barrier. Under the condition $\varepsilon_l < 0$, $|\varepsilon_l| > 1$ we also have $\Gamma_{nl} \ll |\Delta E_{nl}| = |E_{nl} - E_{n-1,l}|$ and δ_l^{res} assumes the form

$$\delta_l^{\text{res}} = \arctan[\Gamma_{nl}/2(E_{nl} - E)] + \pi N_l^*(n), \tag{5.67}$$

where $N_l^*(n)$ is the total number of stationary and quasi-stationary states below the level E_{nl} and the energy E lies close to E_{nl}. The arctan function assumes values within the range 0 to π.

In turn, complex eigenvalues $\mathscr{E}_{nl}$ are to be found from the quantization condition corresponding to the outgoing wave [5.38]. With $\Gamma_{nl} \ll \Delta E_{nl}$ this condition yields two equations, one for E_{nl} and the other for Γ_{nl}:

$$\begin{aligned}&\int_{R^{t'}}^{R^{t''}} (2\mu)^{1/2}[E_{nl} - U - (l+1/2)^2/2\mu R^2]^{1/2}\,dR = (n+1/2)\pi + \phi(\varepsilon_{nl}),\\ &\Gamma_{nl} = (\omega_{nl}/2\pi)\ln\mathscr{R}(\varepsilon_{nl}).\end{aligned} \tag{5.68}$$

Here $\omega_{nl} = \partial E_{nl}/\partial n$ is the classical frequency and $\varepsilon_{nl} = (E_{nl} - U_l^{\text{m}})/\omega_l^{\text{m}}$.

Note that for resonance scattering the phase shift δ_l does not vary smoothly either with E, or with l. The latter generally makes the stationary phase analysis of the scattering amplitude adopted in Sect. 5.1 invalid, and more sophisticated approaches have to be invoked [5.5, 39]. If, e.g. the scattering matrix (5.62) is represented as the sum of exponentials with smoothly varying exponents, the number of terms is infinite.

The quasi-classical phase shifts (5.63), though not quite correct at the limit $E \to 0$, are still consistent with the Levinson theorem, see [5.13, 17],

$$\lim_{E \to 0} \delta_l(E) = \pi N_l, \tag{5.69}$$

where N_l is the number of stationary states (i.e. states with $E_{nl} < 0$) supported by the potential $U_l = U + (l + 1/2)^2/2\,\mu R^2$. This follows from (5.66) with $\lim\limits_{E \to 0} [\delta_l^{>}(E) + \pi/4] = 0$ and $\delta_l^{\text{res}}(0) = \pi N_l$. The scattering phase shifts $\delta_l(E)$ also satisfy other conditions following from the general requirements

$$\lim_{l \to \infty} \delta_l(E) = 0, \tag{5.70a}$$

$$\lim_{E \to \infty} \delta_l(E) = 0. \tag{5.70b}$$

To illustrate the dependence of the scattering phase shifts on l and E (or k), contour maps of $\delta_l(k)$ are sometimes used, the contours corresponding to $\delta_l(k) = (n - 1/2)\,\pi$ (n is an integer). These values of the scattering phase shifts make the factor $1 - S$ in the sum (5.5) a maximum. The contours, called standing-wave Regge trajectories [5.40], extend to negative energies and define the positions of bound states for not necessarily integer values of l. Fig. 5.6 is an example of such a pattern for a simple potential with a minimum and a rigid core [5.40],

$$U(R) = \begin{cases} \infty & R \leqq R_0, \\ -A/R^4 & R > R_0. \end{cases} \tag{5.71}$$

For such a potential the phase shift depends on one parameter $c = (\mu A)^{1/3}/R^0$. The value $c = 27$ corresponds to the approximate interaction potential of $Li^+ + He$. For this system the potential well accomodates eight bound states. The dashed line in Fig. 5.6 defined by the equation $(l + 1/2)^2 = 2kc$ corresponds to the energy E which is equal to the top of the effective potential U^{m} (classical orbiting, see Sect. 5.3).

In this limited part of the plane k, l the phase shift is seen to decrease with increasing l, transition to the limit (5.70a), and tends to the limit (5.69) for $k \to 0$. The limit (5.70b) cannot, in general, be reached for a potential with a rigid core into which the wave can not penetrate. However, this drawback does not manifest itself in the given part of the plane k, l, so that the pattern gives an idea of the typical behaviour of phase shifts for any potential with long-range attraction and short-range repulsion.

Note also the "quasi-crossing" of the level lines (regions $k = 1$, $l = 5$; $k = 2$, $l = 8$, etc.) which corresponds to a sharp increase in the phase shift with a small change in k or l. This increase corresponds to the parameters k, l passing through a resonant state in the effective potential field $U_l(R)$. It is seen that in certain cases when the contour lines approach each other closely, the replacement of quasi-crossing by crossing lines would not change the general pattern far from the quasi-intersection region, but would result in changing the numbering of the curves over n to the left (or to the right). This would be due to a phase increment of π in going through resonance.

More complicated cases of breakdown of the quasi-classical approximation also occur, e.g. the coinciding of three (or more) turning points, as well as the reflection of waves in the region where the potential $U(R)$ tends to the asymptotic limit. These cases have been discussed repeatedly in connection with the quantization of levels in a double potential well [5.13, 22], passing through a double potential barrier (two double turning points) [5.41, 42] and a double barrier with reflection from the wall (two double and one isolated turning point) [5.32, 35, 43] and the threshold behaviour of the cross section due to reflection in the outgoing channel [5.44].

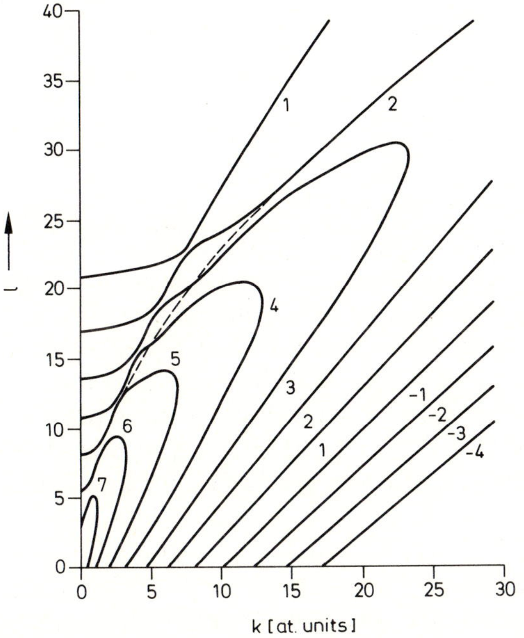

Fig. 5.6. Contour map (—) of the scattering phase shift in the k, l plane for the model potential (5.71); the curves are labeled by values of n. The dashed curve refers to the top of the centrifugal barrier (according to [5.40])

5.3 Classical Scattering

The Hamiltonian for the relative motion of two colliding structureless particle interacting via a central potential $U(R)$ is

$$H = \mathscr{P}_{\mathrm{R}}^2/2 + \frac{\mathscr{J}^2}{2\mu R^2} + U(R), \tag{5.72}$$

where $\mathscr{P}_{\mathrm{R}}$ is the radial momentum and $\mathscr{J}$ is the angular momentum. Since $\mathscr{J}$ is conserved, the trajectory $\boldsymbol{R} = \boldsymbol{R}(t)$ lies in the fixed plane normal to $\mathscr{J}$. Defining two coordinates – the radial distance R and the polar angle ψ – in this plane (Fig. 5.7), the equations of motions are readily solved yielding [5.45, 46]

$$\mp t = \int_{R^t}^{R} v^{-1}[1 - U(R)/E - b^2/R^2]^{-1/2}\, dR. \tag{5.73}$$

The lower integration limit corresponds to the outer turning point R^{t} of radial motion, which is the largest root of the radicand, and the signs minus or plus before t correspond to approaching or receding atoms. Here the total energy is replaced by the relative kinetic energy $E = \mu v^2/2$, and $\mathscr{J}$ is expressed by the relative velocity v and the impact parameter b, $\mathscr{J} = \mu v b$. The rotation angle is

$$\psi(t) = \int_{-\infty}^{t} \left[\frac{\mathscr{J}\, dt}{\mu R^2(t)}\right] dt. \tag{5.74}$$

The trajectory deflection angle η from the direction of the relative velocity $\boldsymbol{v}$ is found from (5.74):

$$\eta(b, E) = \pi - \psi(\infty) = \pi - 2\int_{R^t}^{\infty} [1 - U(R)/E - b^2/R^2]^{-1/2}\, b\,dR/R^2. \tag{5.75}$$

Sometimes it is more convenient to use the rotation angle ψ' referred not to the vector $\boldsymbol{v}$, but to the line equally dividing the angle between the two trajectory asymptotes. Thus $\psi' = \psi - \psi(\infty)/2$ (Fig. 5.7).

In discussing the quantitative dependence of η on b at a fixed energy, we consider various types of trajectories in the field of a typical interatomic potential describing long-range attraction and short-range repulsion. Schematically, the repulsion and attraction subregions are represented by the shaded and unshaded circles corresponding to the effective interaction region (Fig. 5.8).

Since for large R the real potential is not exactly zero, the value $\eta(b)$ is nonzero for large b. However, for $b > b_{\mathrm{M}}$ for instance, the deflection angle becomes so small that the description of scattering in terms of classical

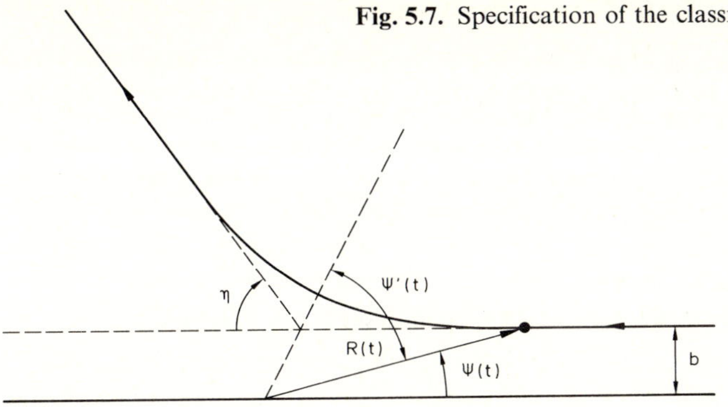

Fig. 5.7. Specification of the classical scattering trajectory

mechanics contradicts the quantum uncertainty principle. Thus for $b > b_M$ the scattering can be described in quantum terms only, and for this reason the quantum region of scattering angles is omitted in the treatment. In the classical scattering angle region, and for sufficiently large b, the atomic interaction is attractive and accordingly $\eta(b)$ is negative (trajectory curved towards the force center). When the interaction is weak, $U(R)/E \ll 1$, then $|\eta(b)| \ll 1$ (scattering through small angles corresponding to the applicability of the high-energy approximation). In this case (5.75) for η reduces to

$$\eta(b, E) = -(2E)^{-1} \int_{-\infty}^{\infty} (dU/dR)\, b\, dz/R, \qquad R = (b^2 + z^2)^{1/2}. \tag{5.76}$$

With a decrease in b, the angle η, though remaining negative, first increases in absolute magnitude, then decreases again and at $b = b_g$ passes through zero. The relevant trajectory, known as the glory trajectory, corresponds

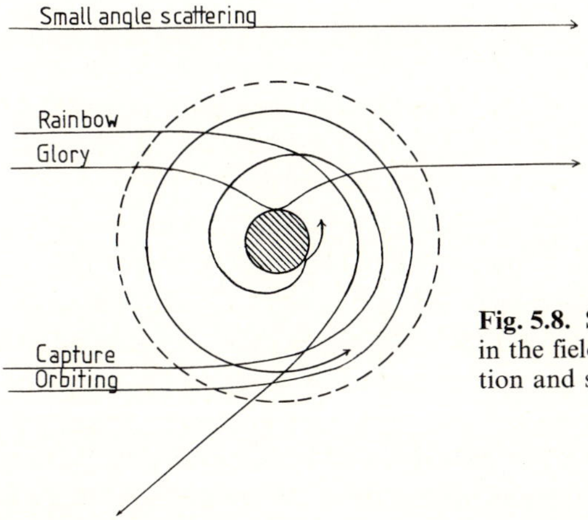

Fig. 5.8. Some characteristic classical trajectories in the field of a potential with long-range attraction and short-range repulsion

to a collision for which the trajectory curvature towards the force center under the action of the attractive forces is completely compensated by the opposite curvature caused by the repulsive forces. In this region ($b_g < b < b_M$) the deflection angle can either attain the extreme value $\eta_{min} = \eta_r$ for $b = b_r$, or become infinite for $b = b_0$, depending upon the relation between the energy E and the depth of the potential well D. The first case is referred to as the rainbow scattering trajectory, and the second as the orbiting trajectory.

The rainbow scattering parameter b_r at fixed E is defined by

$$\partial\eta(b, E)/\partial b|_{b=b_r} = 0. \tag{5.77}$$

The orbiting trajectory parameters are found from the condition of integral divergence in (5.75), i.e. from the condition that a second order root exist in the integrand for $R = R^t$:

$$\begin{aligned} &E - U(R^t) - Eb_0^2/(R^t)^2 = 0, \\ &\partial[U(R) - Eb_0^2/R^2]/\partial R|_{R=R^t} = 0. \end{aligned} \tag{5.78}$$

When the conditions are such that there is orbiting, the relative atomic motion for $b < b_0$ corresponds to a spiral approach until the action of the repulsive forces causes the particles to spiral away. The first stage of such a motion – the approach stage – is the capture of the particle by the center (this notion has a literal meaning when there is no core).

Since the classical trajectory depends on two parameters – the kinetic energy E and the angular momentum $\mathscr{J}$ – each can be specified by a point in the phase plane E, $\mathscr{J}$, the various regions of this plane corresponding to trajectories of different types. Such a classification of trajectories for a potential of well depth D and core radius R_0 is presented in Fig. 5.9. The cross-shaded part of the phase plane for negative energies, delimited by curve 1, corresponds to bound states of a diatom (stable diatomic molecule). The extension of this region (noncrossed shading) to positive energies refers to trajectories of two types: to the bound states inside the centrifugal barrier with turning points $R^{t'}$ and $R^{t''}$ and to the states outside the barrier, i.e. to $R > R^{t'''}$. Curve 2 defined by (5.78) refers to orbiting trajectories, the point with coordinates $E = E^*$, $\mathscr{J} = \mathscr{J}^*$ corresponding to the coinciding of the three roots $R^{t'}$, $R^{t''}$, and $R^{t'''}$. The values of E^* and $\mathscr{J}^*$ are close to the relevant parameters of the potential, $E^* \simeq D$, $\mathscr{J}^* \simeq (2\mu D)^{1/2} R_0$ (μ is the reduced mass). The extension of curve 2 into region $E > E^*$ is the rainbow scattering curve $\eta(E, \mathscr{J}) = \eta_r$, its equation being defined by the condition (5.77). Finally, the same figure presents the curve for glory trajectories $\eta(E, \mathscr{J}) = 0$ separating the positive and negative regions of the deflection function values. The deflection functions for potentials of such

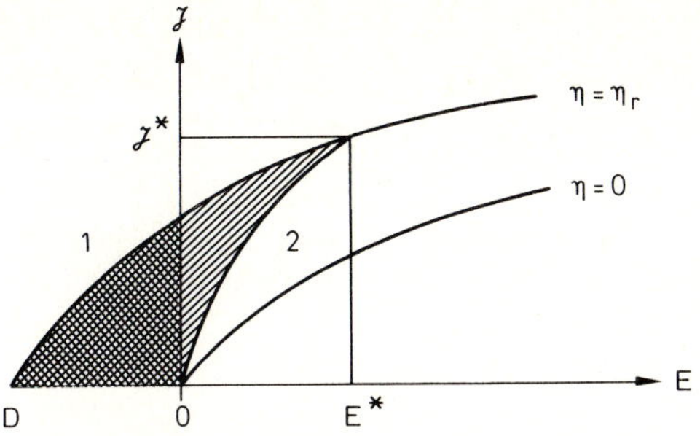

Fig. 5.9. Classification of classical trajectories in the E, $\mathcal{J}$ plane

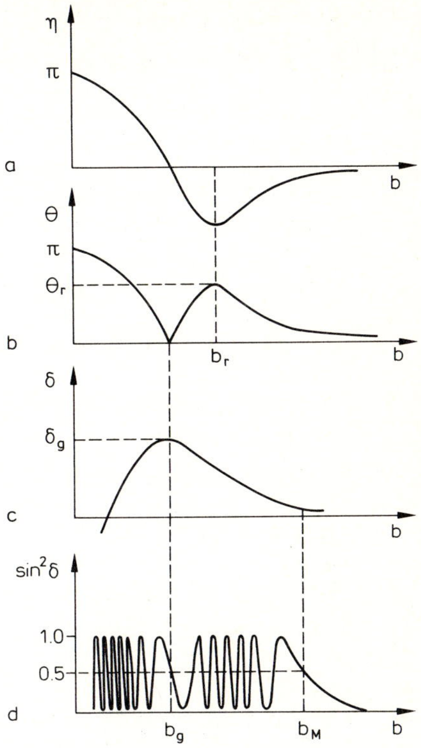

Fig. 5.10. Relation between: (**a**) deflection function η, (**b**) scattering angle θ, (**c**) JWKB scattering phase shift δ and (**d**) $\sin^2 \delta$ ▶

a type, namely for the Lennard-Jones potential, are presented in Fig. 5.10a for the case of rainbow scattering.

Unlike the deflection angle η which acquires different signs depending on b, the observed scattering angle θ, varying by definition over the range $0 \leqq \theta \leqq \pi$, is $\theta = \min |\eta - 2\pi n|$, $n = 0, \pm 1, \pm 2$. The dependence $\theta = \theta(b)$ is shown in Fig. 5.10b. The inverse dependence $b = b(\theta)$ of the impact parameter on the scattering angle is, generally speaking, multi-valued. For instance, three different trajectories with impact parameters b^χ ($\chi = 1, 2, 3$) can correspond to one angle θ for a typical potential with a well. Consistent with this, the classical differential cross section ${}^{\mathrm{cl}}q(\theta)$ equals the sum of contributions from all three branches to scattering through the same angle θ:

$$ {}^{\mathrm{cl}}q(\theta) = \sum_{\chi} {}^{\mathrm{cl}}q^{\chi}(\theta). \tag{5.79} $$

Each contribution is calculated as the ratio of the number of particles scattered through the angles θ, φ into unit solid angle to the incident flux:

$$ {}^{\mathrm{cl}}q(\theta) = |b\,d b\,d\varphi/\sin\theta\,d\theta\,d\varphi|_{b=b^{\chi}(\theta)} = (\sin\theta)^{-1} b^{\chi}(\theta)\,|\partial b^{\chi}/\partial\theta|. \tag{5.80} $$

It is seen that ${}^{\mathrm{cl}}q(\theta)$ defined by (5.80) coincides with ${}^{\mathrm{cl}}q(\theta)$ defined by (5.10) which explains the notation given earlier.

If the contribution from a certain branch dominates, one term only can be retained in (5.79). In particular, for scattering through small angles, $\theta \ll 1$, the main contribution to ${}^{\mathrm{cl}}q(\theta)$ is given by the branch $b^{(1)}(\theta)$ whereas for scattering through angles close to π the branch $b^{(3)}$ often dominates. As an example, consider the potential

$$U(R) = CR^{-s}, \qquad C = 4DR_0^s. \tag{5.81}$$

For repulsive scattering ($C > 0$) the deflection function η close to π is

$$\eta = \pi - g(s)\,(E/4D)^{1/s}(b/R_0), \qquad g(s) = \frac{2\pi^{1/2}\Gamma(1/s)}{s\Gamma(1/s + 1/2)}, \tag{5.82}$$

which gives an angle-independent differential cross section

$${}^{\mathrm{cl}}q(\theta) = R_0^2(4D/E)^{2/s}[g(s)]^{-2}, \qquad \pi - \theta \ll 1. \tag{5.83}$$

For scattering through small angles the deflection function is

$$\eta = -f(s)\,(4D/E)\,(R_0/b)^s, \qquad f(s) = \frac{(s-1)\,\pi^{1/2}\Gamma(s/2 - 1/2)}{2\Gamma(s/2)}. \tag{5.84}$$

With this equation the differential cross section for small angles becomes

$${}^{\mathrm{cl}}q(\theta) = R_0^2(1 - 1/s)\,[f(s)\,4D/E]^{2/s}\,\theta^{-2-2/s}. \tag{5.85}$$

For intermediate scattering angles the series representation of the deflection function is also known [5.47, 48].

Consider now small angle scattering in some detail. The deflection function (5.76) calculated in the high-energy approximation for fixed b is inversely proportional to E. This suggests introducing a reduced deflection function $\eta(b, E)\,E$ and a reduced scattering angle

$$\tau = E\theta. \tag{5.86}$$

In discussing the scattering through small angles it is also convenient to use a reduced differential cross section

$$\bar{q}^{\chi}(\theta, E) = \theta \sin\theta\ {}^{\mathrm{cl}}q^{\chi}(\theta, E). \tag{5.87}$$

According to (5.76) the reduced scattering angle τ is related to the potential U:

$$\tau(b) = w^{\chi} b \int_b^{\infty} \frac{(dU/dR)\,dR}{(R^2 - b^2)^{1/2}}, \tag{5.88}$$

where w^{χ} ($w^{\chi} = \pm 1$) ensures that the rhs of (5.88) is positive.

Equation (5.88) defines different branches

$$b^\chi = b^\chi(\theta). \tag{5.89}$$

When these functions are used to calculate the reduced cross section, the result will be

$$\bar{q}^\chi(\theta, E) = \varrho^\chi(\tau) + O[U(b)/E], \tag{5.90}$$

$$\varrho^\chi(\tau) = (1/2)\, d\,(b^\chi)^2/d \log \tau, \tag{5.91}$$

i.e. at high energies, $U(b) \ll E$, the reduced cross section $\bar{q}^\chi$ depends only on τ and is equal to $\varrho^\chi(\tau)$. This expresses an important scaling principle, see [5.9]. For phases A^χ which define the quasi-classical scattering amplitudes $f^\chi(\theta)$ according to (5.9), the scaling principle assumes the form

$$A^\chi(\theta, E) = (1/v)\, a^\chi(\tau), \tag{5.92}$$

$$a^\chi(\tau) = -2 \int_b^\infty \frac{U(R)\, dR}{(1 - b^2/R^2)^{1/2}} + w^\chi b \tau, \qquad b = b^\chi(\tau). \tag{5.93}$$

Now, when the classical differential cross section is used to calculate the integral cross section, the latter will be infinite. For potentials decaying faster than R^{-2}, this divergence is intrinsically connected with a fundamental breakdown of the classical description of scattering for high impact parameters. This and related topics will be discussed in the next section.

5.4 Integral Cross Sections

The integral cross section for elastic scattering is obtained by integrating over all angles the differential cross section $q(\theta)$. Taking into account (5.5), this gives

$$\sigma = \int |f(\theta)|^2 \sin\theta\, d\theta\, d\varphi = 4\pi k^{-2} \sum_{l=0}^{\infty} (2l+1) \sin^2(\delta_l). \tag{5.94}$$

In the quasi-classical limit, where the summation over l can be replaced by integration, (5.94) becomes

$$\sigma = 2\pi \int_0^\infty |1 - \exp[2i\delta(b)]|^2\, b\, db = 8\pi \int_0^\infty \sin^2[\delta(b)]\, db. \tag{5.95}$$

The value of the integral scattering cross section given by (5.95) is known to depend ultimately on the interference of the wave scattered through a small (diffraction) angle with a nonscattered wave [the amplitudes of these waves in the integrand of (5.95) are $\exp(2i\delta)$ and 1]. If the potential drop is sufficiently sharp, a crude estimate can be obtained either in the random-phase (r. ph.) approximation, replacing the true potential field by an equivalent rigid sphere of radius b_M, or in the Massey-Mohr (MM) approximation, accounting for the contribution of small phase shifts for $b > b_M$, or else in the Landau-Lifshitz-Schiff (LLS) approximation, accurately taking into account the diffuse edge of the scattering boundary by substituting into (5.95) the scattering phase shift in the high-energy approximation.

All these approximations are readily understood by studying the b-dependence of $\delta(b)$ and $\sin^2[\delta(b)]$ for a typical potential with a well (Fig. 5.10c, d).

The random phase approximation corresponds to a replacement of the factor $\sin^2[\delta(b)]$ by the mean value 1/2 for $b < b_M$, completely neglecting the contribution from collisions with $b > b_M$, i.e.

$$\sigma^{\text{r.ph.}} = 2\pi b_M^2, \tag{5.96}$$

b_M usually being obtained from the condition $\sin^2\delta(b_M) = 1/2$.

In the Massey-Mohr approximation, $\sin^2\delta$ is replaced by 1/2 for $b < b_M$, and by $\delta^2(b)$ for $b > b_M$:

$$\sigma^{\text{MM}} = 2\pi b_M^2 + 8\pi \int_{b_M}^{\infty} \delta^2(b)\, b\, db. \tag{5.97}$$

Finally, the Landau-Lifshitz-Schiff approximation uses (5.95) in which $\delta(b)$ is calculated from (5.20). For the potential $U = -C/R^s$, (5.20) gives

$$\delta(b) = \frac{\pi^{1/2}\Gamma(s/2 - 1/2)\, C b^{1-s}}{4v\Gamma(s/2)}. \tag{5.98}$$

For the van der Waals attraction ($s = 6$), (5.95 and 97) give

$$\begin{aligned} \sigma^{\text{MM}} &= 7.547(C/v)^{2/5}, \\ \sigma^{\text{LLS}} &= 8.083(C/v)^{2/5}. \end{aligned} \tag{5.99}$$

The approximation (5.97) is convenient for estimating the contribution from small and large phase shifts to the integral cross section. As the potential slope increases, the main contribution to σ^{MM} in (5.97) shifts to the first term. For the interaction $\sim R^{-3}$ both terms are of the same order; for the van der Waals interaction the ratio of the second to the first is 1/8. For the scattering of rigid spheres with a closest approach distance R_0,

$$\sigma^{\text{r.ph.}} = 2\pi R_0^2. \tag{5.100}$$

This value is twice that of the classical collision cross section for rigid spheres ${}^{\text{cl}}\sigma = \pi R_0^2$. The explanation of this is that half of the quantum scattering cross section is due to isotropic scattering, which is similar to classical scattering, and that the other half is due to scattering through a very small angle (shadow scattering), which is similar to diffraction by an opaque screen.

Taking into account the next-order correction in $1/R_0k$, (5.100) will be replaced by

$$\sigma = 2\pi R_0^2[1 + c(kR_0)^{-2/3}]. \tag{5.101}$$

The exact quantum value of the coefficient c is 0.996 [5.49] and calculation with a quasi-classical phase $\delta(b)$ using (5.94) yields -0.590 [5.50]. The great difference is ascribed to a breakdown of the linear approximation in the turning point region $R \approx R_0$ (Sect. 5.2.3). If the additional phase $\pi/4$ in (5.19) is omitted, then c is 1.024 [5.51] which is already close enough to the exact value.

The above approximation is suitable for a description of the interference phenomena in the region where the scattering phase shifts are of the order of unity and lower. To this approximation, the cross section decreases monotonically with increasing velocity and contains information solely about that part of the potential $U(R)$ which corresponds to distances $R \sim b_M$, where the scattering phase shift monotonically depends on b. However, if there is a region where $\delta(b)$ passes through the extremum, the replacement of $\sin^2\delta(b)$ by its average value for $b < b_M$ is not quite correct. This is the case for a potential with a well for which glory scattering occurs.

In this case, in evaluating the integral (5.95), one has to take into account the contribution of the function $\exp[2i\delta(b)]$ from the point of stationary phase, where the phase shift $\delta(b)$ passes through a maximum in the internal region $b < b_M$ (Fig. 5.10c). As a result of this we obtain, using the quadratic approximation for $\delta(b)$ close to the maximum, the following expression for the cross section, taking into account the glory effect:

$$\begin{aligned} \sigma &= \sigma^{\text{LLS}} + \Delta\sigma, \\ \Delta\sigma &= -4\pi^{-3/2} b_g |d^2\delta/db^2|_{b=b_g}^{-1/2} \cos(2\delta_g - \pi/4), \end{aligned} \tag{5.102}$$

where $\delta_g = \delta(b_g)$ is the maximum scattering phase shift, and $\Delta\sigma$ is assumed to be small compared to σ^{LLS}. It is seen from (5.102) that the correction $\Delta\sigma$ oscillates with energy.

The analysis and interpretation of the oscillations become particularly simple for high E exceeding the depth of the potential well D. In this case $\delta_g(E)$ can be represented in the form of the expansion [5.52],

$$\delta_g(E) = kR_m[a_1(D/E) - A_1(D/E)^2 + A_2(D/E)^3 + \ldots], \tag{5.103}$$

where R_m is the interatomic distance corresponding to the potential minimum and a_1, A_1, A_2 are coefficients depending on the specific form of the potential (for the van der Waals potential $a_1 = 0.4226$, $A_1 = 0.1664$, $A_2 = 0.0842$). In the high-energy approximation the first term only, inversely proportional to the velocity, will be retained in the sum (5.103).

Representing $\delta_g(E)$ and $\Delta\sigma(E)$ as

$$\begin{aligned} \delta_g(E) &= \pi N(E) - 3\pi/8, \\ \Delta\sigma(E) &= C\cos[2\pi N(E)], \quad C > 0, \end{aligned} \tag{5.104}$$

we get the maxima of $\Delta\sigma$ for integer values of N ($N = 1, 2, \dots$) starting from the maximum at the highest velocity, and the minima of $\Delta\sigma$ for half-integer values of N ($N = 3/2, 5/2, \dots$). It is expected from (5.103, 104) that at a high energy the dependence of N on $1/v$ will be linear with an intercept of $-3/8$. Such a dependence has been observed repeatedly in experiment.

It is noted in conclusion that experimental measurement of the integral scattering cross section permits one to evaluate the parameters for the long-range part of the potential (from the absolute value of σ and the dependence of σ on E), the potential well "volume" DR_m (from the oscillation frequencies of $\Delta\sigma$), and the lower limit (or even the accurate value) of the potential well depth (from the number of maxima and minima of σ as a function of E).

For a detailed review of the glory effect in atomic collisions see [5.1, 10, 11, 53].

As the energy decreases, the wavelength of the glory oscillations increases, and at energies at which orbiting becomes possible, resonances due to tunneling or to the singularities of overbarrier reflections might appear in the integral cross section.

It follows from the available estimates of the widths Γ of quasistationary levels of diatoms (from tenths to the order of one and, very infrequently, to tens of cm^{-1}) and from the energy "dispersion" ΔE of beams (of more than 100 cm^{-1}) that experimental observation of "orbiting resonances" in cross sections is very difficult. Indeed, representing the phase shift δ_l in the form, see (5.66),

$$\delta_l(E) = \delta_l^{\mathrm{nonres}} + \arctan[\Gamma_{nl}/2(E_{nl} - E)] \tag{5.105}$$

we see that close to resonance, the partial contribution $\sin^2\delta_l$ from a wave with momentum l is

$$\sin^2\delta_l = \sin^2\delta_l^{\mathrm{nonres}} + \frac{\Gamma_{nl}^2\cos 2\delta_l^{\mathrm{nonres}} + 2\Gamma_{nl}(E - E_{nl})\sin 2\delta_l^{\mathrm{nonres}}}{4(E - E_{nl})^2 + \Gamma_{nl}^2}, \tag{5.106}$$

where δ^{nonres} depends only weakly on energy. Averaging over a narrow energy range such that δ_l^{nonres} can be taken to be unchanging, and Γ_{nl} to be much smaller than ΔE, we get

$$\langle \sin^2 \delta_l \rangle_{\Delta E} = \sin^2 \delta_l^{\text{nonres}} + (\pi \Gamma_{nl}/2\,\Delta E) \cos 2\,\delta_l^{\text{nonres}} . \tag{5.107}$$

Thus, the resonance contribution $\Delta \sigma^{\text{res}}$ to the averaged integral cross section is

$$\langle \Delta \sigma^{\text{res}}(E) \rangle_{\Delta E} = 4\pi k^{-2} \sum_{n,l} (2l+1)(\pi \Gamma_n/2\,\Delta E) \cos 2\,\delta_l^{\text{nonres}} , \tag{5.108}$$

where the summation over n and l is restricted to the quasi-stationary levels within the range ΔE. The difficulty of experimentally detecting such resonances is also caused by the fact that their unambiguous identification is possible only with observation of the characteristic contour against the slowly changing background of the cross section. Up to now, orbiting resonances have been detected for several systems including light atoms. For instance, in the case H – Hg studied over a very wide range of energies [5.54, 55], the resonance states are $n = 4$, $l = 9$, and $n = 3$, $l = 10$ at about the same energy.

5.5 Differential Cross Sections

5.5.1 Scattering Through Classical Angles – Repulsive Potential

Consider now scattering through classical angles using the stationary phase approximation for the scattering amplitude. For a monotonic repulsive potential the deflection function $\eta(b)$ is a monotonic function of b which changes from 0 to π. Then (5.12), rewritten now as

$$\eta(b^{(1)}) = \theta , \tag{5.109}$$

obviously has one root $b^{(1)}$. The corresponding scattering amplitude is given by (5.9). The differential cross section

$$q(\theta) = {}^{\text{cl}}q(\theta) = (\sin\theta)^{-1} b^{(1)}(\theta)\, db^{(1)}(\theta)/d\theta \tag{5.110}$$

monotonically decreases with increasing θ.

However, (5.109) can also have, besides the real root $b = b^{(1)}$, complex roots [5.56]. These roots correspond to trajectories with complex b values deflecting at the same real angle η. Denote these roots by $b^{(1,m)}$ ($m = 0, 1, \dots$) retaining the index 1 as an indication of the existence of one real root, and assuming that $b^{(1,0)} = b^{(1)}$. To find $b^{(1,m)}$ we must consider the general equation (5.12), thus assuming that both exponentials in the square brackets in (5.7) may contribute to the amplitude. Of course, the phases $A^{(1,m)}$ will now be complex, but the contribution will only be from $b^{(1,m)}$, corresponding to negative real values of the exponents.

The scattering for the simple potential (5.81) provides an instructive example of the above [5.57]. For this potential

$$\begin{aligned} b^{(1,m)} &= b^{(1)} \exp[\mathrm{i}(-1)^m \pi m/s], \\ A^{(1,m)} &= A^{(1)} \exp(\mathrm{i}\alpha_m), \end{aligned} \tag{5.111}$$

where α_m denotes certain phases. The stationary phase points $b^{(1,m)}$ ($m = 0, 1, \ldots$) of the first and second exponentials in (5.7) contribute, alternately decreasing, to the scattering amplitude, represented now as

$$f(\theta) = \sum_{m=0} f^{(1,m)}(\theta). \tag{5.112}$$

Thus, the first correction ($m = 1$) to the classical cross section (5.10) stems from the second exponential in (5.7).

The representation of the amplitude in the form of (5.112) implies that the monotonic behavior of the cross section is modulated:

$$q(\theta) = (\sin\theta)^{-1} \left| \sum_m |\partial\eta/b\,\partial b|^{-1/2}_{b=b^{(1,m)}} \exp(\mathrm{i}A^{(1,m)}) \right|^2. \tag{5.113}$$

The modulation frequency depends on the difference $\mathrm{Re}\{A^{(1,m)} - A^{(1,m')}\}$ and its amplitude on $\mathrm{Im}\{A^{(1,m)}\}$. As follows from model calculations, (5.113), representing the scattering amplitude as the sum of contributions from different complex trajectories, reproduces quite satisfactorily quantum results under quasi-classical conditions [5.56, 57]. More than that, this description appears to be valid even in cases when the quasi-classical approximation cannot be justified [5.58].

However, for the general case, accounting for complex stationary phase angular momenta using the JWKB scattering phase shift only is not quite consistent: higher order corrections to δ_l given by (5.18) must also be considered on this level.

5.5.2 Scattering Through Classical Angles – Potential with a Well

For a nonmonotonic potential the function $b = b(\theta)$ has several branches, so that the classical cross section is given by (5.79, 80). An example of constructing ${}^{\mathrm{cl}}q(\theta)$ from the three-valued function $b^{\varkappa}(\theta)$ is shown in Fig. 5.11a, b, which illustrates various contributions to ${}^{\mathrm{cl}}q(\theta)$. When using

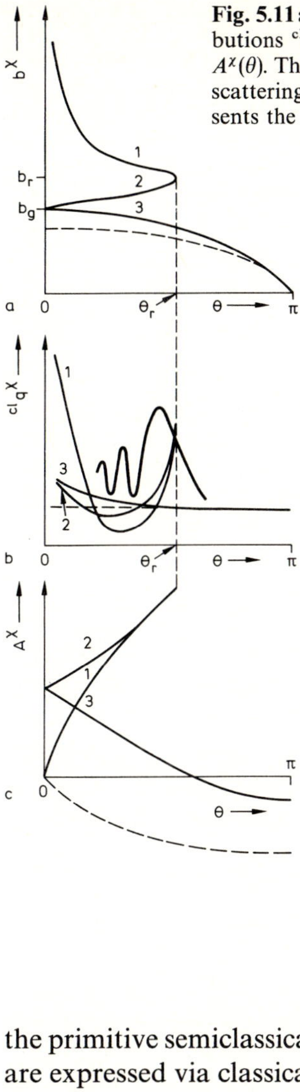

Fig. 5.11 a–c. Relation between: (**a**) branches $b^{\chi}(\theta)$, (**b**) partial contributions ${}^{\mathrm{cl}}q^{\chi}(\theta)$ to the classical differential cross section, (**c**) the phases $A^{\chi}(\theta)$. The curves are labeled by $\chi = 1, 2, 3$. The dashed curves refer to scattering of rigid spheres, and the heavy undulating curve in (**b**) represents the Airy approximation for rainbow scattering

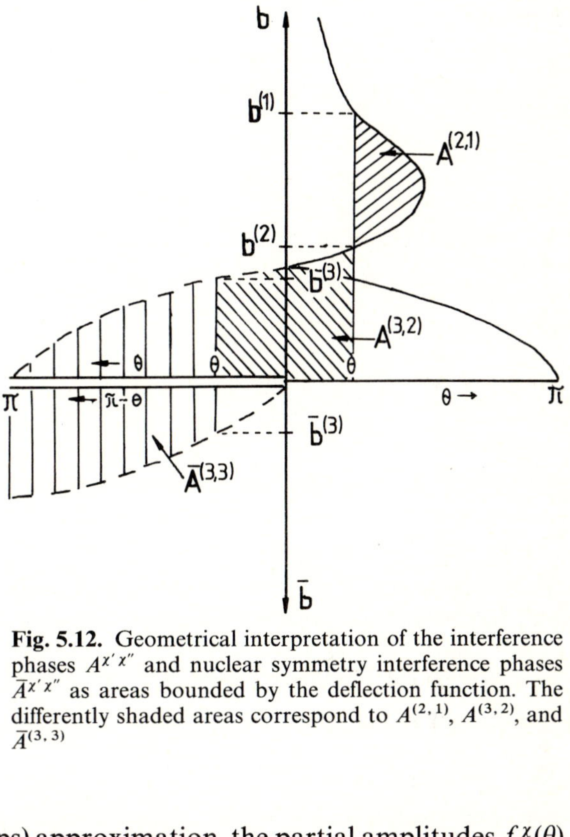

Fig. 5.12. Geometrical interpretation of the interference phases $A^{\chi'\chi''}$ and nuclear symmetry interference phases $\bar{A}^{\chi'\chi''}$ as areas bounded by the deflection function. The differently shaded areas correspond to $A^{(2,1)}$, $A^{(3,2)}$, and $\bar{A}^{(3,3)}$

the primitive semiclassical (ps) approximation, the partial amplitudes $f^{\chi}(\theta)$ are expressed via classical cross sections ${}^{\mathrm{cl}}q$ and appropriate phases A^{χ}. In this way

$$
\begin{aligned}
{}^{\mathrm{ps}}q &= \left|\sum_{\chi} ({}^{\mathrm{cl}}q^{\chi})^{1/2} \exp(\mathrm{i}A^{\chi} + \mathrm{i}\alpha^{\chi})\right|^2 \\
&= {}^{\mathrm{cl}}q + 2 \sum_{\chi',\chi''} ({}^{\mathrm{cl}}q^{\chi'}\,{}^{\mathrm{cl}}q^{\chi''})^{1/2} \cos(A^{\chi'\chi''} + \alpha^{\chi'\chi''}),
\end{aligned}
\tag{5.114}
$$

$$
A^{\chi'\chi''} = A^{\chi'} - A^{\chi''}, \qquad \alpha^{\chi'\chi''} = \alpha^{\chi'} - \alpha^{\chi''}.
$$

The sum over χ', χ'' is the interference contribution from the waves scattered through the same angle by different regions of the potential, and this description is valid, provided $A^{\varkappa'\varkappa''} \gg 1$. The geometrical interpretation of the interference phase is given in Fig. 5.12.

The θ-dependence of $A^{\varkappa}$ (Fig. 5.11c) suggests that for classical scattering angles there are two regions where the primitive semiclassical approximation fails: that of glory scattering and that of rainbow scattering. We shall discuss the latter neglecting the interference contribution from branch 3.

If the Airy uniform approximation is used for the amplitude $f^{(12)}(\theta)$, the corresponding cross section will be

$$\begin{aligned} q^{(12)}(\theta) &= |({}^{\mathrm{cl}}q^{(1)})^{1/2} + ({}^{\mathrm{cl}}q^{(2)})^{1/2}|^2 \pi z^{1/2} \mathrm{Ai}^2(-z) \\ &\quad + |({}^{\mathrm{cl}}q^{(1)})^{1/2} - ({}^{\mathrm{cl}}q^{(2)})^{1/2}|^2 \pi z^{-1/2} [d\,\mathrm{Ai}^2(-z)/dz]^2, \\ z &= [3(A^{(1)} - A^{(2)})/4]^{2/3}. \end{aligned} \tag{5.115}$$

If the second term in (5.115) is neglected, which is often permissible, $q^{(12)}(\theta)$ becomes particularly simple:

$$\begin{aligned} q^{(12)}(\theta) &= (2b_{\mathrm{r}}/k \sin\theta)\,|r|^{-2/3} \mathrm{Ai}^2[|r|^{-1/3}(\theta - \theta_{\mathrm{r}})], \\ r &= 2d^2\theta/k^2 db^2|_{b=b_{\mathrm{r}}} \end{aligned} \tag{5.116}$$

giving the main maximum and other supernumerary extrema at the bright side of the rainbow. The total number of these at a given energy E correlates with the function $N(E)$ describing the glory undulations by (5.104). At the dark side, $q^{(12)}(\theta)$ exponentially decays. The primitive semiclassical approximation is obtained using the asymptotes of the Airy function (5.17),

$${}^{\mathrm{ps}}q^{(12)}(\theta) = {}^{\mathrm{cl}}q(\theta)\, 2\sin^2[2\,|r|^{-1/2}(\theta - \theta_{\mathrm{r}})^{3/2}/3 + \pi/4], \tag{5.117}$$

where the classical cross section near the rainbow angle is

$${}^{\mathrm{cl}}q(\theta) = \begin{cases} \dfrac{b_{\mathrm{r}}(d^2\theta/db^2)^{-1/2}|_{b=b_{\mathrm{r}}}}{(\sin\theta)\, 2^{1/2}(\theta_{\mathrm{r}} - \theta)^{1/2}}, & \theta < \theta_{\mathrm{r}} \\ 0, & \theta > \theta_{\mathrm{r}}. \end{cases} \tag{5.118}$$

An experimental study of scattering in the rainbow region provides information on the potential well depth D which determines θ_{r} (just as the rainbow angle is governed solely by the droplet refractive index), as well as on the interatomic distance in the potential minimum, which together with D is responsible for the supernumerary extrema. Now if the interference of amplitudes $f^{(12)}$ and $f^{(3)}$ is taken into account, there will appear rapid, or secondary oscillations superimposed on the rainbow oscillations. At the dark side of the rainbow these oscillations exponentially disappear.

The structure of the differential cross section becomes even more complicated for the collision of identical atoms (Sect. 2.3.2). For instance in the case of one dominating branch $b^{(3)}(\theta)$ the nuclear symmetry interference contribution ${}^{ps}\Delta q$, calculated in the primitive semiclassical approximation for two identical atoms in the ${}^1S_0$ state with zero nuclear spin, will be

$$\begin{aligned} {}^{ps}\Delta q &= 2[{}^{cl}q^{(3)}(\theta)\, {}^{cl}q^{(3)}(\pi-\theta)]^{1/2} \cos[\bar{A}^{(3,3)}(\theta)], \\ \bar{A}^{(3,3)} &= A^{(3)}(\theta) - A^{(3)}(\pi-\theta). \end{aligned} \tag{5.119}$$

For interpretation of $\bar{A}^{(3,3)}$ see Fig. 5.12.

Thus, in the most general case the differential cross sections of elastic scattering exhibit three types of oscillations, the angular distance $\Delta\theta$ between the extrema being expressed for each type of interference via the impact parameters of appropriate branches using (5.13). These oscillation types in order of decreasing $\Delta\theta$ are:

i) Rainbow undulations: two trajectories pass the scattering center on the same side, $\Delta\theta = (2\pi/k)\,[b^{(1)}(\theta) - b^{(2)}(\theta)]^{-1}$.
ii) Nuclear symmetry undulations: one trajectory corresponds to a projectile, the other to a target, $\Delta\theta = 2\pi/k[b^{\chi}(\theta) + b^{\chi}(\pi-\theta)]$.
iii) Long range – short range undulations: two trajectories pass the scattering center on different sides, $\Delta\theta = 2\pi/k[b^{\chi}(\theta) + b^{(3)}(\theta)]$, $\chi = 1, 2$.

If orbiting occurs instead of rainbow scattering, the undulation pattern becomes even more complicated [5.59].

5.5.3 Scattering Through Small Angles

All features characteristic of differential scattering through classical angles ($\theta \gg \theta^*$) may also appear for scattering through small angles ($\theta \ll 1$). If $\theta^* \ll \theta \ll 1$, the calculation of $q(\theta)$ is similar to that in Sect. 5.5.2 with additional simplification due to the high-energy approximation for the JWKB scattering phase shifts, see (5.20).

Moreover, (5.6) provides for uniform treatment valid for all θ, including the diffraction angles $\theta < \theta^*$. This has been thoroughly studied, see [5.60]. Here we shall consider briefly only the question of extrapolating the small-angle classical cross section ${}^{cl}q(\theta)$, (5.83), for the inverse power potential (5.81) to the region of diffraction angles, $\theta \leqq \theta^* \approx 1/l^*$. Taking the differential cross section to be

$$q(\theta) = F(\theta)\, {}^{cl}q(\theta), \tag{5.120}$$

the function $F(\theta)$ will tend to the asymptotic value $F \to 1$ with increasing angle, provided the contribution from complex trajectories to the scattering amplitude for classical angle is neglected. For a potential in the form (5.81), usually invoked for the interpretation of scattering through small angles, F depends on two parameters s and $\theta^* = (\pi/k^2\sigma)^{1/2}$, i.e. $F = F(\theta/\theta^*, s)$. For the cases of most interest ($s = 4, 5, 6$ and 8) the correction factor F has been obtained numerically [5.61]. The plot for $s = 6$ is presented in Fig. 5.13.

When $\theta/\theta^* < 1$, the product of the function F, which tends to zero as $(\theta/\theta^*)^{2(s+1)/s}$, and of the increasing function ${}^{\mathrm{cl}}q(\theta)$, see (5.85), yields the correct limit – the differential cross section maximum. An alternation of maxima and minima would be expected for increasing θ by analogy with diffraction from a screen. Such a nonmonotonicity of $q(\theta)$ actually follows to a certain extent from Fig. 5.13. However, it is substantially obscured due to the diffusive boundary of the potential field. Consequently it can be conceived to a certain approximation that the first peak of the diffraction maximum smoothly converts to a monotonic function ${}^{\mathrm{cl}}q(\theta)$ defined by (5.85).

This qualitative picture suggests that simple interpolating expressions may describe classical as well as quantum regions of scattering. For instance, see [5.10], the forward diffraction peak (fdp) can be approximated by a simple Gaussian function

$$ {}^{\mathrm{fdp}}q(\theta) = A \exp(-a\theta^2). \tag{5.121} $$

The two constants A and a in the equation can be found from the condition that the Gaussian must smoothly match onto the classical expression (5.83) at $\theta = \theta^*$. For the scattering of rigid spheres, this matching prescription will yield the expected result

$$ 2\pi \int_0^\infty {}^{\mathrm{fdp}}q(\theta)\,\theta\, d\theta = \sigma/2 = \pi R_0^2. \tag{5.122} $$

Note that this approach accounts only for scattering with large impact parameters, i.e. glory scattering is neglected here. This deficiency can be

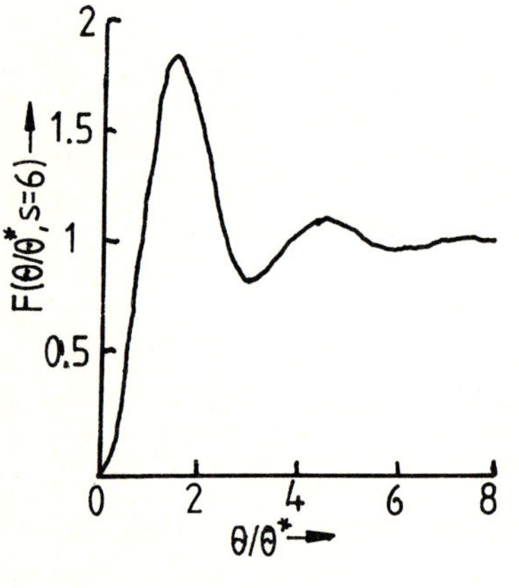

Fig. 5.13. Quantum correction factor $F(\theta/\theta^*, s)$ in (5.120) for $s = 6$ (according to [5.61])

rectified, when needed, as the expression for the forward glory amplitude is known [5.62]. However, the correction is expected to be small, since the glory contribution (5.102) to the integral cross section is small.

Besides forward glory, there can also occur backward glory. The backward scattering corresponding to the head-on collisions does not show any particular features and the differential cross section for this case is given by the classical expression (5.83). The backward scattering for essentially nonzero impact parameters leads to a specific interference pattern [5.62] which becomes quite complicated when the scattering is also of the rainbow type (i.e. the rainbow angle is close to π) [5.63]. In this connection, note the so-called giant glory [5.64]: for classical scattering in a truncated Coulomb potential ($U_G = -1/R$ for $R < R_0$ and $U_G = 0$ for $R > R_0$) at an energy $E = 1/2R_0$ all trajectories with $b < R_0$ are deflected backward. The quasi-classical scattering in a potential $U(R)$ slightly different from U_G and simulating the adiabatic interaction Na–I has been studied in [5.65].

6. Approximate Calculation of a Multichannel Quasi-classical Scattering Matrix

An exact calculation of the scattering matrix calls for the numerical solution of a system of second-order scattering equations formulated using a basis of electronic functions. However, the basis functions and the corresponding interactions are usually known only approximately. The only exception is the interaction at large distances, where the general perturbation approach can be used (Chap. 4).

In such a situation the numerical solution of exact quantum scattering equations of the type (2.80 or 145) is not the best way of handling the whole problem of calculating cross sections for specific processes. One must use additional simplifications (besides the general quasi-classical arguments already discussed in Chap. 2) facilitating the finding of solutions to scattering equations for the given interatomic interaction, and making as clear as possible the correlation between specific properties of the output and the input of the scattering equations: of the S matrix and the interaction. One way would be to reduce the second-order scattering equations to first order, using the notion of the classical atomic trajectory (Sect. 6.1), or to match adiabatic quasi-classical solutions in different regions of R-variation making use of comparison equations (Sect. 6.2). Another approach is to rely on various perturbation treatments, discussed in Sect. 6.3.

To simplify the notations in this and the following chapters the convention adopted in Chap. 2 will be generalized, namely, any basis state used will be labeled by subscripts a, a', a'', etc., each encompassing all the quantum numbers needed. If some quantum numbers are to be singled out, they will appear with a, e.g. a, l; a, v; a, $\tilde{\Omega}$; in this description a will no longer include l, v, $\tilde{\Omega}$. Also, subscripts (or superscripts) denoting the boundary or the initial conditions are usually suppressed in radial wave functions or semiclassical transition amplitudes.

6.1 Common-Trajectory Approach

Under quasi-classical conditions the most intuitively attractive approximation assumes that the nuclei are moving along a certain classical trajectory $\boldsymbol{R}(t)$, thus generating the external time-dependent perturbation which acts on the electrons and induces transitions between their quantum states. This approximation will be called the common-trajectory or the semiclassical approximation. Several other names are also used: the impact-parameter method (for the specific case when $\boldsymbol{R}(t)$ is a rectilinear trajectory), the classical-path approximation, the classical-trajectory approximation.

The semiclassical approximation was first used by *Gaunt* as early as 1927 [6.1]. He considered the excitation of atoms by fast α-particles in terms of first-order perturbation theory. The first general formulation of this approximation was given by *Mott* in 1931 [6.2]. Since then, the semiclassical approximation has been used in numerous papers dealing with atomic collisions (for reviews see, e.g. [6.3–11]) and also with similar problems for inelastic molecular collisions (see, e.g. [6.12–14]) and nuclear collisions (see, e.g. [6.15, 16]).

The derivation and general properties of the semiclassical scattering equations will be considered in Sect. 6.1.1 for the collision of atom A with a structureless particle B. The generalization for two complex atoms is straightforward.

Several questions arise in formulating the common-trajectory approximation as an alternative to the quantum formulation of the collision problem. First, what conditions must be fulfilled for this approximation to be applicable? Second, how do we find the common trajectory $\boldsymbol{R}(t)$? Third, how do we use the semiclassical scattering matrix for the calculation of differential cross sections?

These questions can only be answered provided the semiclassical approximation is derived consistently from the quantum collision picture. The simplest situation here corresponds to a collision energy sufficiently high for the high-energy approximation to be valid, yielding the quantum scattering amplitude in a closed form. The Born approximation and the eikonal approximation fulfill these conditions. It is natural, therefore, that the first reduction of the quantum expression for the inelastic-scattering cross section to the semiclassical one, corresponding to a rectlinear trajectory $\boldsymbol{R}(t)$, was performed by *Frame* in 1931 [6.17] in terms of the first Born approximation. Much later, such a reduction was also done for the Born approximation of arbitrary order [6.18, 19]. These results considerably clarify the problem especially with respect to the relationship between the quantum scattering amplitude $f_{a'm';am}(\theta)$ and the semiclassical scattering matrix obtained as a solution to (6.1) with a rectilinear trajectory. However, the derivation based on the Born expansion seems to be unsatisfactory. It cannot be applied either to the case when the relative motion is considerably disturbed but the coupling between states remains weak, or to the case when the rectilinear trajectory may be used but the coupling between different states is strong. The former can often be adequately described using the distorted-wave approximation. This approximation and the reduction of the quantum scattering matrix to a semiclassical one will be considered in Sect. 6.3. For the latter case the semiclassical approximation was derived by several authors [6.20–28] in terms of the eikonal approximation. This approximation will be considered in Sect. 6.1.2. Three variants of the derivation of the common-trajectory-approximation scattering equations have been proposed for the case when the only simplifying condition is the quasi-classical nature of the nuclear motion. These are the quasi-classical derivation using radial equations [6.10, 15, 29–40], the narrow wave-packet derivation [6.34, 41, 42], and the derivation based on the formal application of the stationary phase method to the calculation of the time-dependent Green's function for multichannel scattering [6.43–45].

In the last two approaches, the aim is to derive a self-consistent set of semiclassical equations with the trajectory defined both by the conventional average potential and by the inelastic scattering amplitudes, thus taking into account the back-coupling of the quantum subsystem to the classical one (in connection with this see also [6.46–48]). However, it is not quite clear so far as to when these complicated self-consistent equations do provide a real improvement to the semiclassical approximation. Moreover, as pointed out in [6.34] the narrow wave-packet derivation demands the fulfillment of too-stringent conditions which, in fact, are not needed for the semiclassical equations to be valid.

Therefore, only the quasi-classical derivation of the semiclassical equations will be considered in Sect. 6.1.3. An attractive feature of this approach, besides the unrestrictive conditions of its applicability, is that it may serve as a convenient first step for generalizations which take into account some quantum effects of the nuclear motion (Sects. 6.2 and 7.2).

6.1.1 Common-Trajectory Scattering Equations

In the common-trajectory approximation the electrons of an atom A are subject to a time dependent perturbation induced by the passing atom B. The electronic transitions in atom A are described by the time-dependent Schrödinger equation for a semiclassical (sc) wave function

$$\mathrm{i}\frac{\partial\, {}^{\mathrm{sc}}\Psi}{\partial t} = (H_{\mathrm{A}} + V_{\mathrm{AB}})\, {}^{\mathrm{sc}}\Psi, \tag{6.1}$$

where H_{A} is the electronic Hamiltonian, and the interatomic interaction V_{AB} given by (2.78) depends on t, since $\boldsymbol{R} = \boldsymbol{R}(t)$. For $t \to \pm\infty$ the interaction V_{AB} tends to zero.

The basic assumption of the common-trajectory approximation is that the classical trajectory is governed by some central-field potential and thus it lies in a fixed plane normal to the relative angular momentum vector $\mathscr{J}$ (referred to hereafter as the collision plane) and is symmetric with respect to the vector $\boldsymbol{R}^{\mathrm{t}}$ specifying the turning point. A justification of these assumptions and the exact physical meaning of $\mathscr{J}$ is revealed by a consistent derivation of the common-trajectory scattering equations (Sect. 6.1.3). The collision plane can be used to define various natural reference frames for the explicit formulation of the semiclassical scattering problem. These frames are shown in Fig. 6.1.

The solution of (6.1) is usually sought in the form

$$^{\mathrm{sc}}\Psi = \sum_{a'} c_{a'} \exp(-\mathrm{i}\phi_{a'})\,|a'\rangle, \tag{6.2}$$

where the $|a'\rangle$ denote electronic wave functions (which might depend on t) forming a complete orthonormal set, and the $\phi_{a'}$ stand for known functions of t. It is assumed also that for $t \to \pm\infty$ the electronic wave functions $|a\rangle$ coincide with the wave functions of a free atom A. Substitution of (6.2) into (6.1) gives for the coefficients $c_{a'}$

$$\mathrm{i}\frac{d c_{a'}}{dt} = \sum_{a''} \mathscr{V}_{a'a''}(t)\, c_{a''}, \qquad \text{where} \tag{6.3}$$

$$\mathscr{V}_{a'a''}(t) = \exp[\mathrm{i}(\phi_{a'} - \phi_{a''})]\Big[-(d\phi_{a'}/dt)\,\delta_{a'a''} + \langle a'|\,-\mathrm{i}\frac{d}{dt}\,|a''\rangle + \langle a'|\,H_{\mathrm{A}} + V_{\mathrm{AB}}\,|a''\rangle\Big]. \tag{6.4}$$

Fig. 6.1 a, b. Coordinate systems used in semiclassical treatment of inelastic collisions. **(a)** Space-fixed systems: standard xyz with z-axis directed along $\mathscr{J}$ and x-axis opposite to $\boldsymbol{v}$; symmetric $x_1\,y_1\,z_1$ obtained from xyz by rotation through Euler angles $\alpha = (\pi - \eta)/2$, $\beta = 0$, $\gamma = 0$; standard $\bar{x}\,\bar{y}\,\bar{z}$ obtained from xyz by rotation through Euler angles $\alpha = \pi$, $\beta = \pi/2$, $\gamma = \pi$. **(b)** Rotating (body-fixed) systems: atomic $x_1'\,y_1'\,z_1'$ obtained from xyz by rotation through Euler angles $\alpha = \psi(t)$, $\beta = 0$, $\gamma = 0$; molecular $x_2'\,y_2'\,z_2'$ obtained from xyz by rotation through Euler angles $\alpha = \psi(t)$, $\beta = \pi/2$, $\gamma = \pi$

Equation (6.3) solved with the initial conditions

$$c_{a'} \xrightarrow[t \to -\infty]{} \delta_{a'a} \tag{6.5}$$

yields the probability amplitudes $c_{a'}(+\infty)$ for the electronic transitions $a \to a'$ caused by the collision, thus defining the semiclassical S matrix ${}^{sc}S_{a'a}$, viz.

$${}^{sc}S_{a'a} = c_{a'}(+\infty). \tag{6.6}$$

These matrix elements depend on the impact parameter b and the initial velocity v which determine the trajectory $\boldsymbol{R}(t)$. The integral inelastic ($a' \neq a$) collision cross section in the common-trajectory approximation is

$$\sigma_{a'a} = 2\pi \int_0^\infty b\,db\, |{}^{sc}S_{a'a}|^2 . \tag{6.7}$$

Now, due to the differences in the electronic basis sets adopted and the coordinate systems used, there will be several different semiclassical S

Table 6.1. Semiclassical scattering matrices for different frames and basis functions

Coordinate system [a]	Basis wave functions [b]	Scattering matrix and coupling
Standard space-fixed x, y, z	Atomic undistorted $\exp[i\nu\psi(t)]\,\|a\nu\rangle$ [c]	${}^{sc}S^{SFN}_{a'\nu';a\nu}$ [d] $\mathscr{V}^{SFN}_{a'\nu';a''\nu''}$
Symmetric space-fixed x_1, y_1, z_1	Atomic undistorted $\exp[i\nu\psi'(t)]\,\|a\nu\rangle$ [e]	${}^{sc}S^{SFS}_{a'\nu';a\nu}$ [f] $\mathscr{V}^{SFS}_{a'\nu';a''\nu''}$
Standard atomic space-fixed $\bar{x}, \bar{y}, \bar{z}$	$\|a\bar{m}\rangle_{st} = \sum_{\nu} \Delta^j_{\bar{m}\nu}$ $\times \exp[i\nu\psi(t)]\,\|a\nu\rangle$	${}^{sc}S^{SF}_{a'\bar{m}';a\bar{m}}$ [d] $\mathscr{V}^{SF}_{a'\bar{m}';a''\bar{m}''}$
Atomic body-fixed x'_1, y'_1, z'_1	Atomic undistorted $\|a\nu\rangle$	${}^{sc}S_{a'\nu';a\nu}$ [f] $\mathscr{V}_{a'\nu';a''\nu''}$
Molecular body-fixed x'_2, y'_2, z'_2	Atomic undistorted $\|a\tilde{\Omega}\rangle_0 = \sum_{\nu} \Delta^j_{\nu\tilde{\Omega}}$ $\times \exp(-i\pi\tilde{\Omega})\,\|a\nu\rangle$	${}^{sc}S^{0}_{a'\tilde{\Omega}';a\tilde{\Omega}}$ [f] $\mathscr{V}^{0}_{a'\tilde{\Omega}';a''\tilde{\Omega}''}$
Molecular body-fixed x'_2, y'_2, z'_2	Molecular, diagonalizing $H_A + V_{AB}$ $\|a\Omega\rangle$	${}^{sc}S_{a'\tilde{\Omega}';a\tilde{\Omega}}$ [f] $\mathscr{V}_{a'\tilde{\Omega}';a''\tilde{\Omega}''}$

[a] The coordinate systems are defined in Fig. 6.1.
[b] Atomic undistorted wave functions $|a\nu\rangle$ in the body-fixed atomic coordinate system x', y', z' are used everywhere except the last line as reference functions
[c] Rotation angle $\psi(t) = \int_{-\infty}^{t} (\mathscr{J}/\mu R^2)\,dt'$ (Fig. 5.7).
[d] The scattering matrix is nonsymmetric
[e] Angle $\psi'(t) = \int_0^t (\mathscr{J}/\mu R^2)\,dt' = \psi(t) - [(\pi - \eta)/2]$ where η is the deflection function for the trajectory $\boldsymbol{R}(t)$ (Fig. 5.7)
[f] The scattering matrix is symmetric

matrices, listed in Table 6.1 together with the notations for the coupling matrices $\mathscr{V}_{a'a''}$ with $\phi_{a'}$ chosen in the form

$$\phi_{a'} = \int_0^t \langle a'|\, H_A + V_{AB}\, |a'\rangle \, dt'. \tag{6.8}$$

The matrix elements $\mathscr{V}_{a'a}$ are given in Table 6.2. If another choice is made,

$$\bar{\phi}_{a'} = \int_0^t \langle a'|\, H_A\, |a'\rangle \, dt', \tag{6.9}$$

the notations $\bar{S}_{a'a}$ and $\bar{\mathscr{V}}_{a'a}$ are used. Two interaction matrices and scattering matrices are simply related as

$$\bar{\mathscr{V}}_{a'a''} = \mathscr{V}_{a'a''} \exp\left[-\mathrm{i} \int_0^t (\langle a'|\, V_{AB}\, |a'\rangle - \langle a''|\, V_{AB}\, |a''\rangle)\, dt' \right] + \delta_{a'a''} \langle a'|\, V_{AB}\, |a'\rangle, \tag{6.10}$$

$$^{sc}\bar{S}_{a'a} = {}^{sc}S_{a'a} \exp\left[-\mathrm{i} \int_0^\infty (\langle a'|\, V_{AB}\, |a'\rangle + \langle a|\, V_{AB}\, |a\rangle)\, dt' \right]. \tag{6.11}$$

Since the scattering matrix $^{sc}S_{a'v';av}$ is most simply related to the quantum scattering matrix $S^J_{a'v';av}$, [see (6.69)], the relationships between the former and all other semiclassical scattering matrices listed in Table 6.1 are given in Table 6.3.

Let us dwell now on the symmetry properties of the semiclassical scattering equations. Consider for definiteness the case of the rotating atomic coordinate system yielding $^{sc}S_{a'v';av}$. Since V_{AB} is invariant with respect to reflection in the collision plane, then

$$\mathscr{V}_{a'v';a''v''} = 0 \quad \text{if} \quad w'w''(-1)^{v'-v''} = -1. \tag{6.12}$$

Therefore, the following selection rules hold for $^{sc}S_{a'v',av}$:

$$^{sc}S_{a'v',av} = 0 \quad \text{if } w'w(-1)^{v'-v} = -1. \tag{6.13}$$

Note that these selection rules coincide with the selection rules (2.88) for the quantum scattering matrix $S^J_{a'v';av}$ obtained on the grounds of parity conservation. Thus in the semiclassical collision picture, invariance with respect to reflection in the collision plane corresponds to invariance with respect to inversion in the quantum collision picture.

The symmetry with respect to the collision plane also ensures that $\langle a'v'|\, V_{AB}\, |a''v''\rangle$ is a real symmetric matrix. Therefore, since $R(t)$ is an even function of t, the coupling matrix $\mathscr{V}_{a'v';av}(t)$ satisfies the relation

$$\mathscr{V}_{a'v';a''v''}(-t) = \mathscr{V}_{a''v'';a'v'}(t). \tag{6.14}$$

It can be shown that this property of the coupling matrix $\mathscr{V}_{a'v';a''v''}$ guarantees the symmetry of $^{sc}S_{a'v';av}$. A detailed discussion of the time-reversal

Table 6.2. Semiclassical coupling matrices for different frames and basis functions

Coupling matrix	Explicit expression
$\mathscr{V}^{\mathrm{SFN}}_{a'v';a''v''}$	$\exp[-\mathrm{i}(v'-v'')\psi(t)]\langle a'v'\|V_{\mathrm{AB}}\|a''v''\rangle\exp\left\{\mathrm{i}\left[\int_0^t(\langle a'v'\|V_{\mathrm{AB}}\|a'v'\rangle + \varepsilon_{a'} - \langle a''v''\|V_{\mathrm{AB}}\|a''v''\rangle - \varepsilon_{a''})\,dt'\right]\right\}(1-\delta_{a'v';a''v''})$ [a]
$\mathscr{V}^{\mathrm{SFS}}_{a'v';a''v''}$	$\exp[-\mathrm{i}(v'-v'')\psi'(t)]\langle a'v'\|V_{\mathrm{AB}}\|a''v''\rangle\exp\left\{\mathrm{i}\left[\int_0^t(\langle a'v'\|V_{\mathrm{AB}}\|a'v'\rangle + \varepsilon_{a'} - \langle a''v''\|V_{\mathrm{AB}}\|a''v''\rangle - \varepsilon_{a''})\,dt'\right]\right\}(1-\delta_{a'v';a''v''})$
$\mathscr{V}^{\mathrm{SF}}_{a'\bar{m}';a''\bar{m}''}$	${}_{\mathrm{st}}\langle a'\bar{m}'\|V_{\mathrm{AB}}\|a''\bar{m}''\rangle_{\mathrm{st}}\exp\left\{\mathrm{i}\left[\int_0^t({}_{\mathrm{st}}\langle a'\bar{m}'\|V_{\mathrm{AB}}\|a'\bar{m}'\rangle_{\mathrm{st}} + \varepsilon_{a'} - {}_{\mathrm{st}}\langle a''\bar{m}''\|V_{\mathrm{AB}}\|a''\bar{m}''\rangle_{\mathrm{st}} - \varepsilon_{a''})\,dt'\right]\right\}(1-\delta_{a'\bar{m}';a''\bar{m}''})$ [b]
$\mathscr{V}_{a'v';a''v''}$	$\exp\left\{\mathrm{i}\left[\int_0^t(\langle a'v'\|V_{\mathrm{AB}}\|a'v'\rangle + \varepsilon_{a'} - \langle a''v''\|V_{\mathrm{AB}}\|a''v''\rangle - \varepsilon_{a''})\,dt'\right]\right\} \times [-\delta_{a'v';a''v''}(v'\mathscr{J}/\mu R^2) + (1-\delta_{a'v';a''v''})\langle a'v'\|V_{\mathrm{AB}}\|a''v''\rangle]$
$\mathscr{V}^{0}_{a'\tilde{\Omega}';a''\tilde{\Omega}''}$	$\exp\left\{\mathrm{i}\left[\int_0^t({}_0\langle a'\tilde{\Omega}'\|V_{\mathrm{AB}}\|a'\tilde{\Omega}'\rangle_0 + \varepsilon_{a'} - {}_0\langle a''\tilde{\Omega}''\|V_{\mathrm{AB}}\|a''\tilde{\Omega}''\rangle_0 - \varepsilon_{a''})\,dt'\right]\right\} \times \{{}_0\langle a'\tilde{\Omega}'\|V_{\mathrm{AB}}\|a''\tilde{\Omega}''\rangle_0 - \delta_{a'a''}(\mathscr{J}/\mu R^2)[(j'+\tilde{\Omega}'+1)^{1/2}(j'-\tilde{\Omega}')^{1/2} \times \delta_{\tilde{\Omega}'';\tilde{\Omega}'+1} + (j'+\tilde{\Omega}')^{1/2}(j-\tilde{\Omega}'+1)^{1/2}\delta_{\tilde{\Omega}'';\tilde{\Omega}'-1}]\}$ [c]
$\mathscr{V}_{a'\tilde{\Omega}';a''\tilde{\Omega}''}$	$\exp\left[\mathrm{i}\int_0^t(U_{a'\tilde{\Omega}'} - U_{a''\tilde{\Omega}''})\,dt'\right]\left\{-\mathrm{i}\frac{dR}{dt}\langle a'\tilde{\Omega}'\|d/dR\|a''\tilde{\Omega}''\rangle\delta_{\tilde{\Omega}'\tilde{\Omega}''} - (\mathscr{J}/\mu R^2)[\langle a'\tilde{\Omega}'\|j_{x_2'}\|a''\tilde{\Omega}'+1\rangle\delta_{\tilde{\Omega}'';\tilde{\Omega}'+1} + \langle a'\tilde{\Omega}'\|j_{x_2'}\|a''\tilde{\Omega}'-1\rangle \times \delta_{\tilde{\Omega}'';\tilde{\Omega}'-1}]\right\}$

[a] The matrix elements $\langle a'v'|V_{\mathrm{AB}}|a''v''\rangle$ are given by (2.84)

[b] The matrix elements

$${}_{\mathrm{st}}\langle a'\bar{m}'|V_{\mathrm{AB}}|a''\bar{m}''\rangle_{\mathrm{st}} = \sum_{v'v''}\exp[-\mathrm{i}(v'-v'')\psi(t)]\,\Delta^{j'}_{\bar{m}'v'}\Delta^{j''}_{\bar{m}''v''}\langle a'v'|V_{\mathrm{AB}}|a''v''\rangle$$

[c] The matrix elements

$${}_0\langle a'\tilde{\Omega}'|V_{\mathrm{AB}}|a''\tilde{\Omega}''\rangle_0 = \sum_{v'v''}\exp[\mathrm{i}\pi(\tilde{\Omega}'-\tilde{\Omega}'')]\,\Delta^{j'}_{v'\tilde{\Omega}'}\Delta^{j''}_{v''\tilde{\Omega}''}\langle a'v'|V_{\mathrm{AB}}|a''v''\rangle$$

are diagonal in $\tilde{\Omega}'$, $\tilde{\Omega}''$ as can be checked using (2.84)

invariance of the semiclassical scattering equations can be found in [6.49–51].

The selection rules and symmetry properties of other semiclassical scattering matrices listed in Table 6.1 can be verified using the relations between these matrices and ${}^{\mathrm{sc}}S_{a'v';av}$ given in Table 6.3. In particular, the selection rules (6.13) lead to similar selection rules for ${}^{\mathrm{sc}}S^{\mathrm{SFN}}_{a'v';av}$ and ${}^{\mathrm{sc}}S^{\mathrm{SFS}}_{a'v';av}$.

Table 6.3. Expression of ${}^{sc}S_{a'v';av}$ through other semiclassical scattering matrices

$\exp[\mathrm{i}v'(\pi-\eta)]\ {}^{sc}S^{SFN}_{a'v';av}$
$\exp[\mathrm{i}(\pi-\eta)(v'+v)/2]\ {}^{sc}S^{SFS}_{a'v';av}$
$\exp[\mathrm{i}v'(\pi-\eta)] \sum_{\bar m \bar m'} \Delta^{j'}_{\bar m' v'}$ $\exp\left[-\mathrm{i}\int_0^\infty ({}_{st}\langle a'\bar m'\vert V_{AB}\vert a'\bar m'\rangle_{st} - \langle a'v'\vert V_{AB}\vert a'v'\rangle)\,dt\right] {}^{sc}S^{SF}_{a'\bar m';a\bar m}$ $\times \exp\left[-\mathrm{i}\int_0^\infty ({}_{st}\langle a\bar m\vert V_{AB}\vert a\bar m\rangle_{st} - \langle av\vert V_{AB}\vert av\rangle)\,dt\right] \Delta^{j}_{\bar m v}$
$\sum_{\tilde\Omega\tilde\Omega'} D^{j'*}_{\tilde\Omega' v'}(0,\pi/2,\pi) \exp\left[-\mathrm{i}\int_0^\infty ({}_0\langle a'\tilde\Omega'\vert V_{AB}\vert a'\tilde\Omega'\rangle_0 - \langle a'v'\vert V_{AB}\vert a'v'\rangle)\,dt\right] {}^{sc}S^{0}_{a'\tilde\Omega';a\tilde\Omega}$ $\times \exp\left[-\mathrm{i}\int_0^\infty ({}_0\langle a\tilde\Omega\vert V_{AB}\vert a\tilde\Omega\rangle_0 - \langle av\vert V_{AB}\vert av\rangle)\,dt\right] D^{j}_{\tilde\Omega v}(0,\pi/2,\pi)$
$\sum_{\tilde\Omega\tilde\Omega'} D^{j*}_{\tilde\Omega' v'}(0,\pi/2,\pi) \exp\left[-\mathrm{i}\int_0^\infty (U_{a'\tilde\Omega'} - \varepsilon_{a'} - \langle a'v'\vert V_{AB}\vert a'v'\rangle)\,dt\right] {}^{sc}S_{a'\tilde\Omega';a\tilde\Omega}$ $\times \exp\left[-\mathrm{i}\int_0^\infty (U_{a\tilde\Omega} - \varepsilon_a - \langle av\vert V_{AB}\vert av\rangle)\,dt\right] D^{j}_{\tilde\Omega v}(0,\pi/2,\pi)$ [a]

[a] This expression is similar to the relation (2.126) between quantum S matrices in the R-helicity and J-representations in the limit of large J. Here the additional phase factors are just those which restore (2.126) when the relationships between semiclassical and quantum S matrices, see (6.67, 69–71), are taken into account.

In the three other cases the symmetry with respect to the collision plane leads to additional symmetry properties for the S matrices, rather than to the selection rules

$$ {}^{sc}S^{SF}_{a'\bar m';a\bar m} = w w'(-1)^{j-j'}\,{}^{sc}S^{SF}_{a'-\bar m';a-\bar m}, \tag{6.15a} $$

$$ {}^{sc}S^{0}_{a'\tilde\Omega';a\tilde\Omega} = w w'(-1)^{j-j'}\,{}^{sc}S^{0}_{a'-\tilde\Omega';a-\tilde\Omega}, \tag{6.15b} $$

$$ {}^{sc}S_{a'\tilde\Omega';a\tilde\Omega} = w w'(-1)^{j-j'}\,{}^{sc}S_{a'-\tilde\Omega';a-\tilde\Omega}. \tag{6.15c} $$

Note that the two last relations are similar to (2.43) for the quantum-scattering matrix elements.

6.1.2 Eikonal and Impact-Parameter Approximations

Consider a high-energy collision for which the inequalities

$$ \bar V \ll \mathscr{E}, \tag{6.16a} $$

$$ \Delta\varepsilon \ll \mathscr{E} \tag{6.16b} $$

hold. Here $\bar V$ is the characteristic value of the interatomic interaction for a collision with energy $\mathscr{E}$, and $\Delta\varepsilon$ is the characteristic excitation energy for states accounted for in the scattering equations. Note that (6.16) does not imply that the actions $\bar{V}\tau_0$ and $\Delta\varepsilon\tau_0$ (τ_0 is the characteristic collision time) associated with $\bar{V}$ and $\Delta\varepsilon$ need be small. This is an important point, showing that the high-energy regime, as defined by (6.16), involves the Born approximation as a limiting case only. For a detailed discussion see [6.27, 52].

Note, first of all, that the scattering amplitude can be exactly represented in the form [6.53, 54]

$$f_{a'm';am} = -\frac{\mu}{2\pi} \sum_{\sigma_1 \dots \sigma_{N_A}} \int \Psi^{0*}_{a'm'} V_{AB} \Psi d^3\boldsymbol{R} d^3\boldsymbol{r}_1 \dots d^3\boldsymbol{r}_{N_A}, \tag{6.17}$$

where

$$\Psi^0_{a'm'} = \exp(\mathrm{i}\boldsymbol{k}_{a'}\boldsymbol{R})\,|a'm'\rangle \tag{6.18}$$

and Ψ is the solution to (2.2) subject to the boundary conditions (2.3). The expression (6.17) shows that in fact it is sufficient to know Ψ only in the region of $\boldsymbol{R}$-variation where the interaction does not vanish. In this region, when the conditions (6.16) hold, Ψ can be considered as an incoming plane wave modulated by a function which varies slowly with $\boldsymbol{R}$ over the wavelength of the nuclear motion. This idea underlies the eikonal approximation, widely used for description of high-energy collisions in nuclear and atomic physics. For a discussion of the different aspects of this approximation see [6.52, 55, 56]. Here we shall consider only how the eikonal approximation permits us to answer the questions stated above about the semiclassical approximation.

Consider a scattering process in the coordinate system $\bar{X}\bar{Y}\bar{Z}$ with the $\bar{Z}$ axis parallel to the initial momentum $\boldsymbol{k}_a$ (Fig. 2.2), and represent Ψ in the form

$$\Psi = \exp(\mathrm{i}k\bar{Z})\,\Phi, \tag{6.19}$$

where $k = (2\mu\mathscr{E})^{1/2}$. According to the above reasoning Φ is assumed to vary much more slowly with $\boldsymbol{R}$ than does $\exp(\mathrm{i}k\bar{Z})$. Substituting the representation (6.19) into (2.2) and neglecting $\nabla_R^2\Phi$ in accord with this assumption, we get a first-order equation for Φ

$$\mathrm{i}\frac{k}{\mu}\partial\Phi/\partial\bar{Z} = H\Phi, \tag{6.20}$$

where $H = H_A + V_{AB}$ is the electronic Hamiltonian.

The inequality (6.16) guarantees that the scattering is confined to small angles only. Therefore, for large negative $\bar{Z}$ corresponding to back scattering, Ψ differs only slightly from the incoming plane wave. Neglecting this difference we arrive at the boundary condition for Φ:

$$\Phi \xrightarrow[\bar{Z}\to-\infty]{} \exp[\mathrm{i}(k_a - k)\bar{Z}]\,|a\bar{m}\rangle. \qquad (6.21)$$

As mentioned above, the eikonal approximation defined by (6.19–21) is valid only for the region of $\bar{Z}$-variation where V_{AB} does not vanish. Indeed, $\partial^2\Phi/\partial X^2$ and $\partial^2\Phi/\partial Y^2$, which were neglected in obtaining (6.20), remain constant at large positive $\bar{Z}$ for this Φ. It can be shown that $\partial^2\Phi/\partial\bar{X}^2$ and $\partial^2\Phi/\partial\bar{Y}^2$ are proportional to $\bar{V}$ if Φ is obtained from (6.20). At the same time, the part of $\partial\Phi/\partial\bar{Z}$ proportional to $\bar{V}$ goes to zero, together with V_{AB}. Therefore, for such $\bar{Z}$, neglecting the terms containing $\partial^2\Phi/\partial\bar{X}^2$ and $\partial^2\Phi/\partial\bar{Y}^2$ compared to the term containing $\partial\Phi/\partial\bar{Z}$ is not legitimate and the eikonal approximation is invalid. The inadequacy of the eikonal approximation at large distances becomes quite clear in the case of elastic potential scattering, when Ψ, given by (6.19) with Φ determined by (6.20), cannot be expressed for large $\bar{Z}$ in the familiar form of the superposition of an incoming plane wave and a spherical outgoing wave [6.52, 57].

Bearing in mind that the coordinates $\bar{X}$ and $\bar{Y}$ appear in (6.20) as parameters only, and setting $\bar{Z} = vt$ ($\boldsymbol{v} = \boldsymbol{k}/\mu$), we get

$$\mathrm{i}\,d\Phi/dt = H[\boldsymbol{R}(t)]\,\Phi \qquad (6.22)$$

with

$$\boldsymbol{R}(t) = \boldsymbol{b} + \boldsymbol{v}t, \qquad (6.23)$$

where $\boldsymbol{b}$ is a vector with the components $\bar{X}$, $\bar{Y}$, 0 normal to $\boldsymbol{v}$. Equation (6.22) coincides with the semiclassical scattering equation (6.1), $\boldsymbol{R}(t)$ being a rectilinear trajectory with impact parameter $\boldsymbol{b}$. Taking into account that $k - k_a \simeq \varepsilon_a/v$ for a high-energy collision, the boundary condition (6.21) reduces to a semiclassical initial condition,

$$\Phi \xrightarrow[t\to-\infty]{} \exp(-\mathrm{i}\varepsilon_a t)\,|a\bar{m}\rangle. \qquad (6.24)$$

Representing Φ in the form

$$\Phi = \sum_{a'\bar{m}'} {}^{\mathrm{eik}}c_{a'\bar{m}'} \exp(-\mathrm{i}\varepsilon_{a'}t)\,|a'\bar{m}'\rangle, \qquad (6.25)$$

we get

$$\mathrm{i}\,d\,{}^{\mathrm{eik}}c_{a'\bar{m}'}/dt = \sum_{a''\bar{m}''} \langle a'\bar{m}'|\,V_{\mathrm{AB}}\,|a''\bar{m}''\rangle \exp[\mathrm{i}(\varepsilon_{a'} - \varepsilon_{a''})\,t]\,{}^{\mathrm{eik}}c_{a''\bar{m}''}, \qquad (6.26)$$

which must be solved with the initial conditions

$${}^{\mathrm{eik}}c_{a'\bar{m}'} \xrightarrow[t\to-\infty]{} \delta_{a'a}\,\delta_{\bar{m}'\bar{m}}, \qquad (6.27)$$

yielding the scattering matrix

$$ {}^{\mathrm{eik}}S_{a'\bar{m}';a\bar{m}}(\boldsymbol{b}) = {}^{\mathrm{eik}}c_{a'\bar{m}'}(+\infty). \tag{6.28} $$

The matrix ${}^{\mathrm{eik}}S_{a'\bar{m}';a\bar{m}}$ is simply related to ${}^{\mathrm{sc}}\bar{S}^{\mathrm{SF}}_{a'\bar{m}';a\bar{m}}$, see Tables 6.1, 2 and (6.10, 11). Indeed, the vector $\boldsymbol{v}$ is directed along the $\bar{z}$ axis of the standard space-fixed coordinate system $\bar{x}\bar{y}\bar{z}$ (Fig. 6.1), and the impact-parameter vector $\boldsymbol{b}$ along the $\bar{y}$ axis of this system. Therefore, from the semiclassical point of view, for the given impact parameter vector $\boldsymbol{b}$ the coordinate system $\bar{x}\bar{y}\bar{z}$ is obtained from the coordinate system $\bar{X}\bar{Y}\bar{Z}$ in which ${}^{\mathrm{eik}}S_{a'\bar{m}';a\bar{m}}$ is defined by rotation through the Euler angles $\alpha = \varphi_0 - (\pi/2)$, $\beta = 0$, $\gamma = 0$, where φ_0 is the azimuthal angle of the vector $\boldsymbol{b}$ in the coordinate system $\bar{X}\bar{Y}\bar{Z}$. Therefore,

$$ {}^{\mathrm{eik}}S_{a'\bar{m}';a\bar{m}}(\boldsymbol{b}) = \exp[\mathrm{i}(\bar{m}-\bar{m}')(\varphi_0 - \pi/2)]\; {}^{\mathrm{sc}}\bar{S}^{\mathrm{SF}}_{a'\bar{m}';a\bar{m}}, \tag{6.29} $$

where ${}^{\mathrm{sc}}\bar{S}^{\mathrm{SF}}_{a'\bar{m}';a\bar{m}}$ depends on b only.

Turn now to the determination of the scattering amplitude in the eikonal approximation. As noted above, this can be done by means of (6.17). Substituting Ψ given by (6.19) into (6.17), we get

$$ f_{a'\bar{m}';a\bar{m}} = -\frac{\mu}{2\pi}\int d^3\boldsymbol{R}\,\exp[\mathrm{i}(\boldsymbol{k}-\boldsymbol{k}_{a'})\boldsymbol{R}]\,\langle a'\bar{m}'|\,V_{\mathrm{AB}}\,|\Phi\rangle, \tag{6.30} $$

where the Dirac brackets denote as usual the electronic matrix elements. For a high-energy collision,

$$ (\boldsymbol{k}-\boldsymbol{k}_{a'})\,\boldsymbol{R} \simeq (\varepsilon_{a'}\bar{Z}/v) - kb\theta\cos(\varphi-\varphi_0), \tag{6.31} $$

where θ, φ are the polar and azimuthal angles of the vector $\boldsymbol{k}_{a'}$ (the scattering angles, see Fig. 2.2). Taking this into account, setting $t = \bar{Z}/v$ and using the expansion (6.25) for Φ, we can represent (6.30) in the form

$$ \begin{aligned} f_{a'\bar{m}';a\bar{m}} = &-\mu v\int_0^\infty b\,db\int_0^{2\pi}\frac{d\varphi_0}{2\pi}\exp[-\mathrm{i}kb\theta\cos(\varphi-\varphi_0)] \\ &\times\int_{-\infty}^{+\infty}dt\sum_{a''\bar{m}''}\exp[\mathrm{i}(\varepsilon_{a'}-\varepsilon_{a''})\,t]\,\langle a'\bar{m}'|\,V_{\mathrm{AB}}\,|a''\bar{m}''\rangle\,{}^{\mathrm{eik}}c_{a''\bar{m}''}. \end{aligned} \tag{6.32} $$

The time integral in (6.32) can be expressed, according to (6.26, 27), as

$$ \begin{aligned} &\int_{-\infty}^{+\infty}dt\sum_{a''\bar{m}''}\exp[\mathrm{i}(\varepsilon_{a'}-\varepsilon_{a''})\,t]\,\langle a'\bar{m}'|\,V_{\mathrm{AB}}\,|a''\bar{m}''\rangle\,{}^{\mathrm{eik}}c_{a''\bar{m}''} \\ &\quad = -\mathrm{i}(\delta_{a'\bar{m}';a\bar{m}} - {}^{\mathrm{eik}}S_{a'\bar{m}';a\bar{m}}). \end{aligned} \tag{6.33} $$

Therefore, using (6.29) and taking into account that

$$\frac{1}{2\pi}\int_0^{2\pi} \exp[-\mathrm{i}n\varphi_0 + \mathrm{i}x\sin\varphi_0]\,d\varphi_0 = \mathrm{J}_n(x), \tag{6.34}$$

where $\mathrm{J}_n(x)$ is the Bessel function, we get

$$f_{a'\bar{m}';a\bar{m}}(\theta,\varphi) = \mathrm{i}\mu v \exp[\mathrm{i}(\bar{m}-\bar{m}')\varphi] \times \int_0^\infty b\,db(\delta_{a'\bar{m}';a\bar{m}} - {}^{\mathrm{sc}}\bar{S}^{\mathrm{SF}}_{a'\bar{m}';a\bar{m}})\,\mathrm{J}_{\bar{m}'-\bar{m}}(kb\theta). \tag{6.35}$$

The quantity $(\delta_{a'\bar{m}';a\bar{m}} - {}^{\mathrm{sc}}\bar{S}^{\mathrm{SF}}_{a'\bar{m}';a\bar{m}})$ in (6.35) can be expressed through ${}^{\mathrm{sc}}S_{a'v';av}$ using (6.11) and Table 6.3:

$$\begin{aligned}\delta_{a'\bar{m}';a\bar{m}}\,{}^{\mathrm{sc}}\bar{S}^{\mathrm{SF}}_{a'\bar{m}';a\bar{m}} &= \sum_{vv'} \exp(-\mathrm{i}\pi v')\,\Delta^{j'}_{\bar{m}'v'} \\ &\times \Bigg\{\delta_{a'v';av}\exp(\mathrm{i}\pi v) - \exp\Bigg[-\mathrm{i}\Bigg(\int_0^\infty \langle a'v'|\,V_{\mathrm{AB}}\,|a'v'\rangle\,dt \\ &+ \int_0^\infty \langle av|\,V_{\mathrm{AB}}\,|av\rangle\,dt\Bigg)\Bigg]\,{}^{\mathrm{sc}}S_{a'v';av}\Bigg\}\,\Delta^{j}_{\bar{m}v}.\end{aligned} \tag{6.36}$$

It was taken into account when obtaining this expression that for the rectilinear trajectory $\eta = 0$.

Comparing (6.35, 36) with the general expression (2.50) for scattering through small angles, we see that they coincide if J is identified with kb, and $T^J_{a'v';av}$ with the term in braces in (6.36). Noting, see (5.20), that at high energies,

$$\int_0^\infty \langle av|\,V_{\mathrm{AB}}\,|av\rangle\,dt = -\,\delta_{av}(b), \tag{6.37}$$

where $\delta_{av}(b)$ is the phase shift of the elastic scattering in the potential $\langle av|\,V_{\mathrm{AB}}\,|av\rangle$ at energy $\mathscr{E}$, the following relation between the quantum scattering matrix $S^J_{a'v';av}$ defined by (2.48, 87) and ${}^{\mathrm{sc}}S_{a'v';av}(b)$ arises in the case of high energy collisions:

$$S^J_{a'v';av} = \exp\{\mathrm{i}[\delta_{a'v'}(b) + \delta_{av}(b)]\}\,{}^{\mathrm{sc}}S_{a'v';av}(b)|_{b=J/k}\,. \tag{6.38}$$

It will be shown in the next section that this relation holds also for a curvilinear trajectory, provided the common-trajectory approximation can be used.

The integral cross section is obtained from $f_{a'\bar{m}';a\bar{m}}$ given by (6.35) in a way similar to that leading from (2.50) to (2.53), and is of the familiar semiclassical form:

$$\sigma_{a'\bar{m}';a\bar{m}} = 2\pi\int_0^\infty b\,db\,|\delta_{a'a}\delta_{\bar{m}'\bar{m}} - {}^{\mathrm{sc}}\bar{S}^{\mathrm{SF}}_{a'\bar{m}';a\bar{m}}|^2. \tag{6.39}$$

The derivation performed here provides answers to the three questions concerning the semiclassical approximation which were put forward at the beginning of this section. First, the conditions (6.16) ensuring the validity of the eikonal approximation are also sufficient to ensure the validity of the common-trajectory approximation. Second, in this case, the rectilinear trajectory with $v = (2\mathscr{E}/\mu)^{1/2}$ is to be used. An answer to the third question concerning differential cross sections is given by (6.35). The derivation shows at the same time that though the results of semiclassical calculations are adequate for high-energy collisions, the semiclassical time-dependent (6.1) must not be interpreted too literally. The thing is, that in the eikonal approximation the motion of nuclei is described by a delocalized modulated plane wave, rather than by the trajectory (6.23), and (6.22) attained its semiclassical form by the formal introduction of a variable with time dimension. To answer the question of whether the classical picture of nuclear motion is actually valid, an additional investigation involving wave packets must be performed [6.34]. Such an investigation shows that the time-dependent picture is valid, provided $(kR^*)^{1/2} \gg 1$, where R^* is the length of the interaction region. This condition is more stringent than the basic quasi-classical condition $kR^* \gg 1$. The inadequacy of the naive interpretation of the semiclassical equations is seen also from the validity of (6.35) for scattering through small diffraction angles, where classical mechanics is certain to fail.

6.1.3 Semiclassical Limit of the Quasi-classical Approximation

For slow atomic collisions the conditions (6.16) under which the eikonal approximation can be applied are often unfulfilled. Most stringent appears to be the condition (6.16a) which allows one to use a rectilinear trajectory to describe the nuclear motion. Therefore a derivation which is not based on this condition is desirable. Such a derivation will be discussed here.

Let there be a group of atom-A states (group G) which is far removed in energy from the other states. The collision occurs with a kinetic energy considerably exceeding the energies $\varepsilon_{a'}$ of all states of group G (the energy of the atom is referred here to the energy of the lowest state of group G), but insufficient for transitions to other states with appreciable probabilities. It can be expected under these conditions that the replacement of the infinite set of equations (2.85 or 2.114) by a finite one taking into account the coupling between states of group G only would not result in a considerable error. We therefore write the sets of equations (2.85) or (2.114) in a unified form considering only the states of this group:

$$\left[\frac{d^2}{dR^2} + \mathscr{K}_{a'}^2(R)\right]\chi_{a'} = 2\mu \sum_{a'' \neq a'} V_{a'a''}\chi_{a''}, \tag{6.40}$$

where

$$\mathscr{K}_{a'}^2(R) = 2\mu\left[\mathscr{E} - \varepsilon_{a'} - \frac{(J+\frac{1}{2})^2}{2\mu R^2} - V_{a'a'}(R)\right] \tag{6.41}$$

and the quantities $V_{a'a''}$ tend to zero for $R \to \infty$. Note that in the case when the molecular basis is used the nondiagonal coupling $V_{a'a''}$ can also contain, besides the functions of R, the operator d/dR (radial nonadiabatic coupling). These equations are solved with the boundary conditions

$$\chi_{a'} \underset{R\to\infty}{\longrightarrow} (k')^{-1/2}\left\{\exp\left[-\mathrm{i}\left(kR - J\frac{\pi}{2}\right)\right]\delta_{a'a} - S^J_{a'a}\exp\left[\mathrm{i}\left(k'R - J\frac{\pi}{2}\right)\right]\right\}. \tag{6.42}$$

To make use of the quasi-classical nature of the nuclear motion, it is convenient to reduce the set of second-order equations (6.40) to an equivalent set of first-order equations representing $\chi_{a'}$, just as in the elastic-scattering case, see (5.29):

$$\chi_{a'} = (\mathscr{K}_{a'})^{-1/2}\{b^-_{a'}(R)\exp[-\mathrm{i}\mathscr{S}_{a'}(R)] + b^+_{a'}(R)\exp[\mathrm{i}\mathscr{S}_{a'}(R)]\}, \tag{6.43}$$

where

$$\mathscr{S}_{a'}(R) = \int_{R^t_{a'}}^{R} \mathscr{K}_{a'}(R')\,dR' \tag{6.44}$$

is the classical radial action for the motion of a particle with energy $\mathscr{E} - \varepsilon_{a'}$ and angular momentum $J + 1/2$ in the potential field $V_{a'a'}(R)$. Since two functions $b^+_{a'}$ and $b^-_{a'}$ are introduced by (6.43) instead of the one function $\chi_{a'}$, some additional condition must be imposed on $b^+_{a'}$ and $b^-_{a'}$. It is convenient to take this condition in a form similar to that used in [6.58] for the one-channel case:

$$\frac{db^+_{a'}}{dR}\exp[\mathrm{i}\mathscr{S}_{a'}(R)] + \frac{db^-_{a'}}{dR}\exp[-\mathrm{i}\mathscr{S}_{a'}(R)] = 0. \tag{6.45}$$

In the one-channel case this condition is trivially fulfilled for a quasi-classical wave function far from the turning points, since then b^+ and b^- are constant. Accounting for this, the set of first-order equations equivalent to (6.40) is readily obtained:

$$\begin{aligned}\mathrm{i}\frac{db^+_{a'}}{dR} &+ \frac{1}{2}\mathscr{K}_{a'}^{-1/2}\left(\frac{d^2}{dR^2}\mathscr{K}_{a'}^{-1/2}\right)[b^+_{a'} + b^-_{a'}\exp(2\mathrm{i}\mathscr{S}_{a'})]\\ &= \mu\sum_{a''\neq a'}\mathscr{K}_{a'}^{-1/2}\exp(-\mathrm{i}\mathscr{S}_{a'})\,V_{a'a''}\exp(\mathrm{i}\mathscr{S}_{a'})\,\mathscr{K}_{a''}^{-1/2}\\ &\quad\times\{b^+_{a''}\exp[\mathrm{i}(\mathscr{S}_{a''} - \mathscr{S}_{a'})] + b^-_{a''}\exp[-\mathrm{i}(\mathscr{S}_{a''} + \mathscr{S}_{a'})]\},\end{aligned} \tag{6.46a}$$

$$\mathrm{i}\frac{db_{a'}^-}{dR}+\frac{1}{2}\mathscr{K}_{a'}^{-1/2}\left(\frac{d^2}{dR^2}\mathscr{K}_{a'}^{-1/2}\right)[b_{a'}^-+b_{a'}^+\exp(2\mathrm{i}\mathscr{S}_{a'})]$$

$$=\mu\sum_{a''\neq a'}\mathscr{K}_{a'}^{-1/2}\exp(\mathrm{i}\mathscr{S}_{a'})\,V_{a'a''}\exp(-\mathrm{i}\mathscr{S}_{a'})\,\mathscr{K}_{a''}^{-1/2}$$

$$\times\{b_{a''}^-\exp[-\mathrm{i}(\mathscr{S}_{a''}-\mathscr{S}_{a'})]+b_{a''}^+\exp[\mathrm{i}(\mathscr{S}_{a''}+\mathscr{S}_{a'})]\}\,. \tag{6.46b}$$

The boundary conditions for these equations follow from a comparison of (6.43) with (6.42), taking (6.44) into account, and read

$$b_{a'}^-\xrightarrow[R\to\infty]{}\delta_{a'a}\exp(-\mathrm{i}\pi/4)\exp(\mathrm{i}\delta_a), \tag{6.47a}$$

$$b_{a'}^+\xrightarrow[R\to\infty]{}-S_{a'a}^J\exp(\mathrm{i}\pi/4)\exp(-\mathrm{i}\delta_{a'}),\qquad\text{where} \tag{6.47b}$$

$$\delta_{a'}=\lim_{R\to\infty}[\mathscr{S}_{a'}(R)-kR]+(J+\tfrac{1}{2})\,\pi/2 \tag{6.48}$$

is the quasi-classical scattering phase shift [see (5.19)].

Consider now the region of R-variation where the motion is quasi-classical in all channels. The exact meaning of this statement is that the inequality

$$\mathscr{K}_{a'}^{-3/2}\,d^2\,\mathscr{K}_{a'}^{-1/2}/dr^2\ll 1 \tag{6.49}$$

holds for all a' [6.58]. This allows one to neglect the second term in the lhs of (6.46). When the molecular basis is used, then it follows from the discussion after (2.113) that the fulfillment of the inequality (6.49) for all channels implies simultaneous fulfillment of the inequality

$$V_{\text{coupl}}\ll\mathscr{K}_{a'}^2/2\mu, \tag{6.50}$$

where V_{coupl} is the characteristic value of the coupling between different states. It is intuitively clear that (6.50) holds also in the atomic basis in the region of R-variation where the quasi-classical condition (6.49) holds simultaneously for both basis sets. Provided the condition (6.50) holds, it is seen from (6.46) that $b_{a'}^\pm$ vary much more slowly than do the rapidly oscillating exponentials $\exp[\pm\mathrm{i}(\mathscr{S}_{a'}+\mathscr{S}_{a''})]$. Therefore according to the averaging principle (see, e.g. [6.59]) the terms proportional to $\exp[\pm\mathrm{i}(\mathscr{S}_{a'}+\mathscr{S}_{a''})]$ in the rhs of (6.46) can be neglected.

Thus we see that in the regions where the motion is quasi-classical in all channels, (6.46a) for $b_{a'}^+$ uncouples from (6.46b) for $b_{a'}^-$. The problem remains, however, for the regions where the quasi-classical conditions fail for one or several channels, i.e. in the turning-point regions. Clearly, if the turning points in different channels are considerably separated, the nuclear motion cannot be described by a single trajectory. With good localization of the regions where the quasi-classical conditions fail and the couplings between different channels are strong, the problem of several sufficiently separated turning points can be treated by the matching method considered in Sect. 6.2.

Here it will be assumed that the turning points are close to each other. This obviously means that there exists a mean potential $U_0(R)$ such that the inequalities

$$|U_0(R) - V_{a'a'}(R) - \varepsilon_{a'}| \ll \mathscr{E} \tag{6.51}$$

hold for all a' and R in the classically allowed region. Introducing also the mean kinetic energy E_0 such that $\mathscr{E} \geqslant E_0 \geqslant \mathscr{E} - \varepsilon_{\max}$ ($\varepsilon_{\max}$ is the highest atomic energy level in the goup G) makes possible the use of the following approximations in (6.46):

$$\mathscr{K}_{a'}(R) \simeq \left\{2\mu\left[E_0 - \frac{(J+\frac{1}{2})^2}{2\mu R^2} - U_0(R)\right]\right\}^{1/2} = \mathscr{K}_0(R), \tag{6.52}$$

$$\begin{aligned} \mathscr{S}_{a'}(R) \simeq \mathscr{S}_0(R) - \int_{R_0^t}^{R} \left\{\frac{2}{\mu}\left[E_0 - \frac{(J+\frac{1}{2})^2}{2\mu R'^2} - U_0(R')\right]\right\}^{-1/2} \\ \times [V_{a'a'}(R') - U_0(R') + \varepsilon_{a'} - \Delta E]\, dR', \end{aligned} \tag{6.53}$$

where $\Delta E = \mathscr{E} - E_0 \ll \mathscr{E}$,

$$\mathscr{S}_0(R) = \int_{R_0^t}^{R} \mathscr{K}_0(R')\, dR' \tag{6.54}$$

and R_0^t is determined from the condition

$$\mathscr{K}_0(R_0^t) = 0. \tag{6.55}$$

We introduce now a new variable

$$t(R) = \int_{R_0^t}^{R} \left\{\frac{2}{\mu}\left[E_0 - \frac{(J+\frac{1}{2})^2}{2\mu R'^2} - U_0(R')\right]\right\}^{-1/2} dR' \tag{6.56}$$

in (6.46a) and $-t(R)$ in (6.46b). Then taking into account that under quasi-classical conditions the operator d/dR which might appear in $\exp(\mathrm{i}\mathscr{S}_{a'})\, V_{a'a} \exp(-\mathrm{i}\mathscr{S}_{a'})$ is replaced by $-\mathrm{i}\mathscr{K}_{a'}(R)$ and omitting all small terms in accord with the previous discussion, we get

$$\mathrm{i}\frac{db_{a'}^-}{dt} = \sum_{a'' \neq a'} V_{a'a''}(t) \exp\left[\mathrm{i}\int_0^t (V_{a'a'} + \varepsilon_{a'} - V_{a''a''} - \varepsilon_{a''})\, dt'\right] b_{a''}^-, \tag{6.57a}$$

$$\mathrm{i}\frac{db_{a'}^+}{dt} = \sum_{a'' \neq a'} V_{a'a''}(t) \exp\left[\mathrm{i}\int_0^t (V_{a'a'} + \varepsilon_{a'} - V_{a''a''} - \varepsilon_{a''})\, dt'\right] b_{a''}^+. \tag{6.57b}$$

Equation (6.57a) refers to $t < 0$ and must be solved with the initial condition

$$b_{a'}^-(-\infty) = \delta_{a'a} \exp(-\mathrm{i}\pi/4) \exp(\mathrm{i}\delta_a) \tag{6.58}$$

following from (6.47a). Equation (6.57b) refers to $t > 0$ and yields the scattering matrix according to the relation

$$b_{a'}^{+}(+\infty) = -S_{a'a}^{J} \exp(\mathrm{i}\pi/4) \exp(-\mathrm{i}\delta_{a'}). \tag{6.59}$$

We have almost arrived at the common-trajectory equations. The only problem remaining is how to match b_a^+ and b_a^- in the turning-point region near $t = 0$, thus setting the initial conditions on (6.57b). It can be seen that the common-trajectory equations valid for the whole range of t-variation $-\infty < t < +\infty$ result from (6.57a, b) solely when the transitions between different states in the turning-point region can be neglected. Then the relations between $b_{a'}^{+}(0)$ and $b_{a'}^{-}(0)$ in different channels are independent and obviously coincide with those for the left turning point in the case of elastic scattering, see (5.49):

$$b_{a'}^{+}(0) = -\mathrm{i} b_{a'}^{-}(0). \tag{6.60}$$

Introducing the function

$$c_{a'}(t) = \begin{cases} b_{a'}^{-}(t) \exp(\mathrm{i}\pi/4) \exp(-\mathrm{i}\delta_a) & \text{for } t < 0, \\ -b_{a'}^{+}(t) \exp(-\mathrm{i}\pi/4) \exp(-\mathrm{i}\delta_a) & \text{for } t > 0, \end{cases} \tag{6.61}$$

we see that taking into account (6.60), the two sets of (6.57a) and (6.57b), each effective within a half of the t-variation range, reduce to one set of semiclassical scattering equations describing the variation of amplitudes over the whole t-variation range:

$$\mathrm{i}\frac{dc_{a'}}{dt} = \sum_{a'' \neq a'} V_{a'a''}(t) \exp\left[\mathrm{i}\int_0^t (V_{a'a'} + \varepsilon_{a'} - V_{a''a''} - \varepsilon_{a''})\, dt'\right] c_{a''}, \tag{6.62}$$

to be solved with the initial conditions

$$c_{a'}(-\infty) = \delta_{a'a}. \tag{6.63}$$

This yields the quasi-classical scattering matrix according to the relation

$$c_{a'}(+\infty) = \exp[-\mathrm{i}(\delta_{a'} + \delta_a)]\, S_{a'a}^{J}. \tag{6.64}$$

The explicit form of (6.62) in the atomic basis will be

$$\begin{aligned} \mathrm{i}\frac{dc_{a'v'}}{dt} = \sum_{a''v'' \neq a'v'} \Bigg\{ & \exp\left[-\mathrm{i}(v' - v'') \int_0^t \frac{(J + \frac{1}{2})}{R^2}\, dt'\right] \\ & \times \exp\left[\mathrm{i}\int_0^t (\langle a'v'|\, V_{\mathrm{AB}}\, |a'v'\rangle + \varepsilon_{a'} - \langle a''v''|\, V_{\mathrm{AB}}\, |a''v''\rangle \right. \\ & \left. - \varepsilon_{a''})\, dt'\right] \langle a'v'|\, V_{\mathrm{AB}}\, |a''v''\rangle \Bigg\} c_{a''v''}. \end{aligned} \tag{6.65}$$

The quantity in braces in the rhs of (6.65) is seen to coincide with the coupling matrix $\mathscr{V}^{\mathrm{SFS}}_{a'v',av}$ from Table 6.2 when the classical trajectory in the potential $U_0(R)$ with $E = E_0$ and $b = (J + 1/2)/k_0\,[k_0 = (2\mu E_0)^{1/2}]$ is used in the matrix. Therefore,

$$c_{a'v'}(+\infty) = {}^{\mathrm{sc}}S^{\mathrm{SFS}}_{a'v';av}|_{b=(J+1/2)/k_0}. \tag{6.66}$$

The same approximation as for $\mathscr{S}_{a'v'}(R)$ [see (6.53)] is used for the phase shifts entering (6.64). Thus the following quantity is substituted for $\delta_{a'}$ in (6.64):

$$\bar{\delta}_{av} = \delta_0 - [\tau_0(\varepsilon_a - \Delta E)/2] - \int_0^\infty (\langle av|\,V_{\mathrm{AB}}\,|av\rangle - U_0)\,dt$$
$$+ \int_0^\infty [v(J+\tfrac{1}{2})/\mu R^2]\,dt \simeq \delta_{av} + v(\pi - \eta_0)/2. \tag{6.67}$$

Here δ_0 is the quasi-classical phase shift for the potential $U_0(R)$ at an energy E_0, η_0 is the deflection function,

$$\tau_0 = 2\partial\delta_0/\partial E_0 \tag{6.68}$$

is the time delay [6.53], and δ_{av} is the elastic-scattering quasi-classical phase shift for the potential $\langle av|\,V_{\mathrm{AB}}\,|av\rangle$ at an energy $\mathscr{E} - \varepsilon_a$. Bearing in mind (6.67) and using the relation between ${}^{\mathrm{sc}}S^{\mathrm{SFS}}_{a'v';av}$ and ${}^{\mathrm{sc}}S_{a'v';av}$ given in Table 6.3, we get

$$S^J_{a'v';av} = \exp[\mathrm{i}(\delta_{av} + \delta_{a'v'})]\,{}^{\mathrm{sc}}S_{a'v';av}|_{b=(J+1/2)/k_0} \tag{6.69}$$

in agreement with the result obtained in the eikonal approximation, see (6.38), and with the general interpretation of $S^J_{a'v';av}$ as the probability amplitude in the rotating coordinate system (Chap. 2).

For the molecular basis, (6.62) coincides with the semiclassical equations with the coupling matrix $\mathscr{V}_{a'\tilde{\Omega}';a\tilde{\Omega}}$ (Table 6.2). Therefore it follows from (6.64) that

$$S^J_{a'\tilde{\Omega}';a\tilde{\Omega}} = \exp[\mathrm{i}(\delta_{a'\tilde{\Omega}'} + \delta_{a\tilde{\Omega}})]\,{}^{\mathrm{sc}}S_{a'\tilde{\Omega}';a\tilde{\Omega}}|_{b=(J+1/2)/k_0}, \tag{6.70}$$

where again $\delta_{a\tilde{\Omega}}$ must be taken in the form

$$\delta_{a\tilde{\Omega}} = \delta_0 - [\tau_0(\varepsilon_a - \Delta E)/2] - \int_0^\infty (U_{a\tilde{\Omega}} - \varepsilon_a - U_0)\,dt. \tag{6.71}$$

The relations (6.69) or (6.70) between the semiclassical and quasi-classical S matrices obviously solve the problem of using the former for calculation of the differential cross sections. A particularly simple situation arises here if ${}^{\mathrm{sc}}S_{a'a}$ varies much more slowly with J than does $\exp[\mathrm{i}(\delta_{a'} + \delta_a)]$. Then (6.69) or (2.126), into which $S^J_{a'\tilde{\Omega}';a\tilde{\Omega}}$ given by (6.70) is substituted, has the form of (2.58) and the general approach to the calculation of differential cross sections for scattering through classical angles based on the representation (2.58) can be applied (see Sect. 2.1.2). In the common-trajectory approximation the expressions for the scattering amplitudes and differential cross sections obtained by this approach are considerably simplified, since all $\eta^{A}_{a'a}(J)$ can be set equal to each other and to $\eta_0(J)$.

Finally, the conditions for the quasi-classical derivation of the common-trajectory approximation will once more be emphasized.

i) The motion in each channel is quasi-classical, except in the well localized turning-point regions. This condition allows one to neglect the elastically reflected waves arising in the classically allowed regions.
ii) The coupling between different channels is small compared to the kinetic energy of the nuclei. This means that the inelastically reflected waves in the classically allowed regions can be neglected.
iii) The elastic trajectories in all channels differ only slightly. This condition makes it possible to introduce a common trajectory in the classically allowed region.
iv) The coupling between channels can be neglected in the turning-point region. This enables one to match in this region the common-trajectory equations for the amplitudes of the incoming and outgoing waves and to obtain equations that would be effective over the entire range $-\infty < t < +\infty$.

The inequalities controlling the fulfillment of conditions (i–iii) have been given above [see (6.49, 50 and 51), respectively]. Consider now in more detail the condition (iv). It corresponds to a replacement of the original scattering problem by a modified one, with interaction between channels vanishing in the region where the quasi-classical approximation fails. It follows immediately that condition (iv), generally speaking, cannot be adopted in the near-adiabatic situation, where even small modifications of the interaction considerably change the transition amplitudes. Note also in this respect that in the common-trajectory approximation there exists an intrinsic uncertainty in E_0 over the range $\mathscr{E}$ to $\mathscr{E} - \varepsilon_{\max}$. Therefore this approximation is adequate only if the results obtained are insensitive to the variation of E_0 over this range. Usually this is not the case under near-adiabatic conditions. Fortunately, under such conditions adiabatic perturbation theory can be applied, avoiding the use of the semiclassical approximation (Sect. 6.2). We assume, therefore, that the coupling between all channels in the molecular basis is strong in the region where the motion is quasi-classical (the Massey parameters for all transitions are of the order of unity). Then the inequality controlling the fulfillment of condition (iv) can be obtained from the following considerations.

The length l_0 of the region where the quasi-classical approximation for motion in the potential $U_0(R)$ fails, is of the order of the first oscillation period of the Airy function describing the nuclear motion near the turning point, i.e.

$$l_0 \sim (\mu F_0)^{-1/3} \sim (R^*/\mu E_0)^{1/3}, \qquad \text{where} \tag{6.72}$$

$$F_0 = -\left.\frac{dU_0}{dR}\right|_{R=R_0^t} \sim E_0/R^*.$$

Note that the effect of the centrifugal potential is neglected here. This is legitimate for order-of-magnitude estimates when considering the scattering through classical angles. The time t_0 taken by the nuclei to pass through this region will be

$$t_0 \sim (l_0\,\mu/F_0)^{1/2} \sim R^*(k_0 R^*)^{-1/3}/(k_0/\mu). \tag{6.73}$$

The switching off the interaction near the turning point for the time interval t_0 will not change the transition amplitudes considerably provided

$$|V_{a'a} t_0| \ll 1. \tag{6.74}$$

The explicit forms of this inequality are

$$|\langle a'\tilde{\Omega}|\, V_{AB}\, |a\tilde{\Omega}\rangle/(U_{a'\tilde{\Omega}} - U_{a\tilde{\Omega}})|\, (k_0 R^*)^{-2/3} \ll 1 \tag{6.75a}$$

for radial coupling in the molecular basis,

$$J(k_0 R^*)^{-4/3} \ll 1 \tag{6.75b}$$

for Coriolis coupling in the molecular basis, and

$$(|\langle a'v'|\, V_{AB}\, |av\rangle|/E_0)\,(k_0 R^*)^{2/3} \ll 1 \tag{6.75c}$$

for the atomic basis.

As seen from the above discussion, condition (iv) is rather stringent. However a more detailed study of the two-state case based on the scattering equations in the momentum representation shows that condition (iv) need not always be satisfied for the semiclassical approximation to be valid [6.33]. It can be replaced by other conditions which will be formulated in Sect. 7.3.1 in connection with a derivation of the common-trajectory approximation for the two-state model. Another example illustrating the superfluous stringency of this condition is provided by scattering through small angles described in the eikonal approximation.

It must be said that conditions (ii) and (iii) are also not obligatory for some specific models. For instance, the two-state linear or similarly constructed model with a slowly varying off-diagonal interaction matrix element allow a semiclassical treatment for any energy (Chaps. 7 and 8).

Note finally that just as in the eikonal approximation the fulfillment of conditions (i–iv) still does not allow a literal interpretation of the functions $c_{a'}(t)$ as probability amplitudes of electronic transitions for nuclei which happen to be at a certain distance at the moment t [6.34].

6.2 Matching Approach

6.2.1 Matching Solution of Scattering Equations

The scattering equations for slow quasi-classical collisions can often be decoupled over some intervals of R by choosing an appropriate basis for the electronic functions. If these intervals are separated by well localized regions, called hereafter $\mathscr{M}$-regions, where either coupling is essential, or the quasi-classical conditions are not fulfilled, or both, the scattering equations must be solved only within these regions. The asymptotes of a solution away from the center of an $\mathscr{M}$-region will have the familiar uncoupled JWKB form which allows the solution of the whole problem to be constructed with scattering boundary conditions by matching the JWKB uncoupled asymptotes over different intervals of R. In a way, the matching approach is a kind of S matrix propagation from small to large R based upon known JWKB solutions for switching from one coupling region to another [6.60–66].

Consider now the scattering equations in the most general form (6.40) for one (say the n-th) $\mathscr{M}$-region. Let R_n be the center of this region, and $R_n^<$ and $R_n^>$ be reasonably estimated boundaries beyond which the radial wave functions $\chi_{a'}(R)$ have reached their uncoupled JWKB asymptotic behavior. To the left and right of R_n, these asymptotics will be of a different form:

$$\chi_{a'}(R) = \mathscr{K}_{a'}^{-1/2}\left[b^-_{<a'}\exp\left(-\mathrm{i}\int\limits_{R_n^<}^{R}\mathscr{K}_{a'}\,dR\right) + b^+_{<a'}\exp\left(\mathrm{i}\int\limits_{R_n^<}^{R}\mathscr{K}_{a'}\,dR\right)\right]$$
$$\text{for } R < R_n^<, \tag{6.76}$$
$$\chi_{a'}(R) = \mathscr{K}_{a'}^{-1/2}\left[b^-_{>a'}\exp\left(-\mathrm{i}\int\limits_{R_n^>}^{R}\mathscr{K}_{a'}\,dR\right) + b^+_{>a'}\exp\left(\mathrm{i}\int\limits_{R_n^>}^{R}\mathscr{K}_{a'}\,dR\right)\right]$$
$$\text{for } R > R_n^>,$$

where $\mathscr{K}_{a'}$ is the local radial momentum for uncoupled motion in the electronic state $|a'\rangle$, as defined by (6.41).

Similar to (5.32) which connects the amplitudes $b^{\alpha'}_{>}$ and $b^{\alpha''}_{<}$ on both sides of the $\mathscr{L}$-region for elastic scattering, we introduce the matrix $\mathscr{M}$ connecting the amplitudes $b^{\alpha'}_{>a'}$ and $b^{\alpha''}_{<a''}$:

$$b^{\alpha'}_{>a'} = \sum_{\alpha'' a''} \mathscr{M}^{\alpha'\alpha''}_{a'a''} b^{\alpha''}_{<a''}, \tag{6.77}$$

where the summation in the rhs of (6.77) is extended over α'', assuming either + or − values, and over all states a'' retained in the scattering equations. Again, similar to (5.33) which introduces the functions

$$\varphi^{\alpha''}_{a''}(R, R_n) = \exp\left(i\alpha'' \int_{R_n}^{R} \mathscr{K}_{a''} dR \right), \tag{6.78}$$

$\chi_{a'}(R)$ can be expressed via $\varphi^{\alpha''}_{a''}$ over each region $R_{n-1} < R < R_n$, and via matrices $\mathscr{M}(R_n)$ for successive distances R_n.

If $\chi_{a'}$ is represented over the interval $R_n < R < R_{n+1}$ as

$$\chi_{a'}(R) = \chi^{-}_{a'}(R) + \chi^{+}_{a'}(R), \tag{6.79}$$

then $\chi^{\alpha'}_{a'}$ can be written in the form

$$\chi^{\alpha'}_{a'}(R) = \mathscr{K}_{a'}^{-1/2} \sum_{\alpha'', \ldots; a'', \ldots} \varphi^{\alpha'}_{a'}(R, R^{>}_{n}) \times \mathscr{M}^{\alpha'\alpha''}_{a'a''}(R_n) \ldots \varphi^{\alpha''\ldots'}_{a''\ldots'}(R^{<}_{k}, R^{t}_{a''\ldots'}) b^{\alpha''\ldots'}_{a''\ldots'}, \tag{6.80}$$

where $b^{\alpha''\ldots'}_{a''\ldots'}$, is the amplitude of the function $\mathscr{K}^{-1/2}_{a''\ldots'} \varphi^{\alpha''\ldots'}_{a''\ldots'}$ in the first uncoupled interval, $R^{t}_{a''\ldots'} < R < R^{<}_{k}$ where $R^{t}_{a''\ldots'}$ is the left turning point and $R^{<}_{k}$ is the left boundary of the first $\mathscr{M}$-region.

Propagating in this way to the right, the wave functions $\chi_{a'}$ for closed channels reach the right turning points, and for open channels $\chi_{a'}$ tend to the free-motion asymptotes for R well beyond the last $\mathscr{M}$-region with its center at R_m.

Now, for closed channels the incoming and outgoing components of the function $\chi_{a''}$ to the left of the right turning point must be related in such a way as to ensure proper JWKB behavior in this region. If there is no coupling near the right turning points, this is achieved by demanding that the relation (5.48) between JWKB amplitudes holds, or, which is the same, the relation between $\chi^{+}_{a''}$ and $\chi^{-}_{a''}$ as such holds at the turning point $R^{t}_{a''}$. Thus, for each molecular state a'' corresponding to a closed channel, the following relation will be valid:

$$\sum_{\alpha''', \ldots; a''', \ldots} \varphi^{-}_{a''}(R^{t}_{a''}, R_m) \mathscr{M}^{-\alpha'''}_{a''a'''}(R_m) \ldots \varphi^{\alpha''\ldots'}_{a''\ldots'}(R^{<}_{k}, R^{t}_{a''\ldots'}) b^{\alpha''\ldots'}_{a''\ldots'}$$
$$= -i \sum_{\alpha''', \ldots; a''', \ldots} \varphi^{+}_{a''}(R^{t}_{a''}, R^{>}_{m}) \mathscr{M}^{+\alpha'''}_{a''a'''}(R_m) \ldots$$
$$\times \varphi^{\alpha''\ldots'}_{a''\ldots'}(R^{<}_{k}, R^{t}_{a''\ldots'}) b^{\alpha''\ldots'}_{a''\ldots'}. \tag{6.81}$$

If there is no coupling close to the turning points as well, the coefficients $b^{+}_{a''\ldots'}$ and $b^{-}_{a''\ldots'}$ satisfy (5.49):

$$b^{+}_{a''\ldots'} = -\mathrm{i}\, b^{-}_{a''\ldots'}. \tag{6.82}$$

For open channels, the wave function $\chi^{\alpha'}_{a'}$ must satisfy the scattering boundary conditions (6.42). For an incoming component this means, as follows from comparing (6.80) and (6.42),

$$\delta_{a'a} = \lim_{R\to\infty} \exp(\mathrm{i}k'R - J\pi/2) \sum_{\substack{\alpha'',\ldots\\ a'',\ldots}} \varphi^{-}_{a'}(R, R^{>}_{m})\, \mathscr{M}^{-\alpha''}_{a'a''}(R_m)\ldots \times \varphi^{\alpha''\ldots'}_{a''\ldots'}(R^{<}_{k}, R^{\mathrm{t}}_{a''\ldots'})\, b^{\alpha''\ldots'}_{a''\ldots'}. \tag{6.83}$$

For a fixed state quantum number a, (6.81–83) constitute a system of $2N$ algebraic equations for the $2N$ unknown coefficients $b^{\alpha''\ldots'}_{a''\ldots'}$, provided the scattering equations are written in a basis of N electronic wave functions: (6.81) involves N_{c} closed channels, (6.82) $N_{\mathrm{c}} + N_{\mathrm{o}} = N$ both closed and open channels, and (6.83) N_{o} open channels. The solution of these equations yields the coefficients $b^{\alpha''\ldots'}_{a''\ldots'}$ which are then used to calculate the S matrix. Each column of the scattering matrix $S_{a'a}$ with a fixed subscript a is defined by the expression which follows from a comparison of the outgoing waves represented by (6.80, 42).

$$S_{a'a} = -\lim_{R\to\infty} \exp(-\mathrm{i}k'R + J\pi/2) \sum_{\alpha'',\ldots;\, a'',\ldots} \varphi^{+}_{a'}(R, R^{>}_{m})\, \mathscr{M}^{+\alpha''}_{a'a''}(R_m)\ldots \times \varphi^{\alpha''\ldots'}_{a''\ldots'}(R^{<}_{k}, R^{\mathrm{t}}_{a''\ldots'})\, b^{\alpha''\ldots'}_{a''\ldots'}. \tag{6.84}$$

We reiterate once more that this approach can be used provided the different $\mathscr{M}$-regions are well separated, so that the uncoupled JWKB asymptotics represent solutions between two successive $\mathscr{M}$-regions. This is usually the case when the phase integrals and their differences for states a' and a'' are great:

$$\left|\int_{R^{>}_{n}}^{R^{<}_{n+1}} \mathscr{K}_{a'}\, dR\right| \gg 1; \quad \left|\int_{R^{>}_{n}}^{R^{<}_{n+1}} \mathscr{K}_{a'}\, dR - \int_{R^{>}_{n}}^{R^{<}_{n+1}} \mathscr{K}_{a''}\, dR\right| \gg 1. \tag{6.85}$$

Two particular cases follow from this general approach.

If the states a'' are not mutually coupled by any nondiagonal interaction, all the $\mathscr{M}$ matrices are diagonal in the subscripts and (6.81, 83) factorize for each $a' = a$. Then each $\mathscr{M}$ matrix becomes the $\mathscr{L}$ matrix discussed in Sect. 5.2 and the S matrix degenerates into an elastic scattering matrix for each independent channel.

If there are no violation regions of the JWKB approximation, except at the turning points, the $\mathscr{M}$ matrices are diagonal in the superscripts. In this case the $\mathcal{M}$ matrix will be called an $\mathcal{N}$ matrix, defined as

$$\mathcal{N}_{a'a''} = \mathcal{M}^{++}_{a'a''} = (\mathcal{M}^{--}_{a'a''})^* . \tag{6.86}$$

with $R_n^<$, R_n, $R_n^>$ coalescing into R_n.

Provided the closed channels are uncoupled, (6.83) for open channels can be solved explicitly making use of (6.86) and the unitarity of $\mathcal{N}$ matrices:

$$\mathcal{N}^{-1}_{a'a''} = \mathcal{N}^*_{a''a'} . \tag{6.87}$$

In this way, we get

$$\begin{aligned} S_{a'a} = {} & \lim_{R\to\infty} \exp(-\mathrm{i}k'R - \mathrm{i}kR + \mathrm{i}\pi J) \\ & \times [A(R, R_m)\, \mathcal{N}(R_m)\, A(R_m, R_{m-1}) \dots A(R_k, R^{\mathrm{t}})\, \mathrm{i} A(R_k, R^{\mathrm{t}}) \dots \\ & \times A(R_m, R_{m-1})\, \mathcal{N}^{\mathrm{tr}}(R_m)\, A(R, R_m)]_{a'a} . \end{aligned} \tag{6.88}$$

Here the A's are diagonal matrices of uncoupled evolution:

$$\begin{aligned} A_{a'a''}(R, R_n) &= \delta_{a'a''} \exp\left(\mathrm{i} \int_{R_n}^{R} \mathcal{K}_{a'}\, dR \right), \\ A_{a'a''}(R, R^{\mathrm{t}}_{a'}) &= \delta_{a'a''} \exp\left(\mathrm{i} \int_{R^{\mathrm{t}}_{a'}}^{R} \mathcal{K}_{a'}\, dR \right), \end{aligned} \tag{6.89}$$

and the $\mathcal{N}(R_n)$ are matrices describing transitions in localized regions centered at $R = R_n$ ($\mathcal{N}$-regions). To find these one has to solve scattering equations in some regions of R encompassing R_n. If to the left of R_n the JWKB solution is of the form

$$\chi_{a'}(R) = \delta_{a'a''} \mathcal{K}_{a'}^{-1/2} \exp\left(\mathrm{i} \int_{R_n}^{R} \mathcal{K}_{a'}\, dR' \right), \qquad R < R_n^< , \tag{6.90}$$

its asymptote to the right of R_n yields a column of the $\mathcal{N}$ matrix,

$$\chi_{a'}(R) = \mathcal{N}_{a'a''} \mathcal{K}_{a'}^{-1/2} \exp\left(\mathrm{i} \int_{R_n^>}^{R} \mathcal{K}_{a'}\, dR \right), \qquad R > R_n^> . \tag{6.91}$$

As the next simplifying step of the matching approach, we introduce the common-trajectory approximation to calculate the $\mathcal{N}$ matrix. This implies the use of a local common trajectory $X(t) = R(t) - R_n$ during some time interval centered at $t = t_n$. The scattering equations (6.40) are now replaced by the time-dependent equations (6.57) which must be solved under the following initial condition at $t < t_n$, i.e. before passing the coupling region:

$$b_{a'}(t) = \delta_{a'a''}, \qquad t < t_n^< . \tag{6.92}$$

The solution for $t > t_n^>$, i.e. after passing the coupling region, directly gives the local common-trajectory $\mathcal{N}$ matrix

$$b_{a'}(t) = \mathcal{N}_{a'a''}, \qquad t > t_n^> . \tag{6.93}$$

Note that in different $\mathcal{N}$-regions, there will be different local common trajectories not necessarily constituting different parts of one common trajectory for the entire collision. Thus the condition (6.51) must be fulfilled locally in each $\mathcal{N}$-region, with optimal choice of $U_0(R)$. Between two successive $\mathcal{N}$-regions the common-trajectory approximation may appear to be invalid. In this respect the representation (6.88) is more general than the common-trajectory approximation, but it suffers from the limitation imposed by the assumption of localized coupling regions.

Now, if the common-trajectory approximation conditions (6.51) are fulfilled for all R and the coupling is localized, (6.88) can be further simplified using the expansion (6.53) in calculating the phase factors of the diagonal matrices A in (6.88). Then (6.88) will represent the connection between a quasi-classical S matrix and one of the semiclassical S matrices discussed in Sect. 6.1.

Since the solution of a time-dependent comparison equation is in general a difficult problem, two limiting situations, where the elements of $\mathcal{N}$ matrices can be calculated explicitly using the near-adiabatic or sudden character of coupling, will be considered in the next sections.

An interesting example of the matching approach is the treatment of the long-range Coriolis coupling between molecular states converging to the same asymptotic limit. Assuming that at large distances, say at $R > R_{\mathrm{M}}$, this type of coupling is the only one of importance, the scattering equations in the molecular basis factorize into blocks, each labeled by an appropriate quantum number j (again, we consider for definiteness the scattering of an atom A on a structureless particle B). Each block describes the evolution of the molecular states $|a\tilde{\Omega}'\rangle$, well approximated by the atomic states $|j\tilde{\Omega}'\rangle$ in the molecular frame, into free atomic states.

The semiclassical comparison equations for this region are just (6.57 a, b) with the following specifications: for each block $\varepsilon_{a'} = \varepsilon_{a''}$, the nondiagonal interactions are the Coriolis coupling matrix elements calculated for the $|j\tilde{\Omega}'\rangle$ basis, and the diagonal interactions are the adiabatic molecular terms $U_{j\Omega'}$. Assume also that the high-energy approximation (rectilinear trajectories) is valid, and introduce the time t_{M} (or the distance R_{M}) as the moment when the Coriolis coupling becomes comparable with the splitting of the molecular terms.

Under these conditions the comparison equations become

$$\mathrm{i}\dot{b}_{j\tilde{\Omega}'} = -\sum_{\tilde{\Omega}''=\tilde{\Omega}'\pm 1} r^j_{\tilde{\Omega}',\tilde{\Omega}''}(vb/R^2)\exp\left\{\mathrm{i}\int_{t_M}^{t}[U_{j\Omega'}(R) - U_{j\Omega''}(R)]\,dt'\right\} b_{j\tilde{\Omega}''},$$

$$r^j_{\tilde{\Omega}',\tilde{\Omega}'\pm 1} = (1/2)\,[(j \pm \tilde{\Omega}' + 1)(j \mp \Omega')]^{1/2}, \qquad R = (b^2 + v^2t^2)^{1/2}. \tag{6.94}$$

For simplicity consider the case of integer j. There will be no coupling between degenerate states in (6.94) (for half-integer j there is the Coriolis coupling between degenerate states $|j\,1/2\rangle$ and $|j - 1/2\rangle$), and for $t < t_M^<$ these equations effectively decouple, as $r^j_{\tilde{\Omega}',\tilde{\Omega}''}(vb/R^2) \ll |U_{j\Omega'} - U_{j\Omega''}|$. This means that the initial conditions for solving comparison equations are

$$b_{j\tilde{\Omega}'}(t) = \delta_{\tilde{\Omega}'\tilde{\Omega}''}, \qquad t < t_M^< . \tag{6.95}$$

For $t > t_M^>$, the equations (6.94) remain coupled, but can be decoupled by transformation from the molecular basis $|j\tilde{\Omega}'\rangle$ to the space-fixed atomic basis $|jv\rangle\exp[\mathrm{i}\psi(t)]$. Thus for $t > t_M^>$, the solution of the comparison equations (6.94) will be

$$b_{j\tilde{\Omega}'}(t) = \sum_v \mathcal{N}^j_{v\tilde{\Omega}'}\exp\left[\mathrm{i}v\psi(t) + \mathrm{i}\pi\tilde{\Omega}' - \int_{t_M}^{t}\langle jv|\,V_{AB}\,|jv\rangle\,dt'\right]. \tag{6.96}$$

Equation (6.96) defines the matrix $\mathcal{N}(t_M)$ which effects a transformation from the molecular to the atomic basis and thus takes into account the long-range Coriolis coupling between converging molecular states.

Suppose now that the scattering solution of molecular scattering equations, with all long-range Coriolis couplings neglected, is known. Let ${}^M\chi$ be such a solution with scattering boundary conditions

$${}^M\chi_{j'\tilde{\Omega}'}\underset{R\to\infty}{\approx}(k')^{-1/2}$$
$$\times[\delta_{j'j}\delta_{\tilde{\Omega}'\tilde{\Omega}}\exp(-\mathrm{i}kR + J\pi/2) - {}^MS^J_{j'\tilde{\Omega}';j\tilde{\Omega}}\exp(\mathrm{i}k'R - J\pi/2)] \tag{6.97}$$

defining the molecular scattering matrix ${}^MS$. For finite R, $R_m < R < R_M$, ${}^M\chi$ will be of the high energy JWKB form

$${}^M\chi_{j'\tilde{\Omega}'}(R) \simeq \mathcal{K}_{0j}^{-1/2}\delta_{j'j}\delta_{\tilde{\Omega}'\tilde{\Omega}}\exp\left[-\mathrm{i}\int_{R_j^t}^{R}\mathcal{K}_{0j}\,dR'\right.$$
$$\left. - \mathrm{i}(\pi/4) + \mathrm{i}\int_0^\infty \mu\,U_{j\Omega}\mathcal{K}_{0j}^{-1}\,dR\right] - \mathcal{K}_{0j}^{-1/2}\,{}^MS^J_{j'\tilde{\Omega}';j\tilde{\Omega}}$$
$$\times\exp\left[\mathrm{i}\int_{R_{j'}^t}^{R}\mathcal{K}_{0j'}\,dR' + \mathrm{i}(\pi/4) - \mathrm{i}\int_R^\infty \mu\,U_{j'\Omega'}\mathcal{K}_{0j'}^{-1}\,dR'\right], \tag{6.98}$$

where

$$\mathcal{K}_{0j'} = [(k')^2 - (J + 1/2)^2/R^2]^{1/2} = [2\mu(\mathcal{E} - \varepsilon_{j'}) - (J + 1/2)^2/R^2]^{1/2}$$

and the R_j^t are defined by the condition $\mathcal{K}_{0j}(R_j^t) = 0$.

Now, the high energy JWKB solution of the scattering equations for $R > R_M$ is defined, by the same arguments, via the matrix elements $S^J_{j'v';jv}$, viz.

$$\chi_{j'v'}(R) \simeq \mathcal{K}_{0j}^{-1/2}\delta_{j'j}\delta_{v'v}\exp\left[-i\int_{R_j^t}^{R}\mathcal{K}_{0j}dR' - i(\pi/4) - iv\int_R^\infty (J+1/2)R^{-2}\mathcal{K}_{0j}^{-1}dR' + i\int_R^\infty \mu\langle jv|V_{AB}|jv\rangle\mathcal{K}_{0j}^{-1/2}dR'\right]$$
$$- \mathcal{K}_{0j'}^{-1/2}S^J_{j'v';jv}\exp\left[i\int_{R_j^t}^{R}\mathcal{K}_{0j'}dR' + i(\pi/4) + iv'\int_0^\infty (J+1/2) \times R^{-2}\mathcal{K}_{0j'}^{-1}dR' - i\int_0^\infty \mu\langle j'v'|V_{AB}|j'v'\rangle\mathcal{K}_{0j'}^{-1}dR'\right]. \quad (6.99)$$

It is seen that the solutions (6.98, 99) can be matched using the time-dependent coefficients $b_{j'\tilde{\Omega}'}$ with t related to R by $R = [b^2 + (k')^2t^2/\mu^2]^{1/2}$. The ultimate result of this matching is an expression of the scattering matrix $S^J_{j'v';jv}$ via the molecular scattering matrix ${}^{M}S^J_{j'\tilde{\Omega}';j\tilde{\Omega}}$ and the matrices $C^j_{v'\tilde{\Omega}'}(J)$ describing the effect of the long-range Coriolis coupling:

$$S^J_{j'v';jv} = \sum_{\tilde{\Omega}'\tilde{\Omega}} C^{j'}_{v'\tilde{\Omega}'}\,{}^{M}S^J_{j'\tilde{\Omega}';j\tilde{\Omega}}C^j_{v\tilde{\Omega}}, \quad \text{where} \quad (6.100)$$

$$C^j_{v\tilde{\Omega}} = \exp(i\pi\tilde{\Omega} + iv\alpha_j - i\delta^{>}_{j\Omega} + i\delta^{>}_{jv})\,\mathcal{N}^j_{v\tilde{\Omega}}. \quad (6.101)$$

Here

$$\alpha_j = \int_{R_M}^{\infty}[(J+1/2)/R^2\mathcal{K}_{0j}(R)]\,dR,$$

$$\delta^{>}_{j\Omega} = \int_{R_M}^{\infty}[\mu U_{j\Omega}(R)/\mathcal{K}_{0j}(R)]\,dR, \quad (6.102)$$

$$\delta^{>}_{jv} = \int_{R_M}^{\infty}[\mu\langle jv|V_{AB}|jv\rangle/\mathcal{K}_{0j}(R)]\,dR.$$

The main implication of (6.100) is that under certain conditions the multichannel S matrix can be constructed from the molecular scattering matrix. This is a considerable simplification, since for slow atomic collisions the calculation of a molecular scattering matrix can often be reduced to a sequence of familiar two-state models of nonadiabatic coupling.

Note also one useful property of the representation (6.100). As the matrices $C^j_{v\tilde{\Omega}}$ are unitary, the following relation holds:

$$\sum_{v'v}|S^J_{j'v';jv}|^2 = \sum_{\tilde{\Omega}'\tilde{\Omega}}|{}^{M}S^J_{j'\tilde{\Omega}';j\tilde{\Omega}}|^2, \quad (6.103)$$

which means that the transition probability from all magnetic substates of the state j to all those of the state j' does not depend on the long-range Coriolis coupling.

6.2.2 Near-Adiabatic Matching

If the N-region corresponds to nonadiabatic coupling between adiabatic terms and the coupling is weak, the $\mathcal{N}$ matrix can be calculated in terms of near-adiabatic perturbation theory.

Considering first the time-dependent formulation and introducing the local common-trajectory approximation, we use the scattering equations for the amplitudes $b_{a'}$ close to the center $t = t_n$ of the nth $\mathcal{N}$-region:

$$\mathrm{i}\dot{b}_{a'} = \sum_{a''} C_{a'a''}(t) \exp\left\{\mathrm{i} \int_{t_n}^{t} [U_{a'}(t) - U_{a''}(t)]\, dt\right\} b_{a''}. \tag{6.104}$$

Here $U_{a'}$ denotes adiabatic terms and $C_{a'a''}$ is a matrix element of the nonadiabatic coupling assumed to be localized around t_n. For a near-adiabatic situation $C_{a'a''}$ is small, and one can attempt to find the solution to lowest order with respect to nonadiabatic coupling. With the initial condition (6.92), the first-order solution is

$$b_{a'}(t) = -\mathrm{i} \int_{t_n^<}^{t} C_{a'a''} \exp\left[\mathrm{i} \int_{t_n}^{t'} (U_{a'} - U_{a''})\, dt''\right] dt', \tag{6.105}$$

where $t_n^<$ is assumed to be sufficiently removed from t_n to ensure $C_{a'a''}(t_n^<) = 0$. For $t \sim t_n$ the integral over t′ in (6.105) can be approximated by the first term of the asymptotic expansion based on rapid oscillation of the integrand exponential:

$$b_{a'}(t)|_{t \sim t_n} \simeq -C_{a'a''}(t)/[U_{a'}(t) - U_{a''}(t)]. \tag{6.106}$$

An important feature of the near-adiabatic limit is that usually $b_{a'}(t_n^>) \ll b_{a'}(t_n)$. This can be seen even from (6.106), showing that $b_{a'}(t)$ tends to zero as $t \to t_n^>$ because $C_{a''a''}(t_n^>) = 0$.

If the first-order solution is used to calculate nondiagonal elements $\mathcal{N}_{a'a''}$, the result will be

$$\begin{aligned} \mathcal{N}_{a'a''} &= -\mathrm{i} \int_{t_n^<}^{t_n^>} C_{a'a''}(t) \exp\left[\mathrm{i} \int_{t_n}^{t} \Delta U_{a'a''}(t')\, dt'\right] dt, \\ \Delta U_{a'a''} &= U_{a'} - U_{a''}. \end{aligned} \tag{6.107}$$

The first contribution to the diagonal elements $\mathcal{N}_{a''a''}$ comes from the second order. The result is obtained by integrating (6.104) with $b_{a''}$ in the rhs taken from (6.105). This yields

$$\begin{aligned} \mathcal{N}_{a''a''} &= 1 - \mathrm{i}\psi_{a''} \simeq \exp(-\mathrm{i}\psi_{a''}), \\ \psi_{a''} &= \int_{t_n^<}^{t_n^>} \left\{\sum_{a'''} |C_{a''a'''}(t)|^2/[U_{a''}(t) - U_{a'''}(t)]\right\} dt. \end{aligned} \tag{6.108}$$

The $\psi_{a''}$ are sometimes referred to as dynamic phases.

An important difference between nondiagonal and diagonal elements of the $\mathcal{N}$ matrix becomes clear if these are expressed via the Massey parameter $\xi = \Delta \bar{U} \Delta R / v$ specifying nonadiabatic coupling (note that v, ΔR and $\Delta \bar{U}$ are the nuclear velocity, the coupling-region length, and the characteristic adiabatic splitting at $t = t_{\mathrm{n}}$). Solution of the integral (6.107) usually gives (see below)

$$\mathcal{N}_{a' a''} \sim \exp(-A_{a' a''} \xi), \qquad a' \neq a'', \tag{6.109}$$

whereas the estimation of $\psi_{a''}$ yields

$$\mathcal{N}_{a'' a''} \sim \exp(-A_{a'' a''}/\xi), \tag{6.110}$$

where $A_{a' a''}$ and $A_{a'' a''}$ are numerical constants.

For near-adiabatic perturbations $1/\xi$ is a small parameter of the perturbation treatment. It is seen then that $\mathcal{N}_{a'' a''}$ can be calculated, at least to the lowest order, using a regular perturbation-theory approach. This is not the case however for nondiagonal elements. For large ξ they are small, as expected, but $\mathcal{N}_{a' a''}$ cannot be represented as a power series over $1/\xi$. In other words, the regular perturbation theory cannot be used to calculate $\mathcal{N}_{a' a''}$.

It was first suggested by *Landau* [6.67] that explicit calculation of the integrals in (6.107) provides a correct exponential factor only, and that for many cases the pre-exponential factor will not differ much from unity.

The exponential factor in the rhs of (6.107) can be calculated simply by extending the integration limits to $\pm \infty$ and displacing the integration contour from the real axis of t, where the integrand is a rapidly oscillating function of t, to the complex plane up to the first singular point $t = t_{\mathrm{s}}$ of the functions $\Delta U_{a' a''}(t)$ or $C_{a' a''}(t)$. The main contribution to integral (6.107) comes from the vicinity of this point, so that the value of the integral is proportional to the integrand exponent at $t = t_{\mathrm{s}}$. This approach gives

$$\mathcal{N}_{a' a''} = B_{a' a''} \exp\left[\mathrm{i} \int_{t_{\mathrm{n}}}^{t_{\mathrm{s}}} \Delta U_{a' a''}(t)\, dt \right], \qquad \mathcal{N}_{a'' a'} = -\mathcal{N}^*_{a' a''}, \tag{6.111}$$

where $B_{a' a''}$ is still undetermined. To calculate $B_{a' a''}$ one has to go beyond the first-order perturbation treatment. One way to accomplish this is to integrate a comparison equation along a contour passing close to $t = t_{\mathrm{s}}$. Again, on this contour the main contribution to the solution of the comparison equation comes from the small region Δt_{s} close to t_{s}. An important point is that the extension Δt_{s} of this region will be generally smaller than the extension Δt_{n} of the $\mathcal{N}$-region on the real t axis. This enables further simplification of the comparison equation to a near-adiabatic form. The latter may be of a simple kind with an analytical solution.

As a practical example, consider the case where a singular point t_s is a branch point t_c of an analytical function $\mathscr{U}(t)$, with two sheets of the latter on the real t axis corresponding to the adiabatic potentials $U_{a'}(t)$ and $U_{a''}(t)$. Consider now the difference $\Delta U_{a'a''}$ for complex t. Then close to t_c [6.68],

$$\Delta U_{a'a''}(t) = g(\Delta t)^{1/2}, \qquad \Delta t = t - t_c. \tag{6.112}$$

It can be shown that the isolated branch-point approximation (6.112) for $\Delta U_{a'a''}(t)$, unambiguously defines $C_{a'a''}(t)$ for this region:

$$C_{a'a''} = 1/2\Delta t. \tag{6.113}$$

For t close to t_c, make a substitution

$$b_{a'} = c_{a'} \exp\left(-\mathrm{i}\int_{t_n}^{t_c} U_{a'}\,dt\right)$$

and note that only the two amplitudes $c_{a'}$ and $c_{a''}$ will then be strongly coupled. The adiabatic comparison equation is of the form

$$\begin{aligned} dc_{a'}/d(\Delta t) &= (\mathrm{i}/2\Delta t)\exp\left[\frac{2\mathrm{i}}{3}g(\Delta t)^{3/2}\right]c_{a''}, \\ dc_{a''}/d(\Delta t) &= (\mathrm{i}/2\Delta t)\exp\left[-\frac{2\mathrm{i}}{3}g(\Delta t)^{3/2}\right]c_{a'}. \end{aligned} \tag{6.114}$$

These equations must be integrated along a contour leading from the real t axis to the vicinity of the point t_c and back to the real axis. The two first-order equations (6.114) have been shown to be equivalent to one second-order equation of the type describing the wave motion in a constant force field [6.68]. This means that analysis of the comparison equation (5.42) remains valid, provided the turning point is redefined. For the adiabatic comparison equation (6.114), the position of the Stokes and anti-Stokes lines is shown in Fig. 6.2a for t_c lying in the upper half-plane of complex t. Now, along the contour $\mathscr{L}$ which does not cross the Stokes lines, the adiabatic approximation holds everywhere except in a small double-circled region, the only one for which the adiabatic comparison equation must be solved. The solution should be found on the anti-Stokes lines where its asymptotics correspond to oscillating adiabatic phase factors propagating in general on both terms. If (6.114) is solved with boundary conditions corresponding solely to motion on the term a'' to the left of $t = t_c$, and to motion on both terms to the right of $t = t_c$, the solution directly gives the transition amplitude for the nonadiabatic transition $a'' \to a'$. Using this approach one finds that in (6.111), $B_{a'a''} = 1$, with t_n

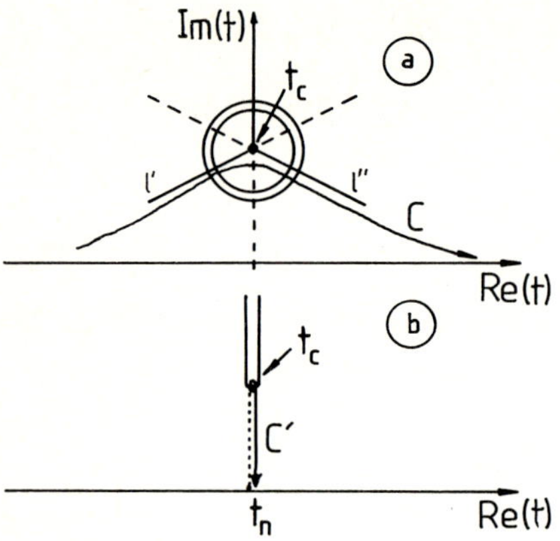

Fig. 6.2. (**a**) Stokes (– – –) and anti-Stokes (—) lines for an isolated branch point of two adiabatic terms as described by (6.114). C is the integration contour passing through the double circled region where the adiabatic approximation is invalid. (**b**) Interpretation of the nonadiabatic transition as a continuous passage from one Riemann sheet to another along the contour C' (dotted line for one sheet and full line for the other) running through the cut

defined as $\mathrm{Re}\{t_c\}$. Then $\mathcal{N}_{a'a''}$ becomes

$$\mathcal{N}_{a'a''} = \exp\left[\mathrm{i}\int_{t_n}^{t_c} \Delta U_{a'a''}(t)\,dt\right]. \tag{6.115}$$

Note that for the square-root behavior of $\Delta U_{a'a''}(t)$ close to t_c the complex exponent in (6.115) can be conceived of as a semiclassical phase integral for motion over a contour $\mathcal{L}_{a'a''}$, starting from the point t_n on one branch of the two-valued function $\mathcal{U}(t)$ up to the branch point t_c, crossing over the branch cut to arrive at another sheet and returning back to t_n on this sheet (Fig. 6.2b). Thus the exponent in (6.115) can be written in the form

$$\mathrm{i}\int_{t_n}^{t_c} \Delta U_{a'a''}(t)\,dt = \mathrm{i}\int_{\mathcal{L}_{a'a''}} \mathcal{U}(t)\,dt\,. \tag{6.116}$$

It is instructive to compare (6.115) with the simple first-order solution of (6.114). If the integration contour is taken to consist of the lines l' and l'' connected by an arc $\alpha = 2\pi/3$ of small radius below t_c (see Fig. 6.2), the contributions from l' and l'' cancel and the pre-exponential factor $B_{a'a''}$ in (6.111) will be $\pi/3$ instead of $B_{a'a''} = 1$ [6.69].

The difference in the pre-exponential factors for $\mathcal{N}_{a'a''}$ calculated by integration of a comparison equation and in terms of the perturbation theory is again due to the fact that the transition amplitude is not an analytical function of $1/\xi$, so that all orders of perturbation contribute to the pre-exponential factor. In the above example, this difference is small, and the factor $\pi/3$ is the first term of an infinite series summed up to $2\sin\pi/6$, i.e. to unity [6.70]. The difference between the two approaches is not great, provided the splitting ΔU near the isolated specific point behaves as $(\Delta t)^\gamma$, $\gamma \neq 1/2$ [6.71, 72]. However, when there are not one, but two closely lying singular points, the pre-exponential factor B can substantially differ from unity. For examples see Sect. 8.3.1 (approach and eventual coalescence of two branching points along a line parallel to the real t axis) and Sect. 9.1.3 (same as above along a line parallel to the imaginary t axis).

Now consider briefly the quasi-classical formulation of near-adiabatic perturbation theory. The basic correction to the semiclassical expression (6.115) for nondiagonal matrix elements of $\mathcal{N}$ is expected to appear in the exponent, rather than in the pre-exponential factor which is equal to unity in the semiclassical approximation.

For the most frequently encountered situation, an isolated branch point, this correction is taken into account by replacing the time integral of $\mathcal{U}(t)$ over the contour $\mathcal{L}_{a'a''}$ by the integral of $\mathcal{K}$ over the contour $\bar{\mathcal{L}}_{a'a''}$, starting from R_n on the branch a' of $\mathcal{K}$, passing through the cut at R_c and ending at R_n on the branch a'' of $\mathcal{K}$. Thus, if the local common-trajectory approximation fails, it is recommended that the following expression for off-diagonal elements of the $\mathcal{N}$ matrix be used:

$$\begin{aligned}\mathcal{N}_{a'a''} &= \exp\left[\mathrm{i}\int_{\bar{\mathcal{L}}_{a'a''}} \mathcal{K}(R)\,dR\right] \qquad (6.117)\\ &= \exp\left[\mathrm{i}\int_{R_\text{n}}^{R_\text{c}} (2\mu)^{1/2}(\mathcal{E}-U_{a'})^{1/2}\,dR - \mathrm{i}\int_{R_\text{n}}^{R_\text{c}} (2\mu)^{1/2}(\mathcal{E}-U_{a''})^{1/2}\,dR\right].\end{aligned}$$

In the case where $\Delta U_{a'a''}$ is small compared to the local kinetic energy $\mathcal{E}-U_0$, the leading term of the expansion of the exponent in (6.117) corresponds to the time-dependent formulation. This is illustrated by the exponential model (Sect. 9.1.2).

The above results correspond to what can be called the first-order near-adiabatic perturbation theory where the transition amplitude is dominated by a path which directly connects the initial and final states. It may happen that account will have to be taken of several paths directly connecting the initial and final states, or the paths that go from the initial to the final Riemann sheet via intermediate states.

The generalization of (6.111) will then be [6.57, 73, 74]

$$\mathcal{N}_{a'a''} = \sum_p B_{a'a''}(\mathcal{L}^p_{a'a''}) \exp\left[-\mathrm{i}\int_{\mathcal{L}^p_{a'a''}} \mathcal{U}(t)\,dt\right], \qquad (6.118)$$

where $\mathcal{L}^p_{a'a''}$ denotes the above-mentioned different paths and where the largest terms only must be retained. Examples of term patterns differing from simple two-valued branching can be found in [6.75, 76]. Note in concluding this section that practical use of the semiclassical (6.115) or of its quasi-classical counterpart (6.117) relies on knowledge of the analytical function $\mathcal{U}(R)$ in the plane of complex R. Though computers enable direct calculation of $\mathscr{U}(R)$ for complex R [6.77, 78], the adiabatic terms $U_{a'}(R)$ are usually known for real R, and very often for a finite number of internuclear distances. In this case the only way of constructing $\mathscr{U}(R)$ is to approximate the entity of terms $U_{a'}(R)$ by a certain multivalued function $\tilde{\mathscr{U}}(R)$, with fitting parameters chosen to make its various branches coincide with $U_{a'}(R)$ at a finite number of interatomic distances [6.79–82]. A comparison of the results thus obtained with the ab initio calculated $\mathscr{U}(R)$ shows that the accuracy of the former strongly depends on the extent to which the analytical features of $\mathscr{U}(R)$ are guessed [6.78]. Of particular interest in this connection are the papers dealing with the general study of the analytical properties of $\mathscr{U}(R)$ (see, e.g. [6.83, 84]).

6.2.3 Near-Sudden Matching

Another important case of localized coupling corresponds to the situation where coupled equations of the type (6.57) can be decoupled away from the center of the $\mathscr{N}$-region by choosing different basis functions. Here the interaction is localized by a switching of the basis, the switching time representing a small parameter in the formulation of the perturbation approach. Using the time-dependent picture, defining t_n as the center of the coupling region and choosing $t_\mathrm{n}^< < t_\mathrm{n}$ and $t_\mathrm{n}^> > t_\mathrm{n}$ as suitable boundaries of this region, assume that at $t_1 < t_\mathrm{n}^<$, (6.57 a, b) are uncoupled in the basis $\{|a'_<\rangle\}$, and for $t_2 > t_\mathrm{n}^>$ in the basis $\{|a'_>\rangle\}$. Call the time interval $\Delta t_\mathrm{n} = t_\mathrm{n}^> - t_\mathrm{n}^<$ the switching time.

The formal representation of the $\mathscr{N}$ matrix can be written via evolution operators $U(t_\mathrm{n}, t_1)$ and $U(t_2, t_\mathrm{n})$ over the intervals $t_1 - t_\mathrm{n}, t_\mathrm{n} - t_2$, and via the overlap matrix $\mathscr{N}^{(0)}_{a'_> a''_<} = \langle a'_> | a''_< \rangle|_{t=t_\mathrm{n}}$.

Explicitly, then,

$$\mathscr{N}_{a'_> a''_<} = \sum_{a'''_>, a''''_<} U_{a'_> a'''_>}(t_2, t_\mathrm{n})\, \mathscr{N}^{(0)}_{a'''_> a''''_<}\, U_{a''''_< a''_<}(t_\mathrm{n}, t_1) . \tag{6.119}$$

It is important to recognize that for sudden matching the evolution operators are close to unity, so that the zero-order approximation with respect to the switching time Δt_n is

$$\mathscr{N}_{a'_> a''_<} = \mathscr{N}^{(0)}_{a'_> a''_<} . \tag{6.120}$$

Some representation of the evolution operator must be used to find higher-order corrections. It is convenient to resort to representations that are unitary at any level of approximation. The most familiar ones are known as the Magnus representation and the Fer representation (see, e.g. [6.85]).

The Magnus representation of the evolution operator is

$$\begin{aligned} U(t, t_0) &= \exp[-\mathrm{i}A(t, t_0)] \\ &= \exp[-\mathrm{i}\,{}^{\mathrm{M}}\!A^{(1)}(t, t_0) - \mathrm{i}\,{}^{\mathrm{M}}\!A^{(2)}(t, t_0) - \mathrm{i}\,{}^{\mathrm{M}}\!A^{(3)}(t, t_0) \dots], \end{aligned} \tag{6.121}$$

where each Hermitian matrix ${}^{\mathrm{M}}\!A^{(n)}$ is expressed via integrals from commutators of the coupling matrices $\mathscr{V}(t)$. The first three terms are

$$\begin{aligned} {}^{\mathrm{M}}\!A^{(1)} &= \int_{t_0}^{t} \mathscr{V}(t_1)\, dt_1, \qquad {}^{\mathrm{M}}\!A^{(2)} = \frac{1}{2}\int_{t_0}^{t} dt_1 \int_{t_0}^{t_1} dt_2 [\mathscr{V}(t_1), \mathscr{V}(t_2)], \\ {}^{\mathrm{M}}\!A^{(3)} &= \frac{1}{6}\int_{t_0}^{t} dt_1 \int_{t_0}^{t_1} dt_2 \int_{t_0}^{t_2} dt_3 \\ &\quad \times \{[[\mathscr{V}(t_1), \mathscr{V}(t_2)], \mathscr{V}(t_3)] + [[\mathscr{V}(t_3), \mathscr{V}(t_2)], \mathscr{V}(t_1)]\}, \end{aligned} \tag{6.122}$$

where $[a, b] = ab - ba$. The Fer representation of the evolution operator is, compare with (6.121),

$$\begin{aligned} U(t, t_0) &= \exp[-\mathrm{i}\,{}^{\mathrm{F}}\!A^{(1)}(t, t_0)] \exp[-\mathrm{i}\,{}^{F}\!A^{(2)}(t, t_0)] \\ &\quad \times \exp[-\mathrm{i}\,{}^{\mathrm{F}}\!A^{(3)}(t, t_0)] \dots, \end{aligned} \tag{6.123}$$

where the Hermitian matrices ${}^{\mathrm{F}}\!A^{(n)}$ are again expressed through time integrals of commutators of $\mathscr{V}$. The first three terms are

$$\begin{aligned} {}^{\mathrm{F}}\!A^{(1)} &= \int_{t_0}^{t} \mathscr{V}(t_1)\, dt_1, \qquad {}^{\mathrm{F}}\!A^{(2)} = \frac{1}{2}\int_{t_0}^{t} dt_1 \int_{t_0}^{t_1} dt_2 [\mathscr{V}(t_1), \mathscr{V}(t_2)], \\ {}^{\mathrm{F}}\!A^{(3)} &= \frac{1}{3}\int_{t_0}^{t} dt_1 \int_{t_0}^{t_1} dt_2 \int_{t_0}^{t_2} dt_3 [\mathscr{V}(t_2), [\mathscr{V}(t_3), \mathscr{V}(t_1)]] \\ &\quad + [\mathscr{V}(t_3), [\mathscr{V}(t_2), \mathscr{V}(t_1)]]. \end{aligned} \tag{6.124}$$

Both representations collapse to one term if $\mathscr{V}(t_1)$ and $\mathscr{V}(t_2)$ commute:

$$U(t, t_0) = \exp\left[-\mathrm{i}\int_{t_0}^{t} \mathscr{V}(t')\, dt'\right]. \tag{6.125}$$

Now, if the switching time is short, the time integrals

$$A^{(1)}_{a'a''}(t_2, t_{\mathrm{n}}) = \int_{t_{\mathrm{n}}}^{t_2} \mathscr{V}_{a'_< a''_<}(t)\, dt, \qquad A^{(1)}_{a'a''}(t_{\mathrm{n}}, t_1) = \int_{t_1}^{t_{\mathrm{n}}} \mathscr{V}_{a'_> a''_>}(t)\, dt, \tag{6.126}$$

are small, and either the first-order Magnus or the first-order Fer approximations can be used to calculate the leading correction which accounts for the final duration of switching [6.86, 87]. Thus, to first order with respect to the switching time, the $\mathcal{N}$ matrix is

$$\mathcal{N}_{a'_{>}\,a''_{<}} = \sum_{a'''\,a''''} \{\exp[-\mathrm{i}A^{(1)}(t_2 t_\mathrm{n})]\}_{a'_{>}\,a'''_{>}}\,\mathcal{N}^{0}_{a'''\,a''''} \times \{\exp[-\mathrm{i}A^{(1)}(t_\mathrm{n} t_1)]\}_{a''''_{<}\,a''_{<}}\,, \tag{6.127}$$

where the exponent matrices are defined by their matrix elements in (6.126).

For a short switching time the first correction is usually important only if the zero-order result gives vanishing or very small amplitudes for certain processess. An example is provided by the $\Sigma - \Pi$ transitions in the united-atom limit (Sect. 9.3.3) where the zero- and first-order approximations compete for small deflection angles η. For a more detailed discussion of the sudden perturbation see [6.88].

As a simple illustration of the matching treatment of the long-range Coriolis coupling, assume that the potentials $U_{j\Omega}(R)$ are very steep compared to the Coriolis coupling. It will be expected then that the sudden approximation can provide a satisfactory solution with an appropriate choice of R_M which can depend on J. The matrix elements $\mathcal{N}^{j}_{\mathrm{v}\tilde{\Omega}}$ are calculated to this approximation as overlap integrals between the functions $|j\tilde{\Omega}\rangle$ and $|j\mathrm{v}\rangle$, giving [6.89, 90]

$$\mathcal{N}^{j}_{\mathrm{v}\tilde{\Omega}} = \exp(-\mathrm{i}\pi\tilde{\Omega})\,\Delta^{j}_{\mathrm{v}\tilde{\Omega}}\,. \tag{6.128}$$

Substitution of $\mathcal{N}^{j}_{\mathrm{v}\tilde{\Omega}}$ from this equation into (6.101) gives, according to (6.100), the expression for the S matrix,

$$S^{J}_{j'\,\mathrm{v}';\,j\,\mathrm{v}} = \exp(\mathrm{i}\mathrm{v}'\alpha_{j'} + \mathrm{i}\mathrm{v}\alpha_j + \mathrm{i}\delta^{>}_{j'\mathrm{v}'} + \mathrm{i}\delta^{>}_{j\mathrm{v}}) \sum_{\tilde{\Omega}'\,\tilde{\Omega}} \exp(-\mathrm{i}\delta^{>}_{j'\Omega'}) \times \Delta^{j'}_{\mathrm{v}'\tilde{\Omega}'}\,{}^{\mathrm{M}}S_{j'\tilde{\Omega}';\,j\tilde{\Omega}}\,\Delta^{j}_{\mathrm{v}\tilde{\Omega}}\exp(-\mathrm{i}\delta^{>}_{j\tilde{\Omega}})\,. \tag{6.129}$$

On general grounds it would be expected that the sudden approximation is the best for collisions with not too large J values. For this limit, $\delta^{>}_{j\mathrm{v}}, \delta^{>}_{j\Omega} < \alpha_j \ll 1$, and

$$S^{J}_{j'\,\mathrm{v}';\,j\,\mathrm{v}} = \sum_{\tilde{\Omega}'\,\tilde{\Omega}} \Delta^{j'}_{\mathrm{v}'\tilde{\Omega}'}\,{}^{\mathrm{M}}S^{J}_{j'\tilde{\Omega}';\,j\tilde{\Omega}}\,\Delta^{j}_{\mathrm{v}\tilde{\Omega}}\,. \tag{6.130}$$

For a large angular momentum the sudden solution of the comparison equation is not expected to be sufficiently accurate. Still with a reasonable choice of R_M, (6.130) would yield a correct result which follows from the perturbation treatment. For peripheric collisions R_M can be identified with the distance of closest approach for rectilinear relative motion. For this case,

$${}^{\mathrm{M}}S^{J}_{j'\tilde{\Omega}';\,j\tilde{\Omega}} = \delta_{j'j}\delta_{\tilde{\Omega}'\tilde{\Omega}}\exp(2\mathrm{i}\delta^{>}_{j\tilde{\Omega}})\,, \qquad \alpha_j = \pi/2\,. \tag{6.131}$$

Putting these values into (6.129) and summing over $\tilde{\Omega}$, $\tilde{\Omega}'$, we get

$$S^J_{j'v';jv} = \delta_{j'j}\delta_{v'v}\exp(i\pi v)\exp(2i\delta^{>}_{jv}), \tag{6.132}$$

which is the Born distorted-wave result for diagonal matrix elements (Sect. 6.3.1). The general expression (6.129) will be discussed once more in Sects. 10.1.1 and 11.2.1.

6.3 Perturbation Approach

The perturbation approach to calculation of the scattering matrix is based on a perturbation solution of the scattering equations. Since these can be formulated in different basis sets using various dynamic approximations, a variety of explicit results can be obtained. This section addresses itself to the first-order solutions of scattering equations in atomic and molecular basis sets. Such approximations known as the distorted-wave Born approximation (DWBA) and the adiabatic distorted-wave approximation (ADWA) are applicable when the coupling is weak. If it is not weak, one must resort to higher-order approximations. It appears, however, that very often direct use of the perturbation series solution is not advantageous compared to the much simpler and virtually more convenient unitarization procedure. This procedure amounts to constructing the unitary S matrix by using the first-order solution only, thus providing a reasonable extrapolation of the first-order scattering matrix to the strong-coupling regime. The unitarization procedure is of course not unique. The one we dwell on is related to the so-called exponential solution of the scattering equations.

6.3.1 First-Order Perturbation Treatment. The Born and Adiabatic Distorted-Wave Approximations

For weak coupling the scattering boundary conditions (6.42) can be satisfied by writing $\chi_{a'}$ as a superposition of the zero- and first-order contributions,

$$\chi_{a'}(R) = {}^0\chi_{a'}(R) + {}^1\chi_{a'}(R), \tag{6.133}$$

which for large R behave as

$$\begin{aligned} {}^0\chi_{a'} &\sim \delta_{a'a}\{\exp[-i(kR - \pi J/2)] - {}^{\mathrm{DWA}}S_{a'a'}\exp[i(k'R - \pi J/2)]\}, \\ {}^1\chi_{a'} &\sim (\delta_{a'a} - 1)\,{}^{\mathrm{DWA}}S_{a'a}\exp[i(k'R - \pi J/2)]. \end{aligned} \tag{6.134}$$

The zero-order functions are solutions to homogeneous equations, so that they describe the elastic scattering in uncoupled channels. These functions are used to construct the Green's function of the lhs of (6.40) and to solve inhomogeneous equations with the rhs containing ${}^0\chi_{a''}$ instead of $\chi_{a''}$. The first order solution yields ${}^1\chi_{a'}$ with the boundary conditions (6.134) ensured by those of the Green's function. The asymptotes of ${}^1\chi_{a'}$ then define ${}^{\mathrm{DWA}}S_{a'a}$.

The ultimate result of these calculations formulated as a distorted-wave scattering matrix is [6.8]

$$\begin{aligned} {}^{\mathrm{DWA}}S_{aa} &= \exp(2\mathrm{i}\delta_a^0), \\ {}^{\mathrm{DWA}}S_{a'a} &= -4\mathrm{i}\exp(\mathrm{i}\delta_{a'}^0 + \mathrm{i}\delta_a^0)\int_0^\infty V_{a'a}(R)\,{}^0\chi_{a'}\,{}^0\chi_a dR, \end{aligned} \tag{6.135}$$

where ${}^0\chi_a$ are standing zero-order distorted waves with the asymptotic behavior

$${}^0\chi_a \xrightarrow[R\to\infty]{} (\mu/k)^{1/2}\sin[kR - \pi J/2 + \delta_a^0]. \tag{6.136}$$

The distorted-wave approximation (6.135) is valid under the general condition

$${}^0\chi_{a'} \gg {}^1\chi_{a'}, \tag{6.137}$$

which must be fulfilled to ensure the replacement of $V_{a'a''}\chi_{a''}$ in the rhs of (6.40) by $V_{a'a''}{}^0\chi_{a''}$. Of course, the condition (6.137) guarantees that $|S_{a'a}|^2 \ll 1$, $a' \neq a$. The reverse might often not be true: if the transition probability is small, this does not necessarily mean that the first-order correction ${}^1\chi_{a'}$ is small compared to the zero-order function ${}^0\chi_{a'}$.

The practical calculations of integrals simplify for quasi-classical scattering conditions. For an isolated turning point in each channel the wave function ${}^0\chi_a$ can be considered in the uniform Airy approximation, combining the Airy function solution in the vicinity of the turning point with correct JWKB asymptotes [6.91–95]:

$$\begin{aligned} {}^0\chi_a(R) &= (\pi\mu/\mathscr{K}_a)^{1/2}(3\mathscr{S}_a/2)^{1/6}\,\mathrm{Ai}[-(3\mathscr{S}_a/2)^{2/3}], \\ \mathscr{S}_a(R) &= \int_{R_a^t}^R \mathscr{K}_a(R)\,dR. \end{aligned} \tag{6.138}$$

If neither the tunneling region nor the classically allowed region close to the turning point make a dominant contribution to the integral in (6.135), the JWKB wave function

$${}^0\chi_a(R) = [\mu/\mathscr{K}_a]^{1/2}\sin\left[\int_{R_a^t}^R \mathscr{K}_a(R)\,dR + \pi/4\right] \tag{6.139}$$

can be used everywhere.

The next simplifying step would be to use the common-trajectory approximation. The product of two JWKB functions is represented as the sum of two oscillating functions, the contribution from the rapidly oscillating term is neglected, and the phase of another much more slowly oscillating term is approximated according to the common-trajectory method:

$$\int_{R_a^t}^{R} \mathcal{K}_a dR' - \int_{R_{a'}^t}^{R} \mathcal{K}_{a'} dR' \simeq \int_{R_0^t}^{R} \mu[(V_{a'a'} - V_{aa})/\mathcal{K}_0]\, dR. \tag{6.140}$$

Now we specify the basis. The scattering matrix in the atomic basis for the common-trajectory distorted-wave Born approximation (CTDWBA), ${}^{\mathrm{CTDWBA}}S$, can be expressed via a first-order semiclassical (FOSC) matrix by an equation similar to (6.64):

$$ {}^{\mathrm{CTDWBA}}S^{J}_{a'v';av} = \exp(\mathrm{i}\bar{\delta}_{a'v'} + \mathrm{i}\bar{\delta}_{av})\, {}^{\mathrm{FOSC}}S^{\mathrm{SFS}}_{a'v';av}(b)|_{b=(J+1/2)/k}. \tag{6.141}$$

Here the $\bar{\delta}_{av}$ are defined by (6.67) and

$$ {}^{\mathrm{FOSC}}S^{\mathrm{SFS}}_{a'v';av}(b) = \delta_{a'a}\delta_{v'v} - \mathrm{i}\int_{-\infty}^{+\infty} \mathscr{V}^{\mathrm{SFS}}_{a'v';av}(t)\, dt, \tag{6.142}$$

where $\mathscr{V}^{\mathrm{SFS}}_{a'v';av}$ is given in Table 6.2. The high-energy (HE) approximation to (6.141) involves the calculation of the phase shifts $\bar{\delta}_{av}$ in (6.67) and the integrals in (6.142) along a rectilinear path, U_0 being taken as zero:

$$ {}^{\mathrm{HE}}\bar{\delta}_{av} = -\mathrm{RP}\int_{0}^{\infty} \langle av|\, V_{\mathrm{AB}}\, |av\rangle\, dt + (\pi v/2), \tag{6.143}$$

$$ {}^{\mathrm{HE\,FOSC}}S^{\mathrm{SFS}}_{a'v';av} = \delta_{a'a}\delta_{v'v} - \mathrm{i}\,\mathrm{RP}\int_{-\infty}^{+\infty} \mathscr{V}^{\mathrm{SFS}}_{a'v';av}(t)\, dt, \tag{6.144}$$

where RP before the integral refers to integration along a rectilinear path. Substitution of the phase shifts and semiclassical scattering matrix into (6.141) gives the high-energy Born distorted-wave scattering matrix, ${}^{\mathrm{HEDWBA}}S$.

Provided all scattering shifts are small, the ${}^{\mathrm{HEDWBA}}S$ matrix reduces to the semiclassical Born scattering matrix ${}^{\mathrm{SCBA}}S$:

$$ {}^{\mathrm{SCBA}}S^{J}_{a'v';av} = \exp\left[\mathrm{i}\frac{\pi}{2}(v'+v)\right]\left[\delta_{a'a}\delta_{v'v} - \mathrm{i}\,\mathrm{RP}\int_{-\infty}^{+\infty} \bar{\mathscr{V}}^{\mathrm{SFS}}_{a'v';av}(t)\, dt\right], \tag{6.145}$$

where $\bar{\mathscr{V}}^{\mathrm{SFS}}_{a'v';av}$ is related to $\mathscr{V}^{\mathrm{SFS}}_{a'v';av}$ by (6.10).

For a molecular basis the common-trajectory adiabatic distorted-wave (CTADWA) scattering matrix is expressed via ${}^{\mathrm{FOSC}}S$ in the same way as in (6.70):

$$ {}^{\mathrm{CTADWA}}S^{J}_{a'\tilde{\Omega}';a\tilde{\Omega}} = \exp[\mathrm{i}(\delta_{a'\tilde{\Omega}'} + \delta_{a\tilde{\Omega}})]\, {}^{\mathrm{FOSC}}S_{a'\tilde{\Omega}';a\tilde{\Omega}}. \tag{6.146}$$

Here the $\delta_{a\tilde{\Omega}}$ are defined by (6.71) and

$$^{\mathrm{FOSC}}S_{a'\tilde{\Omega}';a\tilde{\Omega}} = \delta_{a'a} - \mathrm{i} \int_{-\infty}^{+\infty} \mathscr{V}_{a'\tilde{\Omega}';a\tilde{\Omega}}(t)\,dt, \tag{6.147}$$

where $\mathscr{V}_{a'\tilde{\Omega}';a\tilde{\Omega}}$ is given in Table 6.2. Note that $\mathscr{V}_{a'\tilde{\Omega}';a\tilde{\Omega}}$ involves only radial and rotational nonadiabatic coupling. However, if there is another small perturbation not accounted for in constructing the adiabatic molecular terms, it can also be included into $\mathscr{V}_{a'\tilde{\Omega}';a\tilde{\Omega}}(t)$.

The long-range Coriolis coupling between molecular states converging to the same limit deserves special mention. A partial contribution of this coupling to the FOSC scattering matrix can be crudely estimated for steep interactions by neglecting the differences in molecular potentials starting at some distance $R > R_{\mathrm{M}}$. This contribution can be small only for values of J which correspond to rotation of the molecular axis through small angles, before the diatom enters the adiabatically uncoupled region. Under quasi-classical conditions it would usually correspond to backward scattering. For the opposite case the long-range Coriolis coupling is not weak and cannot be treated by the first-order perturbation procedure. Still, if this long-range coupling is eliminated in the scattering equations by use of the matching approximation, the molecular scattering matrix can be calculated by the adiabatic distorted-wave approximation (ADWA). For the high-energy limit of ADWA, see [6.96].

6.3.2 Unitarized Distorted-Wave Approximation

The first step of the unitarization procedure is to single out large scattering phase shifts, i.e. to represent the $^{\mathrm{DWBA}}S$-matrix in the form

$$^{\mathrm{DWBA}}S = \exp(\mathrm{i}\delta^0)\,(I + \mathrm{i}\beta)\exp(\mathrm{i}\delta^0), \tag{6.148}$$

where $\exp(\mathrm{i}\delta^0)$ is the diagonal matrix

$$[\exp(\mathrm{i}\delta^0)]_{a'a} = \delta_{a'a}\exp(\mathrm{i}\delta^0_a) \tag{6.149}$$

and I is a unit matrix. Then the linear function $I + \mathrm{i}\beta$ is replaced by another reasonable function $F(\beta)$ with the same behavior at small β, but retaining the unitarity. The exponential function is most widely used in the unitarized distorted-wave approximation (UDWA) [6.97]. The scattering matrix is then

$$^{\mathrm{UDWA}}S = \exp(\mathrm{i}\delta^0)\exp(\mathrm{i}\beta)\exp(\mathrm{i}\delta^0). \tag{6.150}$$

To this approximation the solution of the scattering equations is replaced by the matrix exponentiation which is of course a considerable simplification.[1]

For the atomic basis in the common-trajectory approximation, the middle exponential in (6.150) can be identified with the unitarized semiclassical (USC) scattering matrix for the SFS representation

$$ {}^{\mathrm{USC}}S^{\mathrm{SFS}} = \exp({}^{\mathrm{FOSC}}S - I) = \exp(-\mathrm{i}\,W^{\mathrm{SFS}}), \tag{6.151} $$

where the matrix elements of W^{SFS} are

$$ W^{\mathrm{SFS}}_{a'v';av} = \int_{-\infty}^{+\infty} \mathscr{V}^{\mathrm{SFS}}_{a'v';av}\,dt. \tag{6.152} $$

In this way we arrive at the common-trajectory unitarized distorted-wave approximation (CTUDWA),

$$ {}^{\mathrm{CTUDWA}}S^{J}_{a'v';av} = \exp(\mathrm{i}\bar{\delta}_{a'v'} + \mathrm{i}\bar{\delta}_{av})\,\langle a'v'|\exp(-\mathrm{i}\,W^{\mathrm{SFS}})\,|av\rangle|_{b=(J+1/2)/k}. \tag{6.153} $$

For a rectilinear path, (6.153) gives a high-energy (HE) UDWA scattering matrix

$$ {}^{\mathrm{HEUDWA}}S^{J}_{a'v';av} = \exp(\mathrm{i}\,{}^{\mathrm{HE}}\bar{\delta}_{a'v'} + \mathrm{i}\,{}^{\mathrm{HE}}\bar{\delta}_{av})\,\langle a'v'|\exp(-\mathrm{i}\,{}^{\mathrm{HE}}W^{\mathrm{SFS}})\,|av\rangle, $$

$$ {}^{\mathrm{HE}}W^{\mathrm{SFS}}_{a'v';av} = \mathrm{RP}\int_{-\infty}^{+\infty} \mathscr{V}^{\mathrm{SFS}}_{a'v';av}\,dt. \tag{6.154} $$

On the other hand, a physically similar approximation can be formulated by direct exponential unitarization of the semiclassical Born scattering matrix (USCBA) using (6.145):

$$ {}^{\mathrm{USCBA}}S^{J}_{a'v';av} = \exp\left[\mathrm{i}\frac{\pi}{2}(v'+v)\right]\langle a'v';av|\exp(-\mathrm{i}\,{}^{\mathrm{HE}}\bar{W}^{\mathrm{SFS}})\,|av\rangle, $$

$$ {}^{\mathrm{HE}}\bar{W}^{\mathrm{SFS}}_{a'v';av} = \mathrm{RP}\int_{-\infty}^{+\infty} \bar{\mathscr{V}}^{\mathrm{SFS}}_{a'v';av}(t)\,dt. \tag{6.155} $$

Note that for HEUDWA and USCBA the scattering matrices will in general be different, HEUDWA being presumably more accurate. However, if the basis used is that for the degenerate atomic state, both approximations may give comparable results. In practical applications, the use of (6.155) is simpler than that of (6.154) and this explains its wider use. For instance, USCBA has been used in the calculation of the depolarization cross sections of atoms A with different j values in collisions with spherically symmetric atoms B [6.100].

[1] This approach with a uniform approximation for β has been analysed for some two- and three-state problems [6.91–94, 98, 99]

The transformation of the USCBA S matrix from the $|jv\rangle$ to the $|j\bar{m}\rangle$ representation can be carried out using (2.51). The ultimate result will be

$$\begin{aligned} {}^{\mathrm{USCBA}}S^{J}_{a'\bar{m}';a\bar{m}} &= {}^{\mathrm{USC}}\bar{S}^{\mathrm{SF}}_{a'\bar{m}';a\bar{m}}(b)|_{b=(J+1/2)/k} = \langle a'\bar{m}'|\exp(-\mathrm{i}\,{}^{\mathrm{HE}}\bar{W}^{\mathrm{SF}})|a\bar{m}\rangle, \\ {}^{\mathrm{HE}}\bar{W}^{\mathrm{SF}}_{a'\bar{m}';a\bar{m}} &= \mathrm{RP}\int_{-\infty}^{\infty} \bar{\mathscr{V}}^{\mathrm{SF}}_{a'\bar{m}';a\bar{m}}dt. \end{aligned} \tag{6.156}$$

Since ${}^{\mathrm{sc}}\bar{S}^{\mathrm{SF}}$ is related to the eikonal scattering matrix by (6.29), ${}^{\mathrm{USCBA}}S$ can be considered to represent the unitarized first-order solution to eikonal equations.

Note that the USCBA scattering matrices (6.155, 156) are simply related to the first-order Magnus (FOMA) scattering matrix which is obtained from (6.125) in the limit $t_0 \to -\infty$, $t \to +\infty$ with $\bar{\mathscr{V}}^{\mathrm{SFS}}$ or $\bar{\mathscr{V}}^{\mathrm{SF}}$ substituted for the coupling matrix:

$$\begin{aligned} {}^{\mathrm{USCBA}}S^{J}_{a'v';av} &= \exp\left[\mathrm{i}\frac{\pi}{2}(v'+v)\right]{}^{\mathrm{FOMA}}S_{a'v';av}(\bar{\mathscr{V}}^{\mathrm{SFS}})|_{b=(J+1/2)/k}, \\ {}^{\mathrm{USCBA}}S^{J}_{a'\bar{m}';a\bar{m}} &= {}^{\mathrm{FOMA}}S_{a'\bar{m}';a\bar{m}}(\bar{\mathscr{V}}^{\mathrm{SF}})|_{b=(J+1/2)/k}. \end{aligned} \tag{6.157}$$

The FOMA scattering matrix has been discussed for some problems in [6.101, 102].

If the interatomic-interaction-induced perturbation is of a pulse type (PT), i.e. if the collision time τ_0 is short compared to the reciprocal of all eigenfrequencies $\omega_{a'a''}$ (extremely nonadiabatic conditions) the matrix $\bar{\mathscr{V}}^{\mathrm{SF}}$ can be replaced by the interaction matrix V as such, to give

$$\begin{aligned} {}^{\mathrm{PT}}S^{J}_{a'\bar{m}';a\bar{m}} &= \langle a'\bar{m}'|\exp(-\mathrm{i}\,{}^{\mathrm{PT}}W)|a\bar{m}\rangle, \\ {}^{\mathrm{PT}}W_{a'\bar{m}';a\bar{m}} &= \mathrm{RP}\int_{-\infty}^{\infty} V_{a'\bar{m}';a\bar{m}}dt. \end{aligned} \tag{6.158}$$

If the basis used to calculate the matrix $V_{a'\bar{m}';a\bar{m}}$ is considered to be complete, the matrix exponentiation can be accomplished in closed form, yielding the Glauber approximation (GA) (see, e.g. [6.52, 56]):

$$\begin{aligned} {}^{\mathrm{GA}}S^{J}_{a'\bar{m}';a\bar{m}} &= \langle a'\bar{m}'|\exp(-\mathrm{i}W)|a\bar{m}\rangle, \\ W = W(\boldsymbol{r}_i, b) &= \mathrm{RP}\int_{-\infty}^{\infty} V_{\mathrm{AB}}[\boldsymbol{r}_i, \boldsymbol{R}(t)]\,dt. \end{aligned} \tag{6.159}$$

Here W is a function of the electronic coordinates in the standard atomic space-fixed frame $\bar{x}$, $\bar{y}$, $\bar{z}$. The integration of V_{AB} is carried out over nuclear coordinates along the rectilinear trajectory with impact parameter b. This approach can be generalized for bent trajectories [6.103] in which case V_{AB} stands for that part of the interaction responsible for inelastic transitions. For further discussion concerning the relation of the Glauber approximation to the Magnus expansion see [6.104].

7. Two-State Scattering Problem

It was emphasized in the preceding chapter that for quasi-classical slow collisions the general multichannel problem may often be reduced to several two-state problems. Two-state scattering also provides a simple example illustrating the major features peculiar to inelastic scattering. Therefore, the two-state scattering problem will be considered in detail in this and the next two chapters.

This chapter will be concerned with the general formulation and features of the two-state problem. Since many of the general properties of this problem have been discussed in the classic book by *Mott* and *Massey* [7.1], special emphasis will be placed here on aspects specific to slow quasi-classical collisions.

7.1 The Two-State Model. Adiabatic and Diabatic Representations

It seems that from a formal point of view, the two-state model can be used only in the case of two interacting molecular states with $\Omega = 0$, which correlate with spherically symmetric atomic states ($j_1 = j_2 = 0$). This would ensure the absence of the Coriolis interaction which, for large R, strongly couples almost-degenerate states with different values of Ω emerging from any degenerate atomic state. However, if the radial wave functions of the two-state model reach their asymptotic form (2.10) at a value of $\bar{R}$ such that the Coriolis coupling with other quasi-degenerate states becomes important for $R > \bar{R}$ only, then $\bar{R}$ may be formally taken as infinity when solving the two-state radial equations. Then the complete scattering matrix can readily be constructed using the matching procedure and the notion of the dynamic Hund coupling cases (Sects. 3.3.2, 6.2). Keeping this in mind, the two-state model can often be used for states with $\Omega \neq 0$. The degeneracy of such states is of no importance because of the selection rules for radial and Coriolis coupling.

The first step in constructing the two-state scattering model is to retain just two coupled equations for the states of interest in the complete system

(2.145). Strictly speaking, the matrix elements of V_R and V_{Cor} in these equations must be calculated using the general equations (2.146, 147). These expressions contain ill-defined terms responsible for the electron momentum transfer which ensure the proper behavior of the nonadiabatic coupling when $R \to \infty$. However, since for slow quasi-classical collisions the electron momentum-transfer effects are usually unimportant at the finite values of R determining the transition probability (see discussion in Sect. 2.3.1), the corresponding terms may be omitted from V_R and V_{Cor} in constructing the two-state model. If the remaining matrix elements display a spurious asymptotic behavior, then for large R some truncation procedure may be used to ensure the mathematical correctness of the problem. The explicit condition which such truncated coupling should satisfy will be given below. The terms $\langle a' \tilde{\Omega}' | j_{Y'}^2 + j_{X'}^2 | a'' \tilde{\Omega}'' \rangle / 2\mu R^2$ can also be omitted, as follows from the estimates (2.111–113).

Thus, for large J the two-state model scattering equations in the molecular (adiabatic) basis may be written in the form (for simplicity all the indices except a, a', $a'' = 1, 2$ are suppressed):

$$\begin{aligned}
&\left(-\frac{1}{2\mu}\frac{d^2}{dR^2} + \frac{(J+1/2)^2}{2\mu R^2} + U_1 - \mathscr{E}\right)\chi_1^a \\
&\quad = -\left[\langle 1|V_R|1\rangle \chi_1^a + (\langle 1|V_R|2\rangle + \langle 1|V_{Cor}|2\rangle)\chi_2^a\right], \\
&\left(-\frac{1}{2\mu}\frac{d^2}{dR^2} + \frac{(J+1/2)^2}{2\mu R^2} + U_2 - \mathscr{E}\right)\chi_2^a \\
&\quad = -\left[(\langle 2|V_R|1\rangle + \langle 2|V_{Cor}|1\rangle)\chi_1^a + \langle 2|V_R|2\rangle \chi_2^a\right],
\end{aligned} \tag{7.1}$$

where

$$\begin{aligned}
\langle 1|V_R|1\rangle &= \frac{1}{2\mu} G_{11}(R), \\
\langle 2|V_R|2\rangle &= \frac{1}{2\mu} G_{22}(R), \\
\langle 1|V_R|2\rangle &= -\frac{1}{\mu} g(R)\frac{d}{dR} - \frac{1}{2\mu} g'(R) + \frac{1}{2\mu} G_{12}(R), \\
\langle 2|V_R|1\rangle &= \frac{1}{\mu} g(R)\frac{d}{dR} + \frac{1}{2\mu} g'(R) + \frac{1}{2\mu} G_{21}(R), \\
\langle 1|V_{Cor}|2\rangle &= \langle 2|V_{Cor}|1\rangle^* = -\frac{J+1/2}{\mu R^2}\langle 1|j_{X'}|2\rangle,
\end{aligned} \tag{7.2}$$

and

$$G_{a'a''}(R) = \langle (d\langle a'|/dR)|(d|a''\rangle/dR)\rangle = G^*_{a''a'},$$
$$g(R) = \langle 1|d/dR|2\rangle, \tag{7.3}$$
$$g'(R) = dg(R)/dR.$$

As was shown in Sect. 2.2, nonadiabatic coupling between two states occurs either if $\Omega_1 = \Omega_2$ (radial coupling) or if $\Omega_1 = \Omega_2 \pm 1$ (Coriolis coupling). In the absence of an external magnetic field, it is always possible to construct molecular basis functions $|a\rangle$ such that all the matrix elements in (7.1) are real; it will be assumed below that this is the case.

When $\Omega_1 = \Omega_2 \pm 1$, the molecular terms 1 and 2 may cross and therefore from the point of view of the definition of diabatic states given in Sect. 3.3.2, the molecular states $|1\rangle$ and $|2\rangle$ are diabatic.

When the states $|1\rangle$ and $|2\rangle$ have the same symmetry and only radial coupling persists, then (7.1) corresponds to an adiabatic representation. However, in the two-state case considered here, transformation to the diabatic representation can easily be performed [7.2–6]. For a real (7.1) this transformation is

$$\chi^{\mathrm{a}}_{a'} = \sum_{a''} A_{a'a''}(R)\,\chi^{\mathrm{d}}_{a'}, \quad \text{with} \tag{7.4}$$

$$A_{a'a''} = \begin{pmatrix} \cos\vartheta & \sin\vartheta \\ -\sin\vartheta & \cos\vartheta \end{pmatrix}. \tag{7.5}$$

The mixing angle must be chosen such that the terms containing d/dR in the transformed equations will vanish. This corresponds to the condition

$$\langle 1\,\mathrm{d}|\,d/dR\,|2\,\mathrm{d}\rangle = 0, \tag{7.6}$$

where the diabatic basis functions are

$$|a'\mathrm{d}\rangle = \sum_{a''} A_{a''a'}|a''\rangle. \tag{7.7}$$

It follows from (7.6, 7) that

$$\vartheta'(R) = -g(R) \tag{7.8}$$

and therefore

$$\vartheta(R) = -\int_{\infty}^{R} g(R')\,dR' + \vartheta(\infty). \tag{7.9}$$

Almost always $\vartheta(\infty) = 0$ (except for an exact resonance) and for $R \to \infty$ the diabatic and adiabatic representations coincide. This will be assumed in deriving all general expressions in this section. In the case of exact resonance $g(R) = 0$, $G_{12}(R) = 0$ and $\vartheta(\infty) = \pi/4$. Therefore, the diabatic and adiabatic representations are different everywhere and in the latter the problem reduces to two independent elastic scattering problems (Sects. 7.2, 4).

It is seen from (7.9) that the diabatic representation may be properly introduced only if for large R, the radial coupling tends to zero faster than R^{-1}. This is the condition that must be ensured by the truncation procedure.

The scattering equations in the diabatic basis defined by (7.4–9) are

$$\left(-\frac{1}{2\mu}\frac{d^2}{dR^2}+\frac{(J+1/2)^2}{2\mu R^2}+V_{11}+T_{11}-\mathscr{E}\right)\chi_1^{\mathrm{d}}=-(V_{12}+T_{12})\chi_2^{\mathrm{d}},$$

$$\left(-\frac{1}{2\mu}\frac{d^2}{dR^2}+\frac{(J+1/2)^2}{2\mu R^2}+V_{22}+T_{22}-\mathscr{E}\right)\chi_2^{\mathrm{d}}=-(V_{21}+T_{21})\chi_1^{\mathrm{d}}, \tag{7.10}$$

where

$$V_{11}=U_1\cos^2\vartheta+U_2\sin^2\vartheta,$$
$$V_{22}=U_1\sin^2\vartheta+U_2\cos^2\vartheta, \tag{7.11}$$
$$V_{12}=V_{21}=(U_1-U_2)\cos\vartheta\sin\vartheta,$$

and

$$T_{11}=-\frac{\vartheta'^2}{2\mu}+\frac{1}{2\mu}(G_{11}\cos^2\vartheta-2G_{12}\cos\vartheta\sin\vartheta+G_{22}\sin^2\vartheta),$$

$$T_{22}=-\frac{\vartheta'^2}{2\mu}+\frac{1}{2\mu}(G_{11}\sin^2\vartheta+2G_{12}\cos\vartheta\sin\vartheta+G_{22}\cos^2\vartheta), \tag{7.12}$$

$$T_{12}=T_{21}=\frac{1}{2\mu}[G_{12}(\cos^2\vartheta-\sin^2\vartheta)+(G_{11}-G_{22})\cos\vartheta\sin\vartheta].$$

The inverse relations between the adiabatic and diabatic interactions are

$$U_{1,2}=\frac{V_{11}+V_{22}}{2}\pm\frac{1}{2}[(V_{11}-V_{22})^2+4V_{12}^2]^{1/2}, \tag{7.13 a}$$

$$\tan 2\vartheta=2V_{12}/(V_{11}-V_{22}). \tag{7.13 b}$$

Note that the interaction in (7.10) involves two terms: the usual potential matrix $V_{a'a''}$, and the matrix $T_{a'a''}$ reflecting the possible dependence of the diabatic states on R. Indeed, condition (7.6) implies that the diabatic wave functions are independent of R only in the case when the interaction between just two states of separated atoms, $|1\rangle|_{R\to\infty}$ and $|2\rangle|_{R\to\infty}$, is taken into account when constructing the adiabatic states. Then the completeness arguments (Sect. 3.3.2) are applicable and it may be shown that

$$G^{(0)}_{a'a''}=\vartheta'^2\delta_{a'a''}. \tag{7.14}$$

Therefore in this case $T_{a'a''}$ in (7.10) identically vanishes. Thus the quantity $\Delta G_{a'a''} = G_{a'a''} - G^{(0)}_{a'a''}$ characterizes the strength of the radial coupling between the two and other states. However, in terms of the two-state model $\Delta G_{a'a''}/2\mu$ may be formally treated as an additional potential matrix in the partially adiabatic basis obtained by diagonalizing $V_{a'a''}$ given by (7.11). That is, $U_{a'}$ plays the part of $U^{(0)}_{a'\tilde{\Omega}'}$ and $\Delta G_{a'a''}/2\mu$ that of $\langle a'\tilde{\Omega}'|V^{(1)}_{\mathrm{AB}}|a''\tilde{\Omega}''\rangle$ in the general form (2.103) corresponding to the partially adiabatic basis. From this point of view $U_{a'}$ and $V_{a'a''}$ can always be renormalized in such a way that $G_{a'a''} = G^{(0)}_{a'a''}$. Therefore, everywhere below, the two-state-model scattering equations will be written with $G_{a'a''} = G^{(0)}_{a'a''}$ and $T_{a'a''} = 0$. It is noted here that the modification of $U_{a'a''}$ and $V_{a'a''}$ due to $\Delta G_{a'a''} \neq 0$ might be important only if the nuclear motion is essentially quantal, since in the common-trajectory approximation $\Delta G_{a'a''}$ is omitted together with all other second derivative terms (Sect. 6.1).

Another approximation first suggested by *Takayanagi* [7.7] is often used in constructing two-state scattering models. This so-called modified wave-number approximation, is effective when the centrifugal potential $(J+1/2)^2/2\mu R^2$ varies slowly compared with $V_{a'a''}$ or $U_{a'}$ and $g(R)$. In this approximation, the constant value $(J+1/2)^2/2\mu R_0^2$ is substituted for the centrifugal potential. Here R_0 is some average distance corresponding to maximum coupling. For steep interactions the result is expected to depend only weakly on the value of R_0. After this substitution the scattering equations assume the form of the *s*-wave scattering equations with a modified energy,

$$\tilde{\mathscr{E}}(J) = \mathscr{E} - [(J+1/2)^2/2\mu R_0^2]. \tag{7.15}$$

Keeping this in mind, the two-state model equations in the diabatic basis will be written hereafter in the form

$$\begin{aligned} \left(-\frac{1}{2\mu}\frac{d^2}{dR^2} + V_{11} - \mathscr{E}\right)\chi_1^{\mathrm{d}} &= -V_{12}\chi_2^{\mathrm{d}}, \\ \left(-\frac{1}{2\mu}\frac{d^2}{dR^2} + V_{22} - \mathscr{E}\right)\chi_1^{\mathrm{d}} &= -V_{21}\chi_2^{\mathrm{d}}, \end{aligned} \tag{7.16}$$

where, depending on the nature of the interaction, either $V_{a'a''}$ involves the centrifugal potential, or the modified energy (7.15) stands for $\mathscr{E}$. The adiabatic basis equations are

$$\begin{aligned} \left(-\frac{1}{2\mu}\frac{d^2}{dR^2} + U_1 - \mathscr{E}\right)\chi_1^{\mathrm{a}} &= -\frac{1}{2\mu}g^2\chi_1^{\mathrm{a}} + \frac{1}{\mu}\left(g\frac{d}{dR} + \frac{1}{2}g'\right)\chi_2^{\mathrm{a}}, \\ \left(-\frac{1}{2\mu}\frac{d^2}{dR^2} + U_2 - \mathscr{E}\right)\chi_2^{\mathrm{a}} &= -\frac{1}{2\mu}g^2\chi_2^{\mathrm{a}} - \frac{1}{\mu}\left(g\frac{d}{dR} + \frac{1}{2}g'\right)\chi_1^{\mathrm{a}}, \end{aligned} \tag{7.17}$$

where, according to (7.8, 13 b)

$$g = [(V'_{11} - V'_{22})\, V_{12} - (V_{11} - V_{22})\, V'_{12}]\,[(V_{11} - V_{22})^2 + 4\, V_{12}^2]^{-1}\,. \tag{7.18}$$

These equations must be solved with the boundary conditions

$$\chi_{a'}^{a;\,d} \xrightarrow[R \to \infty]{} (k_{a'})^{-1/2} \{\exp[-\mathrm{i}\,(k_a R - \gamma)]\, \delta_{a'a} - S_{a'a} \exp[\mathrm{i}\,(k_{a'} R - \gamma)]\}\,, \tag{7.19a}$$

$$\chi_{a'}^{a;\,d} \xrightarrow[R \to 0]{} 0\,, \tag{7.19b}$$

where $\gamma = 0$ if the modified wave-number approximation is used, and $\gamma = J\pi/2$ otherwise.

The different approximate approaches considered in Chap. 6 can be used to solve (7.16 or 7.17) depending on the character of the interaction and the kinetic energy. If the coupling of the adiabatic terms is well localized and far removed from the turning points, the S matrix can be constructed by the matching method. This will be considered in Sect. 7.2 for the representative case of one coupling region. A practical application of the matching method usually requires knowledge of the semiclassical two-state transition probabilities and phases. Therefore, and also since the semiclassical approximation can be applied in situations when the matching method fails (the couplings are not localized), the two-state semiclassical approximation will be discussed in more detail in Sect. 7.3.

7.2 Construction of the Two-State Quasi-classical *S* Matrix by the Matching Method

Consider first the terms shown in Fig. 7.1. Assume that the nonadiabatic transition is localized in a region of width ΔR_s with center R_s and the local common trajectory approximation is valid (see Sect. 6.2.1). Therefore, the nonadiabatic transition matrix $\mathcal{N}$ is obtained as a solution of the time-dependent equations in the adiabatic basis with time zero associated with R_s (Sect. 7.3.2). The general form of the matrix is

$$\mathcal{N}(R_s) = \begin{pmatrix} (1-P)^{1/2} \exp(-\mathrm{i}\,\varphi) & -P^{1/2} \exp(-\mathrm{i}\,\psi) \\ P^{1/2} \exp(\mathrm{i}\,\psi) & (1-P)^{1/2} \exp(\mathrm{i}\,\varphi) \end{pmatrix}. \tag{7.20}$$

Note that the one-way transition probability P (one-way transit through the nonadiabaticity region) is independent of R_s. The dynamic phase φ also does not depend on R_s, but ψ does.

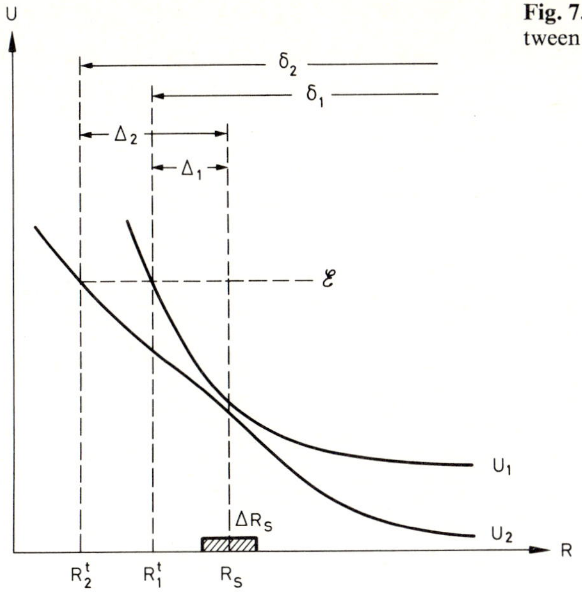

Fig. 7.1. Nonadiabatic transitions between two repulsive adiabatic terms

Substituting (7.20) into the general expression (6.88) for the S matrix, the matching method yields

$$S = \begin{pmatrix} (1 - \Pi^2)^{1/2} \exp(2\mathrm{i}\Gamma_1) & \mathrm{i}\Pi \exp(\mathrm{i}\Gamma_1 + \mathrm{i}\Gamma_2) \\ \mathrm{i}\Pi \exp(\mathrm{i}\Gamma_1 + \mathrm{i}\Gamma_2) & (1 - \Pi^2)^{1/2} \exp(2\mathrm{i}\Gamma_2) \end{pmatrix}, \tag{7.21}$$

with the three parameters Γ_1, Γ_2, and Π expressed via the scattering phase shifts δ_1 and δ_2 for the adiabatic terms U_1 and U_2, the adiabatic phases

$$\Delta_{a'} = \Delta_{a'}(R_s, R^t_{a'}) = \int_{R^t_{a'}}^{R_s} \mathscr{K}_{a'}\, dR, \tag{7.22}$$

and the parameters of the $\mathscr{N}$ matrix

$$\begin{aligned} \Pi &= 2^{1/2}(\bar{\mathscr{P}}_{12})^{1/2} \sin\phi_{12}, \ \bar{\mathscr{P}}_{12} = 2P(1-P), \\ \Gamma_1 &= \delta_1 - \varphi - (1/2)\arctan[P \sin 2\phi_{12}(1 - 2P\sin^2\phi_{12})^{-1}], \\ \Gamma_2 &= \delta_2 + \varphi + (1/2)\arctan[P \sin 2\phi_{12}(1 - 2P\sin^2\phi_{12})^{-1}], \end{aligned} \tag{7.23 a}$$

$$\begin{aligned} \phi_{12} &= \Delta_{12} + \Delta\phi_{12}, \\ \Delta_{12} &= \Delta_1 - \Delta_2, \quad \Delta\phi_{12} = \psi - \varphi. \end{aligned} \tag{7.23 b}$$

Note that the oscillatory factor $\sin\phi_{12}$ responsible for the Stueckelberg oscillations contains not only the difference in adiabatic phases Δ_{12} for $R < R_s$, but also an additional dynamic phase $\Delta\phi_{12}$.

Though the Stueckelberg phase $\phi_{12} = \Delta_{12} + \Delta\phi_{12}$ is independent of the choice of R_s, both contributions Δ_{12} and $\Delta\phi_{12}$ depend on R_s. This ambiguity is implicitly connected with that encountered in choosing the center of the coupling region. Usually for noncrossing terms the three possible choices of R_s are:

i) R_s is identified with the distance R_{max} at which the radial nonadiabatic coupling is a maximum. Then

$$\partial g(R)/\partial R|_{R=R_{max}} = 0, \qquad\qquad (7.24)$$
$$\Delta_{12} = \Delta_1(R_{max}, R_1^t) - \Delta_2(R_{max}, R_2^t).$$

ii) R_s is identified with the distance R_p corresponding to the real part of the complex distance R_c at which the adiabatic term cross. Then

$$U_1(R_c) = U_2(R_c), \quad R_p = \mathrm{Re}\{R_c\}, \qquad\qquad (7.25)$$
$$\Delta_{12} = \Delta_1(R_p, R_1^t) - \Delta_2(R_p, R_2^t).$$

iii) R_s is defined implicitly by identifying Δ_{12} with the real part of the complex phase difference calculated from the tuning points to R_c:

$$\Delta_{12} = \mathrm{Re}\left\{\int_{R_1^t}^{R_c} \mathcal{K}_1(R)\,dR - \int_{R_1^t}^{R_c} \mathcal{K}_2(R)\,dR\right\}$$
$$= \Delta_1(R_s, R_1^t) - \Delta_2(R_s, R_2^t). \qquad\qquad (7.26)$$

Note that definition (iii) ensures that the dynamic correction $\Delta\phi_{12}$ vanishes in the near-adiabatic limit (Sect. 6.2.2).

If the adiabatic terms cross, R_s is taken as the crossing distance R_c.

The basic conditions for representation of the S matrix in the form (7.21) are that absolute values of the phases Δ_1, Δ_2 and their difference be large

$$\Delta_1, \Delta_2, |\Delta_{12}| \gg 1. \qquad\qquad (7.27)$$

These conditions impose certain restrictions on the variation of the channel quantum numbers which parametrize implicitly the matrix elements $S_{a'a}$. For instance, for a fixed energy $\mathscr{E}$ the difference $\Delta_1 - \Delta_2$ usually decreases with increasing J and becomes zero at some $J = J_t$. If the region of J close to J_t heavily contributes to the scattering amplitude, the S matrix in the form (7.21) cannot be used for a calculation of the differential cross section. This occurs in the threshold region and a special treatment must be invoked here (this is discussed briefly for the linear model in Sect. 8.5.3).

However, the conditions (7.27) are sometimes too stringent. This is the case for a resonant two-state process in the collisions of identical atoms (Sect. 2.3.2). Then the $\mathcal{N}$ matrix describes the transformation of the basis $|1d\rangle$, $|2d\rangle$ to the symmetry-adopted (uncoupled) basis of even and odd states. This transformation occurs for $R_s \to \infty$, so that $P = 1/2$, $\varphi = \psi = 0$, $\Delta_{12} = \delta_1 - \delta_2$. Then according to (7.21, 23) the two-state S matrix for a resonant process is

$$S = \exp(\mathrm{i}\,\delta_1 + \mathrm{i}\,\delta_2)\begin{pmatrix} \cos\Delta_{12} & \mathrm{i}\sin\Delta_{12} \\ \mathrm{i}\sin\Delta_{12} & \cos\Delta_{12} \end{pmatrix}. \tag{7.28}$$

This form of the S matrix is correct for any value of the scattering phase shifts. In the uncoupled basis the S matrix is of course diagonal,

$$S = \begin{pmatrix} \exp(2\,\mathrm{i}\,\delta_1) & 0 \\ 0 & \exp(2\,\mathrm{i}\,\delta_2) \end{pmatrix}, \tag{7.29}$$

which corresponds to independent elastic scattering in two channels for the even and the odd states.

Given the S matrix, the transition probability $\mathscr{P}_{12}$ is expressed via S_{12}:

$$\begin{aligned} &\mathscr{P}_{12} = |S_{12}|^2 = 2\,\bar{\mathscr{P}}_{12}\sin^2\phi_{12}\,, \\ &\bar{\mathscr{P}}_{12} = 2\,P(1-P), \qquad \phi_{12} = \Delta_{12} + \Delta\phi_{12}\,. \end{aligned} \tag{7.30}$$

Here $\bar{\mathscr{P}}_{12}$ can be interpreted as the averaged (over many Stueckelberg oscillations) transition probability. It depends, together with the dynamic phase $\Delta\phi_{12}$ on adiabatic terms and on the coupling in a small region ΔR_s near R_s. On the other hand, Δ_{12} depends on adiabatic terms only (but not on coupling) in the extended region $R \leqq R_s$.

The two-state S matrix in the form (7.21) is not always convenient for a calculation of the quasi-classical amplitudes by the stationary phase method (Sect. 2.1), since the parameters Γ_1, Γ_2, and Π are rapidly varying functions of the angular momentum J. For the same reason the form (7.21) is inconvenient for the representation of the results of numerical calculations: the varying values Γ_1, Γ_2, and Π cannot be used for a smooth interpolation over J. In this sense it is more convenient to represent the S matrix via the probability P, the adiabatic phase shifts δ_1 and δ_2, and the phase difference ϕ_{12}. In this parametrization,

$$\begin{aligned} S_{11} &= (1-P)\exp(2\,\mathrm{i}\,\delta_1 - 2\,\mathrm{i}\,\varphi) + P\exp(2\,\mathrm{i}\,\delta_1 - 2\,\mathrm{i}\,\varphi - 2\,\mathrm{i}\,\phi_{12}), \\ S_{22} &= (1-P)\exp(2\,\mathrm{i}\,\delta_2 + 2\,\mathrm{i}\,\varphi) + P\exp(2\,\mathrm{i}\,\delta_2 + 2\,\mathrm{i}\,\varphi + 2\,\mathrm{i}\,\phi_{12}), \\ S_{12} &= S_{21} = -\,P^{1/2}(1-P)^{1/2}\,[\exp(\mathrm{i}\,\delta_1 + \mathrm{i}\,\delta_2 + \mathrm{i}\,\phi_{12}) \\ &\quad - \exp(\mathrm{i}\,\delta_1 + \mathrm{i}\,\delta_2 - \mathrm{i}\,\phi_{12})]\,. \end{aligned} \tag{7.31}$$

Here P, δ_1, δ_2, and ϕ_{12} vary regularly with J and therefore the numerical calculation of these parameters for certain fixed J would enable the construction of smooth functions $P(J)$, $\delta_1(J)$, $\delta_2(J)$ and $\phi_{12}(J)$ by interpolation [7.8].

Turn now to a more complex case (Fig. 7.2). Unlike Fig. 7.1 it contains two separated $\mathcal{M}$-regions: one for nonadiabatic coupling at $R = R_s$ and the other, near the top of the barrier $R = R^m$, for a possible breakdown of the JWKB approximation.

The starting point for the calculation of the scattering matrix at $\mathscr{E} = \mathscr{E}' > U^m$ (classically open channel 1) is the S matrix from (7.21) (denoted now by $\bar{S}_{a'a}$), as well as the matrix $\mathscr{L}^{\alpha'\alpha''}$ which takes into account the overbarrier reflection in the term U_1 (Sect. 5.2.4).

First define the truncated matrix $\bar{s}_{a'a}$ which connects the JWKB amplitudes $b^+_{<,1}$ and $b^-_{<,1}$ to the left of R^m:

$$\begin{aligned} \bar{s}_{a'a} &= \bar{S}_{a'a} \exp(-\mathrm{i}\delta^>_{a'} - \mathrm{i}\delta^>_a), \\ \delta^>_a &= \delta_a - \Delta_a(R^t_a, R^m) - \pi/4 . \end{aligned} \tag{7.32}$$

Then construct the truncated matrix $s_{a'a}$ which accounts for overbarrier reflection at $R = R^m$ and connects the amplitudes $b^+_{>,1}$ and $b^-_{>,1}$ to the right of R^m.

For instance, s_{11} is calculated as

$$\begin{aligned} s_{11} &= -b^+_{>,1}/b^-_{>,1} = -(t b^+_{<,1} + r t^* b^-_{<,1})/(t^* b^-_{<,1} + r^* t b^+_{<,1}) \\ &= (t/t^*)(\bar{s}_{11} + r^*)/(1 + r\bar{s}_{11}), \end{aligned} \tag{7.33}$$

where r and t are elements of the $\mathscr{L}$-matrix (5.54) and two relations are used:

$$\begin{aligned} & rt^* + r^*t = 0, \\ & \quad \bar{s}_{11} = -b^+_{<,1}/b^-_{<,1} . \end{aligned} \tag{7.34}$$

Along this line the full matrix s will be

$$s = (1 + r\bar{s}_{11})^{-1} \begin{pmatrix} (\bar{s}_{11} + r^*)\, t/t^* & \bar{s}_{12} t \\ \bar{s}_{12} t & \bar{s}_{22} + r(\bar{s}_{11}\bar{s}_{22} - \bar{s}^2_{12}) \end{pmatrix} . \tag{7.35}$$

Here the matrix $\bar{s}$ is, according to (7.21, 32),

$$\bar{s} = \begin{pmatrix} (1 - \Pi^2)^{1/2} \exp(2\mathrm{i}\gamma_1) & \mathrm{i}\Pi \exp(\mathrm{i}\gamma_1 + \mathrm{i}\gamma_2) \\ \mathrm{i}\Pi \exp(\mathrm{i}\gamma_1 + \mathrm{i}\gamma_2) & (1 - \Pi^2)^{1/2} \exp(2\mathrm{i}\gamma_2) \end{pmatrix}, \tag{7.36}$$

with γ_a differing from Γ_a in (7.21) in that the δ_a are replaced by $\Delta_a(R^t_a, R^m) + \pi/4$.

Finally the S matrix is calculated as

$$S_{a'a} = s_{a'a} \exp(\mathrm{i}\delta^>_{a'} + \mathrm{i}\delta^>_a). \tag{7.37}$$

The representations (7.35, 37), though derived for $\mathscr{E} > U^m$, are suitable for extension to energies $\mathscr{E} \le U^m$ (e.g., $\mathscr{E} = \mathscr{E}''$, Fig. 7.2). This is accompanied

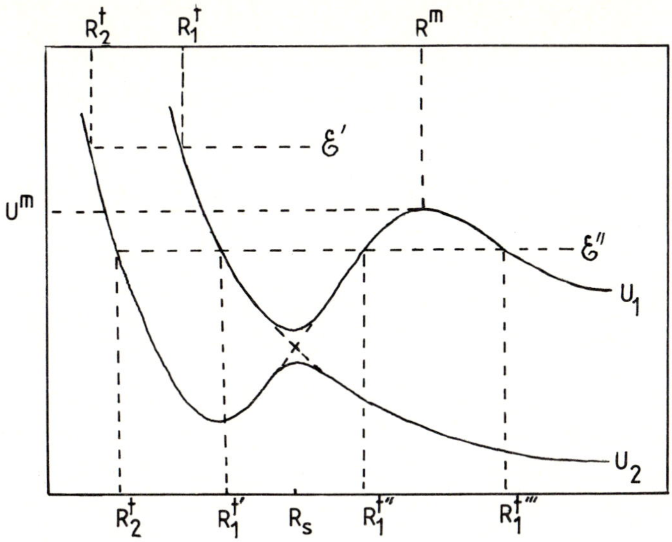

Fig. 7.2. Nonadiabatic transitions between two nonmonotonic adiabatic terms

by replacing Δ_1 (R_1^t, R^m) by a more general expression Δ_1 $(R_1^*, R_1^<)$ and identifying $\delta_1^>$ with the partial contribution to the phase shift from the region $R \geq R_1^>$, according to the one-channel convention (5.60).

The scattering matrix (7.37) with $t = 1$, $r = 0$ reduces of course to the S matrix (7.21). The opposite limiting case corresponds to an impenetrable barrier, i.e., to $t = 0$, $r = -\mathrm{i}$.

In this limit the S matrix (7.37) is diagonal, viz.,

$$S_{11} = (1 - \bar{s}_{11})^{-1}(\bar{s}_{11} + \mathrm{i}) \exp(2\mathrm{i}\delta_1^>) = \exp[2\mathrm{i}(\delta_1^> + \pi/4)], \tag{7.38}$$

$$S_{22} = (1 - \bar{s}_{11})^{-1}[\bar{s}_{22} - \mathrm{i}(\bar{s}_{11}\bar{s}_{22} - \bar{s}_{12}^2)] \exp(2\mathrm{i}\delta_2^>). \tag{7.39}$$

Equation (7.38) describes scattering on the outer part $(R > R_1^{t'''})$ of the potential U_1, and $\delta_1^> + \pi/4$ is simply the appropriate scattering phase shift. Eq. (7.39) describes scattering in the potential field U_2 accompanied by temporary capture in the U_1 potential well (the Feshbach resonance). Substituting $\bar{s}_{a'a}$ from (7.36) into (7.39), we get

$$S_{22} = \frac{(1 - \Pi^2)^{1/2} - \mathrm{i} \exp(2\mathrm{i}\gamma_1)}{1 - \mathrm{i}(1 - \Pi^2)^{1/2} \exp(2\mathrm{i}\gamma_1)} \exp(2\mathrm{i}\Gamma_2). \tag{7.40}$$

If Π is small, the first factor in (7.40) is close to unity for all γ_1 sufficiently removed from the resonant values γ_1^{res}. The latter are obtained from the condition

$$\gamma_1^{\text{res}} + \frac{\pi}{4} = \Delta_1(R_1^{\text{t}}, R_1^{\text{t}''}) + \pi/2 - \varphi$$
$$- (1/2)\arctan[P \sin 2\phi_{12}/(1 - 2P \sin^2\phi_{12})] = \pi n . \quad (7.41)$$

In two limiting cases ($P \ll 1$ and $1 - P \ll 1$), (7.41) coincides with the quantization condition for the energy levels E_n and E_n^{d} in the adiabatic and diabatic potential wells, giving

$$\Delta_1(R_1^{\text{t}} R_1^{\text{t}''}) - \varphi = \pi(n - 1/2), \qquad P \ll 1, \quad (7.42)$$

$$\Delta_2(R_2^{\text{t}'} R_{\text{s}}) + \Delta_1(R_1^{\text{t}'} R_1^{\text{t}''}) - \Delta_1(R_1^{\text{t}'} R_{\text{s}})$$
$$= \Delta_2(R_2^{\text{t}'} R_{\text{s}}) + \Delta_1(R_{\text{s}} R_1^{\text{t}''}) = \pi(n + 1/2), \qquad 1 - P \ll 1 . \quad (7.43)$$

The effect of dynamic phases on the position of resonant energy levels E_n can be seen here. At an energy $\mathscr{E}$ close to E_n,

$$S_{22} \sim (\mathscr{E} - E_n - \mathrm{i}\Gamma_n/2)/(\mathscr{E} - E_n + \mathrm{i}\Gamma_n/2) . \quad (7.44)$$

Here the level width Γ_n is defined by the relations

$$\Gamma_n = 2\nu_n P(E_n) \sin^2\phi_{12}(E_n), \qquad \nu_n = \partial E_n/2\pi\partial n, \; P \ll 1, \quad (7.45)$$

$$\Gamma_n = 2\nu_n^{\text{d}}[1 - P(E_n^{\text{d}})] \sin^2\phi_{12}(E_n^{\text{d}}), \; \nu_n^{\text{d}} = \partial E_n^{\text{d}}/2\pi\partial n, \; 1 - P \ll 1 . \quad (7.46)$$

It will be noted that in both cases the level width is an oscillatory function of energy and of the angular momentum controlling the phase ϕ_{12}. Therefore Γ_n might appear to be very small just when ϕ_{12} is close to $m\pi$, where m is an integer. This has been noted recently in connection with molecular predissociation [7.9, 10]. On the other hand Γ_n can also exhibit very slow oscillations, provided the spacing between the extrema of $\sin^2\phi_{12}(\mathscr{E})$ is not too different from that between the energy levels E_n [7.11].

7.3 Two-State Semiclassical Models

7.3.1 Derivation of Semiclassical Equations

Under quasi-classical conditions, the set of two second-order equations (7.16 or 17) may often be reduced to a set of two first-order equations of the semiclassical (common-trajectory) approximation. Such a reduction has been performed in Sect. 6.1 for the general multistate case. Note that this was done either in the eikonal approximation or under the following assumptions:

i) the nuclear motion is quasi-classical;
ii) the nondiagonal interaction matrix elements and the differences between the diagonal interaction matrix elements are small compared to the nuclear kinetic energy;
iii) the coupling between states can be neglected near the turning points.

Assumption (iii) is often too stringent. It appears however [6.12, 13] that for the two-state model it may be abandoned, provided that in the diabatic representation

iv) the forces $F_1 = -dV_{11}/dR$ and $F_2 = -dV_{22}/dR$ are of the same sign for all R of interest;
v) the inequality

$$|dV_{12}/dR| \ll |F_1|, |F_2| \tag{7.47}$$

holds.

Under these assumptions the two-state model will resemble the linear model discussed in detail in Chap. 8. A peculiar feature of the linear model is the exact equivalence of the quantum and semiclassical formulation when $F_1 F_2 > 0$, provided the trajectory is defined in the proper way. This equivalence has been proved [7.14] using the momentum representation. Therefore it is natural to use the momentum representation in attempting to simplify any two-state model.

Consider the diabatic basis equations (7.16) and define the momentum representation amplitudes $u^{\mathrm{d}}_{a'}(k)$ by the relation

$$\chi^{\mathrm{d}}_{a'} = -\exp(\mathrm{i}\delta_a)\,\mathrm{i}(2\pi\mu)^{-1/2}\int_{-\infty}^{+\infty} u^{\mathrm{d}}_{a'}(k)\exp\left[\mathrm{i}kR - \mathrm{i}\int_0^k \mathscr{R}_{a'}(k')\,dk'\right]dk, \tag{7.48}$$

where δ_a is the phase shift in the diabatic potential V_{aa} corresponding to the incoming state a, $k_a = \{2\mu[\mathscr{E} - V_{aa}(\infty)]\}^{1/2}$ and the $\mathscr{R}_{a'}(k)$ satisfy the equation

$$(k^2/2\mu) + V_{a'a'}[\mathscr{R}_{a'}(k)] = \mathscr{E}. \tag{7.49}$$

The following equations are obtained for u^{d}_a:

$$\begin{aligned}
&\{(k^2/2\mu) + V_{11}[\mathscr{R}_1 + \mathrm{i}(d/dk)] - \mathscr{E}\}\,u^{\mathrm{d}}_1(k)\\
&\quad = -\exp\left[\mathrm{i}\int_0^k(\mathscr{R}_1 - \mathscr{R}_2)\,dk'\right]V_{12}[\mathscr{R}_2 + \mathrm{i}(d/dk)]\,u^{\mathrm{d}}_2(k),\\
&\{(k^2/2\mu) + V_{22}[\mathscr{R}_2 + \mathrm{i}(d/dk)] - \mathscr{E}\}\,u^{\mathrm{d}}_2(k)\\
&\quad = -\exp\left[\mathrm{i}\int_0^k(\mathscr{R}_2 - \mathscr{R}_1)\,dk'\right]V_{21}[\mathscr{R}_1 + \mathrm{i}(d/dk)]\,u^{\mathrm{d}}_1(k).
\end{aligned} \tag{7.50}$$

Equations (7.50) are exact and seem to be more complicated than (7.16) in its coordinate representation. But if the condition (i), which in the momentum representation reads

$$|(k/\mu \mathcal{F}_{a'})(\mathcal{F}_{a'}^{-1}\, d\mathcal{F}_{a'}/d\mathcal{R}_{a'})^2| \ll 1, \tag{7.51}$$

is fulfilled, then $u_{a'}^{\mathrm{d}}$ and $\mathcal{R}_{a'}(k)$ are slowly varying functions and $V_{a'a'}[\mathcal{R}_{a'} + \mathrm{i}(d/dk)]$ can be expanded in powers of $\mathrm{i}(d/dk)$. Note, that here and in equations which follow the square brackets mean "a function of". For instance, $\mathcal{F}_{a'}$ in (7.51) which is a function of k, can be expressed via $F_a(R)$ as $\mathcal{F}_{a'} \equiv \mathcal{F}_{a'}(k) = F_{a'}[\mathcal{R}_{a'}(k)]$, $R = \mathcal{R}_{a'}(k)$. Provided the conditions (ii) and (v) are fulfilled, the coupling terms may be considered to be one order of magnitude higher than the elastic ones. Then, allowing for (7.49) and neglecting the first- and higher-order terms in the V_{12} expansion and second-order terms in the $V_{a'a'}$ expansion, we get the first-order equations

$$\begin{aligned} &\mathrm{i}[-\mathcal{F}_1(du_1^{\mathrm{d}}/dk) + \tfrac{1}{2}(-d\mathcal{F}_1/d\mathcal{R}_1)(d\mathcal{R}_1/dk)\,u_1^{\mathrm{d}}] \\ &\quad = -\exp\left[\mathrm{i}\int_0^k (\mathcal{R}_1 - \mathcal{R}_2)\,dk'\right] V_{12}(\mathcal{R}_2)\,u_2^{\mathrm{d}}, \\ &\mathrm{i}[-\mathcal{F}_2(du_2^{\mathrm{d}}/dk) + \tfrac{1}{2}(-d\mathcal{F}_2/d\mathcal{R}_2)(d\mathcal{R}_2/dk)\,u_2^{\mathrm{d}}] \\ &\quad = -\exp\left[\mathrm{i}\int_0^k (\mathcal{R}_2 - \mathcal{R}_1)\,dk'\right] V_{21}(\mathcal{R}_1)\,u_1^{\mathrm{d}}. \end{aligned} \tag{7.52}$$

Introduce now the amplitudes $c_{a'}^{\mathrm{d}}(k)$ by the relation

$$c_{a'}^{\mathrm{d}}(k) = \mathcal{F}_{a'}^{1/2}\, u_{a'}^{\mathrm{d}}(k). \tag{7.53}$$

Then (7.52) assumes the form

$$\begin{aligned} &\mathrm{i}(\operatorname{sign}\mathcal{F}_1)\,|\mathcal{F}_1\mathcal{F}_2|^{1/2}\, dc_1^{\mathrm{d}}/dk = V_{12}(\mathcal{R}_2)\exp\left[\mathrm{i}\int_0^k (\mathcal{R}_1 - \mathcal{R}_2)\,dk'\right] c_2^{\mathrm{d}}, \\ &\mathrm{i}(\operatorname{sign}\mathcal{F}_2)\,|\mathcal{F}_1\mathcal{F}_2|^{1/2}\, dc_2^{\mathrm{d}}/dk = V_{21}(\mathcal{R}_1)\exp\left[\mathrm{i}\int_0^k (\mathcal{R}_2 - \mathcal{R}_1)\,dk'\right] c_1^{\mathrm{d}}, \end{aligned} \tag{7.54}$$

where

$$\operatorname{sign} x = \begin{cases} +1 & \text{for } x > 0, \\ -1 & \text{for } x < 0. \end{cases} \tag{7.55}$$

The fulfillment of (ii) guarantees that

$$|\mathcal{R}_1 - \mathcal{R}_2|/(\mathcal{R}_1 + \mathcal{R}_2) \ll 1, \tag{7.56}$$

and therefore

$$\bar{\mathcal{R}} = (\mathcal{R}_1 + \mathcal{R}_2)/2 \tag{7.57}$$

can be substituted into $\mathscr{F}_1$, $\mathscr{F}_2$ and V_{12} instead of $\mathscr{R}_1$ and $\mathscr{R}_2$. The set of equations obtained is Hermitian due to assumption (iv).

Define the mean force

$$F = (\text{sign}\, F)\, [F_1(\mathscr{R})\, F_2(\mathscr{R})]^{1/2}, \tag{7.58}$$

where the symbol (sign F) denotes the common sign of F_1 and F_2, and the time t by the relation

$$t = \int_0^k dk'/F(k'). \tag{7.59}$$

This gives the equations

$$\begin{aligned} \mathrm{i}\frac{dc_1^{\mathrm{d}}}{dt} &= V_{12}[R(t)] \exp[\mathrm{i}\Delta(t)]\, c_2^{\mathrm{d}}, \\ \mathrm{i}\frac{dc_2^{\mathrm{d}}}{dt} &= V_{21}[R(t)] \exp[-\mathrm{i}\Delta(t)]\, c_1^{\mathrm{d}}, \qquad \text{where} \end{aligned} \tag{7.60}$$

$$R(t) = \bar{\mathscr{R}}[k(t)], \tag{7.61}$$

$$\Delta(t) = \int_0^t \{\mathscr{R}_1[k(t')] - \mathscr{R}_2[k(t')]\}\, F[k(t')]\, dt' \tag{7.62}$$

and $k(t)$ is defined by (7.59).

Equation (7.60) must be solved with the usual initial conditions

$$c_{a'}^{\mathrm{d}}(-\infty) = \delta_{a'a}. \tag{7.63}$$

Then the two-state quantum S matrix is expressed through the semiclassical scattering matrix ${}^{\mathrm{sc}}S_{a'a}^{\mathrm{d}} = c_{a'}^{\mathrm{d}}(+\infty)$:

$$S_{a'a} = \exp[\mathrm{i}(\delta_{a'}^{\mathrm{d}} + \delta_a^{\mathrm{d}})]\; {}^{\mathrm{sc}}S_{a'a}^{\mathrm{d}} \tag{7.64}$$

where δ_a^{d} denotes the elastic scattering phase shifts in the diabatic potentials V_{aa}.

The intial conditions (7.63) and the relations (7.64) are obtained by substituting $u_{a'}^{\mathrm{d}}(k) = c_{a'}^{\mathrm{d}}(t)\, \mathscr{F}_{a'}$ [where $t = t(k)$] into (7.48) and calculating this integral by the stationary-phase method for large R. The calculation gives $\chi_{a'}^{\mathrm{d}}(R)$ in the form of a superposition of incoming and outgoing waves. Comparing the results with the asymptotics (7.19a) and noting that $k \to -k_a \simeq -k_{a'}$ corresponds to $t \to -\infty$, and $k \to k_{a'} \simeq k_a$ to $t \to +\infty$ we get (7.63) and (7.64).

The equations (7.60) correspond to the usual semiclassical equations (6.3). Therefore $\Delta(t)$ plays the part of $\int_0^t \Delta V(t')\, dt'$, where $\Delta V(t) =$

$V_{11}(t) - V_{22}(t)$. If we neglect the difference between $(F_1 + F_2)/2$ and (sign F) $(F_1 F_2)^{1/2}$, then actually $\Delta(t) = \int_0^t \Delta V(t')\,dt'$.

The semiclassical two-state equations in the adiabatic basis can be obtained by transformation of (7.60), substituting $\int_0^t \Delta V(t')\,dt'$ for $\Delta(t)$. Note that this means that the semiclassical adiabatic and diabatic equations are, generally speaking, equivalent only in the case when we neglect the difference between the trajectories controlled by the mean geometric and the mean arithmetic forces. The resultant adiabatic basis equations are

$$
\begin{aligned}
&\mathrm{i}\frac{dc_1^{\mathrm{a}}}{dt} = C_{12}\exp\left(\mathrm{i}\int_0^t \Delta U\,dt'\right)c_2^{\mathrm{a}},\\
&\mathrm{i}\frac{dc_2^{\mathrm{a}}}{dt} = C_{21}\exp\left(-\mathrm{i}\int_0^t \Delta U\,dt'\right)c_1^{\mathrm{a}}, \qquad \Delta U = U_1 - U_2,
\end{aligned} \tag{7.65}
$$

$$
C_{12} = \langle 1|-\mathrm{i}\partial/\partial t\,|2\rangle = -C_{21} = -\mathrm{i}v_{\mathrm{r}}(t)\,g(R), \qquad R = R(t). \tag{7.66}
$$

Here $v_{\mathrm{r}}(t) = dR/dt$ is the radial velocity of the classical nuclear motion and $g(R)$ is given by (7.3, 18). Note that now the mixing angle

$$
\vartheta(t) = -\mathrm{i}\int_{-\infty}^{t} C_{12}(t')\,dt' = -\int_{-\infty}^{t} v_{\mathrm{r}}(t')\,g(R')\,dt', \qquad R' = R(t'). \tag{7.67}
$$

Equation (7.65) must be solved with the initial conditions

$$
c_{a'}^{\mathrm{a}}(-\infty) = \delta_{a'a}. \tag{7.68}
$$

Since here $t = 0$ corresponds to the turning point, $R(t)$ is an even and $v(t)$ an odd function of t. Therefore, $\vartheta(+\infty) = 0$ and

$$
{}^{\mathrm{sc}}S_{a'a}^{\mathrm{a}} = c_{a'}^{\mathrm{a}}(+\infty) = \exp\left[\mathrm{i}\int_0^\infty (U_a - V_{aa})\,dt + \mathrm{i}\int_0^\infty (U_{a'} - V_{a'a'})\,dt\right]{}^{\mathrm{sc}}S_{a'a}^{\mathrm{d}}. \tag{7.69}
$$

In the approximation adopted here

$$
\int_0^\infty (U_a - V_{aa})\,dt = \delta_a^{\mathrm{d}} - \delta_a^{\mathrm{a}},
$$

where δ_a^{a} are the elastic phase shifts in the adiabatic potentials U_a: therefore the quantum S matrix is expressed through ${}^{\mathrm{sc}}S_{a'a}^{\mathrm{a}}$ in a form completely similar to (7.64). Note that the relation (7.69) is invalid for the case of exact resonance.

7.3.2 Classification of Semiclassical Two-State Models

When the semiclassical approximation is applied to an overall scattering process (usually involving a double passage of the strong-nonadiabatic-coupling region), the interaction satisfies the following conditions in the diabatic basis

$$\Delta V(-\infty) = \Delta V(+\infty), \qquad V_{12}(-\infty) = V_{12}(+\infty) = 0, \tag{7.70a}$$

or in the adiabatic basis,

$$\begin{aligned} &\Delta U(-\infty) = \Delta U(+\infty), \qquad C_{12}(-\infty) = C_{12}(+\infty) = 0, \\ &\vartheta(-\infty) = \vartheta(+\infty) = 0. \end{aligned} \tag{7.70b}$$

These conditions lead to complete equivalence of both basis sets. However it seems useful to consider more general time-dependent models for which the conditions (7.70) are not fulfilled and the adiabatic and diabatic basis sets are, generally speaking, not equivalent. Such models arise naturally in using the matching method to describe one-way nonadiabatic transitions in localized strong-coupling regions (Sects. 6.2 and 7.2). Accordingly, the time $t = 0$ is associated here with the center of the strong nonadiabaticity region R_s, rather than with the turning point of the common trajectory.

For these generalized models it is demanded only that

$$C_{12}(-\infty) = C_{12}(+\infty) = 0, \tag{7.71a}$$

$$\vartheta(-\infty) = 0. \tag{7.71b}$$

The condition (7.71a) permits an unambiguous definition of the semiclassical scattering matrix ${}^{sc}S^{a}_{a'a}$ in the adiabatic basis to be made by the usual relation

$${}^{sc}S^{a}_{a'a} = c^{a}_{a'}(+\infty), \tag{7.72}$$

where the $c^{a}_{a'}$ are obtained solving (7.65) with the initial conditions (7.68). If the model considered describes only one localized strong-coupling region and time $t = 0$ is associated with its center, then ${}^{sc}S^{a}_{a'a}$ coincides with the nonadiabatic two-state transition matrix $\mathcal{N}_{a'a}$ for the region, entering into the general expression for the matching method (Sects. 6.2, 7.2).

In terms of mathematics, the diabatic basis equations are usually simpler than (7.65). More than that, the diabatic basis often naturally arises in choosing certain approximate electronic basis functions which depend only very weakly on the internuclear distance (see Sect. 3.3). However it should be noted that since the condition $\vartheta(+\infty) = 0$ is not demanded now, the coupling between diabatic states may survive in the limit $t \to +\infty$, though the coupling between adiabatic states tends to zero, see (7.71a). The ratio $V_{12}/\Delta V$ tends to a constant value $[\tan 2\vartheta(\infty)]/2$ in this limit. Therefore the solutions to (7.60) will oscillate at large t, thus preventing the matrix ${}^{sc}S^{d}_{a'a}$ being defined by relations similar to (7.72). This is a manifestation of the above-mentioned inequivalence of the two basis sets for models subject to the conditions (7.71) only. Condition (7.71 b) ensures, however, the absence of coupling between diabatic states in the limit $t \to -\infty$, thus enabling one to find ${}^{sc}S^{a}_{a'a}$ using (7.60) with the initial conditions (7.63). The explicit expressions for the matrix elements ${}^{sc}S^{a}_{a'a}$ in terms of $c^{d}_{a'}(t)$ are

$$\begin{aligned}{}^{sc}S^{a}_{1a} = \lim_{t\to\infty} \Bigg\{ &\exp\left(\mathrm{i}\int_0^t U_1\,dt'\right)\left[c^{d}_{1}(t)\exp\left(-\mathrm{i}\int_0^t V_{11}\,dt'\right)\cos\vartheta(t) + c^{d}_{2}(t)\right.\\ &\left.\times\exp\left(-\mathrm{i}\int_0^t V_{22}\,dt'\right)\sin\vartheta(t)\right]\Bigg\}\exp\left[\mathrm{i}\int_{-\infty}^{0}(U_a - V_{aa})\,dt\right],\end{aligned} \tag{7.73}$$

$$\begin{aligned}{}^{sc}S^{a}_{2a} = \lim_{t\to\infty} &\exp\left(\mathrm{i}\int_0^t U_2\,dt'\right)\left[-c^{d}_{1}(t)\exp\left(-\mathrm{i}\int_0^t V_{11}\,dt'\right)\sin\vartheta(t) + c^{d}_{2}(t)\right.\\ &\left.\times\exp\left(-\mathrm{i}\int_0^t V_{22}\,dt'\right)\cos\vartheta(t)\right]\exp\left[\mathrm{i}\int_{-\infty}^{0}(U_a - V_{aa})\,dt\right],\end{aligned}$$

Generally speaking, the two-state semiclassical model is characterized by two functions, ΔV and V_{12} in the diabatic basis, and ΔU and $v_r g$ in the adiabatic basis. However, if V_{12} or $v_r g$ do not change sign as t varies over the range $-\infty$ to $+\infty$ which frequently is the case, the equations can be transformed to a form involving one function only [7.15, 16]. In the diabatic basis this is achieved by introducing another independent variable,

$$s(t) = \int_0^t V_{12}(t')\,dt', \tag{7.74}$$

which is a monotonic function of t, since V_{12} does not change sign. Then the transformed equations read

$$\begin{aligned}\mathrm{i}\frac{dc^{d}_{1}}{ds} &= \exp\left[-\mathrm{i}\int_0^s 2\tau(s')\,ds'\right]c^{d}_{2},\\ \mathrm{i}\frac{dc^{d}_{2}}{ds} &= \exp\left[\mathrm{i}\int_0^s 2\tau(s')\,ds'\right]c^{d}_{1}, \quad \text{where}\end{aligned} \tag{7.75}$$

$$\tau = (V_{22} - V_{11})/2V_{12} = -\cot 2\vartheta. \tag{7.76}$$

Similarly, in the adiabatic basis another variable,

$$S(t) = -\mathrm{i}\int_0^t C_{12}(t')\,dt' = \vartheta(t) - \vartheta(0) \tag{7.77}$$

is introduced and the equations (7.65) assume the form

$$\begin{aligned}\frac{dc^{a}_{1}}{dS} &= \exp\left[-\mathrm{i}\int_0^S T(S')\,dS'\right]c^{a}_{2},\\ \frac{dc^{a}_{2}}{dS} &= -\exp\left[\mathrm{i}\int_0^S T(S')\,dS'\right]c^{a}_{1}, \quad \text{where}\end{aligned} \tag{7.78}$$

$$T = \Delta U / v_r g . \tag{7.79}$$

A discussion of the functions $\tau(s)$ and $T(S)$ for various models which produce analytical solution is given in [7.16].

The gyroscopic interpretation of (7.60, 65) provides still another view on the models available. Setting $a = 1$ in the initial conditions (7.63) and taking the quantities $c_1^{\mathrm{d}} c_1^{\mathrm{d}*} - c_2^{\mathrm{d}} c_2^{\mathrm{d}*}$, $c_1^{\mathrm{d}} c_2^{\mathrm{d}*}$ and $c_1^{\mathrm{d}*} c_2^{\mathrm{d}}$ as spherical components of a unit vector $\boldsymbol{m}$, the system of two equations (7.60) for two complex functions subject to one normalization condition is transformed to a system of three equations for the components of the precessing unit vector $\boldsymbol{m}$,

$$\dot{\boldsymbol{m}} = \boldsymbol{n} \times \boldsymbol{m} . \tag{7.80}$$

Here the vector $\boldsymbol{n}(t)$ is defined by its components

$$n_x = \mathrm{Re}\{2 V_{12}(t)\}, \quad n_y = \mathrm{Im}\{2 V_{12}(t)\}, \quad n_z = \Delta V(t). \tag{7.81}$$

For the case considered, where V_{12} is real, i.e., for $n_y = 0$, the time variation of $\boldsymbol{n}(t)$ can be represented by a curve in the plane n_x, n_z corresponding to the $\boldsymbol{n}$ apex position. Clearly, the length of $\boldsymbol{n}$ is equal to the adiabatic term splitting, and the rate of its rotation is equal to the matrix element for nonadiabatic coupling.

The condition $\vartheta(-\infty) = 0$ and the initial conditions (7.63) in the diabatic basis correspond, for $t \to -\infty$, to the direction of $\boldsymbol{n}$ and $\boldsymbol{m}$ along the z axis. With increasing time, the vector $\boldsymbol{n}$ rotates and the vector $\boldsymbol{m}$ follows it with a delay, suffering precession. In the limit $t \to +\infty$, $\boldsymbol{n}$ tends to a certain vector $\boldsymbol{n}_\infty$, and $\boldsymbol{m}$ precesses around $\boldsymbol{n}$. The frequency of this precession depends on n_∞ and on the angle β between $\boldsymbol{m}$ and $\boldsymbol{n}_\infty$, which specifies the extent of nonadiabaticity when $\boldsymbol{m}$ follows $\boldsymbol{n}$. It will be seen that the nonadiabatic transition probability $P = |{}^{\mathrm{sc}}S_{12}^{\mathrm{a}}|^2$ is expressed through β as

$$P = (1 - \cos\beta)/2 = (1 - \boldsymbol{m} \cdot \boldsymbol{n}_\infty / n_\infty)/2 . \tag{7.82}$$

Other initial conditions for (7.80) imply that at $t \to -\infty$ the vector $\boldsymbol{m}$ is arbitrarily oriented relative to the z axis, and its precession around $\boldsymbol{n}$, with conservation of the angle between $\boldsymbol{m}$ and $\boldsymbol{n}$, corresponds to adiabatic evolution of the system.

Such a gyroscopic interpretation of a two-level system makes it possible to illustrate various models simply, by assigning them certain lines which specify the evolution of the vector $\boldsymbol{n}$ in the space n_z, n_x, t. The projections of these lines on the plane n_z, n_x give curves that contain no information about the rate of $\boldsymbol{n}$-variation, but still represent certain characteristics of the model. Figure 7.3 presents the trajectories of the vector $\boldsymbol{n}$ for the models for which exact solutions are known (Chaps. 8, 9).

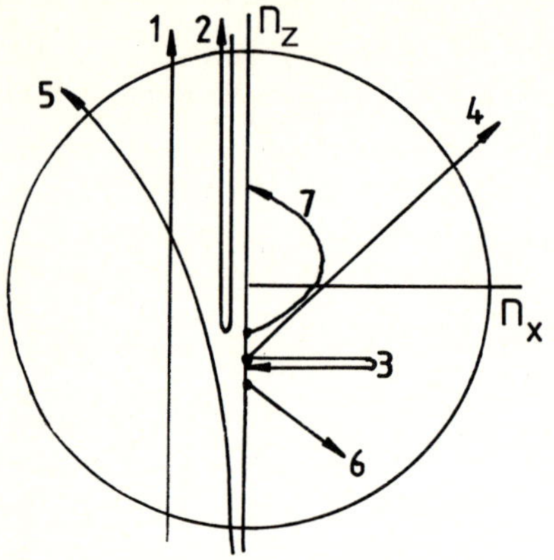

Fig. 7.3. Gyroscopic interpretation of some two-state models of nonadiabatic coupling. (*1*) Original Landau-Zener-Stueckelberg model (Sect. 8.2), (*2*) same model taking the turning point effect into account (Sect. 8.3), (*3*) Rosen-Zener model (Sect. 9.1), (*4*) exponential model (Sect. 9.1), (*5*) linear-exponential model (Sect. 9.2), (*6*, *7*) hypergeometric models (Sect. 9.3.1). The lines outside the circle go to infinity

7.3.3 Approximate Two-State Transition Probabilities

Different variants of the perturbation theory discussed in Chap. 6 can be applied to the two-state semiclassical problem if it contains the relevant small parameters [7.17]. When there are no such parameters, but the different nonadiabatic coupling regions are well separated, the matching method using solvable one-way nonadiabatic coupling models can be applied. When neither of these simplifications is possible, but a not very high accuracy is needed for the transition probability, or else the qualitative dependence of the latter on the collision energy, the impact parameter and the coupling strength is of interest, then the general approximate expressions for the transition probability can be used.

Several expressions of this kind have been suggested since the early thirties. Those referring to the whole collision event (as distinct from the transitions in the localized regions discussed in Sect. 7.2) will be discussed here. Referring to (7.60, 65), we assume that the conditions (7.70 a, b) are fulfilled. It is convenient now to set $\Delta V(t) = \omega + \delta V(t)$, where $\delta V(t)$ vanishes when $t \to \pm\infty$ and ω is a constant, and to introduce the quantities

$$v = \int_{-\infty}^{+\infty} V_{12}(t)\, dt, \tag{7.83}$$

$$w_0 = \int_{-\infty}^{+\infty} V_{12}(t) \exp(\mathrm{i}\omega t)\, dt, \tag{7.84}$$

$$w = \int_{-\infty}^{+\infty} V_{12}(t) \exp\left[\mathrm{i}\int_0^t \Delta V(t')\, dt'\right] dt, \tag{7.85}$$

$$u = \frac{1}{2}\int_{-\infty}^{+\infty} \delta V(t)\, dt, \tag{7.86}$$

$$\mathcal{S}(t) = \int_0^t \{[\Delta V(t')]^2 + 4V_{12}^2(t')\}^{1/2} dt'. \tag{7.87}$$

Since $V_{12}(t)$ is usually an even function of t, w_0 and w are assumed to be real.

Consider first the general perturbation approaches and the symmetric resonance case $\omega = \delta V = 0$. If V_{12} is small, the first order solution of (7.60) reads

$$\mathcal{P} = w^2, \qquad w \ll 1. \tag{7.88}$$

In the near-adiabatic limit (7.65) provides the best starting point. However even with a low transition probability $\mathcal{P}$, the first-order perturbation treatment fails to give a correct result. As discussed in Sect. 6.2.2, the solution of (7.65) using near-adiabatic first-order perturbation theory amounts to simulating (7.65) near singular points by simpler comparison equations, and to their solution. Consider here the simple case with one localized region of nonadiabatic coupling for $R = R_n$ which is not too close to the turning point $R^t = R(0)$. In the time-dependent description the distance R_n is passed twice: on approach (arrival time $-t_n$) and on receding (arrival time t_n).

Let

$$s = s' + \mathrm{i}s'' = \mathcal{S}(t_s), \tag{7.89}$$

where t_s is the singular point for which both s' and s'' are positive. Then using (6.111) and (6.88) we get

$$\mathcal{P} = 4|B| \exp(-2s'') \sin^2(s' + \varphi), \tag{7.90}$$

where the pre-exponential $B = |B| \exp(\mathrm{i}\varphi)$ is still to be determined by solving a proper comparison equation. When t_s is a simple branch point t_c of the function $U(t)$, then according to (6.115), $B = 1$ and

$$\mathcal{P} = 4 \exp(-2s'') \sin^2(s'). \tag{7.91}$$

The quantum version of this expression is obtained when the exponent $2s''$ is replaced by that from (6.117) and the phase s' is replaced by Δ_{12} from (7.26).

Finally, for symmetric resonance, (7.60) with $\Delta V = 0$ is solved to give

$$\mathcal{P} = \sin^2 v. \tag{7.92}$$

We turn now to the approximate expressions for $\mathcal{P}$ describing the transition probability for the general strong-coupling regime.

Chronologically, the first general approximate expression for $\mathcal{P}$ was suggested by *Rosen* and *Zener* in 1932 [7.18]. They proposed an exactly solvable model (see Chap. 9) with $\delta V = 0$ for which $\mathscr{P}$ can be expressed in the form

$$\mathscr{P} = (w_0^2/v^2)\sin^2 v, \tag{7.93}$$

and suggested using this expression for the general case $\delta V = 0$. Another approximate expression giving the exact result for the Rosen-Zener model was suggested much later [7.19] using the notions of adiabatic perturbation theory. This expression reads

$$\mathscr{P} = \sin^2 s'/\cosh^2 s''. \tag{7.94}$$

Several approximate expressions have been proposed in terms of the first-order Magnus approximation (see Sect. 6.2.3). These are readily obtained for the two-state case, because for any 2 × 2 Hermitian traceless matrix $\hat{B}$ the exponentiation is simple:

$$\exp(\mathrm{i}\hat{B}) = \cos[(-\det\hat{B})^{1/2}]\,\hat{I} + \mathrm{i}\hat{B}(-\det\hat{B})^{-1/2}\sin[(-\det\hat{B})^{1/2}], \tag{7.95}$$

where $\hat{I}$ is a unit matrix. In the diabatic representation corresponding to (7.60), the first-order Magnus approximation with the use of (7.95) gives

$$\mathscr{P} = \sin^2 w. \tag{7.96}$$

When using another representation obtained from (7.60) by the transformation

$$c_{a'}^{\mathrm{d}} = \exp\left\{\mathrm{i}\int_0^t [V_{a'a'}(t') - V_{a'a'}(-\infty)]\,dt'\right\}\bar{c}_{a'}^{\mathrm{d}}$$
$$\times \exp\left\{-\mathrm{i}\int_0^{-\infty} [V_{aa}(t') - V_{aa}(-\infty)]\,dt'\right\} \tag{7.97}$$

[see (6.10, 11)], then

$$\mathscr{P} = [w_0^2/(u^2 + w_0^2)]\sin^2(u^2 + w_0^2)^{1/2}. \tag{7.98}$$

In the adiabatic representation the result of the first-order Magnus approximation is

$$\mathscr{P} = \sin^2\left|\int_{-\infty}^{+\infty} C_{12}\exp[\mathrm{i}\mathscr{S}(t)]\,dt\right|. \tag{7.99}$$

For $\delta V = 0$, (7.96) and (7.98) coincide with each other and with the approximate expression proposed in [7.20]. The expression (7.98) for the general case $\delta V \neq 0$ has been discussed in [7.21–23]. The adiabatic representation expression was suggested in [7.24].

One more expression for $\mathscr{P}$ has been obtained in [7.25]. There the initial linear equations in the diabatic basis were transformed to nonlinear equations in such a way that the approximate solution of the latter always gives a value for $\mathscr{P}$ not exceeding unity:

$$\mathscr{P} = \left| \int_{-\infty}^{+\infty} V_{12}(t) \exp[\mathrm{i}\mathscr{S}(t)]\, dt \right|^2 . \tag{7.100}$$

A detailed discussion of the derivation of (7.100) and its use in calculating the integral cross sections can be found in [7.26]. The generalization of (7.100), improving the results for large Massey parameters, was suggested in [7.27, 28].

For other interpolation formulas giving $\mathscr{P}$ not exceeding unity see [7.29–31].

All the approximate expressions for $\mathscr{P}$ considered here satisfy the condition $\mathscr{P} \leq 1$ for any interaction strength. The expressions (7.93, 96, 98, 100) give correct results for a weak interaction ($v \ll 1$) or for exact resonance ($\Delta V = 0$, i.e., $w = w_0$, $u = 0$), but are invalid in the adiabatic limit. The expression (7.99) gives a correct result in the weak-coupling limit, and a virtually correct result in the adiabatic limit, since both the naive and the correct adiabatic perturbation theories lead in fact to close results (Sect. 6.2.2). As for exact resonance, this expression, corresponding to the adiabatic basis, also gives a correct but useless result $\mathscr{P} = 0$, reflecting the absence of transitions between adiabatic states in this case. The expression (7.94) yields a correct result in all limits.

The above approximate equations for $\mathscr{P}$ have been verified in several papers for specific types of interaction. The approximations (7.93, 98) were compared with the results of numerical calculations for the exponential interaction and $\delta V = 0$ in [7.32]. Equation (7.93) gave more accurate results than did (7.98), but the collisions had to be not too adiabatic ($\omega\tau_c < 1/2$, τ_c is the characteristic collision time). A comparison of (7.93) with (7.98) and (7.100) for the dipole – dipole interaction (again $\delta V = 0$) shows [7.33, 34] that (7.100) gives the lowest transition probability under adiabatic conditions (for $\omega\tau_c \gg 1$). Equation (7.100) has also been compared with the exact results [7.35] for the same interaction.

Finally, note two comparisons of the approximate value of $\mathscr{P}$ with the exact results for the linear model, though the latter does not satisfy the conditions (7.70). In [7.24] the expression (7.99) has been verified for the semiclassical one-way model considered in Sect. 8.2. The results were quite satisfactory for a wide variation range of the Landau-Zener parameter δ. A quantum linear model has been considered in [7.36]. When the forces are of the same sign and the model exactly reduces to a semiclassical one (see Sect. 8.3.1), it appears that for the moderate coupling strength β, see (8.6), the expression (7.96) gives very good results, much superior to those obtained by first-order perturbation theory. However, for larger β, the result gets worse.

From general considerations it would be difficult to say which of the above approximations is nearer to the exact result. But the relative simplicity of (7.94) and the correct description it gives for various limiting cases recommend it for the estimation of probabilities in situations where $V_{a'a''}$ satisfies the general condition (7.70 a).

7.4 Differential Cross Sections and Deflection Functions

The general approach outlined in Chap. 2 enables one to calculate quasi-classical scattering amplitudes and differential cross sections for the two-state model, once the scattering matrix is known.

Consider some features of the two-state inelastic scattering problem, looking at symmetric resonant and nonresonant processes separately. For symmetric resonant processes such as excitation and charge transfer, or spin exchange, the scattering matrix is of a simple form. Here elastic and inelastic scattering represent either direct scattering ($q_{11} = q_{22} = q^{\mathrm{d}}$) or state-transfer scattering ($q_{12} = q_{21} = q^{\mathrm{t}}$). Consider for definiteness the excitation transfer between two spinless S-state atoms with the same parities ($j_1 = j_2 = 0, J = l$). Assuming that the nuclei are distinguishable, the amplitudes $f^{\mathrm{d}}(\theta)$ and $f^{\mathrm{t}}(\theta)$ for direct or state-transfer scattering can be expressed via the elastic scattering amplitudes for even or odd molecular terms $U_{\mathrm{g}}(R)$ and $U_{\mathrm{u}}(R)$. According to (2.152), $f^{\mathrm{g}}(\theta)$ and $f^{\mathrm{u}}(\theta)$ are

$$f^{\mathrm{d}} = \tfrac{1}{2}(f^{\mathrm{g}} + f^{\mathrm{u}}), \quad f^{\mathrm{t}} = \tfrac{1}{2}(f^{\mathrm{g}} - f^{\mathrm{u}}). \tag{7.101}$$

Assume for simplicity that there is no rainbow scattering on either potential, and that the amplitudes f^{g} and f^{u} are expressed via classical elastic-scattering cross sections and the corresponding phases by (5.9). Then, for instance, the cross section for inelastic scattering will be

$$q^{\mathrm{t}}(\theta) = (1/4)\,|({}^{\mathrm{cl}}q^{\mathrm{g}})^{1/2}\exp(\mathrm{i}A^{\mathrm{g}} + \mathrm{i}\alpha^{\mathrm{g}}) - ({}^{\mathrm{cl}}q^{\mathrm{u}})^{1/2}\exp(\mathrm{i}A^{\mathrm{u}} + \mathrm{i}\alpha^{\mathrm{u}})|^2. \tag{7.102}$$

Here ${}^{\mathrm{cl}}q^{\mathrm{g}}$ is the differential cross section for elastic scattering in the field of the even potential $U_{\mathrm{g}}(R)$, and A^{g} and α^{g} are given by (5.9, 11). The same applies to ${}^{\mathrm{cl}}q^{\mathrm{u}}$, A^{u} and α^{u}.

Note that for purely repulsive potentials the phases α^{w} are zero and thus do not enter into the exponents. Rather interestingly, the cross section q^{t} oscillates within the limits $[({}^{\mathrm{cl}}q^{\mathrm{g}})^{1/2} \pm ({}^{\mathrm{cl}}q^{\mathrm{u}})^{1/2}]^2$ as a function of the scattering angle θ and in general does not reach zero values at the minima.

Note once more that the cross sections ${}^{cl}q^{g}$ and ${}^{cl}q^{u}$, and the phase differences $A^{g,u} = A^{u} - A^{g}$ responsible for the interference pattern are completely defined by the two deflection functions $\eta_g(b)$ and $\eta_u(b)$ which coincide in the case of symmetric resonance with the deflection functions for elastic scattering. For repulsive interaction these functions are presented schematically in Fig. 7.4. These types of deflection functions describe, for instance, the excitation transfer in the collisions $He(1s2s, {}^1S) + He(1s^2, {}^1S)$ and $He(1s2s, {}^3S) + He(1s^2, {}^1S)$ for kinetic energies lower than the maxima of the potential energy curves for the states $A^1\Sigma_u^+$ and $a^3\Sigma_u^+$ [7.37, 38].

For this type of deflection function, (7.102) is

$$q^{t} = (1/4)\,[{}^{cl}q^{g} + {}^{cl}q^{u} - 2({}^{cl}q^{g}\,{}^{cl}q^{u})^{1/2}\cos A^{g,u}]. \tag{7.103}$$

The phase $A^{g,u}(\theta)$, defined by

$$A^{g,u} = 2\delta_u(b_u) - kb_u\theta - 2\delta_g(b_g) + kb_g\theta, \qquad b_w = b_w(\theta), \tag{7.104}$$

corresponds to the area of the diagonally hatched region in Fig. 7.4. When scattering through small angles is considered and the contribution from impact parameters less than that for rainbow scattering is neglected, an equation similar to (7.104) holds for the case of one attractive and one repulsive term. Then the scattering phase shifts δ_w differ in sign (say, δ_u is negative and δ_g is positive) and the interference phase is

$$A^{g,u} = 2\delta_u(b_u) - kb_u\theta - 2\delta_g(b_g) - kb_g\theta. \tag{7.105}$$

The result (7.105) simplifies if the common-trajectory approximation is adopted. With $\eta = \eta_0(b)$ as the deflection function in the field of the average potential $U_0(R)$, and the difference $\Delta U_w = U_w(R) - U_0(R)$ small com-

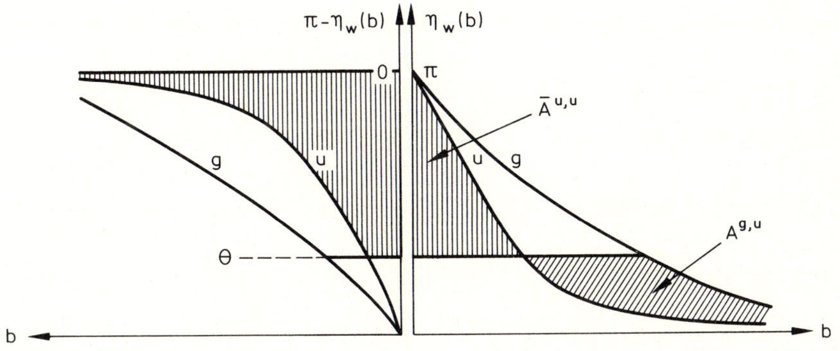

Fig. 7.4. Even g and odd u deflection functions for a symmetric resonant process. Geometrical interpretation of the interference phase $A^{g,u}$ and the nuclear symmetry interference phase $\bar{A}^{u,u}$

pared to the relative kinetic energy, the scattering phase shifts can be approximated as

$$2\delta_{\mathrm{w}}(b_{\mathrm{w}}) = 2\delta_0(b) - \int_{-\infty}^{+\infty} \Delta U_{\mathrm{w}}[R(t)]\,dt - \eta_0(b)\,k(b_{\mathrm{w}} - b), \tag{7.106}$$

where b is the impact parameter for the common trajectory and the time integral is taken along this trajectory, $R = R(b, t)$. This approximation gives for the phase difference

$$A^{\mathrm{g,u}}(b) \simeq A_0^{\mathrm{g,u}}(b) = \int_{-\infty}^{+\infty} \{U_{\mathrm{g}}[R(b,t)] - U_{\mathrm{u}}[R(b,t)]\}\,dt. \tag{7.107}$$

The common-trajectory approximation implies also that ${}^{\mathrm{cl}}q^{\mathrm{g}} = {}^{\mathrm{cl}}q^{\mathrm{u}} = {}^{\mathrm{cl}}q_0$ so that the state-transfer cross section assumes the form

$$q^{\mathrm{t}} = {}^{\mathrm{cl}}q_0(\theta)\sin^2[A_0^{\mathrm{g,u}}(b)/2], \tag{7.108}$$

with $b = b(\theta)$ being an inverse function of $\theta = \eta_0(b)$. Integration of the differential cross section (7.108) over all classical scattering angles will give the "integral" cross section

$$\sigma^{\mathrm{t}} = 2\pi \int_0^{\infty} b\,db\,\sin^2[A_0^{\mathrm{g,u}}(b)/2], \tag{7.109}$$

which can be identified with the true integral cross section, provided the diffraction scattering does not essentially contribute to the inelastic channel. Moreover, for the two-state scattering matrix (7.28), the integral excitation-transfer cross section is given directly by the integral from the transition probability $\sin^2[\delta_{\mathrm{g}}(b) - \delta_{\mathrm{u}}(b)]$. Again, using the expansion

$$\delta_{\mathrm{w}}(b) = \delta_0(b) - \frac{1}{2}\int_{-\infty}^{+\infty} (U_{\mathrm{w}} - U_0)\,dt, \tag{7.110}$$

we get

$$\sigma^t = 2\pi \int_0^{\infty} b\,db\,\sin^2\left[\frac{1}{2}\int_{-\infty}^{+\infty} (U_{\mathrm{g}} - U_{\mathrm{u}})\,dt\right]. \tag{7.111}$$

Comparison of (7.109, 111) shows that for the common-trajectory approximation, the scattering through classical angles accounts for all inelastic transitions. However, this might be different when ${}^{\mathrm{cl}}q^{\mathrm{g}}$ and ${}^{\mathrm{cl}}q^{\mathrm{u}}$ significantly differ.

Now, for identical nuclei the nuclear symmetry effect must be taken into account. The even and odd symmetrized amplitudes have to be constructed first. Then they should be used to calculate the amplitudes for direct and state transfer scattering according to (7.101). According to (2.154), for zero nuclear spin the symmetrized amplitudes ${}^{\mathrm{s}}f^{\mathrm{g}}(\theta)$ and ${}^{\mathrm{s}}f^{\mathrm{u}}(\theta)$

are

$$^{s}f^{g}(\theta) = f^{g}(\theta) + \bar{f}^{g}(\theta), \qquad ^{s}f^{u}(\theta) = f^{u}(\theta) - \bar{f}^{u}(\theta), \tag{7.112}$$

where $\bar{f}^{w}(\theta) = f^{w}(\pi - \theta)$.

Thus, similar to (7.101),

$$f^{d} = (f^{g} + \bar{f}^{g} + f^{u} - \bar{f}^{u})/2, \qquad f^{t} = (f^{g} + \bar{f}^{g} - f^{u} + \bar{f}^{u})/2, \tag{7.113}$$

The state-transfer cross section, for repulsive interactions, for instance, will be

$$q^{t} = (1/4)\,|({}^{cl}q^{g})^{1/2}\exp(\mathrm{i}A^{g}) - ({}^{cl}q^{u})^{1/2}\exp(\mathrm{i}A^{u}) + ({}^{cl}\bar{q}^{g})^{1/2}\exp(\mathrm{i}\bar{A}^{g}) + ({}^{cl}\bar{q}^{u})^{1/2}\exp(\mathrm{i}\bar{A}^{u})|^{2} \tag{7.114}$$

with ${}^{cl}\bar{q}^{w}(\theta) = {}^{cl}q^{w}(\pi - \theta)$ and $\bar{A}^{w}(\theta) = A^{w}(\pi - \theta)$. This expression contains three independent interference phases, two phases being interpreted in Fig. 7.4 as areas bounded by the deflection functions $\eta_{g}(b)$ and $\eta_{u}(b)$.

If one or both potentials cause rainbow scattering (or orbiting) the interference pattern will be more complicated. Yet it can be understood in terms of deflection functions for pure elastic scattering.

Consider now a general case for which the S matrix is given by (7.31) and is suitable for quasi-classical analysis, compare (2.58). To simplify the presentation assume that the interactions U_1 and U_2 correspond to two spinless Σ-states (therefore J coincides with l) and are repulsive (no rainbow scattering), that the contributions from complex impact parameters (Sect. 5.5.1) can be neglected, that the transition probability P depends weakly on l, and that the dynamic phases φ and ψ are much smaller than δ_1 and δ_2. Then, according to the general approach outlined in Sect. 2.1.2, six phases $\delta^{\Lambda}_{a'a}$ and six deflection functions $\eta^{\Lambda}_{a'a} = 2(\partial\delta^{\Lambda}_{a'a}/\partial l)$ can be constructed as is consistent with the number of different exponentials entering into the diagonal and nondiagonal elements of the S matrix.

Introducing the phase shifts δ'_1 and δ'_2 defined by

$$\delta'_1 = \delta_1 - \Delta_1 + \Delta_2, \qquad \delta'_2 = \delta_2 - \Delta_2 + \Delta_1, \tag{7.115}$$

these six phase shifts $\delta^{\Lambda}_{a'a}$ will be

i) for elastic scattering in channel 1:

$$\delta^{(1)}_{11} = \delta_1, \qquad \delta^{(2)}_{11} = \delta'_1, \tag{7.116}$$

ii) for elastic scattering in channel 2:

$$\delta^{(1)}_{22} = \delta'_2, \qquad \delta^{(2)}_{22} = \delta_2, \tag{7.117}$$

iii) for inelastic scattering:

$$\delta^{(1)}_{12} = (\delta_1 + \delta'_2)/2, \qquad \delta^{(2)}_{12} = (\delta'_1 + \delta_2)/2. \tag{7.118}$$

Now, unlike the resonant case, the partial contributions $f^{A}_{a'a}(\theta)$ to the amplitude $f_{a'a}(\theta)$ will also contain dynamic phases, as is evident from the structure of the S matrix (7.31). For instance, two contributions to the inelastic scattering amplitude are

$$\begin{aligned} f^{(1)}_{12}(\theta) &= -\,[{}^{\mathrm{cl}}q^{(1)}_{12}(\theta)]^{1/2}\exp[\mathrm{i}\,A^{(1)}_{12}(\theta) + \Delta\phi_{12}(l^{(1)}_{12})], \\ f^{(2)}_{12}(\theta) &= [{}^{\mathrm{cl}}q^{(2)}_{12}(\theta)]^{1/2}\exp[\mathrm{i}\,A^{(2)}_{12}(\theta) - \Delta\phi_{12}(l^{(2)}_{12})], \end{aligned} \tag{7.119}$$

with

$$\begin{aligned} A^{(1)}_{12}(\theta) &= \delta_1(l^{(1)}_{12}) + \delta'_2(l^{(1)}_{12}) - \theta l^{(1)}_{12}, \\ A^{(2)}_{12}(\theta) &= \delta_2(l^{(2)}_{12}) + \delta'_1(l^{(2)}_{12}) - \theta l^{(2)}_{12}. \end{aligned} \tag{7.120}$$

Here,

$${}^{\mathrm{cl}}q^{A}_{12} = k_2^{-2}\,(\sin\theta)^{-1}\,(l/|\partial\eta^{A}_{12}/\partial l|)|_{l=l^{A}_{12}} \tag{7.121}$$

and $l^{A}_{12} = l^{A}_{12}(\theta)$ are found from the stationary phase condition for repulsive potentials,

$$\eta^{A}_{12}(l^{A}_{12}) = \theta. \tag{7.122}$$

Thus, the inelastic cross section $q_{12}(\theta)$ in the primitive semiclassical approximation will be

$$\begin{aligned} q_{12}(\theta) = \bar{\mathscr{P}}^{(1)}_{12}\,{}^{\mathrm{cl}}q^{(1)}_{12} + \bar{\mathscr{P}}^{(2)}_{12}\,{}^{\mathrm{cl}}q^{(2)}_{12} - 2[\bar{\mathscr{P}}^{(1)}_{12}\bar{\mathscr{P}}^{(2)}_{12}\,{}^{\mathrm{cl}}q^{(1)}_{12}\,{}^{\mathrm{cl}}q^{(2)}_{12}]^{1/2} \\ \times\cos[A^{(1,2)}_{12} + \Delta\phi^{(1,2)}_{12}], \end{aligned} \tag{7.123}$$

where

$$\begin{aligned} \bar{\mathscr{P}}^{A}_{12} = \bar{\mathscr{P}}_{12}(l^{A}_{12}), \quad A^{(1,2)}_{12} = A^{(1)}_{12}(\theta) - A^{(2)}_{12}(\theta), \\ \Delta\phi^{(1,2)}_{12} = \Delta\phi_{12}(l^{(1)}_{12}) - \Delta\phi_{12}(l^{(2)}_{12}). \end{aligned} \tag{7.124}$$

The basic condition for (7.123) to be valid is that the phase difference $A^{(1,2)}_{12}$ be large and the variation of the transition probability $\bar{\mathscr{P}}_{12}(l)$ (compared to the exponential factors) be slow. As l builds up, the radial velocity of nuclei in the region of nonadiabatic coupling decreases and the transition probability drops. When the conditions of near-adiabatic behavior are reached, $\bar{\mathscr{P}}_{12}(l)$ decreases very rapidly with increasing l and this means a breakdown of the primitive semiclassical treatment close to the threshold value $l \simeq l_{\mathrm{th}}$. For two repulsive potentials with a localized coupling region (of the type shown in Fig. 7.1) this threshold implies that the inelastic cross section will markedly increase at some threshold value of the scattering angle $\theta = \theta_{\mathrm{th}}$. The exact value of l_{th} can be defined in different ways. One possibility is to set $A^{(1,2)}_{12}(\theta_{\mathrm{th}})$ equal to the prescribed value, say zero. In turn, $A^{(1,2)}_{12}$ itself can be defined in various ways depending on the choice of the coupling-

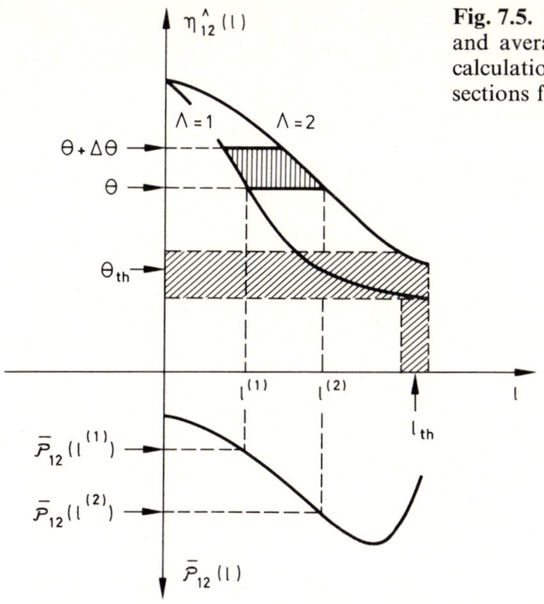

Fig. 7.5. Deflection functions $\eta_{12}^{(1)}(l)$ and $\eta_{12}^{(2)}(l)$ and average transition probability $\mathscr{P}_{12}(l)$ for calculation of semiclassical differential cross sections for inelastic scattering

region center. All this makes ambiguous an exact definition of $A_{12}^{(1;\,2)}$ unless an additional condition is invoked, such as the requirement that even close to the threshold the classical cross sections and the interference phase are determined by two deflection functions. This problem, as well as the threshold behavior of the cross section in the case of pseudocrossing terms is discussed in Sect. 8.5.3 in connection with the Landau-Zener model.

It will only be noted here that the two deflection functions $\eta_{12}^{(1)}$ and $\eta_{12}^{(2)}$ must be matched in some way in a region close to l_{th}, whatever the precise definition of l_{th}. Then the interference phase increment far from the threshold region (diagonally hatched region in Fig. 7.5) in passing from θ to $\theta + \varDelta\theta$ is given by the increment of the area between $\eta_{12}^{(1)}$ and $\eta_{12}^{(2)}$ (vertically hatched region). Thus the upper part of Fig. 7.5 makes possible the calculation of the classical cross sections ${}^{\text{cl}}q_{12}^{(1)}$, ${}^{\text{cl}}q_{12}^{(2)}$ and the interference phase up to a constant term. The lower part shows the l-dependence of the average transition probability $\bar{\mathscr{P}}_{12}$ and the correlation of $\bar{\mathscr{P}}_{12}(l_{12}^{(1)})$ and $\bar{\mathscr{P}}_{12}(l_{12}^{(2)})$ in (7.123) with the scattering angle θ.

Now, if the common trajectory approximation is adopted, i.e., the phase difference ϕ_{12} in (7.31) is assumed to be much smaller than $\delta_1 + \delta_2$, ϕ_{12} can be neglected in determining the stationary phase point in the integral representation of the scattering amplitude. In this approximation

$$ {}^{\text{cl}}q_{12}^{(1)}(\theta) = {}^{\text{cl}}q_{12}^{(2)}(\theta) = {}^{\text{cl}}q_{12}(\theta), \qquad A_{12}^{(1;\,2)}(\theta) = \phi_{12}(l)|_{l=l(\theta)}. \tag{7.125} $$

The cross section will be

$$q_{12}(\theta) = {}^{\text{cl}}q_{12}(\theta)\, \mathscr{P}_{12}(l)|_{l=l(\theta)}, \tag{7.126}$$

where $l = l(\theta)$ corresponds to the common trajectory, and is found from

$$\partial(\delta_1 + \delta_2)/\partial l = \theta. \tag{7.127}$$

$\mathscr{P}_{12}(l)$ is defined by (7.30). This expression for the cross section can be extrapolated to the threshold, provided $\mathscr{P}_{12}(l)$ varies with l less rapidly than does $\exp(\mathrm{i}\delta_1 + \mathrm{i}\delta_2)$. The same description is valid for elastic scattering perturbed by the inelastic channel. Again, the threshold region requires special treatment, which will be exemplified in Sect. 8.5.

Summarizing, we reiterate that the angular Stueckelberg undulations in the differential cross sections represent the main interference phenomenon in two-channel scattering. A special case of this interference is the even-odd undulation phenomenon, when two states can be uncoupled by suitable transformation of the basis. In general these undulations are superimposed on the oscillatory structure of "elastic" scattering characterized by different branches χ of the deflection functions $\eta_{a'a}^{\Lambda\chi}$ with fixed channel quantum numbers a', a, and a fixed "molecular state" quantum number Λ.

8. The Linear Two-State Landau-Zener Model

The two-state nonadiabatic-coupling model discussed in this chapter was the first attempt to describe nonadiabatic coupling over the small range of interatomic separations near to the pseudocrossing or crossing of adiabatic terms. Since the time when *Landau* obtained analytical expressions for the transition probabilities in the weak [8.1] and near-adiabatic [8.2] coupling cases, and *Zener* [8.3] and *Stueckelberg* [8.4] found solutions for strong coupling in the semiclassical approximation, this model has been widely used for the interpretation of various experiments on atomic and ionic collisions [8.5–9]. Moreover, the Landau-Zener formula has often been used beyond the validity range both of the model as such and of the approximations adopted in the solution of nonadiabatic coupling equations. In particular, the solutions found by *Landau*, *Zener* and *Stueckelberg* were obtained making the assumption that the motion of the system in the nonadiabaticity region was quasi-classical. This assumption imposes a lower limit on the nuclear velocity, and the condition that the extent of the interaction region be small imposes an upper limit. However, the range of velocity variations satisfying both the model and the solution constraints appeared to be insufficiently wide for the application of the theory to certain problems. This stimulated the publishing of many papers attempting to clarify the range of applicability of the model, to widen it as much as possible, or even to remove the restrictions imposed by the initial formulation [8.10–12].

8.1 Formulation of the Model

The recent basic advance in the development of the Landau-Zener model seems to be the solution of the quantum problem, with no restrictions on the distance between the turning points and the pseudocrossing (or crossing) point of the adiabatic terms. It was shown that the quantum problem reduces to a semiclassical one if an additional unambiguous definition of the trajectory $R = R(t)$ is made. This removes one of the main restrictions of the semiclassical approximation, which in general requires that the energy of relative motion in the region of nonadiabatic coupling must be much higher than the energy difference between adiabatic terms. This simplification is possible, since the potential-energy matrix of the model two-state Hamiltonian is linear in R. Therefore it is natural to consider the Landau-Zener model as a two-state generalization of the problem of particle motion in a constant force field.

The quantum problem of particle motion in a linear potential field, with its solution expressed by the Airy function, is particularly helpful under quasi-classical conditions when the JWKB asymptotic behavior is attained before the linear approximation for some nonlinear potential becomes invalid. Under these conditions, the quantum solution permits one to find the correct phase of the quasi-classical function (Sect. 5.2.2).

This problem can be generalized for two states, replacing the wave function by a two-component vector, and the linear potential by a matrix V with elements V_{ik} linear in R. The nondiagonal elements of this matrix are responsible for the coupling between electronic states. These states can be either diabatic or adiabatic, depending on the problem. In the first case, the coupling is due to that part of the total interaction which is unaccounted for in the adopted diabatic basis, and the problem is equivalent to nonadiabatic coupling between adiabatic pseudocrossing terms. In the second case, the coupling is of a dynamic nature and can be due, e.g., to the Coriolis interaction between adiabatic crossing terms. Attention will be paid here mainly to pseudocrossing, though crossing will also be considered.

Again, as in the one-state case, the use of the linear model depends on the fulfillment of the quasi-classical conditions: the linear approximation for the matrix of a certain nonlinear potential will become invalid only after the two-component wave function approaches the adiabatic JWKB asymptotes, corresponding to motion in the potential fields $U_1(R)$ and $U_2(R)$ obtained by diagonalizing the potential matrix.

For the pseudocrossing adiabatic terms, the linear model is most simply formulated in the diabatic representation by the nondiagonal matrix V,

$$V_{ik} = [U_{\mathrm{p}} - F_i(R - R_{\mathrm{c}}^{\mathrm{d}})]\,\delta_{ik} + (1 - \delta_{ik})\,a \tag{8.1}$$

where $R_{\mathrm{c}}^{\mathrm{d}}$ is the crossing point of the diabatic linear terms and a is the coupling, assumed to be constant and positive. Note that when the need arises to take the term linear in $R - R_{\mathrm{c}}^{\mathrm{d}}$ into account in the coupling as well, it can be eliminated by an R-independent transformation of the diabatic basis. Introducing the distance $X = R - R_{\mathrm{c}}^{\mathrm{d}}$ we get the following expressions for the adiabatic terms U_1, U_2, the mixing angle ϑ, and the matrix element $g(R)$ (Sect. 7.1):

$$\begin{aligned} U_{1,2} &= U_\mathrm{p} - (F_1 + F_2)\,X/2 \pm \Delta U(X)/2, \\ \Delta U(X) &= (\Delta F^2 X^2 + 4a^2)^{1/2}, \qquad \Delta F = F_1 - F_2, \end{aligned} \tag{8.2}$$

$$\tan 2\vartheta(X) = -2a/\Delta F X, \qquad g(R) = -a\Delta F(\Delta F^2 X^2 + 4a^2)^{-1}. \tag{8.3}$$

It is seen from (8.2) that the adiabatic terms cross at

$$R_\mathrm{c} = R_\mathrm{c}^\mathrm{d} \pm \mathrm{i}(2a/\Delta F), \tag{8.4}$$

so that the pseudocrossing distance $R_\mathrm{p} = \mathrm{Re}\{R_\mathrm{c}\}$ is equal to R_c^d. Moreover, the adiabatic splitting $\Delta U(X)$ is a minimum at this distance and the matrix element $g(R)$ attains its extremal value. Thus, for the linear model all possible choices of the center of the coupling region R_s, as discussed in Sect. 7.2, lead to the same result: $R_\mathrm{s} = R_\mathrm{p}$.

The coupled equations for the nuclear functions χ_1 and χ_2 are

$$\begin{aligned} (2\mu)^{-1} d^2\chi_1/dX^2 + (\mathscr{E} - U_\mathrm{p} + F_1 X)\,\chi_1 - a\chi_2 &= 0, \\ (2\mu)^{-1} d^2\chi_2/dX^2 + (\mathscr{E} - U_\mathrm{p} + F_2 X)\,\chi_2 - a\chi_1 &= 0. \end{aligned} \tag{8.5}$$

The solution to (8.5) has been discussed repeatedly [8.11, 13–15] and so we give below only the basic results needed for finding the applicability regions of various analytical expressions for amplitudes and transition probabilities.

The transformation of (8.5) to the momentum representation yields two first-order equations containing two dimensionless parameters ε and β. The first specifies the kinetic energy of the nuclei at the point R_p, the second, the coupling matrix element [8.16]:

$$\varepsilon = (\mathscr{E} - U_\mathrm{p})\,|\Delta F|/2aF, \qquad \beta = 4a(a\mu/F\,|\Delta F|)^{1/2}, \tag{8.6}$$

where $F = |F_1 F_2|^{1/2}$. In the momentum representation the equations display different properties depending on whether the signs of the slopes F_1 and F_2 are the same or different.

If F_1 and F_2 are of the same sign, then after simple transformations these equations exactly reduce to semiclassical equations with a definite trajectory $X = X(t)$ (Sect. 8.3.1). A typical example of applying the theory developed for this case is the calculation of cross sections for atomic excitation near the threshold (Sect. 8.5.3). If F_1 and F_2 are of different signs, (8.5) has no simple classical analog, since it describes nonadiabatic over-barrier reflection and nonadiabatic tunneling. The theory developed for this case is used, in particular, in calculating the widths and shifts of energy levels of predissociating diatomic molecules [8.17, 18].

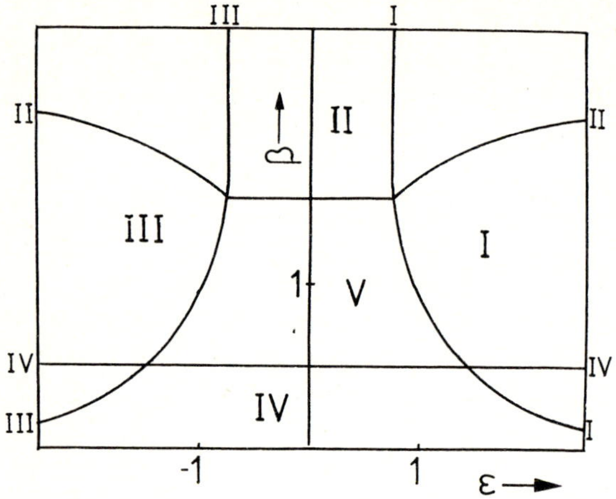

Fig. 8.1. Regions of validity of different analytical expressions for the transition probability, and their boundaries marked by the same numbers

In both cases, the expressions for the transition probabilities can be obtained analytically in certain regions of the ε, β plane. Five such regions can be distinguished (Fig. 8.1). In four of these (regions I, II, III, IV) the probabilities are given by analytical expressions. Since these regions overlap, the probability is given analytically within the entire circular region that circumscribes the region of strong coupling V, where (8.5) must be solved numerically. It is noted also that the values of $|\varepsilon|$ and β are bounded from above by the assumed linear form of the matrix V. This is discussed in Sect. 8.4.

8.2 Nonadiabatic Transitions Far from the Turning Point. Landau-Zener-Stueckelberg Solution

Since the coupling region is assumed to be far from the turning points, the kinetic energy of nuclei near the avoided crossing is large compared with the splitting $\Delta U = U_1 - U_2$, and the quantum equations (8.5) reduce in the usual way (Chap. 6) to semiclassical ones with a classical average trajectory $X = X(t)$.

The semiclassical equations read

$$i\dot{b}_1 = a \exp\left[-i\int^t \Delta F\, X(t)\, dt\right] b_2, \quad i\dot{b}_2 = a \exp\left[i\int^t \Delta F\, X(t)\, dt\right] b_1. \tag{8.7}$$

The nuclear motion in the coupling region is taken to be uniform, i.e.,

$$X(t) = v_p t, \quad v_p = [2(\mathscr{E} - U_p)/\mu]^{1/2}. \tag{8.8}$$

Equations (8.7), taking (8.8) into account, are solved in terms of parabolic cylinder functions.

The probability P was first found by *Zener* [8.3] in the diabatic basis, and then by *Dykhne* [8.19] in the adiabatic basis. Both completely coincided with the results obtained by *Landau* in terms of adiabatic perturbation theory [8.2]. The dynamic phases have been calculated in [8.20–22]. A substitution of appropriate expressions for P, φ, and ψ into the general equation (7.23a) gives a two-channel S matrix (7.21) for the model of pseudocrossing terms. Alternatively, the S matrix can be calculated directly by the method of phase integrals [8.23] and be expressed via the same parameters. In terms of this method, the problem reduces to finding the Stokes coefficients that ensure the unambiguity of the desired quasi-classical functions in the region of their adiabatic asymptotes, bypassing the branch points of adiabatic terms which are treated as analytical functions of a complex variable $Z = X + \mathrm{i}\,Y$. To find an appropriate contour, it is assumed that the branch points R_c and R_c^* are sufficiently far removed from the turning points R_1^t, R_2^t, i.e., that the phases $\Delta_1(R_p, R_1^t)$, $\Delta_2(R_p, R_2^t)$ [as defined by (7.22)] and their difference $\Delta_2 - \Delta_1$ are large, viz.,

$$\Delta_1, \Delta_2, \Delta_{21} \gg \max(1, \pi a^2/|\Delta F|\, v_p), \tag{8.9}$$

where $\pi a^2/|\Delta F|\, v_p$ stands for the phase difference between R_c and R_c^* calculated under condition (8.9).

Unlike the problem with one branch point (see Sect. 6.2.2) corresponding to a near-adiabatic perturbation, that with two branch points [$\mathrm{Im}\{R_c\} = \pm 2a\mathrm{i}/\Delta F$ from (8.4)] needs more careful treatment because of the different possibilities when choosing the cut and the Stokes coefficients (Fig. 8.2).

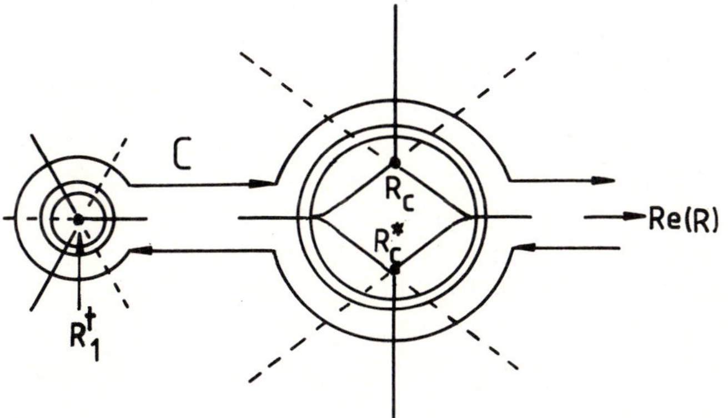

Fig. 8.2. Stokes (---) and anti-Stokes (—) lines in the sheets of two adiabatic terms with two branch points R_c and R_c^* · R_1^t is the turning point in the first sheet. C is a possible circumventing contour for deriving (8.10) by the phase-integral method

The probability P was first calculated using the phase-integral method by *Stueckelberg* [8.4]. Later, various researchers [8.11, 15, 24–26] found not only the probabilities, but also the dynamic phases using the same method. The results for the dynamic phases appeared to be controversial. However, the discrepancies can be removed by proper choice of the Stokes parameters. In particular, the choice of the Stokes parameters in [8.25] gives a dynamic phase differing in sign and constant term from that found in [8.11].

The ultimate result obtained from direct solution of (8.7) and the phase-integral method is

$$\begin{aligned} &P = \exp(-2\pi\delta), && \delta = a^2/|\Delta F|\, v_p = \beta/8\,\varepsilon^{1/2}, \\ &\psi = 0, && \varphi = \pi/4 + \delta\log\delta - \delta - \arg\Gamma(1+\mathrm{i}\delta). \end{aligned} \tag{8.10}$$

The choice of sign in φ corresponds to a positive adiabatic phase difference $\Delta_{21} = -\Delta_{12}$. The latter is calculated using one of (7.24, 25) or (7.26). All three give the same result for the model considered, since for this model the pseudocrossing point of the adiabatic terms coincides with the crossing point of the diabatic terms. When δ increases, the phase φ monotonically decreases from $\pi/4$ to zero as $1/12\delta$, for $\delta \gg 1$.

Thus, the probability of a nonadiabatic transition $\mathscr{P}_{12}$ calculated as $|S_{12}|^2$ from (7.21) is

$$\begin{aligned} &\mathscr{P}_{12} = 2\bar{\mathscr{P}}_{12}\sin^2[\Delta_{21} + \gamma(\delta)], \\ &\bar{\mathscr{P}}_{12} = 2\exp(-2\pi\delta)\,[1 - \exp(-2\pi\delta)], \\ &\gamma(\delta) = \varphi - \psi = \pi/4 + \delta\log\delta - \delta - \arg\Gamma(1+\mathrm{i}\delta). \end{aligned} \tag{8.11}$$

Plots of the functions $\mathscr{P}_{12}(\delta)$ and $\gamma(\delta)$ are presented in Fig. 8.3.

In two limiting cases (8.11) coincides with the results obtained from the distorted-wave approximation and near-adiabatic perturbation theory:

$$\mathscr{P}_{12} = \begin{cases} 8\pi\delta\sin^2(\Delta_{21} + \pi/4), \\ 4\exp(-2\pi\delta)\sin^2\Delta_{21}. \end{cases} \tag{8.12}$$

The generalizations of this expression discussed in the next sections amount to elimination of the conditions (8.9). However it must be noted even here that these conditions, based on general considerations, are too stringent for the linear model; the real constraint applies to the difference $|\Delta_2 - \Delta_1| \gg \max(1, 2\pi\delta)$ rather than to Δ_1 and Δ_2 separately, as will be seen in the next section.

Consider now the transitions between crossing adiabatic terms. The matrix V is now diagonal, with R_c^d replaced by the crossing distance R_c.

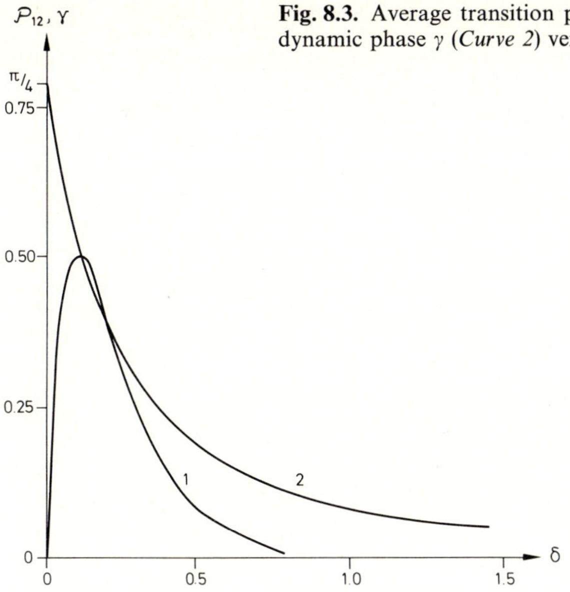

Fig. 8.3. Average transition probability $\bar{\mathscr{P}}_{12}$ (*Curve 1*) and the dynamic phase γ (*Curve 2*) versus δ, see (8.11)

Usually such a situation corresponds to terms of different axial symmetry (see Sect. 3.3.1), in which case we have Coriolis coupling. Equations (8.7) are still valid, provided the coupling matrix element a is replaced by the Coriolis coupling matrix element

$$a \to a_{\mathrm{Cor}} = \dot{\psi}_{\mathrm{c}} r . \tag{8.13}$$

Here $\dot{\psi}_{\mathrm{c}}$ is the angular velocity of the axis rotation and $r = \langle 1 | j_{x_{2}'} | 2 \rangle$ for $R = R_{\mathrm{c}}$. The electronic angular momentum operator is taken with respect to the center of mass of the diatom. However, the choice of the origin is not critical for the case of crossing, as discussed in Sect. 2.3.1.

The transition probability can be derived from (8.10, 11) taking (8.13) into account and identifying the one-way transition probability with $[1 - \exp(-2\pi\delta_{\mathrm{Cor}})]$ rather than with $\exp(-2\pi\delta_{\mathrm{Cor}})$. However, an additional simplification is possible here, namely (as will be shown later in Sect. 8.4), the parameter δ_{Cor} for the Coriolis coupling must not be too large to be compatible with the linear approximation for the crossing adiabatic terms. Thus, retaining the linear term only, we get

$$P = 2\pi\delta_{\mathrm{Cor}}, \quad \mathscr{P}_{12} = 8\pi\delta_{\mathrm{Cor}} \sin^2(\Delta_{21} + \pi/4), \quad \delta_{\mathrm{Cor}} = \dot{\psi}_{\mathrm{c}}^2 r^2 / |\Delta F| v_{\mathrm{c}}, \tag{8.14}$$

where $\Delta_{21} > 0$ is the phase difference between the turning points R_1^{t}, R_2^{t} and the crossing point R_{c}.

8.3 Nonadiabatic Transitions Near the Turning Point

8.3.1 Terms with Slopes of the Same Sign

For terms with slopes of the same sign, i.e., for $F_1 F_2 > 0$, and with an appropriate choice of variables, (8.2) in the momentum representation reduces to the semiclassical (8.7) with a trajectory $X(t)$ corresponding to a uniformly accelerated nuclear motion [8.16, 27, 28], compare Sect. 7.3.1:

$$X(t) = Ft^2/2\mu + \mu v_p^2/2F. \tag{8.15}$$

Figure 8.4 illustrates the exact reduction of the quantal to the semiclassical problem. Here the turning point R^t corresponds to time $t = 0$, and the branch point in the upper (lower) part of the complex plane of R to two branch points t_c in the upper (lower) part of the complex plane t.

Introducing a dimensionless time $\tau = t/t_0$, $t_0 = (2a\mu/F\,|\Delta F|)$, (8.7) with the trajectory (8.15) becomes

$$\begin{aligned} i\,db_1/d\tau &= (\beta/2)\exp\left[i\int_0^\tau \beta(\varepsilon - \tau^2)\,d\tau\right] b_2, \\ i\,db_2/d\tau &= (\beta/2)\exp\left[-i\int_0^\tau \beta(\varepsilon - \tau^2)\,d\tau\right] b_1. \end{aligned} \tag{8.16}$$

This system cannot be solved analytically in the general form. However, various simplifications are possible using a small or a high value of the real (s') or the imaginary (s'') part of the dimensionless action s defined via the adiabatic term splitting by the integral

$$s = s' + i s'' = \beta \int_0^{\tau_c} [1 + (\varepsilon - \tau^2)^2]^{1/2}\,d\tau = \beta[\Delta'(\varepsilon) + i\Delta''(\varepsilon)] \tag{8.17}$$

with $\Delta'(-\varepsilon) = \Delta''(\varepsilon)$. Here $\tau_c = (\varepsilon + i)^{1/2}$ is the root of the radicand for which $s' > 0$, $s'' > 0$. There are four roots in all:

$$\tau_c^{(1)} = \tau_c, \quad \tau_c^{(2)} = -\tau_c^*, \quad \tau_c^{(3)} = -\tau_c, \quad \tau_c^{(4)} = \tau_c^*,$$

located at the apexes of a rectangle (Fig. 8.4). When ε decreases from large positive to large negative values, the rectangle converts from a prolate shape to an oblate one, passing through a square configuration at $\varepsilon = 0$.

Consider first the solution of (8.16), with $|s| \gg 1$, for three distinct cases:

a) $s' \gg s''$, 1 prolate rectangle
b) $s' \approx s'' \gg 1$ nearly square
c) $s'' \gg s'$, 1 oblate rectangle.

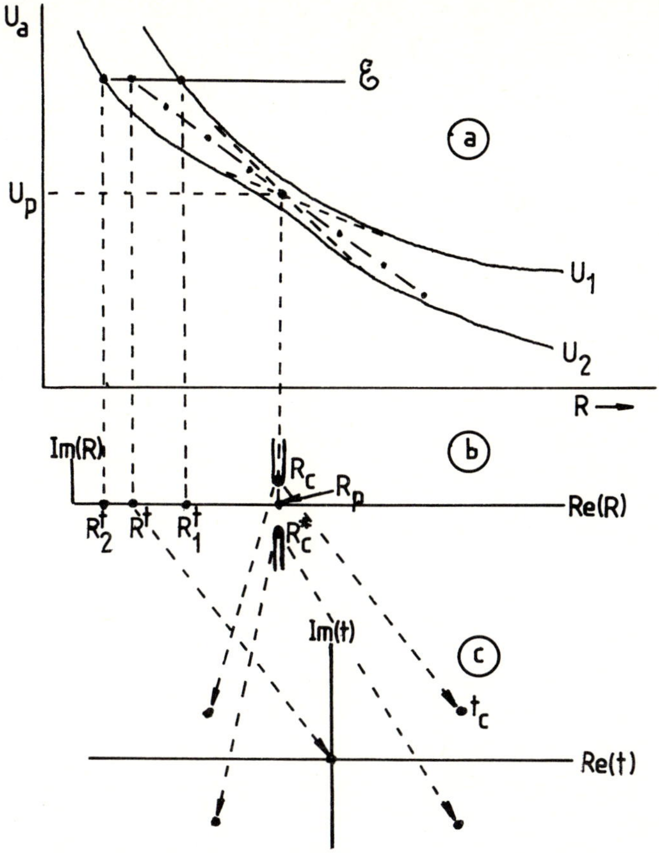

Fig. 8.4. (**a**) Adiabatic (—) and diabatic (---) terms of the linear model for $F_1 F_2 > 0$. The dash-dotted line refers to the linear average potential. (**b**) Analytical structure of adiabatic terms in the complex R-plane. (**c**) Analytical structure of adiabatic terms in the complex t-plane after transformation (8.15)

Since s', s'' are proportional to β, the above conditions impose a lower limit on β.

For $s' \gg s''$, 1 we have

$$\tau_c = \varepsilon^{1/2} + i/2(2\varepsilon)^{1/2}, \qquad s = 2\beta\varepsilon^{3/2}/3 + i\pi\beta/8\varepsilon^{1/2}. \tag{8.18}$$

The condition $s' \gg 1$ means that between two couplings [atoms approaching ($\tau < 0$) and leaving ($\tau > 0$)] the system's behavior reaches its adiabatic asymptotic form. Therefore the solution of (8.16) can be obtained by matching the solutions of two simpler equations. These are obtained from (8.16) making use of the quadratic approximation near two stationary phase points $\tau_p = \pm(\varepsilon)^{1/2}$:

$$i\,db_1/d\tau = (\beta/2)\exp[i\beta\varepsilon^{1/2}(\tau-\tau_p)^2]\,b_2,$$
$$i\,db_2/d\tau = (\beta/2)\exp[-i\beta\varepsilon^{1/2}(\tau-\tau_p)^2]\,b_1. \tag{8.19}$$

These equations coincide with (8.7) for the trajectory (8.8), i.e., the effect of the turning point at $\tau = 0$ is neglected. Matching the two solutions of (8.19) leads, of course, to (8.11).

Now, the conditions $s' \gg s''$ can be rewritten

$$\beta\varepsilon^{3/2} \gg 1, \qquad \varepsilon \gg 1 \tag{8.20}$$

and it is just these conditions that define the boundary of the region I in Fig. 8.1 (here-after, in defining the boundary, the inequalities are replaced by equalities). The conditions (8.20) are much milder than (8.9) and only a high value of the phase difference is required. In particular, if the slope of one term is very small, the Landau-Zener formula (8.11) is valid down to very small velocities and in the limit $F_1 = 0$ it is valid over the whole range of v_p [8.29].

When the condition $s' \gg s''$ is violated, but the inequality $s', s'' \gg 1$ is satisfied, the solution can also be obtained by matching. The comparison equation near each isolated branch point is used. Then (7.91) directly leads to

$$\mathscr{P}_{12} = 4\exp[-2\beta\Delta''(\varepsilon)]\sin^2[\beta\Delta''(-\varepsilon)]. \tag{8.21}$$

The function $\Delta''(\varepsilon)$ can be expressed by complete elliptic integrals [8.30] or by hypergeometric functions [8.31, 32]. A plot of this function is presented in Fig. 8.5, curve 0. The dashed line is for the function $\pi/8\varepsilon^{1/2}$, which gives the constant-velocity approximation for $\Delta''(\varepsilon)$, i.e., for $\varepsilon \gg 1$. The considerable difference between the two plots at low ε is due to the fact that the constant-velocity approximation neglects tunneling: the transition probability is different from zero only when the energy of relative motion is sufficient for the pseudocrossing point to be attained. The contribution from tunneling corrections at high ε can be obtained from the expansion [8.27]:

$$\Delta''(\varepsilon) = (\pi/8\varepsilon^{1/2})(1 - 3/32\varepsilon^2 + \dots) \tag{8.22}$$

demonstrating that $\Delta''(\varepsilon)$ rather rapidly tends to its asymptotic expression for $\varepsilon > 1$. On the other hand, for high negative ε,

$$\Delta''(\varepsilon) = 2|\varepsilon|^{3/2}/3. \tag{8.23}$$

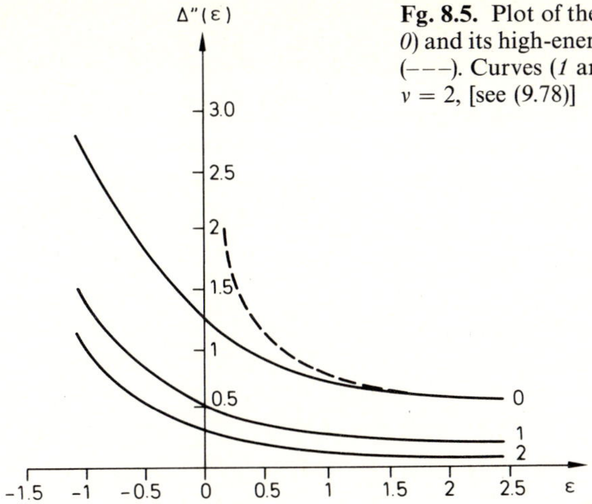

Fg. 8.5. Plot of the function $\Delta''(\varepsilon)$ set by (8.17) (*Curve 0*) and its high-energy (Landau-Zener) approximation (– – –). Curves (*1* and *2*) refer to $\Delta''(\varepsilon, v)$ for $v = 1$ and $v = 2$, [see (9.78)]

Finally, the expansion

$$\Delta''(\varepsilon) = 0.61 - 0.42\,\varepsilon + \dots \tag{8.24}$$

is valid for $|\varepsilon| \ll 1$.

Equation (8.21) is valid provided

$$\beta\,|\varepsilon|^{1/2} \gg 1, \qquad \beta \gg 1. \tag{8.25}$$

These conditions define the boundary of region II in Fig. 8.1.

The relation $s'' \gg s', 1$ is satisfied for high negative ε, i.e., in the region of nonadiabatic tunneling. Then

$$\tau_c = (8\,|\varepsilon|)^{-1/2} + \mathrm{i}\,|\varepsilon|^{1/2}, \qquad s = \pi\beta/4\,|\varepsilon|^{1/2} + 2\mathrm{i}\beta\,|\varepsilon|^{3/2}/3. \tag{8.26}$$

For this case the transition probability is found by integration of the comparison equations which take into account simultaneously the two branch points $\tau_c^{(1)}$ and $\tau_c^{(2)}$ in the upper semiplane located at a distance $|\varepsilon|^{1/2}$ from the real axis (Fig. 8.4). The integration path coincides with the anti-Stokes lines leading to the real axis $\mathrm{Re}\{\tau\}$ (Fig. 8.6).

The solution of this problem yields for the transition probability [8.16]

$$\begin{aligned} \mathscr{P}_{12} &= B(\bar\delta)\exp(-4\beta\,|\varepsilon|^{3/2}/3), \\ B(\bar\delta) &= 2\pi^{-1}\bar\delta[\Gamma(\bar\delta)\exp(\bar\delta - \bar\delta\log\bar\delta)\sin\pi\bar\delta]^2, \end{aligned} \tag{8.27}$$

where now $\bar\delta = \beta/8\,|\varepsilon|^{1/2}$. In the two limiting cases ($\bar\delta \gg 1$ and $\bar\delta \ll 1$),

$$\mathscr{P}_{12} = \begin{cases} 4\sin^2(\pi\bar\delta)\exp(-4\beta\,|\varepsilon|^{3/2}/3) & \text{for } \bar\delta \gg 1, \\ 2\pi\bar\delta\exp(-4\beta\,|\varepsilon|^{3/2}/3) & \text{for } \bar\delta \ll 1. \end{cases} \tag{8.28}$$

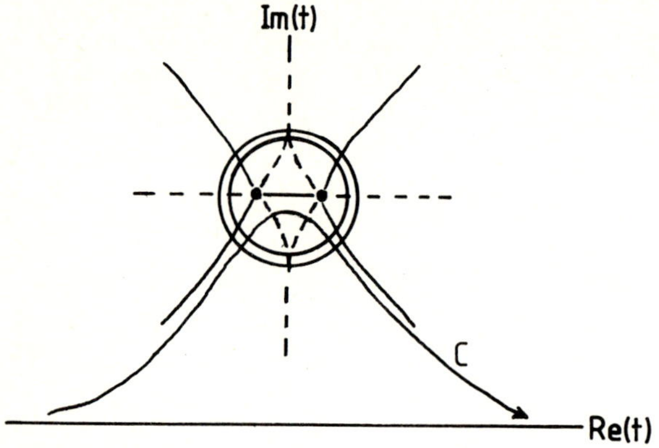

Fig. 8.6. Stokes (---) and anti-Stokes (—) lines for the near-adiabatic case of two branch points of adiabatic terms in the complex t-plane. C is the integration contour passing through the region where the adiabatic approximation is invalid

The small factor before the exponential in the second case implies that two branch points coalesce (the factor $\pi\delta$ is simply proportional to the phase difference between these two points).

The validity conditions for (8.27) are formulated as the inequalities

$$\beta\,|\varepsilon|^{3/2} \gg 1, \qquad |\varepsilon| \gg 1, \tag{8.29}$$

satisfied inside region III (Fig. 8.1).

Turn now to the system (8.19) for low β. The condition $\beta \ll 1$ defining region IV in Fig. 8.1 permits these equations to be solved to first order in β. The transition probability is readily calculated:

$$\mathscr{P}_{12} = \left|(\beta/2)\int_{-\infty}^{\infty} \exp(\mathrm{i}\beta\varepsilon\tau - \mathrm{i}\beta\tau^3/3)\,d\tau\right|^2 = \pi^2\beta^{4/3}\,\mathrm{Ai}^2(-\varepsilon\beta^{2/3}), \tag{8.30}$$

where Ai is the Airy function. The asymptotic representations of this function, see (5.17), yield

$$\mathscr{P}_{12} = \begin{cases} 8\pi\delta\sin^2(2\beta\varepsilon^{3/2}/3 + \pi/4) & \text{for } \beta\varepsilon^{3/2} \gg 1, \quad \varepsilon > 0, \\ 2\pi\delta\exp(-4\,|\varepsilon|^{3/2}/3) & \text{for } \beta|\varepsilon|^{3/2} \gg 1, \ \varepsilon < 0. \end{cases} \tag{8.31}$$

8.3.2 Terms with Slopes of Different Signs

Turn now to the case when the forces F_1 and F_2 have different signs. Then the lower adiabatic term corresponds to the potential barrier, and the upper, to the potential well. The appropriate adiabatic terms are shown in the upper part of Fig. 8.7. The potential well U_1 supports the vibrational quasi-stationary states that decay due to nonadiabatic coupling of the upper with the lower electronic state. The decay results in outgoing waves on the lower electronic term. But when the system originally is on the lower term, the existence of quasi-stationary states in the upper well manifests

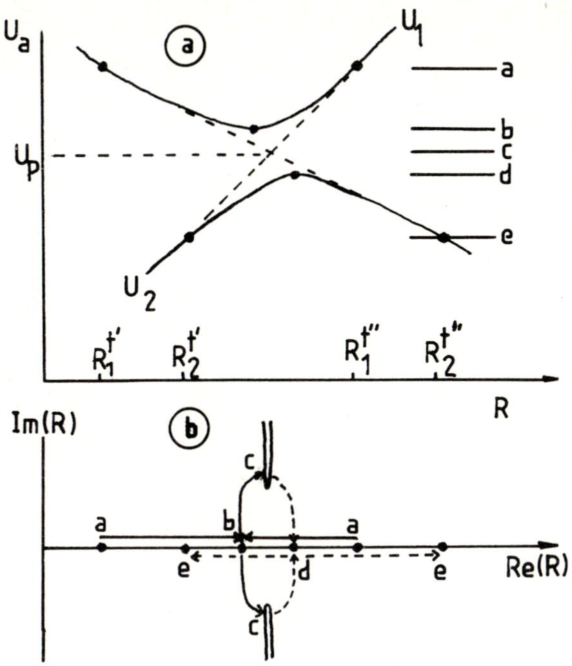

Fig. 8.7. (**a**) Adiabatic terms (—) of the linear model for $F_1 F_2 < 0$. (**b**) Analytical struture of adiabatic terms in the complex R-plane and the motion of turning points with change of energy $\mathscr{E}_a, \ldots, \mathscr{E}_e$

itself as a resonance effect caused by passage of the particle over the potential barrier of the lower adiabatic term. In this case the problem is to calculate the matrix $\mathscr{M}_{22}$ connecting the flux amplitudes at both sides of the barrier, or the unitary matrix $\tilde{\mathscr{M}}_{22}$ expressing the amplitudes of outgoing waves via those of the waves incident on the barrier. As distinct from the case when $F_1 F_2 > 0$, both turning points can now be on the upper, or on the lower term, depending on the energy $\mathscr{E}$. With a continuous variation of $\mathscr{E}$, the turning points pass from the upper to the lower sheet of the complex plane. The motion of the turning points and their correlation with the general form of the adiabatic potentials are shown in the lower part of Fig. 8.7. The positions of the turning points $R_1^{t'}$ and $R_1^{t''}$ correspond to the energy $\mathscr{E}_a$ in the upper part of the figure. As the energy decreases, the turning points approach each other and at position b (energy corresponding to the upper term minimum $\mathscr{E}_b$) depart from the real axis, separating symmetrically to the upper and lower semiplanes of R. Then at $\mathscr{E} = \mathscr{E}_c$ they reach the cuts and pass to the lower Riemann sheet, tending to position d at an energy coinciding with the lower term maximum. Finally, from position d they separate along the real axis to position e corresponding to the energy level $\mathscr{E}_e$ in the upper part of Fig. 8.7.

Consider now the high-energy case similar to that discussed in Sect. 8.2. Assume that the phase between the turning points on the upper term is large, and the splitting of terms in the nonadiabaticity region is so small that the constant velocity approximation is valid near the pseudo-crossing point $R = R_p$, i.e., $\beta\varepsilon^{3/2} \gg 1$, $\varepsilon \gg 1$. In this case the adiabatic functions to the right of the coupling region can be matched with those to the left in terms of the Landau-Zener solution. Appropriate calculations [8.20, 33] give

$$\mathscr{M}_{22} = [1 + (1 - P)\exp(2i\phi_1)]^{-1} \times \begin{pmatrix} P & -2i(1-P)^{1/2}\cos\phi_1 \\ -2i(1-P)^{1/2}\cos\phi_1 & P \end{pmatrix}, \tag{8.32}$$

where P is the Landau-Zener probability (8.10) and ϕ_1 is expressed via the action integral between the turning points of the upper adiabatic term $\Delta_1(\mathscr{E})$ and via the dynamic phase $\gamma(\mathscr{E})$:

$$\phi_1(\mathscr{E}) = \Delta_1(\mathscr{E}) - \gamma(\mathscr{E}). \tag{8.33}$$

Here

$$\Delta_1(\mathscr{E}) = \Delta_1(R_1^{t''}, R_1^{t'}) = \int_{R_1^{t'}}^{R_1^{t''}} [2\mu(\mathscr{E} - U_1)]^{1/2}\, dR,\quad \gamma(\mathscr{E}) = \gamma(\delta),\quad \delta = \delta(\mathscr{E}), \tag{8.34}$$

where $\gamma(\delta)$ is given by (8.11) and $\mathscr{E}$ and ε are related by (8.6).

The transmission $\mathscr{T}$ and reflection $\mathscr{R}$ coefficients for overbarrier passage are readily obtained from (8.32):

$$\mathscr{T} = 4(1-P)\cos^2\phi_1[4(1-P)\cos^2\phi_1 + P^2]^{-1}, \tag{8.35}$$

$$\mathscr{R} = P^2[4(1-P)\cos^2\phi_1 + P^2]^{-1}. \tag{8.36}$$

These expressions allow one to make a simple interpretation for two limiting cases: the nonoverlapping and strongly overlapping resonances.

Consider, for definiteness, the reflection coefficient $\mathscr{R}$. It is seen from (8.36) that for $P \ll 1$ the function $\mathscr{R}(\mathscr{E})$ displays resonance behavior (non-overlapping resonance case). The resonance occurs when $\cos\phi_1(\mathscr{E}_n) = 0$, i.e.,

$$\Delta_1(\mathscr{E}_n) - \gamma(\mathscr{E}_n) = \pi(n + 1/2). \tag{8.37}$$

If $\gamma = 0$, (8.37) defines the JWKB levels $E_{1,n}$ in the upper uncoupled adiabatic potential well $U_1(R)$. The dynamic phase is responsible for the level shift $\Delta E_{1,n}$ which can be found from the expansion

$$\begin{aligned}\Delta_1(\mathscr{E}) &= \Delta_1(E_{1,n}) + (\partial\Delta_1/\mathscr{E})|_{\mathscr{E}=E_{1,n}}(\mathscr{E} - E_{1,n}) \\ &= \Delta_1(E_{1,n}) + (\pi/\omega_{1,n})(\mathscr{E} - E_{1,n}).\end{aligned} \tag{8.38}$$

Substitution of (8.38) into (8.37) gives, to first order,

$$\mathscr{E}_n = E_{1,n} + \omega_{1,n}\gamma(E_{1,n})/\pi. \tag{8.39}$$

Equation (8.39) shows that the quasi-stationary states in the upper well manifesting themselves as resonances in reflection are shifted upwards due to the nonadiabatic coupling. Exact computations [8.34] revealed that this approximation gives satisfactory results even for low-lying resonance states.

To find the shape of $\mathscr{R}(\mathscr{E})$ near an isolated resonance we again use the expansion (8.38) to approximate $\cos^2\phi_1$ near $\mathscr{E}_n$. This yields

$$\mathscr{R} = (\Gamma_n/2)^2\,[(\mathscr{E}-\mathscr{E}_n)^2 + (\Gamma_n/2)^2]^{-1} \quad \text{with} \tag{8.40}$$

$$\Gamma_n = (\omega_{1,n}/\pi)\exp[-2\pi\,\delta(E_{1,n})] \tag{8.41}$$

for $\mathscr{E}$ close to $\mathscr{E}_n$.

If P is sufficiently large, so that the neighboring resonances overlap, one can use the average reflection coefficient $\bar{\mathscr{R}}$ defined by

$$\bar{\mathscr{R}}(\bar{\mathscr{E}}) = (\Delta\mathscr{E})^{-1}\int_{\bar{\mathscr{E}}-\Delta\mathscr{E}/2}^{\bar{\mathscr{E}}+\Delta\mathscr{E}/2} \mathscr{K}(\mathscr{E})\,d\mathscr{E}, \tag{8.42}$$

where the range of $\Delta\mathscr{E}$ accomodates several resonances. Substituting (8.36) into (8.42) we find

$$\mathscr{R} = (2\pi)^{-1}\int_0^{2\pi}[4(1-\bar{P})\cos^2\phi + \bar{P}^2]^{-1}\,\bar{P}^2\,d\phi = \bar{P}/(2-\bar{P}). \tag{8.43}$$

Here $\bar{P}$ will depend on the average energy, viz., $\bar{P} = \exp[-2\pi\,\delta(\bar{\mathscr{E}})]$. Similarly we find

$$\bar{\mathscr{T}} = 2(1-\bar{P})/(2-\bar{P}). \tag{8.44}$$

This expression can also be obtained from simple considerations using the summation of probabilities for capture to and release from the upper potential well for each passage of the pseudocrossing:

$$\bar{\mathscr{T}} = (1-\bar{P}) + \bar{P}^2(1-\bar{P})\sum_{n=0}^{\infty}(1-\bar{P})^{2n} = 2(1-\bar{P})/(2-\bar{P}). \tag{8.45}$$

It will be seen from (8.43, 44) that as the energy increases, the reflection coefficient builds up and the transmission coefficient drops. This dependence is just opposite to that for adiabatic transmission over a simple potential barrier (Sect. 5.2.4).

Consider now the case when $\mathscr{E}$ is less than the height of the lower adiabatic term. When the imaginary action calculated for underbarrier transition is large, i.e., $|\varepsilon| \gg 1$, $\beta\,|\varepsilon|^{3/2} \gg 1$, the transmission coefficient is [8.16]

$$\mathscr{T} = \bar{B}(\bar{\delta})\exp[-2\bar{\Delta}_2(\mathscr{E})]. \tag{8.46}$$

Here

$$\bar{\Delta}_2(\mathscr{E}) = \bar{\Delta}_2(R_2^{t'}, R_2^{t''}) = \int_{R_2^{t'}}^{R_2^{t''}} [2\mu(U_2 - \mathscr{E})]^{1/2}\, dR,$$
$$\bar{B}(\bar{\delta}) = 2\pi\bar{\delta}^{-1}[\Gamma(\bar{\delta})\exp(\bar{\delta} - \bar{\delta}\log\bar{\delta})]^{-2}, \quad \bar{\delta} = \beta/8\,|\varepsilon|^{1/2}, \tag{8.47}$$

and ε and $\mathscr{E}$ are related by (8.6). Now, the exponential in (8.46) is recognized as the transmission coefficient for the adiabatic barrier $U_2(R)$. The factor before the exponential, at the limits of small and large $\bar{\delta}$, assumes the form

$$\bar{B}(\bar{\delta}) = \begin{cases} 2\pi\bar{\delta} & \text{for } \bar{\delta} \ll 1, \\ 1 & \text{for } \bar{\delta} \gg 1. \end{cases} \tag{8.48}$$

It is seen from (8.47, 48) that the probability of nonadiabatic tunneling is always lower than that of adiabatic tunneling.

If the condition $\bar{\delta} \gg 1$ holds close to the top of the barrier, $\mathscr{T}$ can be calculated with the nonadiabatic coupling to the upper term completely neglected. The transmission coefficient will then be given by (5.55) and need not be small. This will be the case for $||\varepsilon| - 1|\,(F/|\Delta F|)^{1/2} \ll 1$. The two cases considered above and valid for $|\varepsilon| \gg 1$, $\beta\,|\varepsilon|^{3/2} \gg 1$ can be bridged between the limits of large negative and large positive ε for the case of small β. In this case the transmission coeffient can be calculated by the perturbation approach using as the zero-order approximation the wave functions corresponding to diabatic motion in the fields of constant forces F_1 and F_2. The result is

$$\mathscr{T} = \pi^2\beta^{4/3}\,\mathrm{Ai}(-\varepsilon\beta^{2/3}). \tag{8.49}$$

Here of course $\mathscr{T} \ll 1$, $\mathscr{R} \approx 1$. Equation (8.49) at the limit $\varepsilon\beta^{2/3} \gg 1$ yields (8.31).

Other interpolating expressions eliminating the restriction $\mathscr{T} \ll 1$ and based on the phase-integral approach have been suggested [8.35].

8.4 Validity of Linear Model and of Analytical Expressions for Transition Probabilities

As stated in Sect. 8.1, the validity limits of the linear model are defined by the condition that the effective width of the nonadiabatic coupling region ΔR be smaller than a certain dimension l, defining the characteristic scale for which the real diabatic terms differ from the linear ones, and for which their interaction cannot be considered as a constant. To find ΔR one must either write down the conditions for the asymptotic behavior of $b_k(t)$ using the known [8.36] asymptotics for a parabolic cylinder function, or (which is, in a certain sense, simpler and more illustrative) evaluate ΔR in the diabatic ($2\pi\delta \ll 1$) or adiabatic ($2\pi\delta \gg 1$) limit. Since in these cases the solution is given analytically in terms of known integrals, it is not difficult to find the width of the region essential for these integrals.

In the adiabatic limit the value ΔR defined as $\Delta_a R$ is equal to the distance from the real axis to the branch point, $\Delta_a R = 2a/|\Delta F|$. In the diabatic limit $\Delta R = \Delta_d R$ is defined by the length at which the phase difference calculated for diabatic terms from the crossing point becomes of the order of unity, $\Delta_d R = 2(v/|\Delta F|)^{1/2}$. Thus the validity conditions of the linear model are formulated as two inequalities,

$$\Delta_a R = 2a/|\Delta F| \ll l, \qquad \Delta_d R = 2(v/|\Delta F|)^{1/2} \ll l. \tag{8.50}$$

Now, the product $|\Delta F|\, l$ can be interpreted as the splitting of the adiabatic terms far from the coupling region. Setting $|\Delta F|\, l = \Delta U_\infty$ and denoting by $\xi_\infty = \Delta U_\infty l/v$ the Massey parameter for nonadiabatic coupling between terms with a splitting ΔU_∞, (8.50) will be rewritten:

$$2a \ll \Delta U_\infty, \qquad 1 \ll \xi_\infty. \tag{8.51}$$

The first inequality means that the adiabatic terms actually reproduce the pseudocrossing pattern (the minimum splitting $2a$ between the adiabatic terms is much smaller than the asymptotic splitting ΔU_∞), and the second, that there is no coupling far from the pseudocrossing (the Massey parameter ξ_∞ is high). The two conditions (8.51) define the outer boundaries of the plane in Fig. 8.1 for the Landau-Zener model. The nonlinear diabatic terms would have to be used to go beyond these limits.

Of most interest here is the case of strong coupling (region I), since for all other regions, the use of common methods of perturbation theory would readily yield the desired solutions. Certain nonlinear models involving the linear model as a particular case under conditions (8.50) or (8.51) will be discussed in the next chapter.

To interpret the conditions (8.50, 51), it is important to realize that the solution of equations describing nonadiabatic coupling display rather peculiar asymptotic behavior. For instance, when using the adiabatic basis, the amplitudes $b_i(t)$ tend to their asymptotics exhibiting an oscillatory behavior. If the probability is low, the slowly decreasing [as $\Delta_a R/(R - R_p)^3$] amplitude of oscillations still remains larger than the asymptotic value $b_i(\infty)$ for the distance $R - R_p$, which is in turn appreciably larger than $\Delta_a R$. This peculiarity found by numerical solution of coupled equations [8.37, 38] can be interpreted, in a sense, as a widening of the coupling region at the near-adiabatic limit, i.e., at low energies. In this limit the transition probability for the linear model is given by (8.21), and (7.91) readily provides generalization for nonlinear models. For a slightly nonlinear model, the exponent $s''_{\rm nonl}$ will be

$$s''_{\rm nonl} = \beta \Delta''(\varepsilon) + \bar{s}''. \tag{8.52}$$

The condition (8.50) ensures only that $\beta \Delta''(\varepsilon) \gg \bar{s}''$, but not that $\bar{s}'' \ll 1$. If $\bar{s}''$ is larger than unity, the near-adiabatic transition probability calculated for linear and slightly nonlinear terms will be substantially different. However, this will happen only when the probability is already very small.

Consider now the implication of using states that are not strictly diabatic. Though for the two-state problem the diabatic states can be rigorously defined, these will not be exactly linear, even for the clear-cut case of pseudocrossing. In other words, the crossing terms are not necessarily diabatic. Thus, besides the potential, the dynamic coupling must also be taken into account. The contribution $\delta_{\rm dyn}$ of the dynamic coupling to δ is [8.39]

$$\delta_{\rm dyn} \approx v/|\Delta F|\, l_0^2, \tag{8.53}$$

where l_0 is the range over which the almost-diabatic states change. As l_0 is expected to be close to l, (8.50) requires that $\delta_{\rm dyn} \ll 1$. This implies that within the range of validity of the Landau-Zener model the dynamic coupling between almost-diabatic states will be weak and as such, cannot be taken as the reason for the substantial change in the form of the velocity dependence of the probability, compared with the traditional Landau-Zener dependence.

The same restrictions arise for the transition probabilities between crossing adiabatic terms. The most typical case concerns terms differing in electronic angular momentum quantum numbers Λ or Ω. The relatively infrequent crossing of terms of the same axial symmetry, possible for specific potential types, for instance, for two Coulomb centers (see Sect. 3.3.3), will not be considered here. For Coriolis coupling, see (8.13),

$$a_{\rm Cor} = rbv/R_{\rm c}^2. \tag{8.54}$$

Assuming that $r \approx 1$ and taking into account the natural condition $l < R_{\rm c}$, we find for $b \sim R_{\rm c}$ that

$$2\pi\delta_{\rm Cor} \sim 2\pi v/|\Delta F|\, R_{\rm c}^2 \ll l^2/R_{\rm c}^2. \tag{8.55}$$

Equation (8.55) implies that $2\pi\delta_{\rm Cor}$ must not be too large, and therefore the transition probability can be calculated to first order in $\delta_{\rm Cor}$. In other words, the linear model is unable to describe strong rotational coupling. If the formula (8.11) were used for large $\delta_{\rm Cor}$, it would predict a decrease in the transition probability with increasing velocity, which is, of course, completely wrong.

Consider now the approximate expressions for $\mathscr{P}_{12}(\varepsilon, \beta)$, valid over a broader range of parameters than are the expressions that are confined to separate regions I, II, III, IV in Fig. 8.1. Several attempts have been made to derive a unified interpolation expression for $\mathscr{P}_{12}(\varepsilon, \beta)$ [8.35, 40–42], but so far these have not been successful. Therefore, we shall consider separately the cases $\beta \ll 1$ and $\beta \gg 1$.

When $\beta \ll 1$, the suggested expression is

$$\mathscr{P}_{12} = 8\pi^2 \delta z^{1/2} \operatorname{Ai}^2(z), \qquad z = (3\Delta_{21}/2)^{2/3}, \tag{8.56}$$

where δ is defined by (8.10) and Δ_{21} is the absolute value of the difference between the adiabatic or the diabatic phases calculated from the turning points to the crossing point of the diabatic terms. The inequalities (8.50) ensure that the Airy function will tend to its asymptotic form faster than the rate at which Δ_{21} becomes essentially different from the value $(2/3)\,\varepsilon\beta^{2/3}$ valid for the linear model. Equation (8.56) is a relatively good approximation even for not too small β. For instance, perturbation theory appears to be valid even for a relatively high probability $\mathscr{P}_{12} = 1/4$ in the maximum of the first oscillation, i.e., for $\beta = 0.16$ [8.43].

When $\beta \gg 1$ the interpolation expression for $\mathscr{P}_{12}$ is

$$\mathscr{P}_{12} = 4B(s'/\pi)\exp(-2s'')\,[1 - \exp(-2s'')]\sin^2[\Delta'_{21} + \gamma(s'/\pi)]. \tag{8.57}$$

Here s' and s'' are defined by (8.17), γ is the dynamic phase, and Δ'_{21} is the real part of the phase difference over the adiabatic terms from the turning points to the complex point of their crossing.

When $\varepsilon \gg 1$, the phase Δ'_{21} virtually coincides with Δ_{21}, and B with unity, so that (8.57) can differ from the Landau-Zener formula only in certain small corrections in the exponent, see (8.22). As ε decreases, the parameter s'' increases and $\gamma \to 0$. The conditions (8.51) ensure here that B appreciably differs from unity only when Δ'_{21} can already be identified with s'. Finally, at large negative ε, (8.57) converts to (8.27).

Simple interpolation equations can be recommended for practical use [8.44]:

$$\begin{gathered} B = 1 + 1/2s', \qquad s' = \beta\Delta''(-\varepsilon), \qquad s'' = \beta\Delta''(\varepsilon), \\ \Delta''(\varepsilon) = \begin{cases} (2/3)\,(0.16 - \varepsilon)^{3/2} + 1.15 & \text{for} \quad \varepsilon \leqq 0, \\ (\pi/8)\,[(1.3 + \varepsilon)^{1/2} - 0.5]^{-1} & \text{for} \quad \varepsilon \geqq 0. \end{cases} \end{gathered} \tag{8.58}$$

A comparison with the results obtained by numerical integration of (8.5) shows that (8.57), together with (8.58), gives a satisfactory approximation for $\beta > 2$. The validity region of (8.57), then, accomodates the values of β for which the transition probability as a function of the dimensionless energy ε oscillates, when $\varepsilon < 0$.

Finally, it is noted that for the ε and β values for which neither of the above expressions is valid, the transition probability has been calculated by numerical solution of (8.16) and tabulated [8.45].

8.5 Cross Sections for the Linear Model

The general expressions for the transition probability obtained in the previous sections enable explicit calculations of the integral and differential cross sections for different types of coupling to be made. This requires parametrization of ε and β in terms of the initial energy E_i and initial angular momentum l_i. In calculating integral cross sections we shall discuss separately the transitions between pseudocrossing and crossing adiabatic terms, and then some specific features of differential cross sections near the threshold for weak coupling between crossing diabatic terms.

8.5.1 Integral Cross Sections – Radial Coupling

Consider two pseudocrossing terms which will be taken, for simplicity, as two spinless Σ-terms. Thus $l_i = l_f = J$. A simple approximation widely used for the calculation of the integral cross section σ_{fi} for the transition $i \to f$ $(i, f = 1, 2)$ is based on the Landau-Zener formula (8.11):

$$\sigma_{fi} = 4\pi \int_0^\infty \bar{\mathscr{P}}_{fi} \sin^2 \phi_{fi} b\, db. \tag{8.59}$$

It is assumed that $\sin^2 \phi_{fi}$ changes with b so rapidly that it can be replaced by 1/2 in the integrand. Moreover, the region of integration over b where (8.11) is invalid, is assumed to be relatively small, so that $\bar{\mathscr{P}}_{fi}$ in (8.59) can be identified with $\bar{\mathscr{P}}_{fi}$ from (8.11), replacing v_p in ε by the radial velocity over all the integration range $0 \leqslant b \leqslant R_p$. Thus, the quantity $\mathscr{E} - U_p$ in (8.6) is expressed via the initial kinetic energy of the atoms E_i and the impact parameter b:

$$\mathscr{E} - U_p = E_i(1 - b^2/R_p^2) - V_{ii}(R_p) + V_{ii}(\infty). \tag{8.60}$$

The value $|\Delta F|$ is defined as the difference in slopes of effective diabatic terms at the point of their crossing:

$$|\Delta F| = |F_1 - F_2| = |\partial V_{11}^{\text{eff}}/\partial R - \partial V_{22}^{\text{eff}}/\partial R|_{R=R_p}. \tag{8.61}$$

Note that due to the centrifugal potential contribution V_{11}^{eff} and V_{22}^{eff} depend on b, but the slope difference is independent of b. The substitution of parametrizations (8.60, 61) into (8.59), taking (8.11) into account, allows the cross section σ_{fi}^{LZ} to be expressed as a function of the parameter y, defined by

$$y = [E_1 + V_{11}(\infty) - V_{11}(R_{\text{p}})]/E^* = [E_2 + V_{22}(\infty) - V_{22}(R_{\text{p}})]/E^*,$$
$$E^* = (\mu/2)\,(2\pi a^2/|\Delta F|)^2, \tag{8.62}$$

provided the pseudocrossing distance R_{p} is classically accessible for all b less than R_{p}, i.e., if there are no limitations imposed on incoming or outgoing fluxes by the potential barriers. This condition is satisfied for terms of the type shown in Fig. 7.1.

Calculation yields

$$\sigma_{fi}^{\text{LZ}} = \pi R_{\text{p}}^2 (E^*/E_{\text{i}})\, f^{\text{LZ}}(y) \quad \text{where} \tag{8.63}$$

$$f^{\text{LZ}}(y) = \int_0^y 2\exp(-x^{-1/2})\,[1 - \exp(-x^{-1/2})]\,dx,$$
$$f^{\text{LZ}}(y) = \begin{cases} 2y^{3/2}\exp(-y^{-1/2}) & \text{for} \quad y \ll 1, \\ 4y^{1/2} & \text{for} \quad y \gg 1. \end{cases} \tag{8.64}$$

Now, for an initial horizontal diabatic term (V_{11} = const), $E_1 = yE^*$, and f^{LZ} is simply related to the reduced cross section Q_{21}^{LZ}:

$$Q_{21}^{\text{LZ}} = \sigma_{21}^{\text{LZ}}/\pi R_{\text{p}}^2 = y^{-1} f^{\text{LZ}}(y). \tag{8.65}$$

The plot of the function $Q_{21}^{\text{LZ}}(y)$ is shown in Fig. 8.8 by the dashed line. The maximum of Q_{21}^{LZ} occurs at $y = 5.6$ and $Q_{21\,\max}^{\text{LZ}} = 0.445$.

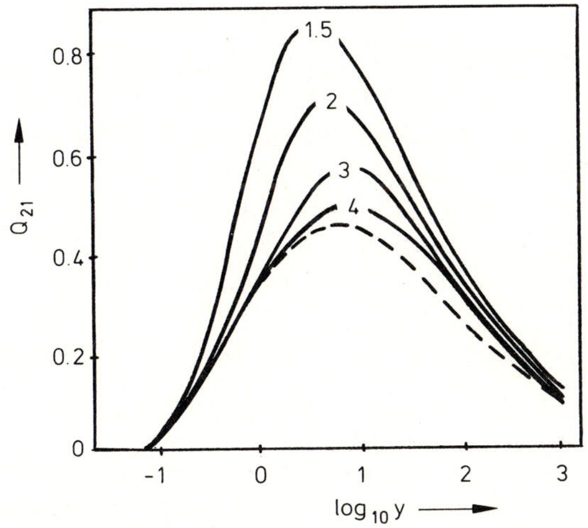

Fig. 8.8. Reduced cross sections $Q_{21}(y, A, B)$ [see (8.69–71)] versus $\log_{10} y$ for $B = 0$ and different A values. The dashed line refers to the Landau-Zener cross section $Q_{21}^{\text{LZ}}(y)$ (8.65)

Equations (8.63, 65), which neglect the undulatory behavior of the probability and tunneling corrections, have been used repeatedly in treating experimental data. Moreover, the cross sections (8.65) have been used for the calculation and tabulation of the average rate-constants of non-adiabatic processes [8.46, 47]:

$$k_{fi}^{\mathrm{LZ}}(T) = (8kT/\pi\mu)^{1/2} \int_0^\infty \sigma_{fi}^{\mathrm{LZ}}(E_\mathrm{i}) \exp(-E_\mathrm{i}/kT)\,(E_\mathrm{i}\,dE_\mathrm{i})/(kT)^2 . \tag{8.66}$$

With a more accurate level of calculation the oscillatory behavior of the transition probability would be retained. In particular, for near-adiabatic collisions when the main contribution to σ^{LZ} comes from small impact parameters, the cross section σ^{LZ} acquires an undulating correction $\Delta\sigma$ [8.48],

$$\Delta\sigma \sim \sin[2\Delta_{21}(E_\mathrm{i}, b=0)] . \tag{8.67}$$

As $\Delta_{21}(E_\mathrm{i}, b=0)$ is the Stueckelberg phase difference for the head-on collision, the undulating structure of the integral inelastic cross section versus energy can provide some information on the adiabatic terms over the range $R_1^\mathrm{t}, R_2^\mathrm{t} - R_\mathrm{p}$.

In a more general case, with the largest b-contribution to the integral cross section coming from the turning-point region, the transition probability should be expressed by equations valid for small ε, rather than by (8.11). Then a definition of the average force

$$F(b) = |(\partial V_{11}^{\mathrm{eff}}/\partial R)\,(\partial V_{22}^{\mathrm{eff}}/\partial R)|_{R=R_\mathrm{p}}^{1/2} \tag{8.68}$$

would have to be added to the parametrizations (8.60) and (8.61). The accuracy of such an approach for weak coupling of diabatic terms ($\beta < 1$) is presented in Fig. 8.9, which refers to the transition probability between two pseudocrossing terms $^2\Sigma$ in the system $\mathrm{He}^+ - \mathrm{Ne}$, also discussed in Sect. 8.5.3 in connection with the calculation of differential cross sections. The full lines refer to the DWBA calculation for nonlinear diabatic terms [8.49], the open points, to close-coupling results. It is seen that for this case, the perturbation treatment is quite satisfactory even for rather high values of $\mathscr{P}_{12}$. The dashed line corresponds to the Landau-Zener formula (8.11) with Δ_{21} calculated for the actual adiabatic terms. Finally, the Airy approximation (8.30) is represented by dots [8.43].

The Airy function approximation is seen to match satisfactorily with the general expression (8.11) and this shows the validity of the interpolation equation (8.56), both for the classical and the quantum regions of the impact parameter ($b < R_\mathrm{p}$ and $b > R_\mathrm{p}$). However, a more careful analysis reveals that this conclusion should be accepted cautiously. A defect,

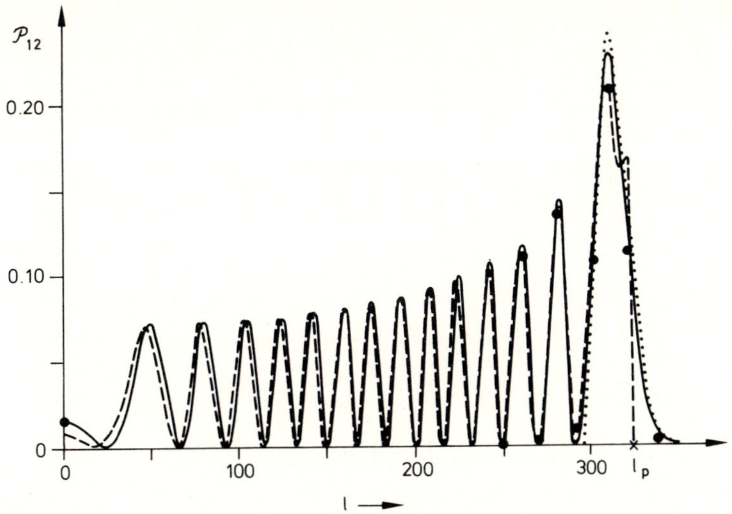

Fig. 8.9. Comparison of different approximations for $\mathscr{P}_{12}$ in the model of the $He^+ - Ne$ system: the full line refers to the DWBA result, the black circles to strong coupling calculations, the dashed line to the Landau-Zener result (8.11), the dots to the Airy approximation (8.30)

indiscernible in the figure, is that the tunneling contribution is overestimated: though the approximation by the Airy function with the parametrizations (8.60, 61, 68) and the curve plotted by the distorted wave method virtually coincide for b slightly higher than R_p, when b increases, the divergence becomes greater. As a result, the calculation of the integral cross section with the conventional parametrizations (8.60, 61, 68) and use of the Airy approximation in the tunneling region yields a divergent integral. This divergence is caused by extending the linear approximation for the diabatic terms near the crossing point to a region far from it [8.50].

This can be illustrated by considering the motion of a representative point in the ε, β plane as a function of b, when integrating over impact parameters (Fig. 8.10). Equation (8.11) is valid to the right of boundary X. The transition probability $\overline{\mathscr{P}}_{fi}$ reaches a maximum on curve Y defined by $2\pi\delta = \ln 2$. The extension of this curve to the region where the Landau-Zener formula is invalid is shown by the dashed line. The region of weak interaction (region I) lies below curve Y, and the region of near-adiabatic coupling above it (region II). Consider a head-on collision ($b = 0$) with the parameters $\varepsilon(0)$ and $\beta(0)$ lying in the validity region of the Landau-Zener formula. As the impact parameter builds up, the point $\varepsilon(b)$, $\beta(b)$ moves along a certain curve. Three types of curves are possible with the parametrizations (8.60) and (8.61): curve 1 intersects line Y in the validity region of the Landau-Zener formula, curve 2 intersects curve Y (more precisely its

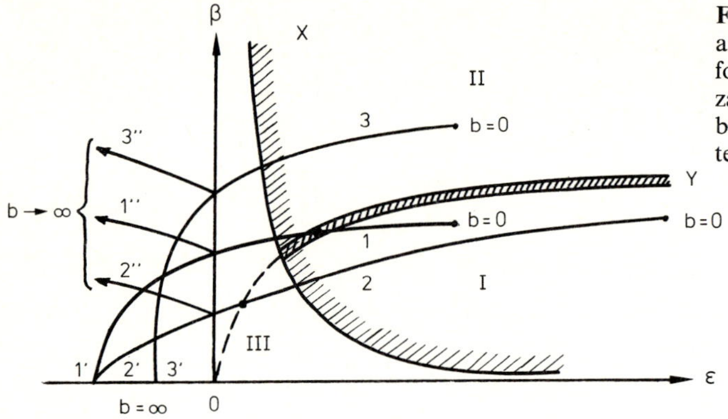

Fig. 8.10. Motion of ε and β with increasing b for different parametrization of the linear diabatic terms (for detail see text)

extension) in the region where the Landau-Zener formula is invalid, curve 3 intersects neither Y nor its extension.

The cross section σ_{fi}^{LZ} is obtained by integrating the probability (8.11) along these curves from a point $b = 0$ to a point $b = R_p$, which corresponds to $\varepsilon = 0$. For curve 1 the cross section will be a maximum, and for curves 2 and 3 the cross sections correspond to values to the right and to the left of the maximum, respectively.

If tunneling is not neglected, but (8.60, 61) are retained, the integration for $\varepsilon < 0$ will be along curves 1′, 2′, 3′. For $b \to \infty$ the probability decrease along these curves is so slow that the cross section becomes infinite in the three cases. This overestimation of tunneling is due to the linear approximation for diabatic terms close to their crossing point. For weak coupling (curve 2), for which an example of the $\mathscr{P}_{12}(b)$ is provided by Fig. 8.9, the divergence sets in only for very large b and therefore it can readily be eliminated simply by truncating the integral. However, in the general case this defect cannot be eliminated in such a simple manner, yet the result can be greatly improved by modifying the parametrization of $\varepsilon(b)$ and $\beta(b)$ in the tunneling region. In this case, the diabatic terms for $\varepsilon < 0$ must be approximated by linear ones, not near the crossing point, but near the nuclei turning points, i.e., in the region providing the largest contribution to the nonadiabatic transition probability $\mathscr{P}_{12}$. For the overbarrier region the linear approximation will be retained near the diabatic-term crossing points. A certain refinement is possible in this case as well, taking into account the fact that as ε increases the transition probability no longer depends on two parameters (ε and β), but on their combination δ only. Thus, the parametrization of ε and β can be altered in such a way that δ remains unchanged, but at the same time the phase Δ_{21} obtained for the linear model becomes as close as possible to the true phase for curvilinear terms [8.51]. For the integral cross section the refinement of the model in the overbarrier region is not very essential, since the positions of the probability extrema have only a weak effect on the integral cross sections.

Thus, in the tunneling region (8.60, 61, 68) are retained for the parameters $\mathscr{E} - U_{\mathrm{p}}$, ΔF and F involved in ε and β, with the only difference being that for $\varepsilon < 0$, R_{p} is replaced by a b-dependent $R_{\mathrm{p}}^{\mathrm{eff}}$ which is the crossing radius of the linear terms tangent to the effective diabatic terms at the turning points [8.51, 52].

With such a parametrization the extensions of curves 1, 2, 3 to the tunneling region are represented by curves 1″, 2″, 3″. The decrease of transition probability along the latter curves is fast enough. In this approximation the integral cross section σ_{fi} can be represented in a form similar to (8.63) [8.52]:

$$\sigma_{fi} = \pi R_{\mathrm{p}}^2 (E^*/E_{\mathrm{i}})\, f(y, A, B), \tag{8.69}$$

which involves, besides y defined by (8.62), two more parameters A and B. The first specifies the crossing radius in the scale of the nonadiabaticity region, and the second the average force for head-on collisions:

$$\begin{aligned} A &= (4R_{\mathrm{p}}\,|\Delta F|/\pi^2 a)^{1/2} = (8\,R_{\mathrm{p}}/\pi^2 \Delta_{\mathrm{a}} R)^{1/2}, \\ B &= F_0 R_{\mathrm{p}}/2E^*, \qquad F_0 = F|_{b=0}. \end{aligned} \tag{8.70}$$

These parameters define the shapes of curves 1″, 2″, 3″ for $b \to \infty$ in Fig. 8.10. Clearly,

$$f^{\mathrm{LZ}}(y) = \lim_{\substack{A \to \infty \\ B \to 0}} f(y, A, B). \tag{8.71}$$

For an initial horizontal diabatic term (hence $B = 0$), examples of reduced cross sections $Q_{21} = \sigma_{21}/\pi R_{\mathrm{p}}^2$ are given in Fig. 8.8. It is seen that for finite values of A the position of the Q_{21} maximum shifts relative to the Q_{21}^{LZ} maximum, and the cross section can be appreciably greater than the maximum value $(Q_{21}^{\mathrm{LZ}})_{\max} = 0.445$. The reason for this difference in the strong-coupling case is seen in Fig. 8.11. The Landau-Zener probability represented by the dash-dotted line reproduces on the average the quantum close-coupling results [8.53] below the threshold, but will eventually tend to zero for $b \to R_{\mathrm{p}}$. However, the outermost maximum for $b \approx R_{\mathrm{p}}$ and the appreciable tunneling for $b > R_{\mathrm{p}}$ contribute quite substantially to the integral cross section. In this region the probability calculated semiclassically for the linear model with a linear approximation near the turning points (dashed line) is close to the quantum result (full line). The discrepancy still existing between the semiclassical and the quantum approach for the deep tunneling region can readily be eliminated, when needed, bypassing the linear model and using the near-adiabatic perturba-

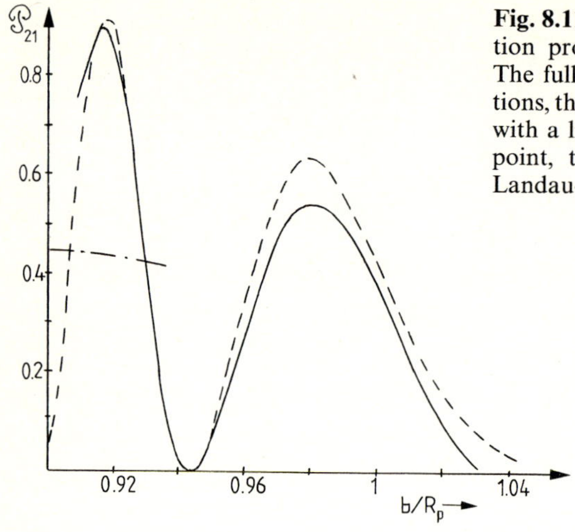

Fig. 8.11. Threshold behaviour of the transition probability for $A = 4$, $B = 0$, $y = 5.56$. The full line refers to close-coupling calculations, the dashed line to the semiclassical result with a linear approximation near the turning point, the dash-dotted line to the average Landau-Zener transition probability (8.11)

tion theory. In practice this is hardly needed, since the transition probability in this region is very small in absolute magnitude. Numerical calculations of the cross sections suggest an approximation for the maximum value of $f(y, A, B = 0)$ [8.52]:

$$f_{\max}(y, A, B = 0) = 0.445\,(1 + 0.92/A^2)^2 . \tag{8.72}$$

Consider now the threshold dependence of the cross section. It is clear from physical considerations that in the threshold region with fixed y the cross section increases with decreasing A (larger size of the nonadiabatic-coupling region) and increasing B (greater slope of terms, which facilitates tunneling). In particular, for $y = 0$ the cross section σ_{fi} is of the order [8.52]

$$\sigma_{fi} = \pi R_p^2 (A/B^{1/2}) \exp(-\,1.24\,AB^{-1/2}). \tag{8.73}$$

To estimate the cross section near the threshold one can use the approximate representations of the transition probability (8.57) with the above-mentioned parametrization of the ε and β dependences on b. This approximation satisfactorily reproduces the exact result (Fig. 8.12). Note that the cross section exhibits shallow undulations caused by incomplete averaging of the Stueckelberg oscillations in the transition probability.

If one or both effective diabatic terms exhibit potential barriers (e.g. of a centrifugal nature) the maximum value of the impact parameter in (8.59) is defined by the condition of overcoming the barrier, rather than by that of vanishing transition probability.

The general method of constructing the scattering matrix by matching (see Sect. 6.2) makes it possible, when needed, to account for the quantum effects of reflection and for penetration through the barrier [8.54]. How-

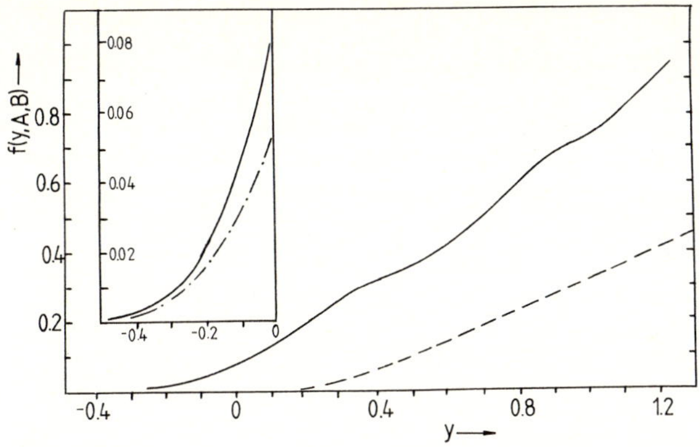

Fig. 8.12. Threshold behaviour of $f(y, A = 2, B = 1)$: full lines refer to the semiclassical result with a linear approximation near the turning points, dashed line to the Landau-Zener function $f^{LZ}(y)$, dash-dotted line in the insert is obtained using interpolations (8.57, 8.58)

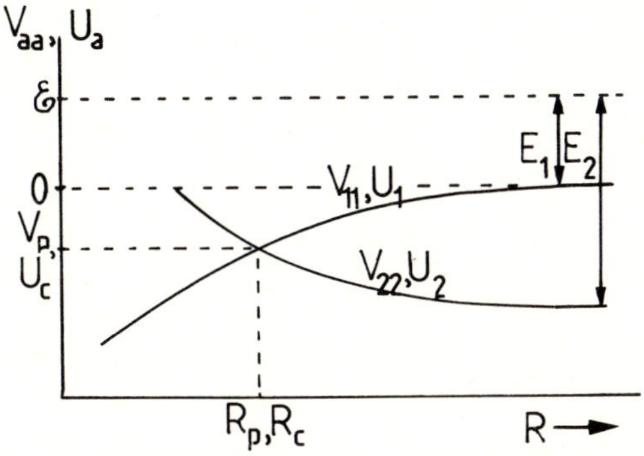

Fig. 8.13. Crossing of attractive and repulsive diabatic (V_{aa}) or adiabatic (U_a) terms

ever, of more importance is the simple classical effect of surmounting the barrier.

For an attractive potential V_{11} (Fig. 8.13) of the form $V_{11}(R) = V_p(R_p/R)^{-s}$, $V_p < 0$, the cross sections σ_{21} for the Landau-Zener transition probability have been calculated [8.55] for different s. The dimensionless cross section $Q_{21} = \sigma_{21}/\pi R_p^2$ depends on the dimensionless velocity $v^* = (E_1/E^*)^{1/2}$ and on two more parameters $V_p^* = V_p/E^*$ and s, which set the form of $V_{11}(R)$. When $s = 6$, the plots of Q_{21} versus v^* are those in Fig. 8.14 for different values of V_p^*. The dotted curve separates the regions where the centrifugal barrier either does (region above the curve) or does not (region below the curve) affect the cross section. The dashed line corresponds to $V_p = 0$, i.e., to a horizontal V_{11}. This is just the Landau-Zener cross section passing through the maximum for $v^* = y_m^{1/2}$. When $V_{11}(R)$ is Coulombic ($s = 1$), there will be no centrifugal barrier and Q_{21} will coincide with $\sigma_{21}^{LZ}/\pi R_p^2$.

The cross sections for the reverse processes σ_{12} can be calculated via σ_{21} using the detailed balance relations

$$\sigma_{12} = (E_1/E_2)\,\sigma_{21}. \tag{8.74}$$

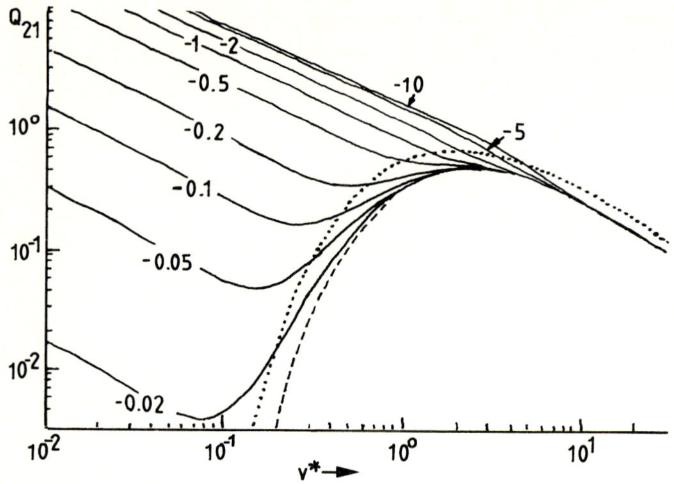

Fig. 8.14. Log-Log plot of the reduced cross section Q_{21} versus v^* for an attractive diabatic term with $s = 6$. The curves are labeled by corredponding V_p^* values. The dashed curve is for a horizontal ($V_p^* = 0$) upper term. The dotted curve separates regions where the centrifugal barrier is important (*above the curve*) or unimportant (*below the curve*) [8.55]

For the attractive Coulomb term $V_{11}(R)$, (8.74) implies an abrupt change of σ_{12} from zero to a finite value at the threshold.

8.5.2 Integral Cross Section – Rotational Coupling

As discussed in Sect. 8.4, the Landau-Zener treatment of rotational coupling between two crossing adiabatic terms U_1 and U_2 is consistent only provided this coupling is not too strong. To calculate the cross section using (8.59), replace v_c in (8.14) by the radial velocity at the point R_c, express $\dot{\psi}$ through the impact parameter from the conservation law for relative angular momentum, and substitute 1/2 for $\sin^2\phi_{21}$.

Introducing the dimensionless energy

$$\begin{aligned}\tilde{y} &= [E_1 + U_1(\infty) - U_1(R_c)]/\tilde{E}^* = [E_2 + U_2(\infty) - U_2(R_c)]/\tilde{E}^*,\\ \tilde{E}^* &= (\mu/2)\,(2\pi r^2/|\Delta F|\,R_c^2)^{-2},\end{aligned} \tag{8.75}$$

and assuming that there are no potential barriers in the incoming and outgoing channels, we obtain for the cross section $\tilde{\sigma}_{fi}^{\mathrm{LZ}}$ for first-order rotational coupling

$$\tilde{\sigma}_{fi}^{\mathrm{LZ}} = \pi R_c^2 (E^*/E_i)\,(8/3)\,\tilde{y}^{3/2} = \frac{16(2\mu)^{1/2}\pi^2 r^2 [E_i + U_i(\infty) - U_i(R_c)]^{3/2}}{3\,|\Delta F|\,E_i}. \tag{8.76}$$

In general, the terms U_1 and U_2 are doubly degenerate so that the square of the coupling matrix element for a two-state model will be obtained by averaging over initial substates and summing over the final ones. This gives, for the $\Omega \to \Omega + 1$ transition,

$$r^2 = \langle a_1, \Omega | j_{x_2'} | a_2, \Omega + 1 \rangle^2 (\delta_{\Omega,0} + 1). \tag{8.77}$$

The appearance of $\delta_{\Omega,0}$ in (8.77) singles out a peculiarity for $\Sigma \to \Pi$ transitions; in this case there is a coupling of the initial state to both final states.

It may be expected that for Coriolis coupling, the use of the transition probability in the form of (8.14), which is invalid for b close to R_c, introduces a larger error into the integral cross section σ_{21} than it does in the case of radial nonadiabatic coupling, since the Coriolis coupling become substantially stronger with increasing b. An equation involving the Airy function would be better instead of (8.14) for a more accurate calculation. The result obtained with a linear approximation for the diabatic terms in the tunneling region is

$$\tilde{\sigma}_{21} = \sigma_{21}^{LZ}[1 - 0.17(\Delta_d R/R_c)^{2/3}]. \tag{8.78}$$

The decrease of the cross section σ_{21} with increasing v, compared to the Landau-Zener model, is due to the fact that the probability oscillations become slower with increasing v_c and the relative contribution from the region where the expression for $\tilde{\sigma}_{21}$ averaged over oscillations is incorrect becomes larger. To understand the limits of the Landau-Zener treatment of rotational coupling, rewrite (8.76) for high energies:

$$\tilde{\sigma}_{21}^{LZ} = \pi R_c^2 (4\pi/3) r^2 (\Delta_d R/R_c)^2. \tag{8.79}$$

As $\Delta_d R \ll R_c$ for the linear model, (8.79) implies that $\tilde{\sigma}_{21}^{LZ} \ll \pi R_c^2$. Attempts are sometimes made to use the Landau-Zener model, or its simple extensions, to describe the situation when $\tilde{\sigma}_{21} \sim \pi R_c^2$. For instance, the linear approximation for $U_1 - U_2$ near $R = R_c$ is retained, but the R-dependence of the coupling is taken into account in the numerical integration of the scattering equations [8.56]. These approaches seem to be beyond the linear model limits and will not be discussed here.

Consider now the barrier effect in one of the channels, assuming that the barrier is due to the centrifugal forces. Let the term pattern be as shown in Fig. 8.13. For the upper term $U_1(R) = U_c(R_c/R)^s$ where $U_c < 0$, calculations have been performed for different values of s [8.55]. For a fixed s the dimensionless cross sections $\tilde{Q}_{21} = \tilde{\sigma}_{21}/\pi R_c^2$ depend on the reduced velocity $\tilde{v}^* = (E_1/\tilde{E}^*)^{1/2}$ and $U_c^* = U_c/E^*$. The results obtained are shown in Fig. 8.15, where the full curves are labeled by U_c^* values and the dotted curve separates regions where the barrier effects are important and un-

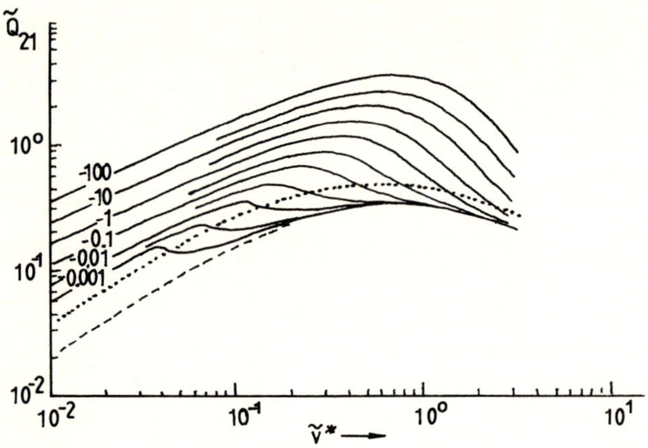

Fig. 8.15. Log-Log plot of the reduced cross section $\tilde{Q}_{21}$ versus $\tilde{v}^*$ for an attractive adiabatic term with $s = 6$. The curves are labeled by corresponding U_c^* values. The dashed curve is for a horizontal ($U_c^* = 0$) upper term. The dotted curve separates regions where the centrifugal barrier is important (*above the curve*) or unimportant (*below the curve*).

important (above or below the curve). Again the dashed curve corresponds to $U_c = 0$, i.e., to the case when there is no barrier effect. Note that the calculations in [8.55] were performed using the full Landau-Zener formula (8.11) which, for high velocities, gives decreasing cross sections. As discussed in Sect. 8.4 in connection with (8.55), this is not correct, so that to the right of the maximum the meaning of the cross sections $\tilde{Q}_{21}$ in Fig. 8.15 is rather obscure.

8.5.3 Differential Cross Sections – Threshold Effects

The primitive semiclassical approximation for differential cross sections for the two-state models discussed in Sect. 7.4 is valid far from the threshold. A more elaborate treatment is needed close to the threshold, which takes into account the interference of two waves with not too different phases and a possible strong l-dependence in the transition probability.

For the Landau-Zener model with weak coupling between diabatic states, the transition probability $\bar{\mathscr{P}}_{12}$ given by (8.30) varies relatively slowly near $l \sim l_p$. This suggests the formulation of a uniform approximation for differential cross sections near the threshold scattering angle θ_{th}. This will be investigated now, making use of the high-energy approximation, i.e., considering scattering through small but classical angles. In this way, the two parameters ε and β of the linear model correspond to rectilinear relative motion, and the time dependence of the trajectory

$R = (b^2 + v^2 t^2)^{1/2}$ is approximated according to (8.15), viz.,

$$X(t) = R(t) - R_p = b - R_p + v^2 t^2/2b,$$
$$\varepsilon(b) = (|\Delta F|\, R_p/4a)\,(R_p^2/b^2 - 1), \qquad \beta(b) = 4a^{3/2}(R_p^3/b^2\,|\Delta F|\, v^2)^{1/2}. \tag{8.80}$$

This parametrization and the weak-coupling condition allow one to identify the scattering phase shifts δ'_1 and δ'_2, as defined by (7.115), with the scattering phase shifts δ_1^d and δ_2^d for the diabatic crossing terms V_{11} and V_{22}. Yet another simplification, also effective in the weak coupling limit, is due to the constant value of the dynamic phase $\Delta\phi_{21} = \pi/4$.

Consider first the inelastic scattering problem. The two high-energy reduced deflection functions are [8.7]

$$\tau_{12}^{(1)}(b) = \bar{\tau}_{12}(b) - \Delta\tau_{12}(b), \qquad \tau_{12}^{(2)}(b) = \bar{\tau}_{12}(b) + \Delta\tau_{12}(b), \tag{8.81}$$

where $\bar{\tau}_{12}(b)$ is the deflection function for the mean (adiabatic or diabatic) potential

$$\bar{U}(b) = (U_1 + U_2)/2 = (V_{11} + V_{22})/2. \tag{8.82}$$

The function $\Delta\tau_{12}(b)$ is given, according to the general (5.88) with $R^2 = b^2 + z^2$, by

$$\Delta\tau_{12} = b \int_0^{(R_p^2 - b^2)^{1/2}} (1/R)\,(d\Delta U/dR)\, dz, \tag{8.83}$$

with ΔU approximated as $-\Delta F(R - R_p)$. It follows that for small $R_p - b$,

$$\Delta\tau_{12}(b) = \begin{cases} |\Delta F|\,[2R_p(R_p - b)]^{1/2} & \text{for } b < R_p, \\ 0 & \text{for } b > R_p. \end{cases} \tag{8.84}$$

The general features of the deflection functions for inelastic scattering are shown in Fig. 8.16a. To simplify notations the subscripts 1, 2 are dropped, so that $\tau^{(1)}(b)$ and $\tau^{(2)}(b)$ represent $\tau_{12}^{(1)}(b)$ and $\tau_{12}^{(2)}(b)$. The diabatic deflection function τ_2^d corresponds to the pure repulsive term V_{22}, and τ_1^d to V_{11}, the latter being slightly attractive for large R. These terms simulate the actual interaction in the He^+–Ne system discussed below. The two deflection functions $\tau^{(1)}$ and $\tau^{(2)}$ join at $b = R_p$, their difference $\Delta\tau(b)$ exhibiting a square root behavior for b close to and less than R_p. Now, with this definition of the deflection functions, the interference phase $A^{(+-)}$ in the semiclassical expression for the inelastic cross section (7.123) will be

$$A^{(+-)}(\tau) = (2/v) \int_{\tau_{th}}^{\tau} [b^+(\tau) - b^-(\tau)]\, d\tau, \tag{8.85}$$

where $b^-(\tau)$ and $b^+(\tau)$ are two branches of the function represented by the

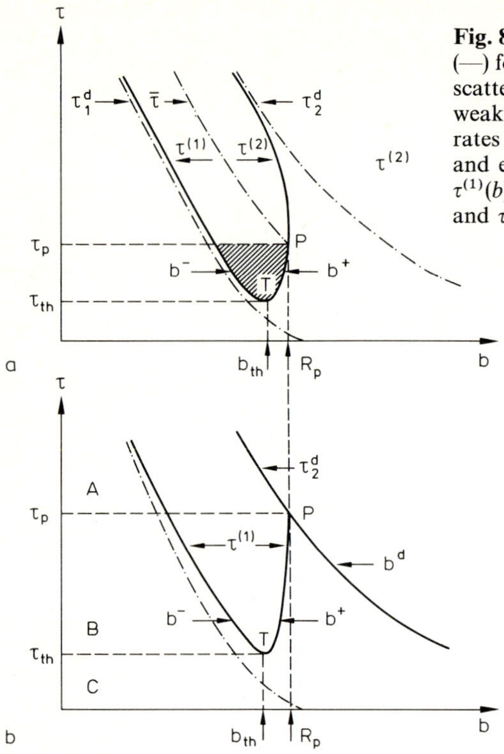

Fig. 8.16 a, b. Threshold deflection functions (—) for (**a**) inelastic $1 \to 2$ and (**b**) elastic $2 \to 2$ scattering for one repulsive and another weakly attractive diabatic term. Point T separates branches $b^-(\tau)$ and $b^+(\tau)$ for the inelastic and elastic cases, point P separates branches $\tau^{(1)}(b)$ and $\tau^{(2)}(b)$ for the inelastic and $\tau^{(1)}(b)$ and $\tau_2^{\mathrm{d}}(b)$ for the elastic case

continuous curve in Fig. 8.16a. The former is obtained by inverting the deflection function $\tau = \tau^{(1)}(b)$ for $b \leqslant b_{\mathrm{th}}$, and the latter by inverting $\tau = \tau^{(1)}(b)$ for $b_{\mathrm{th}} \leqslant b \leqslant R_{\mathrm{p}}$ and $\tau = \tau^{(2)}(b)$ for all b. The integral in (8.85) is taken from the threshold value τ_{th} of the reduced scattering angle corresponding to the minimum of the deflection function $\tau = \tau^{(1)}(b)$. It is seen that the threshold angle τ_{th} is not equal to the scattering angle τ_{p} corresponding to deflection for $b = R_{\mathrm{p}}$.

The high-energy reduced differential cross section $\varrho_{12}(\tau)$ is

$$\varrho_{12}(\tau) = \varrho^+ \bar{\mathscr{P}}_{12}^+ + \varrho^- \bar{\mathscr{P}}_{12}^- - [\varrho^+ \mathscr{P}_{12}^+ \varrho^- \mathscr{P}_{12}^-]^{1/2} \cos[A^{(+-)} + \pi/2], \quad (8.86)$$

where the superscripts + and − correspond to the impact parameter functions $b^+(\tau)$ and $b^-(\tau)$.

A more detailed treatment is needed for the threshold region where $A^{(+-)}$ is not large [8.57]. A general approach would be to use the integral representation of the Airy function without invoking the stationary-phase approximation. The problem amounts to a solution of the two-dimensional integral over the impact parameter b and the variable u, see (5.16), with two coalescent stationary-phase points. Two limiting cases can be distinguished here.

If $|\Delta F|$ is small, the phase $A^{(+-)}$ will become close to unity or less even, when τ is much larger than the difference $\tau_p - \tau_{th}$. This is the case for the common-trajectory approximation characterized by the mean deflection function $\tau = \bar{\tau}(b)$ and the cross section $\bar{\varrho}(\tau)$. With a linear expansion of $\varepsilon(b)$ and $\beta(b)$ over $R_p - b$, the threshold cross section $\varrho_{12}(\tau)$ will be

$$\begin{aligned} \varrho_{12}(\tau) &= \bar{\varrho}(\tau_p)\,\mathscr{P}_{12}(\tau - \tau_p), \qquad \bar{\varrho}(\tau_p) = d\bar{b}^2/d\log\tau|_{\tau=\tau_p}, \\ \mathscr{P}_{12}(\tau - \tau_p) &= \pi^2 \beta_p^{4/3}\,\mathrm{Ai}^2[\beta_p^{2/3}(\partial\varepsilon/\partial b)_p\,|d\tau/db|_p^{-1}\,(\tau - \tau_p)] \\ &= \pi^2 \beta_p^{4/3}\,\mathrm{Ai}^2[A_d^{4/3}\,2^{1/3}(\tau - \tau_p)/\tau_p n]. \end{aligned} \tag{8.87}$$

Here all the quantities with the subscript p refer to $b = R_p$ and

$$n = |R_p\, d\bar{\tau}/\bar{\tau}\, db|_{b=R_p}, \qquad A_d = (|\Delta F|\, R_p^2/v)^{1/2}. \tag{8.88}$$

The parameter n characterizes the steepness of the mean potential $\bar{U}(R)$ for $R = R_p$ and A_d defines the pseudocrossing distance R_p in terms of the dynamic length of the coupling region for the velocity of free motion. Hence, for the Landau-Zener model $A_d \gg 1$. Now, the first maximum of $\varrho_{12}(\tau)$ occurs at $\tau_1 = \tau_p + \Delta\tau_1$, and the condition $\Delta\tau_1 \gg \tau_p - \tau_{th}$ reads

$$A_d^{8/3} \ll l_p \theta_p n \tag{8.89}$$

where $l_p = kR_p$. Note that the rhs of (8.89) is larger than unity, since θ_p is larger than the diffraction angle $1/l_p$ (scattering through classical angles!), and usually $n > 1$. Under condition (8.89) the phase corresponding to the cross-hatched area in Fig. 8.16a is small, so that $\tau - \tau_p$ in (8.87) can safely be replaced by $\tau - \tau_{th}$.

If $|\Delta F|$ is large, and an inequality opposite to (8.89) holds, viz.,

$$A_d^{8/3} \gg l_p \theta_p n, \tag{8.90}$$

the phase corresponding to the cross-hatched region is $\sim A_d^8 l_p^{-3} \theta_p^{-3} n^{-3} \gg 1$ and the undulatory threshold structure of the cross section shows even for τ such that $\tau_{th} < \tau < \tau_p$. Under this condition the threshold behavior of the cross section is similar to that for rainbow elastic scattering. Assuming that the average transition probability $\bar{\mathscr{P}}_{12}$ varies only slightly near the bottom of the deflection function $\tau^{(1)}(b)$, the cross section near the threshold will be [8.58, 59], using the Airy approximation,

$$\begin{aligned} \varrho_{12}(\tau) &= \varrho^{(1)}(\tau - \tau_{th})\,\bar{\mathscr{P}}_{12}(\tau_{th}), \qquad \varrho^{(1)}(\tau - \tau_{th}) = 2\tau b_{th} v^{-4/9}\chi^{-2} \\ &\quad \times \mathrm{Ai}^2[-2\chi v^{-2/3}(\tau - \tau_{th})], \\ \chi &= |d^2\tau^{(1)}/db^2|_{b=b_{th}}^{-1/3}, \\ \bar{\mathscr{P}}_{12}(\tau_{th}) &= 4\pi\delta(b_{th}) = 4\pi a^2/|\Delta F|\, v(1 - b_{th}^2/R_p^2)^{1/2}. \end{aligned} \tag{8.91}$$

An even more exact representation of $\bar{\mathscr{P}}_{12}$ can be formulated using the uniform Airy approximation [8.60]. In both cases, (8.90) guarantees that the Airy function reaches its asymptotic form before τ becomes equal to τ_p, so that for some $\tau_{th} < \tau < \tau_p$, the cross section $\varrho_{12}(\tau)$ will be given by the primitive semiclassical approximation (8.86). Note that, as τ approaches τ_p, the probability $\bar{\mathscr{P}}_{12}(b)$ diverges as $(R_p - b)^{-1/2}$. However, ϱ_{12} goes to zero as $(R_p - b)^{1/2}$ and thus for $b = R_p$ there is no singularity in the cross section. This signifies that, at least for small a, the angular extension of the region near τ_p (where the semiclassical approximation is not quite correct) is also small.

Note finally that the undulations of the cross sections described by (8.87) are quite different from those predicted by (8.91). In the former case the undulation frequency depends on two branches $\tau^{(1)}$ and $\tau^{(2)}$, while in the latter case the frequency is controlled by the curvature of the bottom part of the deflection function $\tau^{(1)}$.

The application of the semiclassical approach to inelastic scattering is exemplified by the process

$$He^+ + Ne(2p^6) \rightarrow He^+ + Ne(2p^5 3s), \tag{8.92}$$

studied experimentally in detail for different energies [8.61, 62]. The initial channel corresponds to the $^2\Sigma$-term of the quasi molecule $HeNe^+$. The final channels involve both doublet and quartet Σ- and Π-terms. Neglecting the spin-orbit and the Coriolis couplings and retaining only the interaction between the initial $^2\Sigma$-term and the two final $^2\Sigma$-terms, the rather complicated term pattern becomes greatly simplified. Analysis of the three-state model shows that the coupling of the initial term to two close-lying terms results in a slow modulation of the differential cross section [8.63]. If this modulation is neglected and attention is paid mainly to the threshold oscillations of ϱ_{12}, then the two-state model can be adopted, at least for not too large values of τ.

The two-state model was used repeatedly for calculation of the cross section for process (8.92), adopting either the semiempirical or the ab initio interaction. Far from the threshold the primitive semiclassical approximation appears to be qualitatively valid [8.22, 49]. Close to the threshold the rainbow description (8.91) of scattering is more adequate than the common-trajectory description [8.59]. The corresponding calculations were based on the Airy approximation [8.59], or on uniform exponential approximations [8.63, 64], or else use was made of the full quantum treatment [8.49]. The most consistent semiclassical treatment used the impact-parameter method to calculate the scattering matrix [8.65], as well as the general integral representation of the scattering amplitude for scattering through small angles (2.50) which is equivalent to the eikonal method (see

Sect. 6.1.2). Though it has been found [8.65] that the basic assumption of the linear model concerning the R-independence of the coupling V_{12} is not quite correct, the linear model, which goes beyond the Landau-Zener-Stueckelberg treatment when accounting for the turning point effect, still satisfactorily describes the main features of the cross section above the threshold. The first maximum of the cross section is at $\tau_{\text{th}} = 1.1$ keV deg, and the two-state approximation appears to be valid up to $\tau = 3$ keV deg. This τ range contains five maxima of $\varrho_{12}(\tau)$. The coupling to other states probably occurs at higher τ [8.65].

Consider now elastic scattering perturbed by inelastic scattering. For scattering in channel 2 the deflection functions are

$$\tau_{22}^{(1)}(b) = \tau_2^{\text{d}}(b) - 2\Delta\tau_{12}(b), \qquad \tau_{22}^{(2)}(b) = \tau_2^{\text{d}}(b), \tag{8.93}$$

where τ_2^{d} is the deflection function for the diabatic term $V_{22}(R)$, and $\Delta\tau_{12}(b)$ is given by (8.83). These functions are shown in Fig. 8.16b where $\tau^{(1)}$ stands for $\tau_{22}^{(1)}$, and b^- and b^+ are two branches of the function $b(\tau)$ inverse to the function $\tau = \tau^{(1)}(b)$. Note that $\tau^{(1)}, b^+, b^-, b_{\text{th}}, \tau_{\text{th}}, \tau_{\text{p}}$ in Fig. 8.16b, having for simplification no subscript 22, are different from the corresponding values in Fig. 8.16a. It is seen from Fig. 8.16b that there are three different regions of scattering angles corresponding to two (A), three (B) or one (C) semiclassical amplitudes. Representing the elastic cross section $\varrho_{22}(\tau)$ as

$$\varrho_{22}(\tau) = \varrho_2^{\text{d}}(\tau) + \Delta\varrho(\tau), \tag{8.94}$$

where $\varrho_2^{\text{d}}(\tau)$ is the elastic scattering cross section for the diabatic term $V_{22}(R)$, consider the behavior of $\Delta\varrho(\tau)$ in the three above-mentioned regions, assuming again that $\bar{\mathscr{P}}_{12} \ll 1$:

(A) $\tau > \tau_{\text{p}}$. The correction $\Delta\varrho$ is

$$\Delta\varrho(\tau) = -\varrho_2^{\text{d}}\bar{\mathscr{P}}_{12}^{\text{d}} - (\varrho_2^{\text{d}}\varrho^-)^{1/2}\bar{\mathscr{P}}_{12}^{-}\sin A^{(\text{d},-)}, \tag{8.95}$$

with the interference phase

$$A^{(\text{d},-)}(\tau) = \frac{2}{v}\int_{\tau_{\text{th}}}^{\tau_{\text{p}}}[b^+(\tau) - b^-(\tau)]\,d\tau + \frac{2}{v}\int_{\tau_{\text{p}}}^{\tau}[b^{\text{d}}(\tau) - b^-(\tau)]\,d\tau, \tag{8.96}$$

which is assumed to be large ($A^{(\text{d},-)} \gg 1$). The first term in (8.95) is the depletion of state 2, and the second represents pure interference.

(B) $\tau_{\text{th}} < \tau < \tau_{\text{p}}$. For this case $\Delta\varrho$ is due to the interference of three scattering amplitudes:

$$\Delta\varrho(\tau) = -(\varrho_2^{\text{d}}\varrho^-)^{1/2}\bar{\mathscr{P}}_{12}^{-}\sin A^{(\text{d},-)} - (\varrho_2^{\text{d}}\varrho^+)^{1/2}\bar{\mathscr{P}}_{12}^{+}\cos A^{(\text{d},+)}. \tag{8.97}$$

The interference phase $A^{(d,-)}$ is given by (8.96) and

$$A^{(d,+)}(\tau) = \frac{2}{v} \int_{\tau}^{\tau_p} [b^d(\tau) - b^+(\tau)]\, d\tau \tag{8.98}$$

which can again be interpreted geometrically.

(C) $\tau < \tau_{th}$. There is no interference here and

$$\Delta\varrho(\tau) = 0. \tag{8.99}$$

Now, (8.95, 97, 99) are not valid near $\tau = \tau_{th}$, where ϱ^- and ϱ^+ become infinitely large (rainbow scattering) and near $\tau = \tau_p$, where $A^{(d,+)}$ becomes small and $\bar{\mathscr{P}}_{12}^+$ large. However with a large phase $A^{(d,-)}(\tau_p)$, different uniform approximations can be formulated for these two regions [8.66].

It can be shown [8.66] that near τ_{th} the uniform approximation for $\Delta\varrho$ is proportional to the product of the rainbow scattering amplitude near the minimum of $\tau^{(1)}$, the scattering amplitude for the potential V_{22} and the oscillatory factor containing a large phase $A^{(d,+)}(\tau)$. When the cross section $\Delta\varrho(\tau)$ approaches the threshold, it oscillates and increases (just as in the dark-side approach to the rainbow angle), and after passing through τ_{th} the fast oscillations are superimposed on those due to the Airy function. In the asymptotic region, where $A^{(d,+)} - A^{(d,-)}$ is large, $\Delta\varrho$ is given by (8.97). As τ approaches τ_p the semiclassical condition breaks down again. In this region the uniform approximation for $\Delta\varrho$ is given by a parabolic cylinder function. This approximation describes the interference of the amplitude corresponding to the branches $b^{(+)}$ and b^d with that for the branch $b^-(\tau)$. The relevant equations are rather cumbersome and will not be given here.

We shall only illustrate the general behavior of the inelastic cross section near the threshold and the elastic cross section near the onset of inelastic scattering for a model two-state problem simulating He^+-Ne scattering at $E = 150$ eV [8.66]. For this model, the deflection functions for inelastic and elastic scattering were chosen in a simple form, ensuring that $\tau_{ik}^{(1)}(b)$ attains its minimum value at the prescribed value of the impact parameter $(b_{th})_{ik}$ and has a vertical tangent at $b = R_p$. The values $(\tau_p)_{ik}$, $(\tau_{th})_{ik}$ and $R_p - (b_{th})_{ik}$ are interrelated through the form of the diabatic terms (assumed to be smooth curves) near the crossing point.

Figure 8.17a presents the inelastic cross section and Fig. 8.17b the elastic one. The latter is calculated using uniform approximations (dashed curves near $\tau = (\tau_{th})_{22}$ and $\tau = (\tau_p)_{22}$ joined by the semiclassical curve (full curve) (8.97). For $\tau > \tau_p$ the parabolic cylinder function uniform approximation is seen to go smoothly to the semiclassical approximation (8.95) (full curve) showing undulations around the monotonic depletion. Finally,

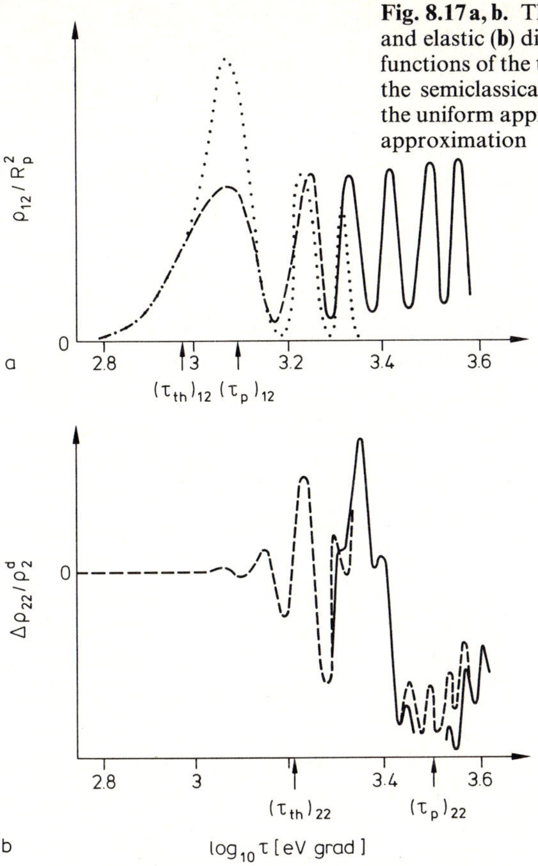

Fig. 8.17 a, b. Threshold behaviour of inelastic (**a**) and elastic (**b**) differential cross sections for deflection functions of the type in Fig. 8.16. The full lines refer to the semiclassical approximation, the dashed line to the uniform approximation, the dotted line is the Airy approximation

for $\tau < (\tau_{th})_{22}$ the ratio $\Delta\varrho/\varrho_2^d$ oscillates with a decreasing amplitude around the semiclassical zero value, see (8.99).

Quantum calculation for the $He^+ - Ne$ system [8.62] shows the same qualitative features for elastic scattering, though for the more realistic diabatic potentials, τ_{th} and τ_p appear to be too close to ensure the applicability of the semiclassical description for $\tau_{th} < \tau < \tau_p$.

It will be mentioned once more that the threshold scattering angles for inelastic scattering $(\theta_{th})_{12}$ and for the onset of inelastic perturbation in elastic scattering $(\theta_{th})_{22}$ [or the corresponding values $(\tau_{th})_{12}$ and $(\tau_{th})_{22}$] are different, implying different values for the threshold impact parameters $(b_{th})_{12}$ and $(b_{th})_{22}$. It is also seen that $(\tau_p)_{22}$ can be estimated approximately from the position of the dip in the elastic scattering cross section.

Analysis of inelastic scattering becomes more complicated when the coupling between diabatic terms is not weak and when interaction corresponds both to attraction and repulsion.

If, for instance, $\Delta\tau_{12}(b)$ is calculated using the general expression (8.2) for the adiabatic splitting $\Delta U(R)$, then over the range $\Delta_a R \sim a/|\Delta F|$ near $b = R_p$, the dependence $\Delta\tau_{12}(b)$ differs from the square root behavior (8.84): the plot shows a bend and extends beyond $b = R_p$, asymptotically tending to zero for $b > R_p$ with a horizontal tangent. This results in additional extrema in all deflection functions τ_{ik}^{Δ}, which in turn lead to additional specific features in the cross sections. Since semiclassical analysis of the cross sections for large a is hindered by the strong dependence of the nonoscillating part of S_{ik} on b, it is interesting to compare directly the exact quantum calculation with a qualitative semiclassical analysis ascribing the basic features of cross sections to some peculiarities or other of the deflection functions. It has been found in this way [8.45, 67] that semiclassical analysis permits a qualitative interpretation of the above features to be made in terms of both adiabatic and diabatic deflection functions, for moderate and large splitting of terms. It has also been found that if diabatic phases only are used for a parametrization of the S matrix, the deflection functions, though giving a correct final result, oscillate around a certain smooth curve [8.45].

Consider now, in brief, a case when the interatomic interaction in one of the channels is substantially attractive. Then the deflection functions for elastic and inelastic scattering extend to the negative region of η, so that for small enough θ, not two, but four amplitudes interfere. Figure 8.18 presents as an example the deflection functions $\eta^{(1)}$ and $\eta^{(2)}$ for inelastic scattering in the chemoionization process

$$A + B \rightarrow A^+ + B^- . \tag{8.100}$$

In the two-state approximation the adiabatic terms of such a system are represented approximately by a truncated Coulomb potential (a covalent

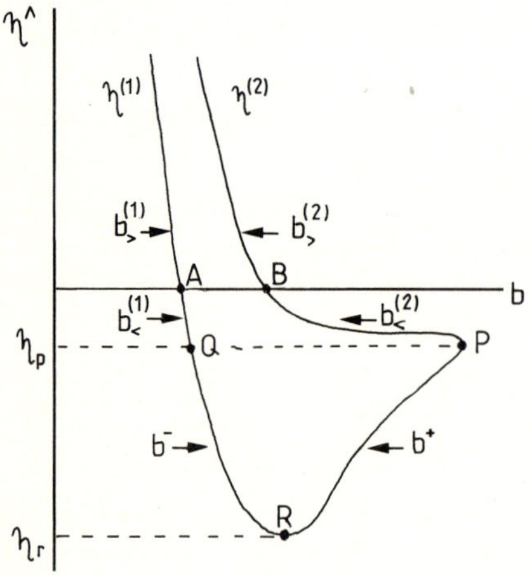

Fig. 8.18. The inelastic-scattering deflection function for one strongly attractive diabatic term (for detail see text)

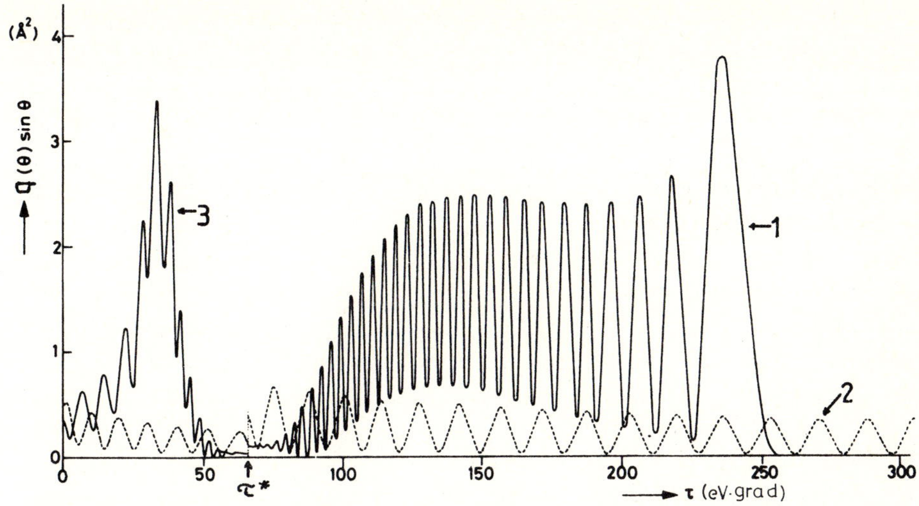

Fig. 8.19. Low frequency contributions to the differential cross section for a model Na + I collision. Curves 1, 2, 3 account for the interference between different pairs of branches: $b^- - b^+$, $b_>^{(1)} - b_>^{(2)}$, $b_<^{(1)} - b_<^{(2)}$, respectively. For $\tau < \tau^* = 65$ eV grad the curves are divided by 3 [8.68]

potential for $R > R_p$, relatively weakly dependent on R, and a Coulomb potential for $R < R_p$, modified at small interatomic separations due to the repulsive core) and by the long-range Coulomb tail for $R > R_p$, which for $R < R_p$ goes over to a weakly attractive covalent potential. These adiabatic potentials lead to deflection functions $\eta_{12}^{(1)}$ and $\eta_{12}^{(2)}$ which join at the pseudocrossing point R_p. There will be several branches $b = b(\theta)$, each between pairs of points (Fig. 8.18). The minimum of $\eta_{12}^{(1)}$ corresponds to rainbow scattering on the lower adiabatic potential, and the negative value of θ_p is due to the weak attraction at the pseudocrossing distance R_p. For $\theta > |\eta_r|$ two amplitudes contribute to the inelastic cross section. For $\theta < |\eta_r|$ there will be four interfering amplitudes corresponding to the branches b^-, b^+, $b_>^{(1)}$, $b_<^{(2)}$ for $|\eta_p| < \theta < |\eta_r|$ and to the branches $b_<^{(1)}$, $b_<^{(2)}$, $b_>^{(1)}$, $b_>^{(2)}$ for $\theta < |\eta_p|$.

The differential cross section of process (8.100) with A = Na, B = I was calculated [8.68] in the semiclassical approximation, using the Landau-Zener formula for the probability P at $E = 13.1$ eV.

The rather complicated interference structure becomes somewhat simpler if the rapidly oscillating interference terms are neglected. For instance, if only the contribution from the b^- and b^+ branches is taken into account, the conventional uniform Airy approximation can be used. Curve 1 in Fig. 8.19 was obtained in this way. Accounting for interference from branches $b_>^{(1)}$ and $b_>^{(2)}$ separately yields curve 2 with a slowly changing amplitude. Finally, accounting for interference from branches $b_<^{(1)}$ and $b_<^{(2)}$ gives curve 3 with a maximum at $\tau = 30$ eV deg. A summation of the three contributions gives an approximate differential cross section involving no high-frequency oscillations. Note that for this case, the cross section is larger to the left of the threshold $\tau_{\text{th}} \approx 250$ eV deg, i.e., different from the repulsive scattering case.

9. Nonlinear Two-State Models of Nonadiabatic Coupling

This chapter deals with various nonadiabatic-coupling models different from the Landau-Zener model. Since diabatic terms are not supposed to be linear in R and the coupling matrix element is taken to be R-dependent, all these models can be called nonlinear. Some new physical effects can also be considered, associated with the finite splitting of adiabatic terms away from the coupling region. This finite splitting (not infinitely large as in the Landau-Zener model) means that the extent of the coupling region will be greater than for the linear model, and thus more information on the coupling and on adiabatic terms is required. In this sense any nonlinear model is more specific and hence more restrictive. Moreover, an unique feature of the linear model concerning the one-to-one correspondence between the quantum and the semiclassical problems is lost for nonlinear models. Nevertheless some nonlinear models are general enough and deserve a detailed discussion.

The semiclassical approach is used in this chapter for certain nonlinear models classified according to the R-dependence of the diabatic two-state Hamiltonian. We start with the exponential model (Sect. 9.1) which seems to be the most widely used. A model of this type was first discussed by *Rosen* and *Zener* [9.1]. Compared to the Landau-Zener model it represents, in a certain sense, another limiting case, since in the diabatic basis the difference in the diagonal Hamiltonian elements was taken to be time independent, and the nondiagonal element was approximated by a bell-shaped function H_{12} [$H_{12} \sim \cosh^{-1}(t/\tau_0)$]. Since in the limit $t \to \pm\infty$ the interaction H_{12} disappears, this model can be used, in principle, to describe nonadiabatic processes over the entire collision duration. Explicit specification of $H_{12}(t)$ over the whole t range imposes a superfluous restriction on the model when the nonadiabaticity regions are far from the turning point. For this reason the model was not applied to specific problems for a long time. The use of matching for the construction of the S matrix from nonadiabatic transition matrices (Sect. 6.2) allows one to resort to more flexible models, valid only for a single passage of the nonadiabatic coupling region. One of these models approximating the matrix elements of a two-level semiclassical Hamiltonian by exponential time functions has been discussed by *Nikitin* [9.2, 3] and was proposed for a description of nonresonant charge exchange [9.4]. A particular case of this model has been studied in detail by *Demkov* [9.5] for charge exchange with a small resonance defect. The exponential model has recently been widely used for nonadiabatic coupling in many processes with variation of one of the model parameters, the resonance defect $\Delta\varepsilon$, in the range from several thousands (transition between inner atomic shells [9.6–9]) to thousandths of eV [transitions between components of atomic fine structure, (Chap. 11), or between states split by the isotope shift [9.10]]. For a particular set of parameters this is the linear semiclassical model.

Another model attracting increasing attention is a generalization of the Landau-Zener approximation that implies that the matrix element of interaction between diabatic states is not constant, but exponentially dependent on R (linear-exponential model). This is important for many cases, particularly near the adiabatic threshold of a process (Sect. 9.2).

Section 9.3 is devoted to other nonlinear models. The generalization of the linear and exponential models is provided by hypergeometric models for which the transition probability depends on three parameters, rather than two (exponential model) or one (linear model).

Finally the power models are discussed. One set describes transitions at large interatomic distances and is used mainly for collisional processes at low energies. Another applies to transitions in the united-atom limit and is therefore suitable for processes occurring over a medium energy range.

9.1 Exponential Model

9.1.1 Formulation of the Model

A two-state exponential model in the diabatic basis is described by an interaction matrix of the form

$$V_{ik}(R) = \delta_{ik}[V_0(R) \mp \Delta\varepsilon/2 \pm (A/2)\cos 2\vartheta_0 \exp(-\alpha R)] + (1-\delta_{ik})(A/2)\sin 2\vartheta_0 \exp(-\alpha R), \quad (9.1)$$

where $V_0(R)$ is a certain function of R. The matrix involves four parameters: $\Delta\varepsilon$, A, ϑ_0, and α. Two parameters (A and α) disappear with an appropriate choice of the origin and with scale transformation. Setting

$$A \exp(-\alpha R_\mathrm{p}) = \Delta\varepsilon, \quad (9.2\,\mathrm{a})$$

$$x = \alpha(R - R_\mathrm{p}), \quad (9.2\,\mathrm{b})$$

we obtain

$$V_{ik}(x) = \delta_{ik}\{V_0(x) \mp (\Delta\varepsilon/2)[1 - \cos 2\vartheta_0 \exp(-x)]\} + (1 - \delta_{ik})(\Delta\varepsilon/2)\sin 2\vartheta_0 \exp(-x). \tag{9.3}$$

The diagonalization of this matrix by the general transformation (7.4) yields adiabatic terms $U_k(x)$, the mixing angle $\vartheta(x)$ and the matrix element $g(R)$ (Sect. 7.1):

$$\begin{aligned} U_{1,2}(x) &= V_0(x) \pm [\Delta U(x)/2], \\ \Delta U(x) &= \Delta\varepsilon[1 - 2\cos 2\vartheta_0 \exp(-x) + \exp(-2x)]^{1/2}, \end{aligned} \tag{9.4}$$

$$\begin{aligned} \tan 2\vartheta(x) &= \frac{\sin 2\vartheta_0 \exp(-x)}{\cos 2\vartheta_0 \exp(-x) - 1}, \\ g(R) &= -(1/2)\sin 2\vartheta_0 \exp(-x) \\ &\quad \times [1 - 2\cos 2\vartheta_0 \exp(-x) + \exp(-2x)]^{-1}. \end{aligned} \tag{9.5}$$

The function $g(R)$ in (9.5) has a symmetric bell-shaped form with a maximum at $x = 0$. The complex crossing points of the adiabatic terms, obtained from the condition $\Delta U(x_c) = 0$ are

$$R_c = R_p + x_c\alpha^{-1}, \quad x_c = \mathrm{i}(\pm 2\vartheta_0 \pm 2\pi n), \quad n = 0, \pm 1 \ldots . \tag{9.6}$$

Here the signs $\pm$ before ϑ_0 and the positive or negative values of n must be taken independently. Taking this into account, the region of ϑ_0 variation can be confined to the range $0 \leqq \vartheta_0 \leqq \pi/2$.

Thus, for the exponential model, just as for the linear one, the nonadiabatic coupling maximum is attained at point R_p corresponding to the real part of the complex distance R_c where the terms cross. The distance R_p defined by (9.2 a) is taken to be the center of the nonadiabaticity region. As a measure of the width of the coupling region we take

$$\Delta_a R = (2/\alpha)\sin\vartheta_0 . \tag{9.7}$$

The splitting of the adiabatic terms for $R = R_p$ is

$$\Delta U_p = \Delta U(R_p) = 2\Delta\varepsilon \sin\vartheta_0 . \tag{9.8}$$

However, as distinct from the Landau-Zener model, this point does not correspond, generally speaking, to the splitting minimum. The latter exists only for a limited range of ϑ_0 variation $(0 < \vartheta_0 < \pi/4)$:

$$U_{\min} = \Delta U(R_m) = \Delta\varepsilon \sin 2\vartheta_0, \quad R_m = R_p - \alpha^{-1}\log(\cos 2\vartheta_0). \tag{9.9}$$

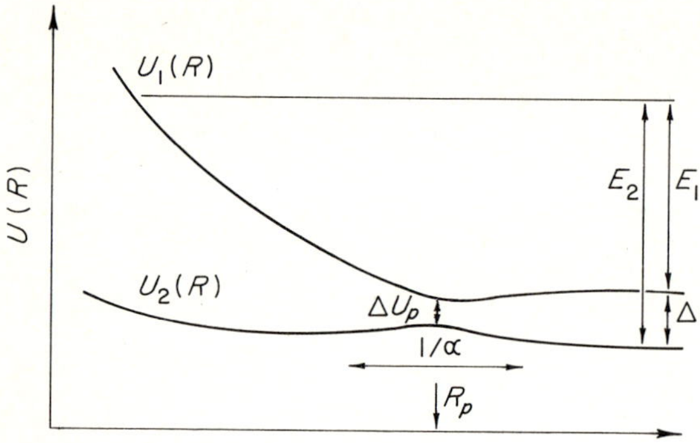

Fig. 9.1.

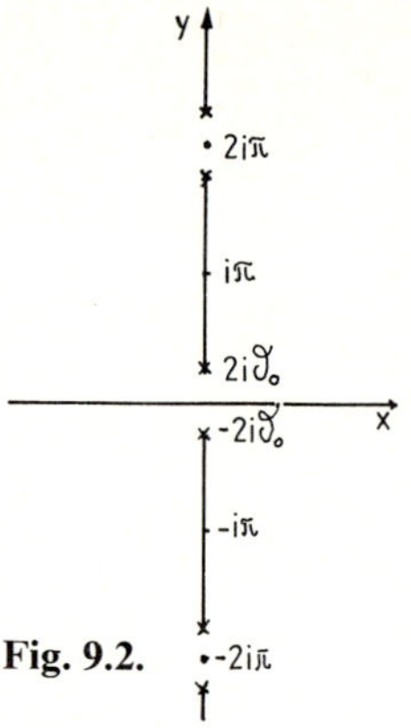

Fig. 9.2.

Fig. 9.1. Adiabatic terms for the exponential model

Fig. 9.2. Branch points (x) and cuts for the two-sheet analytical function $\Delta U(z)$ of the exponential model

The matrix (9.1) or the equations (9.4, 5) completely define the model, both in the diabatic and in the adiabatic basis. The adiabatic terms $U_k(R)$ are presented in Fig. 9.1, and the positions of branch points and cuts for $\Delta U(z)$ on a two-sheet complex plane $z = \alpha(R - R_p)$ in Fig. 9.2.

9.1.2 Transition Probability and Dynamic Phases

Consider first the case when the coupling region is far from the turning point. The simplest trajectory $R = R(t)$ for the semiclassical exponential model corresponds to uniform nuclear motion in the nonadiabaticity region (local-rectilinear-motion approximation):

$$R(t) = R_p + v_p t. \tag{9.10}$$

Such a parametrization of the trajectory gives the exponential time dependence of the matrix elements V_{ik} and allows the analytical properties of the function $\Delta U(t)$ to be identified with those of $\Delta U(x)$.

As stated in Chap. 6, the semiclassical description is valid under two conditions, namely, when the wavelength $1/k$ is small compared to the characteristic size of the nonadiabaticity region $1/\alpha$ ($\mu v_p/\alpha \gg 1$) and when the term splitting $\Delta\varepsilon$ is small compared to the kinetic energy ($2\Delta\varepsilon/\mu v_p^2 \ll 1$). Taking the Massey parameter ξ for this model as

$$\xi = \Delta\varepsilon/\alpha v_p, \tag{9.11}$$

we find the region of ξ variation:

$$\Delta\varepsilon/\mu v_p^2 \ll \xi \ll \mu v_p/\alpha. \tag{9.12}$$

Since the upper limit in (9.12) is much greater than unity and the lower limit much smaller, the permissible range of ξ variation includes the value $\xi = 1$.

Another important parameter of the exponential model is the Massey parameter ξ_p for the transition region, defined via the term splitting ΔU_p and the effective width $\Delta_a R$:

$$\xi_p = \Delta U_p \Delta_a R / 4 v_p = (\xi/2)(1 - \cos 2\vartheta_0). \tag{9.13}$$

When $2\vartheta_0$ varies within the range $0 \leqq 2\vartheta_0 \leqq \pi$, the ξ_p value varies within the limits $0 \leqq \xi_p \leqq \xi$.

The diabatic semiclassical equations of the exponential model written for a dimensionless time $\tau = \alpha v_p t$ and containing the parameters ξ and ξ_p are

$$\begin{aligned} i\,db_1/d\tau &= \xi_p^{1/2}(\xi - \xi_p)^{1/2}\exp(-\tau) \\ &\quad \times \exp\left\{ i\int_0^\tau [\xi - (\xi - 2\xi_p)\exp(-\tau)]\,d\tau \right\} b_2, \\ i\,db_2/d\tau &= \xi_p^{1/2}(\xi - \xi_p)^{1/2}\exp(-\tau) \\ &\quad \times \exp\left\{ -i\int_0^\tau [\xi - (\xi - \xi_p)^{1/2}\exp(-\tau)]\,d\tau \right\} b_1, \quad 0 \leqq \xi_p \leqq \xi. \end{aligned} \tag{9.14}$$

These equations are solved in terms of confluent hypergeometric functions [9.2, 3] yielding analytical expressions for the three parameters specific to solution of the comparison equation (9.14) (Chap. 7), namely, the one-way transition probability P [9.11] and the dynamic phases φ and ψ [9.12]:

$$P = \exp(-\pi\xi_p)\sinh(\pi\xi - \pi\xi_p)/\sinh \pi\xi, \tag{9.15a}$$

$$\begin{aligned} \varphi &= \gamma(\xi_p) - \gamma(\xi), \\ \psi &= \gamma(\xi - \xi_p) - \gamma(\xi) - 2\left[(\xi\xi_p)^{1/2} - \frac{(\xi - \xi_p)}{2}\log\frac{\xi^{1/2} - \xi_p^{1/2}}{\xi^{1/2} + \xi_p^{1/2}}\right]. \end{aligned} \tag{9.15b}$$

Here the signs of φ and ψ are consistent with the positive value of $\Delta_{21} = \Delta_2 - \Delta_1$ and $\gamma = \gamma(x)$ is defined by (8.11).

The transition probability $\mathscr{P}_{12}$ is (compare Sect. 7.2)

$$\begin{aligned} \mathscr{P}_{12} &= 2\bar{\mathscr{P}}_{12}\sin^2(\Delta_{21} + \Delta\phi_{21}), \\ \bar{\mathscr{P}}_{12} &= 2\exp(\pi\xi\cos 2\vartheta_0)\sinh[(\pi/2)\,\xi(1 - \cos 2\vartheta_0)] \\ &\quad \times \sinh[(\pi/2)\,\xi(1 + \cos 2\xi_0)]\sinh^{-2}\pi\xi. \end{aligned} \tag{9.16}$$

For the exponential model, unlike for the linear one, the dynamic phase $\Delta\phi_{21} = -\Delta\phi_{12}$ depends on whether Δ_{21} is defined according to (7.24, 25) or (7.26). If the definitions (7.24) or (7.25) are accepted, then [9.12]

$$\Delta\phi_{21} = \varphi - \psi = \gamma(\xi_p) - \gamma(\xi - \xi_p) + 2\left[(\xi \cdot \xi_p)^{1/2} - (1/2)(\xi - \xi_p)\log\frac{\xi^{1/2} - \xi_p^{1/2}}{\xi^{1/2} + \xi_p^{1/2}}\right]. \tag{9.17}$$

On the other hand, if Δ_{21} is defined according to (7.26), then [9.13]

$$\Delta\phi_{21} = \gamma(\xi_p) - \gamma(\xi - \xi_p). \tag{9.18}$$

It will be noted that only with the definition (9.18) does the dynamic phase go to zero at the near-adiabatic limit.

The contour lines of the average transition probability $\bar{\mathscr{P}}_{12}$ in the plane $2/\pi\xi = 2\alpha v_p/\pi\Delta\varepsilon$ and $\Delta U_p/\Delta\varepsilon = 2\sin\vartheta_0$ are given in Fig. 9.3. Note that the maximum $\bar{\mathscr{P}}_{12} = 1/2$ is attained along the curve $\exp(\pi\xi\cos 2\vartheta_0) = \cosh(\pi\xi)$.

Up to now the effect of the turning point has been completely neglected. To account for it, one must start with quantum equations. It will be assumed that $V_0(R)$ is also given by the exponential function, so that the two-state scattering equations are

$$\begin{aligned}
&\mu^{-1}d^2\chi_1/dR^2 + [2\mathscr{E} - \Delta\varepsilon - (2B - A\cos 2\vartheta_0)\exp(-\alpha R)]\,\chi_1 \\
&\qquad - A\sin 2\vartheta_0\exp(-\alpha R)\,\chi_2 = 0, \\
&\mu^{-1}d^2\chi_2/dR^2 + [2\mathscr{E} + \Delta\varepsilon - (2B + A\cos 2\vartheta_0)\exp(-\alpha R)]\,\chi_2 \\
&\qquad - A\sin 2\vartheta_0\exp(-\alpha R)\,\chi_1 = 0.
\end{aligned} \tag{9.19}$$

Consider first the case when $(A/2)\sin 2\vartheta_0 \ll B + (A/2)\cos 2\vartheta_0$, $B - (A/2)\cos 2\vartheta_0$. Then the transition probability can be calculated in terms of first-order perturbation theory. The distorted waves ${}^0\chi$ are ex-

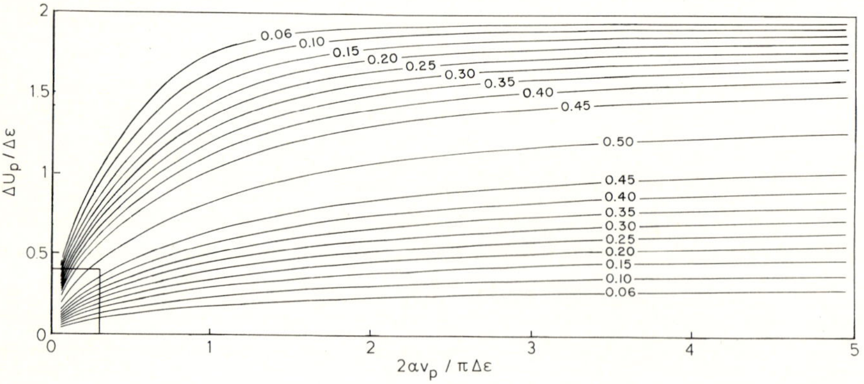

Fig. 9.3. Contour map for the average transition probability (9.16) in the plane $2\alpha v_p/\pi\Delta\varepsilon$, $\Delta U_p/\Delta\varepsilon$

pressed via the Bessel functions and the transition amplitude is calculated analytically, provided the boundary conditions for $R = 0$ are shifted to $R = -\infty$ (this is permissible for a large classically forbidden region near $R = 0$). The transition probability was first calculated for $2\vartheta_0 = \pi/2$ [9.14], and then for an arbitrary $2\vartheta_0$ [9.15]:

$$\mathscr{P}_{12} = \left(\frac{A \sin 2\vartheta_0}{2B - A\cos 2\vartheta_0}\right)^2 \frac{\pi^2(q_2^2 - q_1^2)^2}{4} \frac{\sinh \pi q_1 \sinh \pi q_2}{(\cosh \pi q_1 - \cosh \pi q_2)^2}$$
$$\times |\mathrm{F}[1 + (\mathrm{i}/2)(q_1 + q_2); 1 + (\mathrm{i}/2)(q_1 - q_2); 2; z]|^2, \tag{9.20}$$

$$q_1 = 2(2\mu)^{1/2}\left(\mathscr{E} + \frac{\Delta\varepsilon}{2}\right)^{1/2}\alpha^{-1}, \quad q_2 = 2(2\mu)^{1/2}\left(\mathscr{E} - \frac{\Delta\varepsilon}{2}\right)^{1/2}\alpha^{-1},$$
$$z = 1 - \frac{2B + A\cos 2\vartheta_0}{2B - A\cos 2\vartheta_0}, \tag{9.21}$$

and $\mathrm{F}[a; b; c; z]$ is the hypergeometric function.

Equation (9.20) simplifies under semiclassical conditions (Sect. 6.1.3): $q_1, q_2 \gg 1$; $\Delta\varepsilon \ll \mathscr{E}$; $B \gg A\cos 2\vartheta_0$. Then

$$\frac{4 \sinh \pi q_1 \sinh \pi q_2}{(\cosh \pi q_1 - \cosh \pi q_2)^2} \simeq \sinh^{-2}(\pi\Delta\varepsilon/\alpha v),$$
$$v = (2\mathscr{E}/\mu)^{1/2}, \qquad z \simeq -A(\cos 2\vartheta_0)/B, \tag{9.22}$$

and the hypergeometric function in (9.20) becomes confluent hypergeometric function

$$\Phi\left[1 + \mathrm{i}\frac{\Delta\varepsilon}{\alpha v}, 2, -\frac{2\mathrm{i}\mu v A\cos 2\vartheta_0}{\alpha B}\right].$$

Thus, the final semiclassical result is

$$\mathscr{P}_{12} = \left(\frac{A(\sin 2\vartheta_0)\,\mu\Delta\varepsilon}{B\alpha^2}\right)^2 [\sinh(\pi\Delta\varepsilon/\alpha v)]^{-2}$$
$$\times \left|\Phi\left[1 + \mathrm{i}\frac{\Delta\varepsilon}{\alpha v}, 2, -\frac{2\mathrm{i}\mu v A\cos 2\vartheta_0}{\alpha B}\right]\right|^2. \tag{9.23}$$

This probability can be obtained also as a first-order solution of semiclassical equations, for the interaction matrix $V_{ik}(R)$ given by (9.1) [9.3]

$$\mathscr{P}_{12} = \left|\left(\frac{A}{2}\sin 2\vartheta_0\right)^{1/2} \int_{-\infty}^{\infty} \exp(-\alpha R)\right.$$
$$\left.\times \exp\left[\mathrm{i}\Delta\varepsilon t - \mathrm{i}A\cos 2\vartheta_0 \int_0^t \exp(-\alpha R)\,dt'\right] dt\right|^2, \qquad R = R(t). \tag{9.24}$$

Here the classical trajectory $R = R(t)$ is to be found from the equation of motion

$$\mu \dot{R}^2/2 + B\exp(-\alpha R) = \mathscr{E}, \tag{9.25}$$

which gives

$$R(t) = R^{\mathrm{t}} + (2/\alpha)\log[\cosh(\alpha v t/2)], \qquad R^{\mathrm{t}} = \log(B/\mathscr{E}). \tag{9.26}$$

The transition probability given by (9.23) can be correlated with $\mathscr{P}_{12}$ from (9.16), if the distance $R_{\mathrm{p}} - R^{\mathrm{t}}$ is large enough and ϑ_0 is small.

Turn now to near-adiabatic conditions. Here the problem amounts to a determination of the first quantum correction to the semiclassical exponent. The exponential factor in the nonadiabatic transition probability for near-adiabatic conditions is, according to the quantum version of (7.91),

$$\mathscr{P}_{12} \sim \exp\left[-2(2\mu)^{1/2}\,\mathrm{Im}\left\{\int_{R_2^{\mathrm{t}}}^{R_{\mathrm{s}}}(\mathscr{E} - U_2)^{1/2}\,dR - \int_{R_1^{\mathrm{t}}}^{R_{\mathrm{s}}}(E - U_1)^{1/2}\,dR\right\}\right], \tag{9.27}$$

where R_{s} is the stationary phase point corresponding to the lowest value of the exponent. Using U_1 and U_2 from (9.4), we find the equations defining the possible R_{s}:

$$U_1(R_{\mathrm{s}}^{(1)}) - U_2(R_{\mathrm{s}}^{(1)}) = 0, \qquad U_1(R_{\mathrm{s}}^{(2)}) = U_2(R_{\mathrm{s}}^{(2)}) = \infty. \tag{9.28}$$

The solutions to (9.28) are

$$R_{\mathrm{s}}^{(1)} = R_{\mathrm{p}} \pm \mathrm{i}\alpha^{-1}2\vartheta_0, \qquad R_{\mathrm{s}}^{(2)} = -\infty. \tag{9.29}$$

Now, $R_{\mathrm{s}}^{(1)}$ is seen to coincide with R_{c} from (9.6). This means that the calculation using (9.27) with $R_{\mathrm{s}} = R_{\mathrm{c}}$ will give the exponent, which for high energy $\mathscr{E}$ reduces to the semiclassical result. To obtain this result, expand the square roots for $R = R_{\mathrm{p}}$ retaining the terms linear in

$$\Delta U(R)\,[\mathscr{E} - V_0(R_{\mathrm{p}})]^{-1} \quad \text{and in } [V_0(R) - V_0(R_{\mathrm{p}})]\,[\mathscr{E} - V_0(R_{\mathrm{p}})]^{-1}.$$

Thus we obtain

$$\mathscr{P}_{12} \sim \exp\left[-(2\Delta\varepsilon/\alpha v_{\mathrm{p}}) \times \mathrm{Im}\left\{\int_{R_{\mathrm{p}}}^{R_{\mathrm{c}}}(\Delta U/\Delta\varepsilon)\,dR + \int_{R_{\mathrm{p}}}^{R_{\mathrm{c}}}(\Delta U\,\Delta V_0/\Delta\varepsilon\,\mu v_{\mathrm{p}}^2)\,dR\right\}\right], \tag{9.30}$$

where $v_{\mathrm{p}} = (2/\mu)^{1/2}[\mathscr{E} - V_0(R_{\mathrm{p}})]^{1/2}$ and $\Delta V_0 = V_0(R) - V_0(R_{\mathrm{p}})$. Calculation of the first term in (9.30) gives the expected result

$$(2\Delta\varepsilon/\alpha v_{\mathrm{p}})\,\mathrm{Im}\left\{\int_{R_{\mathrm{p}}}^{R_{\mathrm{c}}}(\Delta U/\Delta\varepsilon)\,dR\right\} = (\pi\Delta\varepsilon/\alpha v_{\mathrm{p}})(1 - \cos 2\vartheta_0) = 2\pi\xi_{\mathrm{p}}. \tag{9.31}$$

The second term in (9.30) provides the correction due to the nonuniform relative motion of atoms in the coupling region. Its specific form depends on the average potential $V_0(R)$. If the latter is taken to be the exponential function $V_0(R) = B\exp(-\alpha R)$ and if $V_0(R_p)$ is small enough compared to $\mathscr{E}$, we obtain

$$\mathscr{P}_{12} \sim \exp\{-(\pi\Delta\varepsilon/\alpha v_p)(1-\cos 2\vartheta_0) \times [1 + (B\Delta\varepsilon/2A\mu v_p^2)(1-\cos 2\vartheta_0)]\}. \quad (9.32)$$

This expression is valid only when the correction in square brackets is small compared to unity. However its absolute contribution to the probability can be substantial, since the correction is multiplied by the large factor before the brackets.

As the ratio $B\Delta\varepsilon/A\mu v_p^2$ increases, the turning point effect becomes greater, since the transition region approaches the turning points. When the correction term in brackets becomes of the order of unity, expansion (9.30) becomes invalid, and it can be expected that under these conditions the main contribution to the transition probability is from the stationary phase point $R_s^{(2)}$. In this case the calculation of the exponent for $2\vartheta_0 = \pi/2$ is simple. The result reads [9.16]

$$\begin{aligned}\mathscr{P}_{12} \sim \exp\Bigg\{&2(2\mu)^{1/2}\int_{-\infty}^{R_1^t}[B\exp(-\alpha R) - E_1]^{1/2}dR \\ &- 2(2\mu)^{1/2}\int_{-\infty}^{R_2^t}[B\exp(-\alpha R) - E_2]^{1/2}dR\Bigg\} \\ = \exp[-\pi(k_2 - k_1)/\alpha] &= \exp[-\pi\Delta\varepsilon/\alpha(v_1 + v_2)], \end{aligned} \quad (9.33)$$

where v_1, k_1, E_1 and v_2, k_2, E_2 are the initial and final velocities, wave vectors and kinetic energies, respectively, and $E_2 - E_1 = \Delta\varepsilon > 0$.

In conclusion we shall mention another exponential model for the nonadiabatic coupling of two states, which can be solved in terms of confluent hypergeometric functions [9.17]. The time-dependent Hamiltonian V_{ik} for this model is

$$V_{ik}(t) = \delta_{ik}\Delta\varepsilon_1[1 - \exp(-2\alpha v_p t)] + (1-\delta_{ik})\Delta\varepsilon_2\exp(-\alpha v_p t). \quad (9.34)$$

The one-way transition probability P is

$$P = \frac{\exp(-\pi\Delta\varepsilon_1/\alpha v_p) + \exp(-\pi\Delta\varepsilon_2/\alpha v_p)}{1 + \exp(-\pi\Delta\varepsilon_1/\alpha v_p)} \quad (9.35)$$

and depends, just as the probability in (9.15a) on two dimensionless parameters.

9.1.3 Specific Cases of the Exponential Model – Probabilities and Cross Sections

The variation of the parameter $2\vartheta_0$ in the exponential model leads to a qualitatively different form of the adiabatic splitting $\Delta U(R)$, varying from the pseudocrossing ($\vartheta_0 \ll 1$) to almost parallel terms ($|\vartheta_0 - \pi/2| \ll 1$) with the important intermediate case $\vartheta_0 = \pi/4$. These limiting cases which are of interest for certain specific applications will now be discussed in detail.

Consider first the pseudocrossing case. The pseudocrossing condition, formulated as

$$\Delta U_p \ll \Delta\varepsilon, \tag{9.36}$$

corresponds to $\vartheta_0 \ll 1$. The general equation (9.15a) with the first two terms of the expansion in ϑ_0 gives

$$P = \exp(-\pi\xi\vartheta_0^2)\sinh[\pi\xi(1-\vartheta_0^2)]\sinh^{-1}\pi\xi. \tag{9.37}$$

With the additional condition

$$\xi = \Delta\varepsilon/\alpha v_p \gg 1, \tag{9.38}$$

(9.37) converts to the Landau-Zener formula (8.10) for which $a = \Delta\varepsilon\vartheta_0$, $|\Delta F| = \Delta\varepsilon\alpha$:

$$P = \exp(-2\pi\delta), \qquad \delta = a^2/|\Delta F|\, v_p = \xi\vartheta_0^2. \tag{9.39}$$

It will be seen that within the limits of the model discussed, the conditions (9.36) and (9.38) correspond to those defining the validity of the linear model, (8.51). The region of the $2\alpha v_p/\pi\Delta\varepsilon$, $U_p/\Delta\varepsilon$ plane, where conditions (9.36) and (9.38) are satisfied is shown as the left lower rectangle in Fig. 9.3. It corresponds to the first quadrant of the ε, β plane in Fig. 8.1. To estimate the correction to the Landau-Zener probability for a small deviation of the diabatic terms from the linear terms, one has to take into account the next term in the expansion of $\cos 2\vartheta_0$ in ϑ_0. For this case

$$P = \exp(-2\pi\delta)\,[1 + (2\pi/3)(\delta^2/\xi) + \ldots], \tag{9.40}$$

where the second term in brackets is assumed to be smaller than unity, whence the relative correction to the probability when $2\pi\delta \sim 1$ is of the order of $1/\xi$. This correction increases with δ. For $2\pi\delta \gg 1$ it can be obtained using adiabatic perturbation theory, which yields, for $\xi(1 \pm \cos 2\vartheta_0)^{1/2} \gg 1$,

$$P = \exp(-2\pi\xi_p), \tag{9.41}$$

where ξ_p is defined by (9.13).

Taking into account two terms of the expansion in ϑ_0 and expressing the result via $\Delta_d R = 2(v_p/\Delta\varepsilon\alpha)^{1/2}$ and $\Delta_a R = 2\vartheta_0/\alpha$, we have, in the near-adiabatic limit,

$$P = \exp[-2\pi(\Delta_a R)^2/(\Delta_d R)^2 + \pi(\Delta_a R)^4\alpha^2/6(\Delta_d R)^2 + \dots]. \qquad (9.42)$$

The first expansion term in the exponent coincides with the Landau-Zener parameter $2\pi\delta$, the second is the correction due to the deviation of the terms from linearity, the value $1/\alpha$ representing the characteristic length l introduced in Sect. 8.4. It is seen from (9.42) that the condition for the validity of the linear model formulated in (8.50) ensures the smallness of the second term compared to the first one in the exponent, but does not guarantee that its absolute contribution to the probability is small. For this to be so, the more stringent condition

$$\Delta_d R \gg (\Delta_a R)^2\alpha \qquad (9.43)$$

must be satisfied, which is, however, violated at low velocities. This violation can be interpreted as an effective broadening of the transition region in the limit $v_p \to 0$, actually found in [9.18].

In the limit considered, $\vartheta_0 \ll 1$, $\xi \gg 1$, the general expression (9.17) yields for the dynamic phase

$$\phi_{21} = \gamma(\delta), \qquad (9.44)$$

which coincides, as expected, with the dynamic phase of the linear model (Sect. 8.2).

In the case of parallel diabatic terms ($2\vartheta_0 = \pi/2$), (9.15) gives

$$P = \exp(-\pi\xi)\,[1 + \exp(-\pi\xi)]^{-1}. \qquad (9.45)$$

The dynamic phase $\Delta\phi_{21}$ calculated from (9.17) is

$$\Delta\phi_{21} = \xi[2^{1/2} - \log(2^{1/2} + 1)]. \qquad (9.46)$$

Thus, taking (9.16) into account, the transition probability becomes

$$\mathscr{P}_{12} = \sin^2\phi_{21}\cosh^{-2}(\pi\Delta\varepsilon/2\alpha v_p), \quad \text{where} \qquad (9.47)$$

$$\phi_{21} = \Delta_{21} + \Delta\phi_{21} = \int_{R^t}^{R_p} [(\Delta\varepsilon)^2 + 4V_{12}^2(R)]^{1/2}\,dt + \Delta\phi_{21}. \qquad (9.48)$$

Assume now that the exponential approximation for the nondiagonal matrix element V_{12} is valid not only in the nonadiabaticity region of width $1/\alpha$ near R_p, but everywhere for $R > R_p$. The integral in (9.47) and the dynamic phase can then be combined to yield an alternative expression for ϕ_{21}:

$$\phi_{21} = \int_{-\infty}^{\infty} V_{12}(R)\, dt . \tag{9.49}$$

The integration over time in (9.49) is carried out along the entire classical trajectory $R = R(t)$ and $V_{12}(R)$ is assumed to be exponentially dependent on R for $R \sim R_p$ and $R > R_p$. Equation (9.47), valid for $\phi_{21} \gg 1$, can be compared with the exact expression for the transition probability $\mathscr{P}_{12}^{RZ}$ found by *Rosen* and *Zener* [9.1] for the model Hamiltonian:

$$V_{ik}^{RZ}(t) = \pm\, \delta_{ik}\, \Delta\varepsilon/2 + (1 - \delta_{ik})\,(A/4)\cosh^{-1}(\alpha v_p t) . \tag{9.50}$$

For this Hamiltonian, the transition probability $\mathscr{P}_{12}^{RZ}$ is

$$\begin{aligned} \mathscr{P}_{12}^{RZ} &= \sin^2 \phi_{21}^{RZ} \cosh^{-2}(\pi \Delta\varepsilon/2\alpha v_p),\\ \phi_{21}^{RZ} &= \int_{-\infty}^{\infty} V_{12}^{RZ}(t)\, dt = \pi A/2\alpha v_p , \end{aligned} \tag{9.51}$$

which is valid for arbitrary phase of the oscillatory factor. Comparing (9.51) with (9.47) we see that the exponential model reproduces the exact results obtained by Rosen and Zener, provided the phase ϕ_{21} is calculated according to (9.49), where $V_{12}[R(t)]$ is put equal to $V_{12}^{RZ}(t)$. This result, which is expected for $\phi_{21} \gg 1$, remains valid for $\phi_{21} \gtrsim 1$.

Thus, there exists a case for which the matching of the exponential model solutions provides an exact result for all values of the phase, and not only for large ones, as originally assumed in deriving the basic (9.16). This is important when calculating the integral cross sections (see below).

The case of almost parallel adiabatic terms ($\pi - 2\vartheta_0 = 2\bar{\vartheta}_0 \ll 1$) will be considered in order to illustrate the effect of two coalescent branch points in the upper halfplane of τ on the transition probability, under near-adiabatic conditions, $\xi \approx \xi_p \gg 1$. The transition probability can be calculated using the theory of near-adiabatic perturbations (Sect. 6.2.2), which must account for two branch points. This differs from the case described in Sect. 8.2 only in that the two points $\tau_c^{(1)}$ and $\tau_c^{(2)}$ lie on a straight line parallel to the imaginary axis, whereas for the tunneling case in the linear model these points lie on a line parallel to the real axis. At the same time, this transition probability is expected to correlate in a certain sense with the Landau-Zener probability for a single passage of the coupling region for which the branch points are essential (Sect. 8.2).

In the limit $\bar{\vartheta}_0 \ll 1$, $\xi \gg 1$ the general expression (9.15a) yields

$$P = \exp(-2\pi\xi + 2\pi\xi\bar{\vartheta}_0^2)\,[1 - \exp(-2\pi\xi\bar{\vartheta}_0^2)] . \tag{9.52}$$

The first factor in (9.52) is identified with the exponential factor appearing in the near-adiabatic perturbation theory, and the second accounts for the existence of two close-lying branch points. It can readily be checked that $2\pi\vartheta_0^2$ equals the phase between the points $\tau_c^{(1)}$ and $\tau_c^{(2)}$. For this reason this factor is similar in structure to the Landau-Zener survival probability $(1-P)$ when the interaction region is traversed along the contour lying close to $\tau_c^{(1)}$ and $\tau_c^{(2)}$.

Turn now to the cross section for the exponential model. The possibility of retaining an arbitrary asymptotic splitting for the adiabatic terms suggests using this model to describe nonresonant processes, provided the resonance defect $\Delta\varepsilon$ is small compared to the spacing from the initial and final levels to all other atomic levels. This is the case, e.g., for quasi-resonant charge exchange,

$$A^+ + B \rightarrow A + B^+ + \Delta\varepsilon, \qquad (9.53)$$

for which the interaction between atomic states is due to electron transfer and thus can actually be approximated by the exponential function within the region of nonadiabatic coupling.

The simplest case corresponds to two molecular Σ-terms. Taking the initial and final atomic s-states as diabatic, assuming that there are no first-order corrections to the diabatic terms $V_{11}(R)$ and $V_{22}(R)$, and approximating $V_{12}(R)$ close to $R = R_p$ by the exponential function, we get

$$V_{11} - V_{22} = \Delta\varepsilon, \qquad V_{12} = (\Delta\varepsilon/2)\exp(-\alpha R + \alpha R_p). \qquad (9.54)$$

Here R_p is evaluated as usual with the actual adiabatic terms $U_1(R)$ and $U_2(R)$, and α is expressed via the orbital exponent of the transferred electron and via the distance R_p [9.19, 20]. Equation (9.54) specifies the particular case of the exponential model with $\vartheta_0 = \pi/4$. The transition probability will be given by (9.16), provided the approximation of local rectilinear motion (9.10) is accepted. The accuracy of this approximation depends on the impact parameter and is best for small b. Naturally, for $b \sim R_p$ this approximation fails because the two coupling regions coalesce. In this case the Rosen-Zener-model Hamiltonian will provide a better approximation [9.21].

However, with the condition $\alpha R_p \gg 1$ the transition probability will be given by (9.47) over a wide range of b, and the region $b \sim R_p$ makes a negligible contribution to the integral cross section [9.19].

Thus, the integral inelastic cross section σ_{21} calculated in the high-energy approximation ($\Delta\varepsilon \ll E_1, E_2$) is

$$\sigma_{21}(v) = 2\pi \int_0^{R_p} \sin^2\phi_{21}(b,v)\cosh^{-2}[\pi\Delta\varepsilon/2\alpha v(1-b^2/R_p^2)^{1/2}]\,b\,db. \qquad (9.55)$$

Now, the oscillatory factor $\sin^2\phi_{21}$ can be dealt with using the random phase approximation, i.e., by replacing $\sin^2\phi_{21}$ by 1/2 for $b < b^*$ and by zero for $b > b^*$. According to *Firsov* [9.22], b^* is found from

$$\phi_{21}(b^*, v) = 1/\pi. \tag{9.56}$$

Again, using the high-energy approximation for calculations at large impact parameters, we get, according to (9.49, 54),

$$\begin{aligned}\phi_{21}(b, v) &= \int_{-\infty}^{\infty} V_{12}[(b^2 + z^2)^{1/2}]\, dz/v \\ &= (\Delta\varepsilon/\alpha v) \exp(-\alpha b + \alpha R_{\mathrm{p}}) (\pi b \alpha/2)^{1/2}.\end{aligned} \tag{9.57}$$

Combining (9.55–57) we get

$$\sigma_{21}(v) = \pi \int_0^{\min(b^*, R_{\mathrm{p}})} \cosh^{-2}[\lambda/(1 - b^2/R_{\mathrm{p}}^2)^{1/2}]\, b\, db \tag{9.58}$$

with $\lambda = \pi \Delta\varepsilon/2\alpha v$ and b^* found from

$$\lambda \exp(-\alpha b^* + \alpha R_{\mathrm{p}}) (2 b^* \alpha/\pi)^{1/2} = 1/\pi. \tag{9.59}$$

Equation (9.58), together with (9.56), implies that σ_{21} first increase with increasing velocity as

$$\sigma_{21}(v) = \pi R_{\mathrm{p}}^2 (2/\lambda) \exp(-2\lambda), \qquad \lambda \gg 1, \tag{9.60}$$

then reaches a broad maximum

$$\sigma_{21}^{\max}(v) \approx \pi R_{\mathrm{p}}^2/2 \tag{9.61}$$

for λ somewhat less than unity, and then slowly decreases

$$\sigma_{21}(v) = \pi (b^*)^2/2 = (\pi/2\alpha^2) \log^2(v_0/v), \qquad \lambda \ll 1. \tag{9.62}$$

Here v_0 depends weakly (logarithmically) on v and can be calculated by (9.59) substituting $\alpha^{-1} \log(v_0/v)$ for b^*.

The slow decrease in the cross section at high energies is connected with the progressively smaller contribution from high impact parameters, for which the phase difference ϕ_{21} is low. This behavior is typical for resonant processes. A more accurate treatment of resonant and quasi-resonant charge-exchange processes, which goes beyond the simple exponential model and the random-phase approximation, can be found in [9.19, 20, 23]. The application of the exponential model to differential cross section calculations can be found in [9.24–27].

A similar approach can be used for arbitrary ϑ_0. This problem will be discussed in Chap. 11 in connection with intramultiplet mixing.

The exponential model is also used for interpreting the undulatory behaviour of the integral excitation cross section for close-lying atomic states. This phenomenon, studied both experimentally and theoretically (see, e.g., the reviews [9.28–30]) consists in coherent population of the close-lying molecular states $|k\rangle$, caused by nondiabatic transitions at small distances, from the same molecular state correlating, for $R \to \infty$, with the ground state of the colliding partners. When nonadiabatic transitions between terms $U_k(R)$ are possible in the region of transformation of molecular to atomic functions, the interference effects due to different population routes for a certain atomic state persist even in the integral cross section [9.31]. Though in the cases studied experimentally the cross sections must be calculated in terms of a multilevel, rather than a two-level model, yet the exponential model was the first invoked for a qualitative interpretation of this effect [9.32]. It is essential here that the exponential model provides flexibility in the transition probability at high velocities. As follows from (9.15a) for $\xi \ll 1$, $P = \cos^2 \vartheta_0$ and does not necessarily tend to unity as in the Landau-Zener model.

9.2 Linear-Exponential Model

9.2.1 Formulation of the Model

This section deals with a linear-exponential model of two-state nonadiabatic coupling, which is an example of the linear model generalization, and which makes it possible to account correctly for transitions near the turning point, as discussed in Sect. 8.3.

Assume that the coupling matrix element in the nonadiabaticity region is exponentially dependent on R, and the diabatic terms are linear. Such a model gives a fair description, for example, of the chemoionization process [9.33]

$$\mathrm{A + B \to A^+ + B^-}, \tag{9.63}$$

induced by nonadiabatic transitions of the system from the covalent to the ionic term. The model Hamiltonian is

$$V_{11} - V_{22} = -\Delta F(R - R_0), \qquad V_{12} = a \exp[-\alpha(R - R_0)]. \tag{9.64}$$

If a dimensionless variable

$$x = \alpha(R - R_0) \tag{9.65}$$

is introduced, the splitting of the adiabatic terms obtained as a result of the matrix (9.64) diagonalization is

$$\Delta U(x) = (|\Delta F|/\alpha)\,[x^2 + v^2 \exp(-2x)]^{1/2}, \tag{9.66}$$

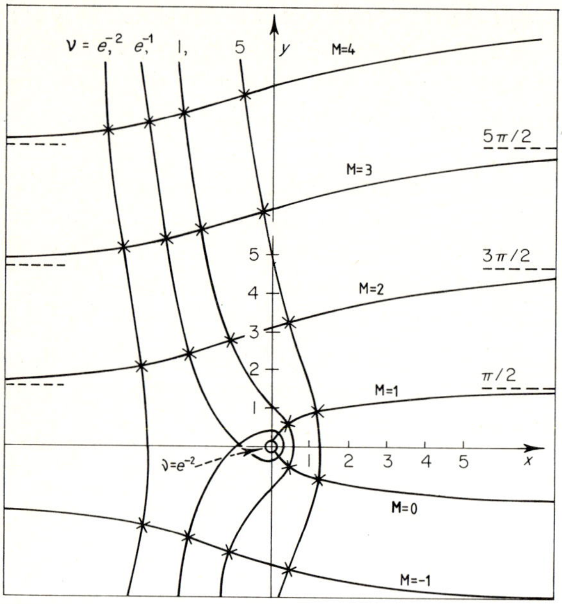

Fig. 9.4. Positions of branch points (intersection of curves) for $\Delta U(z)$ of the linear-exponential model

where v is a parameter representing the ratio of the nonadiabaticity range for the linear model $\Delta_a R$ to the characteristic length $1/\alpha$ of V_{12}:

$$v = \alpha \Delta_a R = 2a\alpha/|\Delta F|. \tag{9.67}$$

The correlation of the linear-exponential model with the linear (Chap. 8) and the exponential (Sect. 9.1) ones is most readily found by considering the analytical features of the function $\Delta U(z)$ in the complex plane z, and specifically the positions of the branch points. All branch points z_c lie at intersections of curves defined by the relations

$$\arg(\mathrm{i} z \exp z) = \pi M, \qquad M \text{ is an integer,} \tag{9.68}$$

$$|z \exp(z)| = v. \tag{9.69}$$

These curves are shown in Fig. 9.4. Curves of one family defined by (9.68) are labeled by values of M. Equation (9.69) defines the family of single (for $v > 1/\mathrm{e}$) or double (for $v < 1/\mathrm{e}$) curves which are labeled by values of v. For small v ($v \ll 1$), (9.68, 69) give two sets of branch points z_c^a and z_c^b:

$$z_c^a = v^2 \pm \mathrm{i} v, \quad z_c^b = \log v \pm \mathrm{i}(\pi/2 - \pi/2 \log v + n\pi), \quad n = 0, 1, \ldots . \tag{9.70}$$

The first pair of points corresponds to a situation similar to that for the Landau-Zener model. Consequently, the transition probability can be calculated using the linear model equations with certain redefinition of the parameters. The second set of branch points approximately corresponds to a particular case of the exponential model for $\vartheta_0 = \pi/4$, and thus the transition probability can be estimated by (9.47).

As v increases, the distance between the pair of branch points nearest to the real axis increases, and for $v \gg 1$ it becomes equal to $\pi/2$. All other branch points lie at a distance $\pm i\pi n$ from the above two sets, namely,

$$z_c = \log v \pm i(\pi/2 - \pi/2 \log v + \pi n). \tag{9.71}$$

Such a pattern of branch point positions resembles that in Fig. 9.2 and therefore for $v \gg 1$ it is similar in nature to coupling in the exponential model.

Thus, for $v \ll 1$ the model (9.64) has, in principle, two nonadiabaticity regions, one near $R_a \approx R_0$, the second near $R_b \approx R_0 + (1/\alpha) \log v < R_a$.

The semiclassical equations describing transitions are

$$\begin{aligned} i\dot{b}_1 &= a \exp(-\alpha X) \exp\left(-i\int_0^t \Delta F X\, dt\right) b_2, \\ i\dot{b}_2 &= a \exp(-\alpha X) \exp\left(i\int_0^t \Delta F X\, dt\right) b_1, \end{aligned} \tag{9.72}$$

with $X = R - R_0$. Depending on whether the coupling region is far from or near to the turning point of the classical trajectory, a different time dependence of X is assumed. Consequently, it is expedient to consider these two limiting cases.

9.2.2 Transition Probabilities and Cross Sections

If the coupling region is far from the turning point, the local-rectilinear-motion approximation can be adopted. However, with $X = v^* t$, (9.72) cannot be solved analytically and numerical integration must be resorted to [9.34]. Integration gives the one-way transition probability P which defines the averaged transition probability $\bar{\mathscr{P}}_{12}$. The plot of $\bar{\mathscr{P}}_{12}$ versus $1/2\delta$, $\delta = a^2/|\Delta F|\, v^*$, is shown in Fig. 9.5. Two features deserve mention.

First, in the low velocity region ($\delta \gg 1$), the transition probability $\bar{\mathscr{P}}_{12}$ for $v > 0$ increases faster with $1/2\delta$ than it does for the linear model ($v = 0$). This is explained in terms of near-adiabatic perturbation theory (Sect. 6.2.2) giving

$$\begin{aligned} \bar{\mathscr{P}}_{12} &= 2 \exp[-2\pi\delta(v)] \\ &= 2 \exp\left\{-8\delta \operatorname{Im}\left\{\int_0^{z_c} [z^2 + \exp(-2vz)]^{1/2}\, dz\right\}\right\}. \end{aligned} \tag{9.73}$$

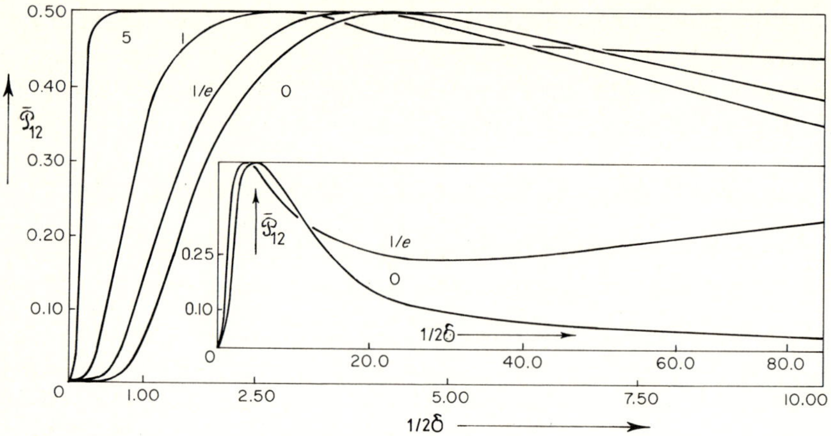

Fig. 9.5. Average transition probability $\bar{\mathscr{P}}_{12}$ for the linear-exponential model. The curves are labeled by v values

Here $\delta(v)$ is smaller than δ, viz., for small v, $\delta(v) = \delta(1 - 2v^2)$. Since $\delta \gg 1$, even a small change in $\delta(v)$ results in a substantial change of $\bar{\mathscr{P}}_{12}$.

Second, at high velocities ($\delta \ll 1$), the transition probability $\bar{\mathscr{P}}_{12}$ tends to a constant nonzero value displaying a dip at intermediate velocities that can be quite pronounced for small v. The high velocity limit is reproduced by invoking the result for the exponential model with

$$\bar{\mathscr{P}}_{12} = 1/2 \cosh[(\pi/2)\,\xi], \qquad \xi = \Delta F \log(1/v)/\alpha^2 v^*. \tag{9.74}$$

Note that for small v the maximum of the Landau-Zener-like portion of $\bar{\mathscr{P}}_{12}$ and the steep rise of the Rosen-Zener-like portion correspond to quite different collision energies, as can be inferred from the ratio of the two Massey parameters:

$$4\pi\delta/\pi\xi \approx v^2/\ln(1/v). \tag{9.75}$$

Moreover, when two nonadiabaticity regions are sufficiently separated, the Landau-Zener formula with slightly modified parameters to account for the exponential dependence of V_{12} on R reproduces the quantum calculation result [9.35].

The dip in the velocity dependence of $\bar{\mathscr{P}}_{12}$ also appears in the dependence of $\bar{\mathscr{P}}_{12}$ on the impact parameter b with the usual parametrization $v^* = v\,(1 - b^2/R_0^2)^{1/2}$. This may eventually result in two maxima in the dependence of the cross section on velocity, one associated with the Landau-Zener maximum (strong coupling at the pseudocrossing $R \approx R_a$ and adiabatic behavior for $R \approx R_b$), the other with a maximum similar to that for quasi-resonant charge exchange (diabatic behavior for $R \approx R_a$ and strong coupling for $R \approx R_b$).

The region of the Landau-Zener-like maximum will now be discussed in more detail, taking into account the effect of the turning point. Equation (8.15) will be used for the trajectory, and the parameters ε, β, t_0 of the linear model (Chap. 8) will be introduced. Then (9.72) reads

$$\begin{aligned} \mathrm{i}\,db_1/d\tau &= (\beta/2)\exp[-\nu(\tau^2-\varepsilon)]\exp\left[-\mathrm{i}\beta\int_0^\tau(\tau^2-\varepsilon)\,d\tau\right]b_2,\\ \mathrm{i}\,db_2/d\tau &= (\beta/2)\exp[-\nu(\tau^2-\varepsilon)]\exp\left[\mathrm{i}\beta\int_0^\tau(\tau^2-\varepsilon)\,d\tau\right]b_1. \end{aligned} \tag{9.76}$$

An analytical solution of this system can be obtained for weak coupling between diabatic or adiabatic states. For the former case use is made of the semiclassical (or quantum) variant of the distorted-wave approximation and the transition probability is expressed via the Airy function [9.36] with the maximum somewhat shifted relative to that for the linear model. In the asymptotic representation of the Airy function this shift is given by the additional phase $\sim(\alpha\Delta_d R)^2$. The transition probability also involves an additional factor very close to unity [9.36].

For near-adiabatic conditions the transition probability is given to exponential accuracy [9.37] by

$$\bar{\mathscr{P}}_{12} = 2\exp[-2\beta\Delta''(\varepsilon,\nu)], \tag{9.77}$$

where $\Delta''(\varepsilon,\nu)$ is expressed by the integral

$$\Delta''(\varepsilon,\nu) = \mathrm{Im}\left\{\int_0^{\tau_c}[(\tau^2-\varepsilon)^2+\exp(-2\nu\tau^2+2\nu\varepsilon)]^{1/2}\,d\tau\right\}. \tag{9.78}$$

For $\nu = 0$ the function $\Delta''(\varepsilon,\nu)$ coincides with $\Delta''(\varepsilon)$ introduced in Sect. 8.3. In the general case, the plot of $\Delta''(\varepsilon,\nu)$ is similar to that of $\Delta''(\varepsilon)$; however, the entire curve lies lower. The curves for $\nu = 1$ and $\nu = 2$ are presented in Fig. 8.5. The use of (9.77, 78) for the energy dependence of cross sections near the threshold has been discussed in [9.37].

For the linear-exponential model the inelastic integral cross section will depend on parameters previously defined for the linear model (Sect. 8.5.1) and also on ν. In this way

$$\sigma_{fi}(E_i) = \pi R_0^2(E^*/E_i)\,f(y, A, B, \nu), \tag{9.79}$$

with y and E^* defined by (8.62) and A and B by (8.70), where R_p is replaced by R_0. Clearly, $f(y, A, B, 0) = f(y, A, B)$ defined by (8.69). Plots of reduced cross sections $Q_{21} = \sigma_{21}/\pi R_0^2$ for the initial horizontal term (hence $B = 0$), $A = 4$, and different ν are presented in Fig. 9.6. It is seen that in the

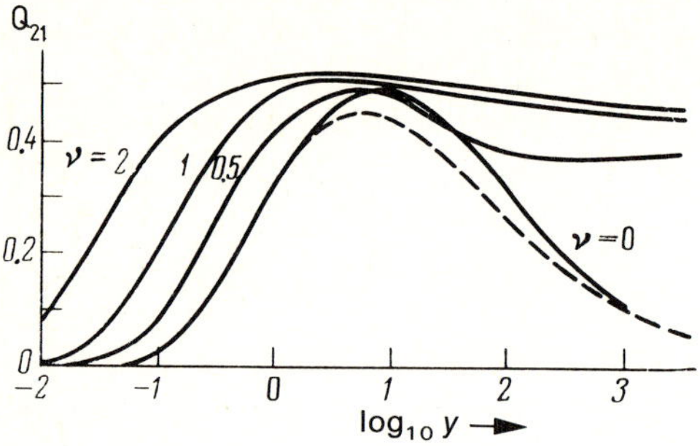

Fig. 9.6. Reduced cross sections Q_{21} for the linear-exponential model versus $\log_{10} y$ for different v values and $A = 4$, $B = 0$. The dashed line refers to the Landau-Zener cross section

near-adiabatic region the cross sections are larger for higher values of v. Also, the cross section maximum is displaced to lower energies.

To the left of the maximum, the function $f(y, A, 0, v)$ for $0 \leqq v \leqq 2$ can be expressed by a simple interpolation equation based on numerical results [9.38]:

$$f(y, A, 0, v) = f[y \exp(1.54\, v), A, 0]. \tag{9.80}$$

When A is sufficiently high (in particular, when $A > 4$), then approximately

$$f(y, A \gg 1, 0, v) = f^{\mathrm{LZ}}[y \exp(1.54\, v)], \tag{9.81}$$

i.e., the cross section for the linear-exponential model is obtained by a scale transformation of the cross section for the Landau-Zener model.

In high energy regions the cross sections for the linear-exponential model decrease very slowly due to the effect of the second transition region. Two maxima separated by a minimum at $y \approx 100$ can be seen for $v = 0.5$.

9.3 Other Nonlinear Models

Other nonlinear models for the two-state nonadiabatic coupling will be considered here. Two models represent a generalization of the exponential model and are formulated in terms of the time-dependent interaction $V_{ik}(t)$ without specifying the trajectory (Sect. 9.3.1). Then follows a discussion on power models. This kind of interaction ($V_{ik} \sim R^n$) between atoms is usually operative either for very large or very small internuclear separations. The former (for n negative, see Sect. 4.2), for which the rectilinear trajectory $R = R(t)$ can safely be assumed, is discussed in Sect. 9.3.2. The latter ($n = 2$, see Sect. 3.2.3), for which the effect of the trajectory distortion can be considerable, is dealt with in Sect. 9.3.3.

9.3.1 Hypergeometric Models

Comparison equations for the linear and exponential models with a local-rectilinear-motion trajectory are solved analytically in terms of parabolic cylinder functions (or Weber functions), or confluent hypergeometric functions. Correspondingly, the transition amplitudes depend asymptotically (far from the coupling region) on one (δ) or two (ξ, ξ_p) parameters. It appears, however, that even more general models can also be constructed which make it possible to find analytical solutions to semiclassical problems, including the linear and exponential models as particular cases. This is possible since the most general second-order equation to which the system of two first-order equations reduces and the solution of which is expressed by well-known higher transcendental functions, depend on three parameters. This is a hypergeometric equation, hence – the hypergeometric models. Two such models will be discussed [9.39] and will be formulated in the diabatic basis which decouples at $t \to \infty$.

Model 1. The Hamiltonian matrix is

$$\begin{aligned} V_{11} - V_{22} &= [(\Delta\varepsilon_1 + \Delta\varepsilon_3) + (\Delta\varepsilon_1 - \Delta\varepsilon_3)\tanh(\alpha v t)]/2, \\ V_{12} &= \Delta\varepsilon_2/2\cosh\alpha v t. \end{aligned} \tag{9.82}$$

The nonadiabatic transition probability P has a different form depending on the sign of the product $\Delta\varepsilon_1\Delta\varepsilon_3$. Introducing $\lambda_k = \pi\,|\Delta\varepsilon_k|/2\alpha v$, one gets for $\Delta\varepsilon_1\Delta\varepsilon_3 > 0$ (noncrossing diabatic terms):

$$\begin{aligned} P = {} & \sinh\{(\lambda_1 - \lambda_3)/2 + [(\lambda_1 - \lambda_3)^2/4 - \lambda_2^2]^{1/2}\} \\ & \times \sinh\{(\lambda_1 - \lambda_3)/2 - [(\lambda_1 - \lambda_3)^2/4 - \lambda_2^2]^{1/2}\}\cosh^{-1}\lambda_1\cosh^{-1}\lambda_3, \end{aligned} \tag{9.83}$$

and for $\Delta\varepsilon_1\Delta\varepsilon_3 < 0$ (crossing diabatic terms):

$$\begin{aligned} P = {} & \cosh\{(\lambda_1 + \lambda_3)/2 + [(\lambda_1 + \lambda_3)^2/4 - \lambda_2^2]^{1/2}\} \\ & \times \cosh\{(\lambda_1 + \lambda_3)/2 - [(\lambda_1 + \lambda_3)^2/4 - \lambda_2^2]^{1/2}\}\cosh^{-1}\lambda_1\cosh^{-1}\lambda_3. \end{aligned} \tag{9.84}$$

These equations are valid for all values of the three parameters λ_k. For $|\Delta\varepsilon_1 - \Delta\varepsilon_3| < \Delta\varepsilon_2$ the radicands in (9.83) and (9.84) are negative and the arguments of the hyperbolic functions are complex. After transforming these functions to functions of real arguments, P will involve an oscillatory factor appearing as the result of interference caused by the existence of two nonadiabaticity regions.

In a particular case, when $\lambda_1 = \lambda_3$, (9.83) converts to the Rosen-Zener formula (9.51):

$$P = \mathscr{P}_{12}^{\mathrm{RZ}} = \sin^2\lambda_2\cosh^{-2}\lambda_1. \tag{9.85}$$

In another limiting case, λ_1, λ_3, $|\lambda_1 - \lambda_3| \to \infty$, when $\lambda_2^2/|\lambda_1 - \lambda_3|$ is limited, the Landau-Zener formula is recovered:

$$P = \exp(-2\lambda_2^2/|\lambda_1 - \lambda_3|). \tag{9.86}$$

These limiting cases can be readily conceived of in terms of the gyroscopic interpretation of the two-state models (Sect. 7.3).

Model 2. The Hamiltonian matrix is

$$\begin{aligned} V_{11} - V_{22} &= \Delta\varepsilon_1 - 2^{-1}\cos 2\vartheta_0\,\delta\varepsilon[1 - \tanh(\alpha v t/2)], \\ V_{12} &= 4^{-1}\sin 2\vartheta_0\,\delta\varepsilon[1 - \tanh(\alpha v t/2)]. \end{aligned} \tag{9.87}$$

The probability of a nonadiabatic transition is

$$\begin{aligned} P &= \sinh[\lambda_3 + \lambda_2 - \lambda_1 + (\lambda_3^2 + \lambda_2^2 - \lambda_1^2)^{1/2}] \\ &\quad \times \sinh[\lambda_3 - \lambda_2 + \lambda_1 - (\lambda_3^2 + \lambda_2^2 - \lambda_1^2)^{1/2}]\sinh^{-1}2\lambda_1\sinh^{-1}2\lambda_3, \\ \lambda_k &= \Delta\varepsilon_k\pi/2\alpha v. \end{aligned} \tag{9.88}$$

Here the energy parameter $\Delta\varepsilon_1$ is defined by the Hamiltonian (9.87) and $\Delta\varepsilon_2$ and $\Delta\varepsilon_3$ by the relations

$$\Delta\varepsilon_2 = \Delta\varepsilon_1\cos 2\vartheta_0, \qquad \Delta\varepsilon_3 = (\Delta\varepsilon_1^2 - \Delta\varepsilon_1\,\delta\varepsilon + \delta\varepsilon^2)^{1/2} \tag{9.89}$$

under the assumption that $\Delta\varepsilon_3 > \Delta\varepsilon_1 > \Delta\varepsilon_2$. It will be seen that $\Delta\varepsilon_1$ and $\Delta\varepsilon_3$ are adiabatic term splittings for $t \to \infty$ and $t \to -\infty$, respectively. The exponential model is obtained from (9.87) in the limit $\delta\varepsilon \to \infty$, and the parameters $\Delta\varepsilon_1$ and $\cos 2\vartheta_0$ coincide with $\Delta\varepsilon$ and $\cos 2\vartheta_0$ of the Hamiltonian (9.1). In the limit $\lambda_3 \to \infty$ in (9.88), we obtain the transition probability of the exponential model (9.15 a), viz.,

$$P = \exp(-\lambda_1 + \lambda_2)\sinh(\lambda_1 + \lambda_2)\sinh^{-2}2\lambda_1. \tag{9.90}$$

Other limiting cases of P follow from (9.90).

9.3.2 Power Models – Large Interatomic Separations

In the mathematical sense the exponential function can be taken to be the limit of the power function with a high exponent n. Thus it is natural to suggest that the two-state exponential model can appear to be a useful approximation for the description of a situation when the matrix elements are power functions of R. Consider a simple Hamiltonian,

$$V_{11} - V_{22} = \Delta\varepsilon, \qquad V_{12} = CR^{-n}. \tag{9.91}$$

Under near-adiabatic conditions the transition probability is, see (7.91),

$$\mathscr{P}_{12} = 4\sin^2\phi_{21}\exp\left[-\operatorname{Im}\left\{\int_{R_s^*}^{R_s} \Delta U(R)\,dR/v_R\right\}\right]. \tag{9.92}$$

Here the adiabatic term-splitting corresponding to the interaction (9.91) is

$$\Delta U(R) = (\Delta\varepsilon^2 + 4C^2R^{-2n})^{1/2} \tag{9.93}$$

and the dependence of the radial velocity v_R on R for the rectilinear trajectory is $v_R = v(1 - b^2/R^2)^{1/2}$.

The stationary phase point nearest to the real axis is found from the condition $\Delta U = 0$:

$$R_s = R_0\exp(\pm i\pi/2n), \qquad R_0 = (2C/\Delta\varepsilon)^{1/n}. \tag{9.94}$$

To calculate the exponent in (9.92) expand the integrand in powers of b/R_0 and keep the first two terms. Integration gives [9.40]

$$\operatorname{Im}\left\{\int_{R_s^*}^{R_s} \Delta U(R)\,dR/v_R\right\} = (\Delta\varepsilon R_0/vn)\,[M(n) + M(-n)\,b^2/2R_0^2 + \dots], \tag{9.95}$$

where $M(n)$ denotes the integral

$$\begin{aligned} M(n) &= \operatorname{Im}\left\{\int_{-i}^{i} y^{1-1/n}(1+y^2)^{1/2}\,dy\right\} \\ &= n\sin(\pi/2n)\,\Gamma(1-1/2n)\,\Gamma(1/2)/\Gamma(3/2-1/2n). \end{aligned} \tag{9.96}$$

The corresponding expression for the exponential model in the near-adiabatic case can be obtained from (9.47), replacing the velocity v_p by the radial velocity $v(1 - b^2/R_p^2)^{1/2}$. Expanding $(1 - b^2/R_p^2)^{1/2}$ in powers of b^2/R_p^2 we have

$$\pi\xi = (\pi\Delta\varepsilon/\alpha v)\,(1 - b^2/R_p^2)^{1/2} = (\Delta\varepsilon/\alpha v)\,[\pi + \pi b^2/2R_p^2 + \dots]. \tag{9.97}$$

Let us now show that the result for the exponential model (9.97) provides a good approximation to (9.95) when the power potential is approximated by the exponential. Assuming

$$\begin{aligned} V_{12}(R) &= CR^{-n} = (\Delta\varepsilon/2)\,[1 - (R - R_0)/R]^n \\ &\approx (\Delta\varepsilon/2)\exp[-n(R - R_0)/R_0], \end{aligned} \tag{9.98}$$

we find that for the exponential model α must be identified with n/R_0 of the power model. The difference between (9.95) and (9.97) arising in integration still persists. However for $n \gg 1$ this difference is small. Expanding the Γ-functions in (9.96) in powers of $1/n$ we get

$$M(n) = \pi[1 + (1 - \log 2)/n] = \pi(1 + 0.307/n). \tag{9.99}$$

In the most unfavorable, but physically interesting case of $n = 3$, we find $M(3) = 1.1\pi$ and $M(-3) = 0.9\pi$, which differ by only 10% from the value π for exponential coupling in (9.97).

It will be seen from (9.95) that under adiabatic conditions the basic contribution to the transition probability occurs at small b values. As b increases, the transition probability decreases and finally for $\phi_{21} < 1$ perturbation theory with respect to the coupling $V_{12}(R)$ becomes valid. Equation (9.92) with the oscillatory factor remains valid as long as there is an adiabatic region between the branch points R_s and the turning point $R^t = b$.

Thus, the following conclusions hold for the near-adiabatic limit:

i) The transition probability is a maximum for trajectories with $b \ll R_0$. This is simply caused by the fact that for head-on collisions, i.e., for $b = 0$, the radial velocity is a maximum and the Massey parameter is a minimum.
ii) When $b < R_0$ the main contribution to the transition probability is from the region $R \sim R_0$, rather than from the region of closest approach.
iii) When $b > R_0$ and $\phi_{21} < 1$, the main contribution to $\mathscr{P}_{12}$ is from the region of closest approach.
iv) The preexponential factor in the expression for $\mathscr{P}_{12}$ can never exceed the value 4.

In one of the first papers on the theory of nonadiabatic transitions *Stueckelberg* [9.41] used perturbation theory to calculate the transition probability at large impact parameters. He obtained an expression similar to (9.92), but with ϕ_{21}^2 replacing $\sin^2 \phi_{21}$. This is a gross overestimate of the probability for $\phi_{21} > 1$, which illustrates that a low value of the transition probability is not a validity criterion for the Born approximation. *Bates* [9.42] corrected Stueckelberg's error by taking into account the oscillatory factor. However, an approximate expression (7.93) was used for the transition probability, which erroneously ascribed a great importance to the region of closest approach. Naturally, under adiabatic conditions, the transition probability (7.93) is critically dependent either on the distance of closest approach or on the truncation radius of the interaction. The incorrectness of this approach is evident (see also [9.43]).

Turn now to the Hamiltonian in a more general form

$$V_{11} - V_{22} = \Delta\varepsilon - BR^{-n}, \qquad V_{12} = CR^{-n}. \tag{9.100}$$

Nonadiabatic coupling has been studied for this case in [9.44] and the numerical solution of the coupled equations for certain parameters B, C was compared to the approximate solution (7.98). The latter gives the same results, independently of the sign of the product $B \Delta\varepsilon$ at the same values of $|B|$ and $\Delta\varepsilon$. In terms of the exponential model this means that the approximation (7.98) erroneously does not discriminate between $\cos 2\vartheta_0$ and $-\cos 2\vartheta_0$. This error could be avoided by direct use of the exponential model to approximate the power interaction (9.100).

Note also the particular case of (9.100) for $n = 1$ which can be solved analytically for head-on collisions [9.45], i.e., for $R(t) = vt$. The solution is expressed via confluent hypergeometric functions, and the transition probability is

$$P = \frac{\exp\{[\pi(B^2 + 4C^2)^{1/2} + B]/v\} - 1}{\exp[2\pi(B^2 + 4C^2)^{1/2}/v] - 1}. \tag{9.101}$$

9.3.3 Power Models – Small Interatomic Separations

In the united-atom limit certain molecular terms degenerate to one atomic term, and in the general case their splitting at small R is proportional to R^2 (Sect. 3.2.3). The nonadiabatic interaction between terms for $R \to 0$ is the Coriolis coupling between components of a split atomic state, differing in projections Λ. This enables one to formulate a relatively general problem which has been studied in detail for the coupling of Σ- and Π-states [9.46–49].

For this case the equations for the amplitudes of the Σ-state and one of the Π-states in the molecular (adiabatic) basis are

$$\begin{aligned} i\dot{a}_1 &= (bv/R^2)\exp\left(iB\int_0^t R^2\,dt\right)a_2, \\ i\dot{a}_2 &= (bv/R^2)\exp\left(-iB\int_0^t R^2\,dt\right)a_1, \end{aligned} \tag{9.102}$$

where B is a certain constant specifying the splitting of Σ- and Π-terms; the function $R = R(t)$ depends on the trajectory. For a rectilinear trajectory, after introducing a dimensionless time $\tau = vt/b$ and the Massey parameter $\gamma = Bb^3/2v$, (9.102) is rewritten in the form

$$\begin{aligned} i\,da_1/d\tau &= (1+\tau^2)^{-1}\exp\left[2i\gamma\int_0^\tau (1+\tau^2)\,d\tau\right]a_2, \\ i\,da_2/d\tau &= (1+\tau^2)^{-1}\exp\left[-2i\gamma\int_0^\tau (1+\tau^2)\,d\tau\right]a_1. \end{aligned} \tag{9.103}$$

Equations (9.103) have been studied in the near-adiabatic and in the sudden limits.

In the near-adiabatic limit, $\gamma \gg 1$, (9.103) can be solved by a transformation to comparison equations. Now, the case encountered most often has two separated branch points for which the transition probability is given by (7.91). However, in the present case, the singular points in the upper half-plane of τ are the pole $\tau_0 = \mathrm{i}$ and four branch points surrounding it at a distances $|\Delta\tau| \sim \gamma^{-1/2}$. The comparison equations in the vicinity of the pole are solved in terms of confluent hypergeometric functions, and integration along the anti-Stokes lines yields the coefficient of the exponential for transition amplitudes [9.48]. In this way, for $\gamma \gg 1$,

$$\mathscr{P}_{12} = 2\exp\left[-4\mathrm{i}\gamma\int_0^{\mathrm{i}}(1+\tau^2)\,d\tau\right] = 2\exp(-8\gamma/3). \qquad (9.104)$$

Note that although for $\gamma \gg 1$ the branch points tend to coalesce, the coefficient of the exponential in (9.104) remains constant. This behavior is different from that in the case of the linear and exponential models, for which the exponential coefficients tend to zero when two close branch points of ΔU coalesce. This difference can be attributed to the fact that for (9.103) there is a pole between the close branch points.

Note also that the value 2 for the coefficient in (9.104) is not caused by averaging the oscillatory transition probability as is the case for near adiabatic transitions on passing through the coupling region twice, but results from passing through only once.

For $\gamma \ll 1$ the solution of (9.103) is most readily obtained using first-order perturbation theory, passing from the molecular (adiabatic) to the atomic (diabatic) basis. In this basis,

$$\begin{aligned} \mathrm{i}\,db_1/d\tau &= 2\gamma\tau\exp\left[2\mathrm{i}\gamma\int_0^{\tau}(\tau^2-1)\,d\tau\right]b_2, \\ \mathrm{i}\,db_2/d\tau &= 2\gamma\tau\exp\left[-2\mathrm{i}\gamma\int_0^{\tau}(\tau^2-1)\,d\tau\right]b_1. \end{aligned} \qquad (9.105)$$

For small γ the second term in the integral is neglected, so that the transition probability is

$$\mathscr{P}_{12} = \left|\int_{-\infty}^{\infty} 2\gamma\tau\exp(2\mathrm{i}\tau^3/3)\,d\tau\right|^2 = \pi^2(2/3)^{8/3}\Gamma^{-2}(4/3)\,\gamma^{2/3}. \qquad (9.106)$$

Analytical expressions are also obtained for a parabolic Coulomb trajectory for large and small Massey parameters [9.47]. This parameter for a trajectory with a scattering angle θ for the Coulomb potential $Z_A Z_B/R$ is

$$\gamma_c = BZ_A^3 Z_B^3[\cot(\theta/2)]^3\,[\cos(\theta/2)]^4/2\mu^3 v^7 \qquad (9.107)$$

with θ related to the impact parameter b by

$$b = Z_A Z_B \cot(\theta/2)/\mu v^2 . \tag{9.108}$$

The transition probability for the near-adiabatic case ($\gamma_c \gg 1$) is

$$\begin{aligned} \mathscr{P}_{12} &= 2\exp(-c_1\gamma_c), \qquad \gamma_c \gg 1, \\ c_1 &= 4\cos^{-6}(\theta/2)\,\{2/3 + (11/6)\sin^2(\theta/2) \\ &\quad + (1/2)(\pi+\theta)\,[3/2 + \sin^2(\theta/2)]\tan(\theta/2)\}, \end{aligned} \tag{9.109}$$

whereas for the sudden limit ($\gamma_c \ll 1$),

$$\begin{aligned} \mathscr{P}_{12} &= (\sin\theta + c_2\gamma_c^{1/3})^2, \qquad \gamma_c \ll 1, \\ c_2 &= \pi[(2/3)\cos(\theta/2)]^{4/3}(\cos\theta)/\Gamma(4/3). \end{aligned} \tag{9.110}$$

For the intermediate case ($\gamma_i \approx 1$), the transition probability is calculated either by solving the scattering equations numerically, or by using some unitarization procedure, or else by resorting to one of the interpolation expressions discussed in Sect. 7.3.3. In this connection we shall mention the Magnus approximation expansion (6.121) which was found [9.50] to give a satisfactory result when calculated to third order. For the semiempirical interpolation, alongside that based on (7.100) [9.51], a simple rule of thumb gives a rather accurate description of the approximation of $\mathscr{P}_{12}$ [9.47]:

$$\mathscr{P}_{12} = 2\exp(-c_1\gamma_c)\frac{(\sin\theta + c_2\gamma_c^{1/3})^2}{2 + c_2^2\gamma_c^{2/3}} . \tag{9.111}$$

This is also valid for a rectilinear trajectory if θ in c_1 and c_2 is taken to be zero, and γ_c is replaced by γ.

An interesting peculiarity of the probability for the $\Sigma - \Pi$ transitions in the united-atom limit is that the probability as a function of the scattering angle can exhibit two maxima – one corresponding to the scattering angle close to $\pi/2$ (when the initial Σ-function strongly overlaps the final Π-function after the trajectory turns through $\pi/2$) and the other to small scattering angles.

In applying the above equations to specific processes one must take into account the fact that sometimes the united-atom limit does not adequately describe the real situation [9.52]. In that case the transition probability includes more parameters [9.53, 54] and needs more complicated interpolation expressions for P [9.55]. Moreover, for heavy elements (large Z_A, Z_B) the spin-orbit splitting of the united-atom P state must be taken into account [9.56].

10. Multistate Models of Nonadiabatic Coupling

Some multichannel scattering problems of general interest will be discussed in this chapter. Strong coupling between several molecular states in the same region of R variation is common to all these problems.

Section 10.1 is concerned with transitions between states of degenerate atomic terms resulting in depolarization, resonant excitation transfer, or mixing of different L, M_L states in a hydrogen-like atom. Section 10.2 deals with the general approach to the calculation of transition probabilities between close-lying atomic states under conditions when correspondence principle arguments and classical perturbation theory are valid. Finally, the generalization of the linear model is considered in Sect. 10.3. This refers to transitions between several states perturbed by one state, to transition from one discrete state to a continuum, and to nonadiabatic transitions between two states coupled to continua.

10.1 Transitions Between Degenerate States

The transitions between degenerate states of an atom are a particular problem. The most important specific feature of this problem is that the main contribution to integral cross sections comes from large impact parameters where the interaction is of a rather general form (Sects. 3.2.2 and 4.2) and the atomic functions are only slightly disturbed. This makes the scattering equations for large interatomic distances relatively simple. As a rule, these equations cannot be integrated analytically, but nevertheless the dependence of the cross section on interaction parameters, velocity, etc., can be found explicitly, and only constant factors are obtained as a result of numerical integration.

10.1.1 Collisional Depolarization of an Isolated Atomic State

Consider the depolarizing collisions

$$A(j, m) + B(^1S_0) \rightarrow A(j, m') + B(^1S_0) \tag{10.1}$$

in the heat bath of atoms B and assume an isotropic distribution of relative velocities. Here m, m' are projections of j onto the z axis of some space-fixed reference frame. The depolarization cross sections $\sigma_{jm';jm}$ or the multipole relaxation cross sections σ_{χ}^{jj} for this process were defined in Sect. 2.4 through the scattering matrix $S^J_{jv';jv}$. To formulate the scattering equations the atomic wave functions $|jm\rangle$ are assumed to be only slightly disturbed when forming molecular states $|j\tilde{\Omega}\rangle$, although the Hund couplings d or e may change during collisions to couplings b or c (Sect. 3.3.2). In the first case, the spin-orbit coupling is neglected, and in the second it is assumed to be so strong that the fine-structure splitting greatly exceeds the splitting of the molecular terms emerging from the given atomic state. Now, the condition for slight admixing of other atomic states into state $|jm\rangle$ means that the matrix elements of the Coriolis coupling can be calculated using atomic functions quantized on the molecular axis. The interaction will be either of the Coulomb multipole type (if B is an ion) or of a dispersion type (if B is neutral), i.e., it may be either first or second order (Sect. 4.2). Let $U_{j\Omega}(R)$ be represented by

$$U_{j\Omega} = C g(j\Omega)\, R^{-s}, \tag{10.2}$$

where the dimensionless coefficients $g(j\Omega)$ depend on the type of interaction.

The molecular-basis scattering equations in the impact-parameter approximation are given by (6.94) which should now be integrated along the entire trajectory. Identifying t_{M} in (6.94) with the time corresponding to the turning point $R^{\mathrm{t}} = b$ and introducing the rotation angle $\psi' = \arctan[v(t - t_{\mathrm{M}})/b]$ instead of the time variable t, we get

$$\begin{aligned} \mathrm{i}\, dc_{j\tilde{\Omega}'}/d\psi' &= - \sum_{\tilde{\Omega}'' = \tilde{\Omega}' \pm 1} (r^{j}_{\tilde{\Omega}'\tilde{\Omega}''}) \exp\left(\mathrm{i}\, B_{\Omega'\Omega''} \int_{-\pi/2}^{\psi'} \sin^{s-2}\psi\, d\psi\right) c_{j\tilde{\Omega}''}, \\ B_{\Omega'\Omega''} &= [g(j\Omega') - g(j\Omega'')]\, C b^{1-s}/v. \end{aligned} \tag{10.3}$$

Equation (10.3) should be integrated over the interval $-\pi/2 \leqq \psi' \leqq \pi/2$. It contains one dynamic parameter $B(b) = C v^{-1} b^{1-s}$ and numerical coefficients $g(j\Omega)$ depending on the type of interaction.

Equation (10.3) cannot be solved analytically except in the case $s = 2$. This value of s corresponds to the interaction of an ion with an atom possessing a dipole moment. As nonzero matrix elements of the dipole moment exist only between states of different parity, which is not the case for an isolated atomic state $|j\rangle$; this will not be considered here. The most frequent cases are $s = 3$ (interaction of an ion with an atom possessing a quadrupole moment) and $s = 6$ (dispersion interaction) (see, however, Sect. 10.1.3 for the specific case of hydrogen-like degeneration with $s = 2$).

Table 10.1 Average relaxation cross sections of atoms A with electronic angular momentum j in collisions with spherical symmetric atoms B for the dispersion interaction [in units of $\pi(|C_1 - C_0|/\bar{v})^{2/5}$]

Relaxation process and cross section	$j = 1$	$j = 3/2$	$j = 2$
Relaxation of orientation, $\langle\sigma_1^{jj}\rangle$	0.776	0.770	0.610
Relaxation of alignment, $\langle\sigma_2^{jj}\rangle$	0.689	0.957	0.833
Relaxation of octupole moment, $\langle\sigma_3^{jj}\rangle$	—	0.848	0.862
Relaxation of hexadecapole moment, $\langle\sigma_4^{jj}\rangle$	—	—	0.664

For $s = 6$, (10.3) has been solved numerically for $j = 1$ [10.1, 2], $j = 3/2$ [10.3, 4], $j = 2$ [10.5] in calculations of the integral cross sections for multipole relaxation of atoms A in the heat bath of atoms B. The relaxation rates Γ_χ^{jj} as defined by the general (2.172) can be written as $\Gamma_\chi^{jj} = N_B \bar{v} \langle\sigma_\chi^{jj}\rangle$, where $\langle\sigma_\chi^{jj}\rangle$ is the velocity-averaged cross section. As the velocity dependence of σ_χ^{jj} follows directly from dimensional analysis, $\sigma_\chi^{jj} \sim v^{-2/(s-1)}$, the average cross section $\langle\sigma_\chi^{jj}\rangle$ can be rather simply expressed via $\sigma_\chi^{jj}(\bar{v})$, viz., $\langle\sigma_\chi^{jj}\rangle = A_s \sigma_\chi^{jj}(\bar{v})$. Here A_s is a numerical coefficient (e.g., $A_6 = 0.98$) and $\bar{v}$ is the average thermal velocity $\bar{v} = (8kT/\pi\mu)^{1/2}$. The calculated values of $\langle\sigma_\chi^{jj}\rangle$ are given in Table 10.1 for the dispersion interaction. Note that for a diatom $A(j) + B$ this interaction involves two parameters, as seen from (4.54). In particular, the following parametrization can be adopted:

$$U_{j\Omega}(R) = [(j^2 - \Omega^2)\, C_1 - (j^2 - \Omega_0^2)\, C_0]/(j^2 - \Omega_0^2)\, R^6, \tag{10.4}$$

where Ω_0 is the smallest projection of j on the molecular axis: $\Omega_0 = 0$ for integer j and $\Omega_0 = 1/2$ for half-integer j. The cross sections $\langle\sigma_\chi^{jj}\rangle$ are expressed in units of $\pi(|C_1 - C_0|/\bar{v})^{2/5}$.

For qualitative interpretation of these numerical calculations we resort to the matching method (Sect. 6.2) and consider that at some critical distance $R = R_M$ the change in the coupling schemes d → b and e → c is instantaneous [10.6] [1]. This implies free atomic motion for $R > R_M$, and adiabatic diatomic motion for $R < R_M$. We take into account the fact that for the molecular term $\Omega = 1/2$, two degenerate components with $\tilde{\Omega} = 1/2$ and $\tilde{\Omega} = -1/2$ are coupled by the Coriolis interaction, and the coefficients $c_{j,\pm 1/2}$ satisfy the equations

$$i\, dc_{j,\pm 1/2}/d\psi' = -r^j_{1/2,-1/2}\, c_{j,\mp 1/2}\,. \tag{10.5}$$

The solution to these equations reads

$$c_{j\tilde{\Omega}}(\psi'_2) = \sum_{\tilde{\Omega}''} d^{1/2}_{\tilde{\Omega}'\tilde{\Omega}''}(r^j_{1/2,-1/2}\,\Delta\psi)\, c_{j\tilde{\Omega}''}(\psi'_1), \qquad \tilde{\Omega}', \tilde{\Omega}'' = \pm 1/2, \tag{10.6}$$

[1] The accuracy of this approximation can be inferred, e.g., from [10.7]

where $\Delta\psi = \psi'_2 - \psi'_1$ is the rotation angle of the molecular axis for the diatomic motion from R_1 to R_2.

Turn now to the scattering matrix (6.129). If the interaction at $R > R_M$ is neglected, the phases $\delta^>_{j\Omega}$ and $\delta^>_{j\nu}$ will be zero. Then the scattering matrix will be

$$S^J_{j\nu';j\nu} = \exp[i(\nu + \nu')\alpha_j] \sum_{\tilde{\Omega}',\tilde{\Omega}} \Delta^j_{\nu'\tilde{\Omega}'} \, {}^{M}S^J_{j\tilde{\Omega}';j\tilde{\Omega}} \Delta^j_{\nu\tilde{\Omega}}, \tag{10.7}$$

where ${}^{M}S$ is the molecular scattering matrix.

The molecular scattering matrix is diagonal for all terms except the term $\Omega = 1/2$. Using (10.6) we get

$$\begin{aligned} {}^{M}S^J_{j\tilde{\Omega}';j\tilde{\Omega}} &= \delta_{\tilde{\Omega}'\tilde{\Omega}} \exp(2i\delta^J_{\Omega}), \qquad \tilde{\Omega}',\tilde{\Omega} \neq \pm 1/2, \\ {}^{M}S^J_{j\tilde{\Omega}';j\tilde{\Omega}} &= \begin{pmatrix} \cos(r^j\Delta\psi) & i\sin(r^j\Delta\psi) \\ i\sin(r^j\Delta\psi) & \cos(r^j\Delta\psi) \end{pmatrix} \exp(2i\delta^J_{1/2}), \qquad \tilde{\Omega}',\tilde{\Omega} = \pm 1/2. \end{aligned} \tag{10.8}$$

Here $\Delta\psi$ is the rotation angle of the molecular axis during classical diatomic motion at $R < R_M$, r^j stands for $r^j_{1/2,-1/2}$, and the phases δ^J_{Ω} can be identified with the elastic scattering phase shifts for molecular terms $U_{j\Omega}(R)$.

Now, (10.7) can be rewritten in the form (2.58) where summation is carried out over Ω (which replaces Λ). This representation yields

$$S^J_{j\nu';j\nu} = \sum_{\Omega} s^{Jj}_{\nu'\nu\Omega} \exp(2i\delta^J_{\Omega}), \tag{10.9}$$

where the $s^{Jj}_{\nu'\nu\Omega}$, accounting for the symmetry properties of $\Delta^j_{\nu\tilde{\Omega}}$, are

$$\begin{aligned} s^{Jj}_{\nu'\nu\Omega} &= [1 + (-1)^{\nu'-\nu}] \exp[i(\nu' + \nu)\alpha_j] \\ &\quad \times \exp[-i(-1)^{j+\nu}\delta_{1/2,\Omega}\phi^J_j] \, \Delta^j_{\nu'\Omega}\Delta^j_{\nu\Omega}(1 + \delta_{0\Omega})^{-1}. \end{aligned} \tag{10.10}$$

Here

$$\phi^J_j = 2\int_{R^j_{1/2}}^{R_M} \frac{r^j_{1/2,-1/2}(J + 1/2)(2\mu)^{-1/2}\,dR}{R^2[E - U_{1/2}(R) - (J + 1/2)^2/2\mu R^2]^{1/2}} \tag{10.11}$$

and α_j is defined by (6.102).

Equation (10.10), where $\delta_{1/2,\Omega}$ and $\delta_{0\Omega}$ are the δ-symbols, clearly displays the factorization of the S matrix according to the total parity (for quantum treatment) or to the symmetry character of the electronic wave function on reflection in the collision plane (for semiclassical treatment).

Note that (10.9) is valid for $J < kR_M$ when the phase differences are large. For $J > kR_M$ the matching method in its simple form is invalid, but the S matrix can be calculated to first order with respect to the interatomic interaction, assuming the pure coupling case d or e.

For the calculation of integral cross sections, one is interested in the S matrix with J ranging from 0 to $J_M = kR_M$. The probability of multipole relaxation $P_\chi^{jj}(J)$ is obtained by substitution of the S matrix from (10.9) into (2.186), and the average transition probability $\bar{P}_\chi^{jj}$ is obtained directly from (2.186) using (2.188).

As an illustration consider the case $j = 1$ in more detail. There are two relaxation cross sections: σ_1^{11} (for orientation) and σ_2^{11} (for alignment). The corresponding transition probabilities $\bar{P}_\chi^{11}$ are [10.8]

$$\bar{P}_1^{11}(b) = 1 + \tfrac{1}{3}\cos(2\alpha_1 + \eta_0), \tag{10.12a}$$

$$\bar{P}_2^{11}(b) = \tfrac{4}{5} + \tfrac{1}{5}\cos(2\alpha_1 + \eta_0) - \tfrac{1}{10}\cos(4\alpha_1 + 2\eta_0) - \tfrac{1}{10}\cos(4\alpha_1 + 2\eta_1). \tag{10.12b}$$

Note that since the matching approximation does not use the straight-path approach, both $\bar{P}_1^{11}$ and $\bar{P}_2^{11}$ also depend, together with $\alpha_1(b)$, on the deflection functions $\eta_0(b)$ and $\eta_1(b)$ in the field of the adiabatic potentials $U_{1,0}$ and $U_{1,1}$ (referred to below as U_0 and U_1).

The matching radius R_M for $s = 6$ (more generally for $s > 3$) is defined by the condition that the adiabatic half-splitting of terms U_0 and U_1 be equal to the matrix element of the Coriolis coupling [10.6]:

$$|U_1(R_M) - U_0(R_M)| = v/R_M. \tag{10.13}$$

To interpret the numerical results obtained from the scattering equations (10.3), assume a straight-line trajectory ($\eta_0 = \eta_1 = 0$) in (10.12a, b). With $\alpha_1(b) = \arcsin(b/R_M)$ the calculation yields

$$\sigma_1^{11} = 2\pi \int_0^{R_M} \bar{P}_1^{11}(b)\, b\, db = \pi R_M^2, \tag{10.14a}$$

$$\sigma_2^{11} = 2\pi \int_0^{R_M} \bar{P}_2^{11}(b)\, b\, db = \tfrac{13}{15}\pi R_M^2. \tag{10.14b}$$

A comparison of the results obtained by the matching method and by numerical solution of the impact parameter scattering equation is given in Table 10.2. The results of the first-order Magnus approximation are also given (Sect. 6.3.2). The matching method is seen to provide more accurate results compared to those obtained by the FOMA, as the former partially accounts for time-ordering of the general solution in the exponential form.

Equations (10.12a, b) also show under what conditions the adiabatic approximation is valid during the entire collision, and how the trajectory deflection affects the transition probabilities.

Table 10.2 Comparison of relaxation cross sections for orientation and alignment for the state $j = 1$ calculated in impact parameter approximation [in units of $\pi(|C_1 - C_0|/v)^{2/5}$]

Cross sections	Numerical integration of coupled equations	Matching approach	First order Magnus approximation
σ_1^{11}	0.785	0.77	0.97
σ_2^{11}	0.705	0.67	0.58
$\sigma_1^{11}/\sigma_2^{11}$	1.11	1.15	1.67

The adiabatic approximation is valid when there is no appreciable transformation of the basis, i.e., when $\alpha_1(b) \ll 1$. Since R_M is roughly defined by $B_{10}(R_M) \simeq 1$, the condition $\alpha_1(b) \sim b/R_M \ll 1$ becomes

$$|B_{10}(b)|^{1/(s-1)} \gg 1, \tag{10.15}$$

which is consistent with the adiabatic Born-Fock criterion for degenerate states [10.9]. The adiabatic transition probabilities for a rectilinear trajectory are ${}^{(a)}\bar{P}_1^{11} = 4/3$ and ${}^{(a)}\bar{P}_2^{11} = 4/5$, whereas $\bar{P}_1^{11}$ and $\bar{P}_2^{11}$ with nonzero $\alpha_1(b)$ vary from the adiabatic values for $b = 0$, to $\bar{P}_1^{11} = 2/3$ and $\bar{P}_2^{11} = 2/5$ at the boundary $b = R_M$.

To illustrate the effect of the trajectory deflection consider collisions with small impact parameters ($\alpha_1 \ll 1$) for two limiting scattering cases: the forward scattering ($\eta_0, \eta_1 \ll 1$) and the backward scattering ($\eta_0, \eta_1 \simeq \pi$). From (10.12) in the former case we get $\bar{P}_1^{11} = 4/3$, $\bar{P}_2^{11} = 4/5$, whereas for the latter case $\bar{P}_1^{11} = 2/3$, $\bar{P}_2^{11} = 2/5$. The depolarization probability is seen to be greater for forward scattering, which is due to additional dephasing between the degenerate components of the $\Omega = 1$ state with rotation of the molecular axis.

Consider now the total integral cross sections for a state $j = 1$. Unlike the depolarization cross sections, the total cross sections depend on U_0 and U_1 as such, rather than on the difference potential $U_0 - U_1$. If the difference $U_0 - U_1$ is small, $|U_0 - U_1| \ll |U_0|, |U_1|$, the total scattering cross sections can markedly exceed the depolarization cross sections

For a quantization axis directed along the vector $\boldsymbol{k}$, the total cross section $\sigma_{j\bar{m}}^{\text{tot}}$ describing attenuation of the state-selected atomic beam is defined by (2.31). In this equation we specify $a = j$ and rewrite it in terms of $T^J_{j'v';jv}$ from (2.48). Realizing that the value of the integral cross section is controlled by large relative angular momenta where the interaction energy is small compared to the kinetic energy, we use the impact-parameter approximation. The matrix $S^J_{j'v';jv}$ is replaced by ${}^{\text{sc}}S_{j'v';jv}(b)$ using (6.69) and then transformed to ${}^{\text{sc}}S^{\text{SF}}_{j'\bar{m}';j\bar{m}}$ according to Table 6.3. The Clebsch-Gordan coefficients in (2.31) are then expressed in the asymptotic form, which results in cancelling out all the matrices Δ^j_{mv}. This ultimately gives

$$\sigma_{j\bar{m}}^{\text{tot}} = 4\pi \int_0^\infty b\,db \left[1 - \text{Re}\left\{\exp\left(-\text{i}\int_{-\infty}^{\infty} {}_{\text{st}}\langle j\bar{m}|\, V_{\text{AB}}\, |j\bar{m}\rangle_{\text{st}}\, dt\right) {}^{\text{sc}}S^{\text{SF}}_{j'\bar{m}';j\bar{m}}\right\}\right], \tag{10.16}$$

where j is taken to be equal to 1, and $\bar{m}$ to 0 or ± 1.

To calculate the relevant phase factors and ${}^{\text{sc}}S^{\text{SF}}_{j'\bar{m}';j\bar{m}}$ it is convenient to parametrize the adiabatic molecular potentials for the pure coupling case b (spinless P-state) or e (isolated fine structure state $j = 1$) as

$$U_0(R) = CR^{-6}(1 + 2q/5), \qquad U_1(R) = CR^{-6}(1 - q/5). \tag{10.17}$$

The matrix elements ${}_{\text{st}}\langle j\bar{m}'|\, V_{\text{AB}}\, |j\bar{m}\rangle_{\text{st}}$ entering the exponents in (10.16) and defining the time-dependent equations for the calculation of ${}^{\text{sc}}S^{\text{SF}}_{j\bar{m}';j\bar{m}}$ can be expressed via $U_0(R)$, $U_1(R)$ and ψ' using Table 6.2, footnotes b and c. This gives

$$\begin{aligned} {}_{\text{st}}\langle 10|\, V_{\text{AB}}\, |10\rangle_{\text{st}} &= CR^{-6}\left[1 + \frac{q}{10}(1 - 3\cos 2\psi')\right], \\ {}_{\text{st}}\langle 1 \pm 1|\, V_{\text{AB}}\, |1 \pm 1\rangle_{\text{st}} &= CR^{-6}\left[1 - \frac{q}{20}(1 - 3\cos 2\psi')\right], \end{aligned} \tag{10.18}$$

where the time dependence of R and ψ' is of course controlled by motion along the straight-line trajectory.

Now note that the right hand side of the semiclassical scattering equation for ${}^{\text{sc}}S^{\text{SF}}$ is proportional to the parameter q. Hence these equations will give the second- and higher-order corrections to the zero-order values of the diagonal elements ${}^{\text{sc}}S^{\text{SF}}_{j\bar{m};j\bar{m}} = 1$. This means that to first order in q, the diagonal matrix elements in (10.16) can be assumed to be equal to unity. In this way, using (10.18) to calculate the phase factors in (10.16), we get

$$\begin{aligned} \sigma_{10}^{\text{tot}} &= 4\pi \int_0^\infty b\,db\{1 - \cos[2\delta(b)(1 - q/10)]\}, \\ \sigma_{1\pm1}^{\text{tot}} &= 4\pi \int_0^\infty b\,db\{1 - \cos[2\delta(b)(1 + q/20)]\}, \end{aligned} \tag{10.19}$$

where $\delta(b)$ is the high energy JWKB scattering phase shift for a central potential $U(R) = CR^{-6}$. Using the Landau-Lifshitz-Schiff approximation to solve the integrals in (10.19) (Sect. 5.4) and retaining only the first-order correction in q, we get

$$\begin{aligned} \sigma_{10}^{\text{tot}} &= \sigma^{\text{LLS}}(1 - q/10)^{2/5} \simeq \sigma^{\text{LLS}}(1 - 0.04q), \\ \sigma_{1\pm1}^{\text{tot}} &= \sigma^{\text{LLS}}(1 + q/20)^{2/5} \simeq \sigma^{\text{LLS}}(1 + 0.02q), \qquad \sigma^{\text{LLS}} = 8.083\,|C/v|^{2/5}. \end{aligned} \tag{10.20}$$

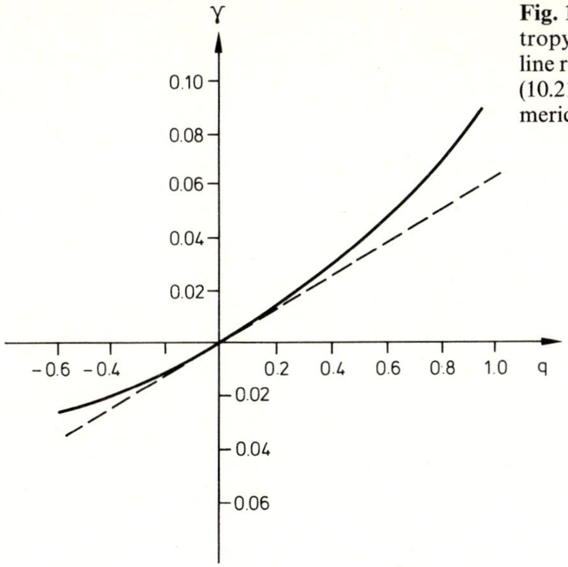

Fig. 10.1. Relation between anisotropy parameters γ and q. The dashed line refers to the perturbation approach (10.21), the full line represents the numerical impact parameter result

Thus the dimensionless anisotropy parameter γ characterizing the difference between the total cross sections for scattering of polarized atoms is

$$\gamma = (\sigma_{1\pm 1}^{\text{tot}} - \sigma_{10}^{\text{tot}})/\sigma_{10}^{\text{tot}} = 0.06\,q. \tag{10.21}$$

Figure 10.1 demonstrates how this first-order result is substantiated by the impact-parameter calculations [10.10]. Note that for small q the total cross sections are linear in q, and the cross sections for depolarizations are proportional to $q^{2/5}$.

Turn now to the depolarization of an atom in the state $j = 1/2$. The scattering matrix for this case can be found from (10.5) which can be solved not only for constant $r_{1/2-1/2}^{1/2}$, but also for any function $r_{1/2-1/2}^{1/2}(\psi')$. Moreover, the matching approximation given by (10.9–11) provides the exact result.

The scattering matrix is of the form

$$S_{1/2\,v';\,1/2\,v}^{J} = \exp(2\mathrm{i}\delta_{1/2}^{J}) \times \begin{pmatrix} \exp\left(\dfrac{\mathrm{i}}{2}\eta_{1/2}^{J} - \mathrm{i}\beta^{J} - \mathrm{i}\pi/2\right) & 0 \\ 0 & \exp\left(\dfrac{\mathrm{i}}{2}\eta_{1/2}^{J} + \mathrm{i}\beta^{J} + \mathrm{i}\pi/2\right) \end{pmatrix} \tag{10.22}$$

where $\eta_{1/2}^{J}$ is the deflection function for the adiabatic term $U_{1/2}(R)$ and β^{J}

(independent now of R_m) is

$$\beta^J = \phi^J_{1/2} + \alpha^J_{1/2} - \tfrac{1}{2}(\pi - \eta^J_{1/2})$$
$$= \int\limits_{R^t_{1/2}}^{\infty} \frac{[2r^{1/2}_{1/2\,-1/2}(R) - 1]\,(J + 1/2)\,dR}{R^2(2\mu)^{1/2}[E - U_{1/2}(R) - (J + 1/2)^2/2\mu R^2]^{1/2}}. \tag{10.23}$$

It is seen from (10.23) that if $r^{1/2}_{1/2\,-1/2}$ is equal to its value for a free atom, i.e., $r^{1/2}_{1/2\,-1/2} = 1/2$, β^J is zero. Identifying then $2\delta^J_{1/2} \pm \eta^J_{1/2}/2$ with $2\delta^l_{1/2}$, $l = J \pm 1/2$ and changing from the indexes v', v to l', l the S matrix (10.22) takes the form

$$S^J_{jl';jl} = \delta_{l'l} \exp(2\mathrm{i}\delta^l_{1/2}), \qquad l = J \pm 1/2. \tag{10.24}$$

This means, of course, pure elastic scattering and no depolarization. For $\beta^J \neq 0$, we get from (10.22, 2.186) an expression for the probability of orientation relaxation:

$$P_1^{1/2\,1/2} = (4/3)\sin^2\beta(b). \tag{10.25}$$

The calculation of the differential cross section $q_{1/2\bar{m};1/2-\bar{m}}$ for scattering with angular momentum reorientation is also simple for a one branch function $b = b(\theta)$:

$$q_{1/2\,1/2;1/2\,-1/2} = \sin^2\beta(b)\,q_{1/2}(\theta), \qquad b = b(\theta), \tag{10.26}$$

where $q_{1/2}(\theta)$ is the differential cross section for elastic scattering in a potential $U_{1/2}(R)$. With β^J small, the transition probability $P_1^{1/2\,1/2}$ and the cross section $\sigma_{1/2\,-1/2}$ are also small. This is the case for collisional depolarization of an atom in an isolated ^2S state, where $r^{1/2}_{1/2\,-1/2}$ differs only slightly from its asymptotic value 1/2 during the collision [10.11]. On the other hand, if $r^{1/2}_{1/2\,-1/2}$ becomes zero for all $R < R_0$ and the change from the asymptotic value 1/2 occurs suddenly, (10.25) yields

$$P_1^{1/2\,1/2} = (4/3)\cos^2(\alpha_{1/2} + \tfrac{1}{2}\eta_{1/2}), \tag{10.27}$$

where $\alpha_{1/2} = \arcsin(b/R_0)$. This is the case for depolarization of the adiabatically isolated state $^2P_{1/2}$ which will be discussed in detail in Chap. 11.

10.1.2 Resonant Excitation Transfer by Dipole-Dipole Interaction

Another example of transitions between degenerate states is the resonant excitation transfer. Of all types of interactions between neutrals, the most long-range one is the dipole-dipole interaction which couples states differing in parity and in j by 0 or ± 1. This interaction is given by (4.35).

Consider the process

$$\mathrm{A}'(j_1) + \mathrm{A}''(j_2) \rightarrow \mathrm{A}'(j_2) + \mathrm{A}''(j_1), \tag{10.28}$$

where the prime and the double prime denote different (but energetically identical) atoms, and j_1 and j_2 the excited and unexcited states coupled by the dipole-dipole interaction. The semiclassical wave function of a diatom A′A″ in the coupled representation of the diabatic basis is

$$\Psi = \sum_{j'v'} c^{(\mathrm{d})}_{j_1 j_2 j' v'}(t)\, |j_1 j_2 j' v'\rangle + \sum_{j'v''} c^{(\mathrm{d})}_{j_2 j_1 j' v'}(t)\, |j_2 j_1 j' v'\rangle, \tag{10.29}$$

where the diabatic functions $|j_a j_b j' v'\rangle$ are constructed from the atomic functions $|j_1 v_1\rangle$, $|j_2 v_2\rangle$ according to (2.131):

$$\begin{aligned} |j_1 j_2 j' v'\rangle &= \sum_{v_1, v_2} \begin{bmatrix} j_1 & j_2 & j' \\ v_1 & v_2 & v' \end{bmatrix} |j_1 v_1\rangle_{\mathrm{A}'} |j_2 v_2\rangle_{\mathrm{A}''}, \\ |j_2 j_1 j' v'\rangle &= \sum_{v_1, v_2} \begin{bmatrix} j_1 & j_2 & j' \\ v_1 & v_2 & v' \end{bmatrix} |j_1 v_1\rangle_{\mathrm{A}''} |j_2 v_2\rangle_{\mathrm{A}'}. \end{aligned} \tag{10.30}$$

The scattering equations for $c^{(\mathrm{d})}_{j_a j_b j' v'}$ for the dipole-dipole interaction are of the form

$$\begin{aligned} \mathrm{i}\dot{c}^{(\mathrm{d})}_{j_1 j_2 j' v'} = &- 12(6\pi/5)^{1/2}(\gamma/R^3)(-1)^{j_1 - j_2} \\ &\times \sum_{q j'' v''} (-1)^{j'' + v''}\, Y_2^q\left(\frac{\pi}{2}, \psi'\right) (2j' + 1)^{1/2} (2j'' + 1)^{1/2} \\ &\times \begin{Bmatrix} j' & j'' & 2 \\ j_1 & j_2 & 1 \\ j_2 & j_1 & 1 \end{Bmatrix} \begin{bmatrix} j'' & j' & 2 \\ v'' & v' & q \end{bmatrix} c^{(\mathrm{d})}_{j_2 j_1 j'' v''}, \end{aligned} \tag{10.31}$$

where γ is expressed via the oscillator strength f_{12} in the absorption with transition frequency ω_{12},

$$\gamma = (2j_1 + 1) f_{12}/4\omega_{12}. \tag{10.32}$$

Equations (10.31) are already block factorized according to the different symmetry of Ψ upon reflection in the collision plane. Each block can be further factorized by introducing a new coefficient with a certain electronic parity w:

$$c^{(\mathrm{d})}_{w, j' v'} = \sqrt{1/2}\,(c^{(\mathrm{d})}_{j_1 j_2 j' v'} + w c^{(\mathrm{d})}_{j_2 j_1 j' v'}). \tag{10.33}$$

For a rectilinear trajectory the time variable t is conveniently expressed via the rotation angle ψ' [compare with (10.3)]. In particular, for the $0 - 1$

transition, the scattering equations are

$$i\,dc^{(d)}_{w,10}/d\psi' = 2wB\cos\psi'\,c^{(d)}_{w,10}, \tag{10.34a}$$

$$i\,dc^{(d)}_{w,11}/d\psi' = -wB\cos\psi'[c^{(d)}_{w,11} - 3\exp(-2i\psi')\,c^{(d)}_{w,1-1}], \tag{10.34b}$$

$$i\,dc^{(d)}_{w,1-1}/d\psi' = -wB\cos\psi'[c^{(d)}_{w,1-1} - 3\exp(2i\psi')\,c^{(d)}_{w,11}],\ B = \gamma/vb^2.$$

Equations (10.34) have been solved numerically by several authors [10.12–14] in calculations of the excitation transfer cross section $\sigma_0^{A'A''}$. Solutions to the general equation (10.31) have been obtained [10.15] for different values of j_1 and j_2 and used for the calculation of multipole relaxation cross sections $\sigma_\chi^{A'A'}$ and multipole transfer cross sections $\sigma_\chi^{A'A''}$. All these cross sections are of the form

$$\sigma_\chi^{A'A''} = g_\chi^{A'A''}\,2\pi\gamma/v, \qquad \sigma_\chi^{A'A'} = g_\chi^{A'A'}\,2\pi\gamma/v. \tag{10.35}$$

The numerical coefficients $g^{A'A'('')}$ are given in Table 10.3. For qualitative interpretation of these results turn again to (10.34) and consider the population transfer. A solution to the uncoupled equation (10.34a) is obtained analytically. The two equations (10.34b) can be solved approximately in an appropriate basis. It turns out [10.16] that for the dipole-dipole interaction a substantial phase difference between even and odd terms builds up even at interatomic distances where the Hund coupling cases d or e prevail. Thus, the factor $\exp(\pm 2i\psi')$ in the rhs of (10.34b) rapidly oscillates and

Table 10.3 Coefficients g for cross sections in (10.35) [10.15]

Process	$j_1 = 1$, $j_2 = 0$	$j_1 = 1/2$, $j_2 = 1/2$	$j_1 = 3/2$, $j_2 = 1/2$	$j_1 = 1$, $j_2 = 1$
Population relaxation of A′, $g_0^{A'A'}$	2.242	1.681	1.279	1.267
Orientation relaxation of A', $g_1^{A'A'}$	3.014	2.711	1.643	1.673
Alignment relaxation of A′, $g_2^{A'A'}$	3.134	–	1.793	1.688
Octupole moment relaxation of A′, $g_3^{A'A'}$	–	–	1.925	–
Population transfer A′→A″, $g_0^{A'A''}$	−2.242	−1.681	−1.279	−1.267
Orientation transfer A′→A″, $g_1^{A'A''}$	0.607	0.304	0.308	0.171
Alignment transfer A′→A″, $g_2^{A'A''}$	−0.132	–	−0.017	−0.028
Octupole moment transfer A′→A″, $g_3^{A'A''}$	–	–	0.011	–

therefore decouples the equations in the diabatic basis. In this approximation the problem reduces to solving six uncoupled equations, two (even and odd) in each channel $jv \to jv$ ($j = 1, v = -1, 0, 1$). The transition probabilities are given by simple two-state resonance equations, viz.,

$$\begin{aligned} P_{10,10} &= \sin^2 \int_{-\pi/2}^{\pi/2} 2B \cos\psi' d\psi' = \sin^2(4\gamma/vb^2), \\ P_{11,11} &= P_{1-1,1-1} = \sin^2 \int_{-\pi/2}^{\pi/2} B \cos\psi' d\psi' = \sin^2(2\gamma/vb^2). \end{aligned} \tag{10.36}$$

Using the Firsov criterion (9.56) to choose b_{jv}^*, the excitation transfer cross section will be

$$\sigma_0^{A'A''} = \tfrac{1}{3}[\pi(b_{10}^*)^2 + \pi(b_{11}^*)^2 + \pi(b_{1-1}^*)^2] = (2\pi/3)\, 2\pi\gamma/v. \tag{10.37}$$

The numerical coefficient in brackets $(2\pi/3) \simeq 2.09$ is seen to be close to the exact value $g_0^{A'A''} = 2.242$ (Table 10.3). The excitation transfer cross section can also be calculated analytically by using the decoupling of scattering equations in the atomic basis for the general case $j_1, j_2 \gg 1$ [10.17].

In conclusion to this section note that when atoms A′ and A″ are not identical, i.e., the energy release $\Delta\varepsilon$ is not zero, the resonance approximation is valid only for very small $\Delta\varepsilon$ satisfying the Massey criterion $\Delta\varepsilon b^*/3v \leqq 1$. Since b^* is of the order of $(\gamma/v)^{1/2}$, $\Delta\varepsilon$ must satisfy the condition

$$\Delta\varepsilon \leqq v^{3/2}\gamma^{-1/2}. \tag{10.38}$$

For thermal energies ($E \approx 10^{-3}$), ordinary atoms ($\mu \approx 10^4$), and high oscillator strength ($f_{12} \sim 1$), (10.38) implies that $\Delta\varepsilon$ does not exceed several cm^{-1}. For quasi-resonant collisions the relaxation cross sections for $S - P$ transitions have been calculated in [10.18] as a function of $\Delta\varepsilon$.

10.1.3 Transitions Between Degenerate Hydrogen States in Collisions with Ions

An excited nonrelativistic hydrogen atom with principal quantum number n possesses n^2 degenerate states with different angular momenta such that $0 \leqq L \leqq n - 1$. Since these degenerate states exhibit the linear Stark effect in an external electric field, the transitions between these states in distant collisions with ions B (charge Z) are induced predominantly by the charge – dipole interaction. Neglecting the coupling between states with different n and using the charge – dipole interaction, we get an analytically solvable multistate model referred to as the H-ion model [10.19, 20]. The applicability of this model to a real collisional process is restricted to intermediate impact parameters: at large impact parameters this model is affected by finite relativistic splittings of the hydrogen atom levels, and at small impact parameters by the contribution from higher-order multipoles. These restrictions are discussed in detail in [10.19, 20] (see also [10.21, 22]), and will not be considered here.

The interaction matrix in the radial scattering equations (2.85) for the H-ion model is

$$V_{L'v';L''v''} = -R^{-2}(Jv'/\mu)\,\delta_{v'v''} + ZR^{-2}r_{L'L''}\,\Delta^{1}_{v'-v'';0}(2L''+1)^{-1/2}\begin{bmatrix} L' & 1 & L'' \\ v' & v''-v' & v'' \end{bmatrix}, \quad (10.39)$$

where

$$r_{L'L''} = \int_0^\infty r\,\varphi_{nL'}(r)\,\varphi_{nL''}(r)\,r^2\,dr. \quad (10.40)$$

An exact analytical solution of the scattering equations is based here on the possibility of diagonalizing the matrix $V_{L'v';L''v''}$ by an R-independent transformation. Moreover, this transformation can be done explicitly for arbitrary n, due to the fact that $V_{L'v';L''v''}$ formally coincides with the perturbation matrix of a hydrogen atom in the constant perpendicular electric and magnetic fields. The first term in (10.39) corresponds to a magnetic field $\mathscr{H}$ with components $\mathscr{H}_x = \mathscr{H}_y = 0$, $\mathscr{H}_z = -2Jc/\mu R^2$ (c is the velocity of light), and the second term to the electric field $\mathscr{E}$ with components $\mathscr{E}_y = \mathscr{E}_z = 0$, $\mathscr{E}_x = -Z/R^2$. The problem of a hydrogen atom in perpendicular electric and magnetic fields has been solved in [10.23] with use of the additional dynamic symmetry of the Coulomb interaction. In this particular case the results obtained in [10.23] can be formulated as follows. The hydrogen atom states which diagonalize $V_{L'v';L''v''}$ are characterized by quantum numbers λ_1 and λ_2, which assume the values $-I$, $-I+1, \ldots I$, where $I = (n-1)/2$. These states $|\lambda_1\lambda_2\rangle$ are expressed through states $|Lv\rangle$:

$$|\lambda_1\lambda_2\rangle = \sum_{v,L} A_{Lv;\lambda_1\lambda_2}|Lv\rangle, \quad (10.41)$$

where

$$A_{Lv;\lambda_1\lambda_2} = \sum_{i_1=-I}^{I}\sum_{i_2=-I}^{I}\begin{bmatrix} I & I & L \\ i_1 & i_2 & v \end{bmatrix} d^{I}_{\lambda_1 i_1}(\beta)\,d^{I}_{\lambda_2 i_2}(-\beta) \quad (10.42)$$

and

$$\tan\beta = -3Zn\mu/2J, \qquad \pi/2 < \beta < \pi. \quad (10.43)$$

The corresponding eigenvalues $V_{\lambda_1\lambda_2}$ are

$$V_{\lambda_1\lambda_2} = R^{-2}f(J)(\lambda_1+\lambda_2), \quad (10.44)$$

where

$$f(J) = [(3Zn/2)^2 + (J/\mu)^2]^{1/2}. \tag{10.45}$$

The transformed wave functions $\chi_{\lambda_1' \lambda_2'; \lambda_1 \lambda_2}$, defined by the relations

$$\chi_{L'v';Lv} = \sum_{\lambda_1', \lambda_2'} \sum_{\lambda_1, \lambda_2} A_{L'v'; \lambda_1' \lambda_2'} A^*_{L'v; \lambda_1 \lambda_2} \chi_{\lambda_1' \lambda_2'; \lambda_1 \lambda_2} \tag{10.46}$$

satisfy the uncoupled equations with the usual boundary conditions

$$\chi_{\lambda_1' \lambda_2'; \lambda_1 \lambda_2} \xrightarrow[R \to \infty]{} (k')^{-1/2} \left\{ \exp\left[-\mathrm{i}\left(kR - \frac{J\pi}{2} \right) \right] \delta_{\lambda_1' \lambda_2'} \delta_{\lambda_1 \lambda_2} - S^J_{\lambda_1' \lambda_2'; \lambda_1 \lambda_2} \right.$$
$$\left. \times \exp\left[\mathrm{i}\left(kR - \frac{J\pi}{2} \right) \right] \right\}. \tag{10.47}$$

The scattering matrix $S^J_{\lambda_1' \lambda_2'; \lambda_1 \lambda_2}$ is diagonal:

$$S^J_{\lambda_1' \lambda_2'; \lambda_1 \lambda_2} = \delta_{\lambda_1' \lambda_1} \delta_{\lambda_2' \lambda_2} \exp[2\mathrm{i}\delta_{\lambda_1 \lambda_2}(J)]. \tag{10.48}$$

Here the coefficients of the exponentials are the δ-symbols and the exponents (e.g., [10.24]),

$$\delta_{\lambda_1 \lambda_2}(J) = (\pi/2) \{(J + 1/2) - [(J + 1/2)^2 + 2\mu f(J)(\lambda_1 + \lambda_2)]^{1/2}\}, \tag{10.49}$$

are the phase shifts for elastic scattering in the potentials $V_{\lambda_1 \lambda_2}$. Note here that the H-ion model is mathematically correct only for J large enough to prevent collapse in the most attractive potential $V_{\lambda_1 \lambda_2}$ with $\lambda_1 = \lambda_2 = -(n-1)/2$. An explicit condition can be obtained here demanding that the corresponding phase shift is real:

$$(J + 1/2)^2 > 2\mu(n-1) f(J). \tag{10.50}$$

Taking into account (10.46) we get the following expression for the scattering matrix $S^J_{L'v';Lv}$ in the usual representation:

$$S^J_{L'v';Lv} = \sum_{\lambda_1 \lambda_2} A_{L'v'; \lambda_1 \lambda_2} A^*_{Lv; \lambda_1 \lambda_2} \exp[2\mathrm{i}\delta_{\lambda_1 \lambda_2}(J)]. \tag{10.51}$$

This expression solves the dynamic problem and can be readily used for calculation of the differential cross sections (the integral cross sections in the H-ion model diverge). Since (10.51) is similar to (2.58), the approach to scattering through classical angles formulated in Sect. 2.1.2 is straightforward here. This is illustrated by scattering through small classical angles in the case $n = 2$. In this case small classical angles correspond to J given by

$$3\pi\mu Z \gg J \gg (3\pi\mu Z)^{1/2}. \tag{10.52}$$

The quantity $3\pi\mu Z$ is of the order of $10^4 - 10^5$ and therefore the range of small classical angles is rather wide. When the inequalities (10.52) hold,

$$\delta_{\lambda_1\lambda_2}(J) \simeq -\pi\mu(\lambda_1+\lambda_2)\,3Z/2J, \tag{10.53 a}$$

$$\eta_{\lambda_1\lambda_2}(J) = 2d\delta_{\lambda_1\lambda_2}(J)/dJ \simeq \pi\mu(\lambda_1+\lambda_2)\,3Z/J^2. \tag{10.53 b}$$

In the case under consideration, $I = 1/2$ and λ_1, λ_2 can assume the values $\pm 1/2$. Since $\delta_{1/2\,-1/2} = \delta_{-1/2\,1/2} = 0$, the corresponding terms in (10.51) will not contribute to scattering through classical angles. Introducing the notation

$$\begin{aligned} s^{+}_{L'\nu';L\nu}(J) &= A_{L'\nu';1/2\,1/2}A^{*}_{L\nu;1/2\,1/2}, \qquad \delta^{+}(J) = \delta_{1/2\,1/2}(J), \\ s^{-}_{L'\nu';L\nu}(J) &= A_{L'\nu';-1/2\,-1/2}A^{*}_{L\nu;-1/2\,-1/2}, \qquad \delta^{-}(J) = \delta_{-1/2\,-1/2}(J), \end{aligned} \tag{10.54}$$

the matrix will be of the form

$$S^{J}_{L'\nu';L\nu} = s^{+}_{L'\nu';L\nu}(J)\exp[2i\delta^{+}(J)] + s^{-}_{L'\nu';L\nu}(J)\exp[2i\delta^{-}(J)]. \tag{10.55}$$

Thus the S matrix in (10.55) is of the form (2.58) with Λ assuming two values $+$ and $-$. Therefore the general approach of Chap. 2, including (2.59 a, b) can be used to calculate the cross section. For small angle scattering (2.59) can have only one root with $M = 0$ and if it is assumed for definiteness that $Z > 0$, it will be seen from (10.53 b) that the plus sign must be taken in (2.59 a) for $\Lambda = +$, and the minus sign for $\Lambda = -$. Therefore in (2.61–63), $w^{+} = 1$, $w^{-} = -1$. At the same time it is seen from (10.53 b) that $v^{+} = -1$ and $v^{-} = 1$, and it follows from (10.53 b) that

$$J^{+}(\theta) = J^{-}(\theta) = (3Z\pi\mu/\theta)^{1/2}, \tag{10.56}$$

$${}^{\mathrm{cl}}q^{+}(\theta) = {}^{\mathrm{cl}}q^{-}(\theta) = 3Z\pi/4E\theta^3. \tag{10.57}$$

Thus all the quantities in (2.62, 64) are defined, and simple algebra gives the result found in [10.20]

$$\begin{aligned} q_{2p;2s}(\theta) &= \sum_{\overline{m}} |f^{+}_{1\overline{m};00} + f^{-}_{1\overline{m};00}|^2 = (3Z\pi/4E\theta^3)\sin^2\beta(1+\cos^2\beta) \\ &\quad \times \sin^2(12\pi Z\mu\theta)^{1/2}, \end{aligned} \tag{10.58}$$

$$\begin{aligned} q_{2s;2s}(\theta) &= |f^{+}_{00;00} + f^{-}_{00;00}|^2 \\ &= (3Z\pi/4E\theta^3)\sin^4\beta\cos^2(12\pi Z\mu\theta)^{1/2}, \end{aligned} \tag{10.59}$$

where

$$\tan\beta = -(3Z\mu\theta/\pi)^{1/2}. \tag{10.60}$$

10.2 Transitions Between Highly Excited States

The Magnus series representations (6.121, 122) of the evolution operator and the scattering matrix truncate if the matrices $\mathscr{V}(t)$ commute at different times. This is always the case if the matrix elements $\mathscr{V}_{a'a}$ depend on the difference $\Delta a = a' - a$ only, rather than on a and a' separately (any two matrices possessing this property will commute). On the other hand, the same property of the matrix elements of a wide class of operators is known to be approximately true for states with large quantum numbers a', a. Namely, in the limit $a', a \gg \Delta a, 1$ the matrix elements $V_{a'a''}$ are strongly dependent on Δa but only weakly dependent on $\bar{a} = (a'' + a')/2$. Moreover, the correspondence principle states that for an operator V, depending on the coordinates q of the quantum system, the matrix element $V_{a'a}(t)$ in this limit becomes the Fourier component of the classical function $V[q(t)]$ [10.25]. For one coordinate $q(t)$,

$$V_{a'a} \simeq V_{\Delta a}(\bar{a}) = \frac{\bar{\omega}}{2\pi} \int_0^{\bar{T}} V[q(t')] \exp(\mathrm{i}\Delta a \bar{\omega} t')\, dt', \tag{10.61}$$

where $\bar{\omega}$ is the angular frequency of the classical motion in the state with an average quantum number $\bar{a}$, and $\bar{T} = 2\pi/\bar{\omega}$. The relation (10.61) is of crucial importance in obtaining an explicit expression for the semiclassical S matrix.

It is assumed in applying the correspondence principle to the scattering problem that the time-dependent Hamiltonian H is partitioned to a zero order Hamiltonian H_0 and to an interaction $V(t)$ which vanishes before $(t \to -\infty)$ and after $(t \to +\infty)$ a collision.

Consider first the case of a system with one coordinate q. For such a system,

$$H(q, t) = H_0(q) + V(q, t). \tag{10.62}$$

The spectrum of H_0 is assumed to be homogeneous in a region centered at $\bar{a}$ with frequency $\bar{\omega}$, and the matrix elements of V are assumed to possess the property expressed by (10.61). According to the basic formula (6.125) the problem of calculating the S matrix is now reduced to the matrix exponentiation

$$S = \exp(-\mathrm{i}A). \tag{10.63}$$

The exponent matrix A is defined by its matrix elements

$$A_{a'a} = \int_{-\infty}^{\infty} V_{a'a}(t) \exp[\mathrm{i}(a' - a)\,\omega t]\, dt. \tag{10.64}$$

The exponentiation (10.63) is easy, since the eigenvectors of the matrix A are known. Because of the translational invariance, these are

$$l_{a'}(\alpha) = (2\pi)^{-1/2} \exp(\mathrm{i}\alpha a'). \tag{10.65}$$

The eigenvalues $A(\alpha)$ of the matrix A are evaluated using the eigenvectors and the matrix elements $A_{a'a}$:

$$\begin{aligned} A(\alpha) &= \sum_{a',a} l^*_{a'}(\alpha)\, A_{a'a}\, l_a(\alpha) \\ &= \int_{-\infty}^{\infty} dt \sum_{a',a} (2\pi)^{-1} V_{a'a}(t) \exp[\mathrm{i}(a'-a)(\bar{\omega} t - \alpha)]. \end{aligned} \tag{10.66}$$

Now, as seen from (10.61), the integrand sum represents the classical interaction $V(q, t)$ at time t for the classical trajectory $q = \bar{q}(t, \alpha)$. This trajectory is defined by the Hamiltonian function $H_0(p, q)$, the conserved average action variable $\bar{a}$, and the initial phase α. The time integral from $V[\bar{q}(t, \alpha); t]$ taken along an undistorted trajectory relates the first-order correction $\mathscr{S}^{(1)}(\bar{a}, \alpha)$ (taken with a minus sign) to the classical action associated with the perturbation V. Thus

$$A(\alpha) = -\mathscr{S}^{(1)}(\bar{a}, \alpha) = \int_{-\infty}^{\infty} V[\bar{q}(t, \alpha); t]\, dt. \tag{10.67}$$

This finally yields the expression for the scattering matrix,

$$\begin{aligned} S_{a'a} &= (2\pi)^{-1} \int_0^{2\pi} d\alpha\, l_{a'}(\alpha) \exp[-\mathrm{i} A(\alpha)]\, l_a(\alpha) \\ &= (2\pi)^{-1} \int_0^{2\pi} d\alpha \exp[-\mathrm{i}(a'-a)\alpha + \mathrm{i}\mathscr{S}^{(1)}(\bar{a}, \alpha)]. \end{aligned} \tag{10.68}$$

This equation establishes the relation between the correspondence principle method and the classical S matrix method. Equation (10.68) is the initial angle representation of the classical S matrix calculated to the first-order approximation of classical mechanics [10.26, 27].

It is seen from (10.68) that besides the time dependence of the interaction required by the common-trajectory approach, one needs to choose a classical trajectory $q(t)$ which generates quasi-classical matrix elements, according to (10.61). This implies one more level of approximation compared to the semiclassical formulation of the scattering problem.

The specification of the initial action variable $\bar{a}$ leaves some freedom, and this can be used to make the correspondence-principle transition probabilities more accurate [10.28, 29]. With $\bar{a}$ fixed the S matrix (10.67) is unitary and symmetric. If $\bar{a}$ is taken to be some average value between a' and a, i.e., if $\bar{a}$ is made dependent, though weakly, on Δa, the individual transition probabilities can be made more exact, but then the S matrix is no longer unitary. Other corrections have to be introduced to rectify this deficiency.

For classically allowed transitions the stationary phase approximation can be used to calculate the integral (10.67). The possible stationary angles $\alpha^\chi = \alpha^\chi(a', a)$ are found from

$$a' - a = \left[\frac{\partial}{\partial\alpha}\mathscr{S}^{(1)}(\bar{a}, \alpha)\right]\bigg|_{\alpha=\alpha^\chi} \tag{10.69}$$

and the S matrix assumes the form of the primitive semiclassical approximation,

$$^{\text{ps}}S_{a'a} = \sum_\chi |B^\chi_{a'a}|^{1/2} \exp\left(\mathrm{i}A^\chi_{a'a} + \mathrm{i}\frac{\pi}{4}\operatorname{sign} B^\chi_{a'a}\right), \tag{10.70}$$

where $B^\chi_{a'a}$ and $A^\chi_{a'a}$ are calculated for the stationary trajectory $\bar{q}^\chi(t) = \bar{q}(t, \alpha^\chi)$:

$$\begin{aligned} B^\chi_{a'a} &= (\partial^2\mathscr{S}^{(1)}/\partial\alpha^2)|_{\alpha=\alpha^\chi(a'a)}, \\ A^\chi_{a'a} &= [(a' - a)\,\alpha - \mathscr{S}^{(1)}(\bar{a}, \alpha)]\,|_{\alpha=\alpha^\chi(a'a)}. \end{aligned} \tag{10.71}$$

The independent contribution of each stationary phase point α^χ to the scattering matrix implied by (10.70) is valid when the phase differences are large:

$$|A^\chi_{a'a} - A^{\chi'}_{a'a}| \gg 1. \tag{10.72}$$

Otherwise, different kinds of uniform approximations must be invoked, which is usually the case for the transitions $a \to a'$ with the final state a' close to the boundary of the classically allowed region.

In the uniform approximation, this boundary can be crossed and the transition probability calculated for transition which are classically not strictly forbidden. For transitions which are strictly forbidden classically, the stationary phase approach is again valid. In this case α^χ defined by (10.69) will generally be complex, and only one dominating term need be retained in the sum (10.70) corresponding to the phase $A^\chi_{a'a}$ with the smallest imaginary part.

All the above equations make a simple generalization possible for several degrees of freedom of the Hamiltonian H_0. The unperturbed classical trajectory of this system is then the N-dimensional vector which depends on time t, on the initial (or otherwise chosen) action variable $\bar{\boldsymbol{a}}$ $(\bar{a}_1, \bar{a}_2 \ldots \bar{a}_N)$ and on the corresponding initial angles $\boldsymbol{\alpha}$ $(\alpha_1, \alpha_2 \ldots \alpha_N)$. Direct generalization of (10.68) yields

$$S_{\boldsymbol{a}'\boldsymbol{a}} = (2\pi)^{-N} \int d^N\boldsymbol{\alpha} \exp[-\mathrm{i}(\boldsymbol{a}' - \boldsymbol{a})\,\boldsymbol{\alpha} + \mathrm{i}\mathscr{S}^{(1)}(\bar{\boldsymbol{a}}, \boldsymbol{\alpha})], \tag{10.73}$$

where

$$\mathscr{S}^{(1)}(\bar{\boldsymbol{a}}, \boldsymbol{\alpha}) = -\int_{-\infty}^{\infty} V[\bar{\boldsymbol{q}}(t, \boldsymbol{\alpha}), t]\, dt. \tag{10.74}$$

Again, the stationary phase approximation for (10.73) reads

$${}^{\mathrm{ps}}S_{\boldsymbol{a}'\boldsymbol{a}} = \sum_{\chi} \mathrm{Det}(B^{\chi}_{\boldsymbol{a}'\boldsymbol{a}}) \exp[\mathrm{i} A^{\chi}_{\boldsymbol{a}'\boldsymbol{a}} + \mathrm{i}\phi(B^{\chi}_{\boldsymbol{a}'\boldsymbol{a}})]. \tag{10.75}$$

Here $\phi(B^{\chi}_{\boldsymbol{a}'\boldsymbol{a}})$ is a certain phase which depends on the signs of the diagonal elements $(B_{\boldsymbol{a}'\boldsymbol{a}})_{nn}$ of the matrix $B^{\chi}_{\boldsymbol{a}'\boldsymbol{a}}$. Its matrix elements $(B^{\chi}_{\boldsymbol{a}'\boldsymbol{a}})_{n'n}$ and the exponent $A^{\chi}_{\boldsymbol{a}'\boldsymbol{a}}$ are calculated for the stationary trajectory $\boldsymbol{q}(t)$ as

$$\begin{aligned} (B^{\chi}_{\boldsymbol{a}'\boldsymbol{a}})_{n'n} &= (\partial^2 \mathscr{S}^{(1)}/\partial\alpha_{n'}\partial\alpha_n)|_{\boldsymbol{\alpha}=\boldsymbol{\alpha}^{\chi}(\boldsymbol{a}'\boldsymbol{a})}, \\ A^{\chi}_{\boldsymbol{a}'\boldsymbol{a}} &= [(\boldsymbol{a}' - \boldsymbol{a})\,\boldsymbol{\alpha} - \mathscr{S}^{(1)}(\bar{\boldsymbol{a}}, \boldsymbol{\alpha})]\,|_{\boldsymbol{\alpha}=\boldsymbol{\alpha}^{\chi}(\boldsymbol{a}'\boldsymbol{a})}. \end{aligned} \tag{10.76}$$

The components α^{χ}_n of the vector $\boldsymbol{\alpha}^{\chi} = \boldsymbol{\alpha}^{\chi}(\boldsymbol{a}'\boldsymbol{a})$ are found from the stationary phase conditions

$$\boldsymbol{a}' - \boldsymbol{a} = \frac{\partial}{\partial\boldsymbol{\alpha}} \mathscr{S}^{(1)}(\bar{\boldsymbol{a}}, \boldsymbol{\alpha})\,|_{\boldsymbol{\alpha}=\boldsymbol{\alpha}^{\chi}(\boldsymbol{a}'\boldsymbol{a})}. \tag{10.77}$$

The scattering matrix (10.73) can be simplified if only it is the total probability of transitions to specified final states which is of interest. Let these states be specified by quantum numbers $\boldsymbol{a}_1$, and let $\boldsymbol{a}_2$ be complementary quantum numbers (i.e. the set $\boldsymbol{a}_1, \boldsymbol{a}_2$ is the set $\boldsymbol{a}$). The total transition probability, from the initial state $\boldsymbol{a}_1\boldsymbol{a}_2$ to all final states with fixed $\boldsymbol{a}'_1$, is

$$P_{\boldsymbol{a}'_1\boldsymbol{a}_1}(\boldsymbol{a}_2) = \sum_{\boldsymbol{a}'_2} P_{\boldsymbol{a}'_1\boldsymbol{a}'_2;\boldsymbol{a}_1\boldsymbol{a}_2} = \sum_{\boldsymbol{a}'_2} |S_{\boldsymbol{a}'_1\boldsymbol{a}'_2;\boldsymbol{a}_1\boldsymbol{a}_2}|^2. \tag{10.78}$$

Summation over $\boldsymbol{a}'_2$ gives

$$\begin{aligned} P_{\boldsymbol{a}'_1\boldsymbol{a}_1}(\boldsymbol{a}_2) &= (2\pi)^{-N_2} \int d^{N_2}\boldsymbol{\alpha}_2\, \bar{P}_{\boldsymbol{a}'_1\boldsymbol{a}_1}(\bar{\boldsymbol{a}}, \boldsymbol{\alpha}_2), \\ \bar{P}_{\boldsymbol{a}'_1\boldsymbol{a}_1}(\boldsymbol{a}_2, \boldsymbol{\alpha}_2) &= |\bar{S}_{\boldsymbol{a}'_1\boldsymbol{a}_1}(\boldsymbol{a}_2, \boldsymbol{\alpha}_2)|^2, \end{aligned} \tag{10.79}$$

where the reduced scattering matrix is

$$\begin{aligned} \bar{S}_{\boldsymbol{a}'_1\boldsymbol{a}_1}(\boldsymbol{a}_2, \boldsymbol{\alpha}_2) = {} & (2\pi)^{-N_1} \int d^{N_1}\boldsymbol{\alpha}_1 \\ & \times \exp[-\mathrm{i}(\boldsymbol{a}'_1 - \boldsymbol{a}'_1)\,\boldsymbol{\alpha}_1 + \mathrm{i}\mathscr{S}^{(1)}(\boldsymbol{a}_1\boldsymbol{a}_2; \boldsymbol{\alpha}_1\boldsymbol{\alpha}_2)]. \end{aligned} \tag{10.80}$$

Thus the total transition probability is found by averaging the transition probability $\bar{P}_{\boldsymbol{a}'_1\boldsymbol{a}_1}(\boldsymbol{a}_2\boldsymbol{\alpha}_2)$ obtained for fixed initial phases of the degrees of freedom that are of no interest here.

In concluding this section we note that (10.73) bridges the gap between the first-order perturbation treatment in classical mechanics and the higher-order perturbation treatment in the Rayleigh-Schrödinger perturbation theory [10.30, 31]. Moreover, (10.73) remains basically correct if $\mathscr{S}^{(1)}$ is replaced by the exact increment in the classical action due to interaction during the collision. This is the main idea of the classical S matrix theory [10.26, 27].

On the other hand, (10.73) can be looked upon here as an explicit expression for exponential unitarization. The first-order (FO) scattering matrix, formally found from (10.73) is

$$ {}^{\mathrm{FO}}S_{\boldsymbol{a}'\boldsymbol{a}}(\bar{\boldsymbol{a}}) = \delta_{\boldsymbol{a}'\boldsymbol{a}} + \mathrm{i}(2\pi)^{-N} \int d^N \boldsymbol{\alpha}\, \mathscr{S}^{(1)}(\bar{\boldsymbol{a}}, \boldsymbol{\alpha}) \exp[-\mathrm{i}(\boldsymbol{a}' - \boldsymbol{a})\, \boldsymbol{\alpha}]. \tag{10.81} $$

The inverse transformation defines $\mathscr{S}^{(1)}$ via ${}^{\mathrm{FO}}S$,

$$ \mathscr{S}^{(1)}(\bar{\boldsymbol{a}}, \boldsymbol{\alpha}) = -\mathrm{i} \sum_{\boldsymbol{a}'\boldsymbol{a}''} [{}^{\mathrm{FO}}S_{\boldsymbol{a}'\boldsymbol{a}''}(\bar{\boldsymbol{a}}) - \delta_{\boldsymbol{a}'\boldsymbol{a}''}] \exp[\mathrm{i}(\boldsymbol{a}' - \boldsymbol{a}'')\, \boldsymbol{\alpha}], \tag{10.82} $$

which finally gives the correspondence-principle exponential unitarization

$$ \begin{aligned} S_{\boldsymbol{a}'\boldsymbol{a}} = (2\pi)^{-N} \int d^N \boldsymbol{\alpha} \exp\Big\{ &-\mathrm{i}(\boldsymbol{a}' - \boldsymbol{a})\, \boldsymbol{\alpha} \\ &+ \sum_{\boldsymbol{a}''\boldsymbol{a}'''} [{}^{\mathrm{FO}}S_{\boldsymbol{a}''\boldsymbol{a}'''}(\bar{\boldsymbol{a}}) - \delta_{\boldsymbol{a}''\boldsymbol{a}'''}] \exp[\mathrm{i}(\boldsymbol{a}'' - \boldsymbol{a}''')\, \boldsymbol{\alpha}] \Big\}. \end{aligned} \tag{10.83} $$

The correspondence-principle approach has been repeatedly used to calculate the cross sections for transitions between highly excited atomic states, either in terms of the classical perturbation treatment [10.32], or with the correspondence principle discrimination of states [10.33, 34], or else for directly restoring the scattering amplitude from the Born scattering amplitudes [10.35]. A review on transitions between Rydberg states induced by collisions with charged particles is given in [10.36], and the generalization for non-hydrogen-like states is presented in [10.37]. Collisions of highly excited atoms with neutrals are discussed in [10.38, 39].

10.3 Generalizations of the Linear Model

10.3.1 Interaction of a Diabatic Term with a Set of Parallel Diabatic Terms and a Continuum

The interaction of one linear diabatic term with two mutually uncoupled linear parallel diabatic terms was first discussed by *Ovchinnikova* [10.40] in the semiclassical approximation. An arbitrary number of mutually uncoupled parallel diabatic terms each interacting with one crossing linear diabatic term has been considered by *Osherov* [10.41, 42] and *Demkov* [10.43].

Later on, by analogy with the two-state case [10.44] it was found that a quantum solution was possible for parallel terms with zero slope [10.45, 46]. The relation between this problem and the general class of stationary and nonstationary problems solvable by the contour-integral method has also been obtained [10.47].

The Hamiltonian of interest is

$$\begin{aligned} &V_{00} = -F_0(R-R_0), \quad V_{nn} = u_n - F(R-R_0), \quad V_{0n} = a_n, \\ &V_{n'n} = 0, \quad n', n > 0, \quad n = 1, 2, \dots N. \end{aligned} \tag{10.84}$$

It represents N parallel running diabatic terms V_{nn} with energies u_n for $R = R_0$, crossed by one linear diabatic term V_{00}. The latter is coupled to each of the terms V_{nn} by constant but otherwise arbitrary interactions a_n.

If a linear time-dependent parametrization similar to (8.8) is adopted, i.e.,

$$R - R_0 = vt, \tag{10.85}$$

the equations for the diabatic scattering amplitudes $c_n^{(d)}$ become

$$\begin{aligned} &\mathrm{i}\dot{c}_0^{(d)} = \sum_{n=1}^{N} a_n \exp(\mathrm{i}\Delta F v t^2/2 - \mathrm{i}u_n t)\, c_n^{(d)} \\ &\mathrm{i}\dot{c}_n^{(d)} = a_n \exp(-\mathrm{i}\Delta F v t^2/2 + \mathrm{i}u_n t)\, c_0^{(d)} \end{aligned} \tag{10.86}$$

with $\Delta F = F - F_0$.

The general solution to (10.86) is represented by a contour integral containing the Green's function matrix of the rhs of (10.86) at $t = 0$. The solutions for different initial conditions are obtained by choice of the integration contour (for details, see [10.47]). Only the final result will be given here: the probability $P_{n'n}$ of a transition from the initial n to the final n' state can be obtained as the probability of independent pair transitions each localized at the crossing point of the terms V_{00} and V_{nn}. The probability of pair transitions is given by the Landau-Zener formula

$$p_n = \exp(-2\pi a_n^2/\Delta F v), \tag{10.87}$$

with ΔF assumed to be positive. Thus, for a single transit of the nonadiabaticity region, we have

$$\begin{aligned} &P_{n'n} = 0, \quad \text{when } n' < n, \\ &P_{n'n} = (1-p_{n'})\, p_{n'-1} \cdots p_{n+1}(1-p_n) \quad \text{when } n' > n > 0, \\ &P_{n'0} = (1-p_{n'})\, p_{n'-1} \cdots p_3 p_2 p_1 . \end{aligned} \tag{10.88}$$

In particular, the cumulative transition probability $P(m)$ from state 0 to all states $n' \leqq m$ is

$$P(m) = \sum_{n'=1}^{m} P_{n'0} = 1 - \exp\left(-2\pi \sum_{n'=1}^{m} a_{n'}^2/\Delta F v\right). \tag{10.89}$$

Now, the transition from a discrete state $|0\rangle$ to a continuum $|n\rangle$ can be described using (10.89). With this aim one has to take the limit $u_n - u_{n-1} \to 0$, and replace summation by integration. Then the function $|u\rangle$ is taken to represent the continuum functions normalized to $\delta(u' - u)$. Taking into account that dn and du are related as $dn = \varrho(u)\, du$, where $\varrho(u)$ is the density of states with energy u, we have

$$P(u) = 1 - \exp\left[-(2\pi/\Delta F v) \int_{u_1}^{u} |V_{0u'}|^2 \varrho(u')\right], \tag{10.90}$$

where $P(u)$ is the cumulative transition probability from state 0 to all continuum states with energy $u' \leqq u$ at $R = R_0$. Correspondingly, the distribution function $W(u)$ over the continuum states is $W(u) = dP(u)/du$. At this point it would be simpler to consider the problem with a horizontal continuum boundary, i.e., with $F = 0$ in (10.84). Then u can be interpreted as the continuum-state energy for $R \to \infty$ and $P(u)$ as the cumulative distribution of final states in energy.

Now, (10.90) represents the solution to (10.86) for an infinite time interval. Nevertheless it is still possible, under some restrictive conditions and in spirit of the semiclassical treatment, to interpret the distribution $P(u)$ as arising from consecutive population of continuum states during the motion of a diatom over the diabatic term $V_{00}(R)$. This interpretation suggests that the relation between the energy u and the arrival time t at this energy level is

$$V_{00}[R(t)] = u. \tag{10.91}$$

With the parametrization (10.85) and the relation (10.91), (10.90) now takes the form

$$P[u(t)] = 1 - \exp\left[-\int_{t_0}^{t} \Gamma(t')\, dt'\right], \tag{10.92}$$

where t_0 is the arrival time at the bottom of the continuum. Here the time-dependent width $\Gamma(t)$ of the discrete state embedded in the continuum is just the width $\Gamma(R)$ of the state $|0\rangle$ for clamped nuclei a distance R apart. The nuclei are then allowed to move according to (10.85). The width $\Gamma(R)$ is calculated in terms of the Fermi golden rule,

$$\Gamma(R) = 2\pi |V_{0u}|^2 \varrho(u), \tag{10.93}$$

where u is related to R by $V_{00}(R) = u$.

In addition to the continuum-state population probability, we define the survival probability $P_0(t)$, i.e., the probability that an atom will remain in the initial state $|0\rangle$:

$$P_0(t) = 1 - P(t) = \exp\left[-\int_{t_0}^{t} \Gamma(t')\,dt'\right]. \tag{10.94}$$

The result (10.94) and a more detailed treatment [10.47] show that the time dependence of a diabatic state coupled to the continuum is described by the amplitude

$$c_0(t) = \exp\left\{-\int_{t_0}^{t} \mathscr{U}[R(t)]\,dt\right\}, \tag{10.95}$$

where $\mathscr{U}(R)$ is the complex potential energy

$$\mathscr{U}(R) = V_{00}(R) - \mathrm{i}\Gamma(R)/2. \tag{10.96}$$

On the other hand, (10.95) evidently represents a semiclassical generalization of the adiabatic approximation for stationary states to quasi-stationary states. This provides an important link between two different types of description of transitions to the continuum: either as a coupling of a stationary diabatic state to continuum states, or as a decomposition of the quasi-stationary state with complex potential energy. For certain conditions this correspondence goes beyond the linear model [10.47].

Note that the use of a local and energy-independent imaginary correction $-\mathrm{i}\Gamma(R)/2$ to the adiabatic energy $U(R)$ in describing the decay of a quantum state provides quite a definite picture of the process: the wave function of a quasi-stationary state remains localized, but its amplitude decreases exponentially with time. However a different decay type is possible: the wave function of the initially localized state begins to spread when the continuum is reached. A description of this behavior resorts to other models of interaction of an initially bound state with the continuum, such as the zero-range potential models, discussed in detail elsewhere [10.47–49].

The linear model with parametrization (10.85) describes the population of continuum states as a result of the stationary state entering the continuum. The situation is different when the diabatic term enters the continuum and then leaves it. For this frequently occurring case, the same energy state is populated at least twice and there arises the possibility of interference. In the general case the population distribution $W(u)$ over continuum states is calculated as [10.50]

$$W(u) = \left|\int_{t_0}^{\infty} [\Gamma(t)/2]^{1/2}\, c_0(t)\exp(\mathrm{i}ut)\,dt\right|^2 \tag{10.97}$$

with $c_0(t)$ given by (10.95).

The stationary-phase approximation to (10.97) reads

$$W(u) = \left| \sum_n \frac{1}{2} \left[\frac{-\Gamma}{d\mathscr{U}/dt} \right]^{1/2}_{t=t_n} \exp\left[-\mathrm{i} \int_{t_0}^{t_n} \mathscr{U}(t)\, dt + \mathrm{i} u t_n \right] \right|^2, \quad (10.98)$$

where the stationary-phase points t_n are found from

$$\mathscr{U}(t_n) = u. \quad (10.99)$$

With Γ tending to zero some t_n move towards the real time-axis and others stay complex. The former correspond to classically allowed transitions (real value for the arrival time, see above), and the latter to classically inaccessible transitions.

As an example, consider the linear Hamiltonian (10.84) with $F = 0$, a_n all equal, and the quadratic time parametrization similar to (8.15):

$$R - R_0 = Ft^2/2\mu - \mu v^2/2F. \quad (10.100)$$

The time-dependent representation of this process is shown in Fig. 10.2. With the complex potential $\mathscr{U}(t)$ in the form

$$\mathscr{U}(t) = u^* - bt^2 - \mathrm{i}\Gamma/2, \quad (10.101)$$

the distribution over continuum states is calculated explicitly to give [10.50]

$$W(u) = C\,|\mathrm{Ai}\,[b^{-1/2}(u - u^*) + \mathrm{i}\Gamma/2]|^2, \quad (10.102)$$

where Ai is the Airy function and C is the normalization factor. For $u > u^*$ the continuum states are populated by tunneling and $W(u)$ decays rapidly with increasing u, showing no interference. For $u < u^*$ the continuum states are populated classically, and the distribution $W(u)$ oscillates as a result of the interference. Unlike the Stueckelberg oscillations in a two-state system, these ocillations are of a smaller amplidute, because only part

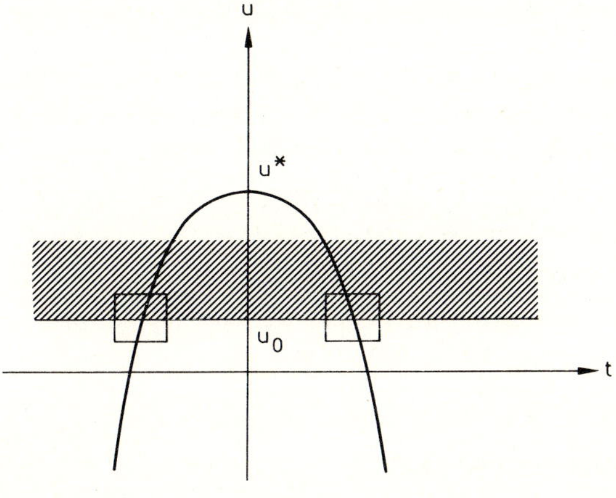

Fig. 10.2. Quadratic approximation (10.101) for transition to the continuum. The squares refer to regions where the linear approximation is valid

of the system remains in the initial state until the second arrival. If the decomposition of the initial state is neglected, i.e., Γ is taken to be zero, (10.102) will represent the first-order solution to the coupled equations (10.86) where the quadratic time-dependence of the exponent is replaced by a cubic dependence [10.51–53].

Note finally that the explicit solution of the coupling of one well-localized state to many close-lying discrete and continuum states allows the formulation of a unified theory for Penning ionization and the excitation of Rydberg states of an atom [10.54].

10.3.2 Nonadiabatic Coupling Between Two Quasi-Stationary States

If there are two quasi-stationary adiabatic states coupled by a nonadiabatic interaction, the semiclassical equations for the probability amplitudes can be derived from those for the coupling of two stationary states (Chap. 7) by adding imaginary parts to the diagonal and nondiagonal elements of the Hamiltonian matrix. This correction can be introduced both in the adiabatic and diabatic representations. The imaginary parts of the diagonal matrix elements describe the decomposition of adiabatic and diabatic states, respectively, and those of nondiagonal elements represent the interaction via the continuum. It will be realized that this interaction depends on the overlap of two continua relevant to the states $|1\,\mathrm{d}\rangle$ and $|2\,\mathrm{d}\rangle$. If these states decompose to the same continuum and have widths Γ_1 and Γ_2, the imaginary part of the nondiagonal matrix element will be $-(1/2)(\Gamma_1\Gamma_2)^{1/2}$, but if these states decompose to physically different continuum states, there will be no imaginary correction to the nondiagonal matrix element. In general, the imaginary part of a nondiagonal matrix element of the two-state Hamiltonian can be represented as $-(1/2)\,S_{12}(\Gamma_1\Gamma_2)^{1/2}$, where S_{12} is the overlap of two continua. Thus, the generalizations of two-state models for quasi-stationary states require the introduction of three independent functions Γ_1, Γ_2, and S_{12}. Of the models so far studied, we consider the case of linear quasi-stationary crossing terms with constant coupling, constant widths, and S_{12} either zero or unity.

The Hamiltonian in the diabatic basis reads [compare with (8.1)]

$$V_{nm} = (E_0 - F_n X - \mathrm{i}\Gamma_n/2)\,\delta_{nm} + (1-\delta_{nm})\,[a - \mathrm{i}(S_{12}/2)(\Gamma_1\Gamma_2)^{1/2}],$$
$$n, m = 1, 2. \tag{10.103}$$

The diagonalization of this Hamiltonian can be carried out by a unitary matrix B of the form

$$B = \begin{pmatrix} \exp(\mathrm{i}\varphi)\cos\vartheta & \exp(\mathrm{i}\psi)\sin\vartheta \\ -\exp(-\mathrm{i}\psi)\sin\vartheta & \exp(-\mathrm{i}\varphi)\cos\vartheta \end{pmatrix}, \tag{10.104}$$

containing not one, as in (7.5), but three, real parameters φ, ϑ and ψ. The adiabatic wave functions $|1\rangle$, $|2\rangle$ and the adiabatic quasi-stationary terms $\mathcal{U}_1(R)$ and $\mathcal{U}_2(R)$ can be used for representation of the total function Ψ in the form

$$\Psi = c_1(t)\,|1\rangle \exp\left[-\mathrm{i}\int_{t_0}^{t} \mathcal{U}_1(t')\,dt'\right] + c_2(t)\,|2\rangle \exp\left[-\mathrm{i}\int_{t_0}^{t} \mathcal{U}_2(t')\,dt'\right]. \tag{10.105}$$

In the limit $t \to +\infty$ both exponential factors singled out in (10.105) disappear, but the coefficients c_n tend to a constant limit. This allows one to introduce the notion of the transition probability, defined in the usual way, by analogy with the transition probability for nondecaying states.

The semiclassical problem for the Hamiltonian (10.103) with parametrization (10.85) is solved analytically in terms of parabolic cylinder functions, and the transition probability is different depending on S_{12}.

If the states $|1\,\mathrm{d}\rangle$ and $|2\,\mathrm{d}\rangle$ decompose to different continua ($S_{12} = 0$), the transition probability is given by the Landau-Zener formula [10.55]

$$P = \exp(-2\pi a^2/|\Delta F|\,v). \tag{10.106}$$

That means that the exponential decay of diabatic states does not change the transition probability compared to the case of stationary states. However, for decaying states the transition region is more extended than for stationary states [10.55].

If the states $|1\,\mathrm{d}\rangle$ and $|2\,\mathrm{d}\rangle$ decay to the same continuum ($S_{12} = 1$), the transition probability is [10.56]

$$P = \exp[-2\pi(a^2 - \Gamma_1\Gamma_2/4)/|\Delta F|\,v]. \tag{10.107}$$

It is seen that for the same a the transition probability (10.107) is greater than (10.106). Moreover, for $a^2 < \Gamma_1\Gamma_2/4$ the transition probability becomes greater than unity. This simply means that nonadiabatic transitions slow down the decay of adiabatic states.

A similar analysis has been carried out for the exponential [10.57] and the hypergeometric [10.58] models yielding much more complicated expressions for the transition probabilities. An important feature of the exponential model is the opportunity given to introduce time-dependent widths and this widens its application to collision problems.

Alongside the calculation of the transition probability between decaying states, another problem is the calculation of the distribution over continuum states. For linear and exponential models this has been done in [10.59] and will not be discussed here.

11. Case Study – Intramultiplet Mixing and Depolarization of Alkalis in Collisions with Noble Gases

The collisions of excited alkali atoms M* in a resonant substate ^2P_j with noble gas atoms X result both in depolarization

$$\mathrm{M}^*\,(^2P_j, m) + \mathrm{X} \rightarrow \mathrm{M}^*\,(^2P_j, m') + \mathrm{X} \tag{11.1}$$

and in intramultiplet mixing

$$\mathrm{M}^*\,(^2P_j) + \mathrm{X} \rightarrow \mathrm{M}^*\,(^2P_{j'}) + \mathrm{X} + \Delta\varepsilon_{jj'}\,. \tag{11.2}$$

Below the energy of several eV these are the only processes which occur with appreciable probability. In the last twenty years considerable effort has been directed towards the study of such processes occurring in bulk, in beams, and under laser excitation [11.1–13].

Theoretically the M* + X collisions are appealing in that the atomic interaction is fairly well known and the scattering equations are rather simple, since they refer to three-channel scattering. On the other hand, a three-state problem differs in many respects from the two-state problem traditionally invoked to illustrate the general approach.

This chapter aims to demonstrate the validity of different approaches in the calculation of adiabatic diatomic terms (Chap. 4), the reduction of a three-state problem to a sequence of two two-state problems (Chap. 6), the use of the exponential model (Chap. 9) to solve a two-state problem, the matching approximation for long-range Coriolis coupling (Chap. 10), and finally the accuracy of the approximations used. Some exact quantum results reported in many papers [11.14–24] will be used for the latter purpose. Accordingly this chapter is organized as follows.

Section 11.1 covers the scattering equations and the basic information concerning the M*–X interaction. Section 11.2 describes the calculation of the S matrix and the relevant comparison equations. Section 11.3 deals in brief with the transition probabilities and cross sections for two inelastic processes in the $^2P_{1/2}$ substate of M*.

11.1 Formulation of the M*–X Scattering Problem

11.1.1 Scattering Equations and Couplings

An alkali atom M possesses one valence electron outside closed shells (see Sect. 4.1.2). The ground state $^2S_{1/2}$ of this electron is specified by the quantum numbers n and $L = 0$, and the first excited state ^2P_j by n and $L = 1$. As X approaches M* (^2P_j), one molecular term $A\ 1/2$ with $\Omega = 1/2$ originates from the atomic substate $^2P_{1/2}$, while two molecular terms, $B\ 1/2$ with $\Omega = 1/2$ and $A\ 3/2$ with $\Omega = 3/2$, originate from the atomic substate $^2P_{3/2}$. The first excited doublet ^2P_j of alkali atoms is separated from the nearest levels with different n or L by an energy gap ΔE of the order of one eV. The terms $A\ 1/2$, $A\ 3/2$ and $B\ 1/2$ can cross or pseudocross the molecular terms originating from these levels only at rather small displacements where the potential energy is very high. Therefore, at not very high collision energies, only coupling between six states corresponding to the above three molecular terms can be taken into account.

The molecular-basis scattering equations (2.114) valid under quasi-classical conditions will be written here in the form (V^1_{AB} is set equal to zero)

$$\left(-\frac{1}{2\,\mu}\frac{d^2}{dR^2} + \frac{(J+1/2)^2}{2\,\mu\,R^2} + U_a - \mathscr{E}\right)\chi_{a'\tilde{\Omega}'} = -\sum_{a''\tilde{\Omega}''} V_{a'\tilde{\Omega}';a''\tilde{\Omega}''}\,\chi_{a''\tilde{\Omega}''}\,, \quad (11.3)$$

where a', a'' are $A\ 1/2$, $A\ 3/2$, $B\ 1/2$ and

$$\begin{aligned} V_{a'\tilde{\Omega}';a''\tilde{\Omega}''} = &-\delta_{\tilde{\Omega}'\tilde{\Omega}''}\,\mu^{-1}\left\langle a'\,\tilde{\Omega}'\left|\frac{\partial}{\partial R}\right|a''\,\tilde{\Omega}'\right\rangle\frac{d}{dR} \\ &-[(J+1/2)/\mu R^2)]\,[\delta_{\tilde{\Omega}'';\tilde{\Omega}'+1}\,\langle a'\,\tilde{\Omega}'\,|j_{x_2'}|\,a'\,\tilde{\Omega}'+1\rangle \\ &+\delta_{\tilde{\Omega}'';\tilde{\Omega}'-1}\,\langle a'\,\tilde{\Omega}'\,|j_{x_2'}|\,a''\,\tilde{\Omega}'-1\rangle]\,. \end{aligned} \quad (11.4)$$

Bearing in mind that the semiclassical collision picture will be used below, the X' axis of the body-fixed coordinate system $X'\,Y'\,Z'$, used in the quantum scattering problem (Fig. 2.1 b), is identified with the x_2' axis of the molecular body-fixed coordinate system $x_2'\,y_2'\,z_2'$ used in the semiclassical collisions problem (Fig. 6.1). In discussing the scattering equations and the couplings, we will use simplified notations for molecular states:

$$|A1/2, \pm 1/2\rangle = |\pm A\rangle, \qquad |A\,3/2, \pm\,3/2\rangle = |\pm A'\rangle$$

$$|B1/2, \pm\,1/2\rangle = |\pm B\rangle\,. \quad (11.5)$$

The molecular terms will be denoted simply as A, A' and B, or Σ and Π (for no spin-orbit coupling). The fine-structure splitting $\Delta\varepsilon_{3/2,\,1/2}$ in (11.2) will be denoted by $\Delta\varepsilon$.

Consider now the nonadiabatic coupling in more detail. Taking the symmetry properties of the nonadiabatic coupling (Sect. 2.2.2) into account, (11.3) will contain, generally speaking, nine different functions of R: three molecular terms U_A, $U_{A'}$, and U_B, one radial coupling matrix element

$$g(R) = \langle + A|\partial/\partial R| + B\rangle = - \langle - A|\partial/\partial R| - B\rangle \tag{11.6}$$

and five different matrix elements $\langle a'\Omega'|j_{x'_2}|a''\Omega''\rangle$, namely, those between states $+A$, $-A$; $+A$, $+A'$; $+A$, $-B$; $+A'$, $+B$; $+B$, $-B$.

However, for the system of interest, the number of independent functions of R in (11.3) can be drastically reduced. To this end, neglect for a moment the spin-orbit coupling. Then two terms only ($A^2\Pi$ and $B^2\Sigma$) arise from the ^2P state. The term $B^2\Sigma$ exhibits early exponential repulsion, since the density distribution of the valence electron of M in this state has a maximum in the direction of the repulsive (from the Pauli principle) atom X. At the same time, in the $A^2\Pi$ state the node of the density distribution of the valence electron corresponds to the direction to atom X. Therefore at large and intermediate distances there is no repulsion in this state. At small R both terms exhibit strong repulsion due to the interaction of the M core with atom X.

Taking into account now the spin-orbit coupling, note first that this coupling varies slowly with R (Sect. 4.5). This means that with the LS-coupling condition in a free M atom, the region where the spin-orbit coupling becomes comparable to the coupling of the orbital motion of the valence electron of M to the molecular axis is located at a rather large interatomic distance. Take this distance R_1 as

$$\Delta\varepsilon = U_\Sigma(R_1) - U_\Pi(R_1) = \Delta U_{\Sigma\Pi}(R_1) \tag{11.7}$$

which is a naive condition for changing the Hund coupling case c (valid for $R > R_1$) to case a (valid for $R < R_1$) for the nonrotating molecular axis. It will be shown later (Sect. 11.2.2) that (11.7) is a rigorous definition of the center of the radial nonadiabaticity region, provided the exponential modelling of $\Delta U_{\Sigma\Pi}$ is adopted.

Thus it is expected that for part of the region of the pure Hund coupling case a, where the distortion of the spatial parts of the valence electron wave function can be significant, $g(R)$ is very small for any reasonable R-dependence of the spin-orbit coupling. The Coriolis coupling in this region is produced solely by the electron spin coupling to the molecular axis and therefore is independent of the spatial wave functions' distortion.

From these considerations the following model, with three additional assumptions, seems to be adequate for the system of interest. The assumptions are:

i) For nonadiabatic coupling the wave functions $|a'\Omega'\rangle$ can be constructed only from the unperturbed atomic wave functions pertaining to the doublet ^2P of M* and to the closed shell of X.

ii) The spin-orbit coupling can be taken in the form

$$V_{\text{s.o.}} = f(r, R)\,\boldsymbol{L}\cdot\boldsymbol{S}, \tag{11.8}$$

where $\boldsymbol{r}$ is the distance from the electron to the alkali atom nucleus, and $f(r, R)$ is a slowly varying function of R which can significantly differ from the atomic $f^0(r) = f(r, \infty)$ only in the region of coupling case a. In all practical calculations which have been performed up to now it was assumed that $f(r, R) = f^0(r)$.

iii) The best available molecular terms U_Σ and U_Π must be used in calculating the coordinate matrix elements of the molecular Hamiltonian matrix.

Under these assumptions the adiabatic terms will be

$$\begin{aligned} U_A &= (1/2)\,[U_\Sigma + U_\Pi + (3/2)\,\zeta(R) - \Delta U_{AB}], \\ U_{A'} &= U_\Pi + (3/2)\,\zeta(R), \qquad\qquad (11.9\,\text{a}) \\ U_B &= (1/2)\,[U_\Sigma + U_\Pi + (3/2)\,\zeta(R) + \Delta U_{AB}], \end{aligned}$$

$$\Delta U_{AB} = [(9/4)\,\zeta^2(R) + \zeta(R)\,\Delta U_{\Sigma\Pi} + (\Delta U_{\Sigma\Pi})^2]^{1/2}, \tag{11.9 b}$$

where

$$\begin{aligned} \zeta(R) &= \int_0^\infty f(r, R)\,\varphi_{np}^2(r)\,r^2\,dr, \\ \zeta(\infty) &= (2/3)\,\Delta\varepsilon. \end{aligned} \tag{11.10}$$

The relevant adiabatic wave functions are

$$\begin{aligned} |\pm A\rangle &= |1/2, \pm 1/2\rangle \cos\vartheta \pm |3/2, \pm 1/2\rangle \sin\vartheta, \\ |\pm A'\rangle &= |3/2, \pm 3/2\rangle, \\ |\pm B\rangle &= \mp |1/2, \pm 1/2\rangle \sin\vartheta + |3/2, \pm 1/2\rangle \cos\vartheta, \end{aligned} \tag{11.11}$$

where $|j\tilde{\Omega}\rangle$ are the free-atom wave functions quantized on the molecular axis and

$$\tan 2\vartheta = 4(2)^{1/2}\,\Delta U_{\Sigma\Pi}/(9\,\zeta + 2\,\Delta U_{\Sigma\Pi}). \tag{11.12}$$

The radial coupling $g(R)$ is expressed via ϑ by (7.8). Bearing in mind that in the region of R variation where the radial coupling is large, $\zeta(R)$ can be considered as a constant, we get

$$g(R) = -\,\partial\vartheta/\partial R = -\,\zeta(d\,\Delta U_{\Sigma\Pi}/dR)/(2)^{1/2}\,\Delta U_{AB}^2. \tag{11.13}$$

Table 11.1 Coupling Matrix $V_{a'\tilde{\Omega}';a''\tilde{\Omega}''}$

$a''\tilde{\Omega}''$ / $a'\tilde{\Omega}'$	$-A$	$-A'$	$-B$	$+B$	$+A'$	$+A$
$-A$	0	$\frac{(J+1/2)}{\mu R^2}\frac{(3)^{1/2}}{2}\sin\vartheta$	$\mu^{-1}g(R)\frac{d}{dR}$	$\frac{(J+1/2)}{\mu R^2}\frac{3}{4}\sin 2\vartheta$	0	$-\frac{(J+1/2)}{\mu R^2}\frac{1}{4}(3\cos 2\vartheta-1)$
$-A'$	$\frac{(J+1/2)}{\mu R^2}\frac{(3)^{1/2}}{2}\sin\vartheta$	0	$-\frac{(J+1/2)}{\mu R^2}\frac{(3)^{1/2}}{2}\cos\vartheta$	0	0	0
$-B$	$-\mu^{-1}g(R)\frac{d}{dR}$	$-\frac{(J+1/2)}{\mu R^2}\frac{(3)^{1/2}}{2}\cos\vartheta$	0	$-\frac{(J+1/2)}{\mu R^2}\frac{1}{4}(3\cos 2\vartheta+1)$	0	$-\frac{(J+1/2)}{\mu R^2}\frac{3}{4}\sin 2\vartheta$
$+B$	$\frac{(J+1/2)}{\mu R^2}\frac{3}{4}\sin 2\vartheta$	0	$-\frac{(J+1/2)}{\mu R^2}\frac{1}{4}(3\cos 2\vartheta+1)$	0	$-\frac{(J+1/2)}{\mu R^2}\frac{(3)^{1/2}}{2}\cos\vartheta$	$\mu^{-1}g(R)\frac{d}{dR}$
$+A'$	0	0	0	$-\frac{(J+1/2)}{\mu R^2}\frac{(3)^{1/2}}{2}\cos\vartheta$	0	$-\frac{(J+1/2)}{\mu R^2}\frac{(3)^{1/2}}{2}\sin\vartheta$
$+A$	$-\frac{(J+1/2)}{\mu R^2}\frac{1}{4}(3\cos 2\vartheta-1)$	0	$-\frac{(J+1/2)}{\mu R^2}\frac{3}{4}\sin 2\vartheta$	$-\mu^{-1}g(R)\frac{d}{dR}$	$-\frac{(J+1/2)}{\mu R^2}\frac{(3)^{1/2}}{2}\sin\vartheta$	0

The Coriolis coupling matrix elements are expressed via $\vartheta(R)$. The calculated result with the wave functions defined by (11.11) is summarized in Table 11.1.

It is difficult to calculate ab initio the function $\zeta(R)$ (Sect. 4.5) and one has to rely on recent experimental information about the R dependence of ζ which has been obtained for the diatoms Na*–Ne and Na*–Ar [11.8, 25,

26]. It appears that in the region where the $A^2\Pi$ state is a minimum, $\zeta(R)$ is only about 10–20% larger than the free-atom value. This favours the naive assumption that ζ can be taken as a constant, $\zeta = 2\Delta\varepsilon/3$. This approximation will be used hereafter, thus reducing the formulation of the collision problem to the determination of only two potentials U_Σ and U_Π.

In the next section the quantities U_Σ and U_Π are considered in more detail.

11.1.2 M* – X Interaction

Many experimental [11.3, 25–33] and theoretical [11.23, 34–45] studies have been performed to determine U_Σ and U_Π for different M* – X pairs. Only the most widely studied systems Na* + He, Ne, Ar will be considered here. These systems are representative because the cases X = He, Ne and X = Ar correspond to weak and strong dispersion interactions, and for given X the qualitative features of U_Σ and U_Π are the same for all M*.

For He and Ne the situation is simple and the multiple scattering method discussed in Sect. 4.3.2 can be applied here. This method yields analytical expressions for the one-electron terms [11.40]. First, we note that the radial wave function of Na* for $3 < r < 10$ can be adequately represented by a Slater-type function,

$$\varphi_{3p}(r) = \gamma^2(\gamma/24)^{1/2}\, r \exp(-\gamma r/2) \tag{11.14}$$

with $\gamma = 0.85$. Adding the core-core repulsion $U^{\mathrm{HF}}_{\mathrm{M^+X}}$ and the long-range attraction, the expressions for the total U_Σ and U_Π will be

$$U_\Sigma = U^{\mathrm{HF}}_{\mathrm{M^+X}} + (3/2)\, L_{\mathrm{X}}\, \varphi^2_{3p}(R) + U^{(\mathrm{d})}_\Sigma, \tag{11.15 a}$$

$$U_\Pi = U^{\mathrm{HF}}_{\mathrm{M^+X}} + U^{(\mathrm{d})}_\Pi. \tag{11.15 b}$$

Here L_{X} is the e — X scattering length, and the second-order electrostatic interactions $U^{(\mathrm{d})}_\Sigma$ and $U^{(\mathrm{d})}_\Pi$ are expressed as

$$\begin{aligned} U^{(\mathrm{d})}_\Sigma = &-(\alpha_{\mathrm{X}}/2)R^{-4} + (\alpha_{\mathrm{X}}/32)\gamma^4 z[(3/2 + 2z^{-2} + 3z^{-4})\psi_- \\ &-(z/2 - z^{-1} - 3z^{-3})\psi_+ - 2z^{-1} \\ &-6z^{-3} + 32z^{-5} + 1152z^{-7} - Q_\Sigma(z)\exp(-z)], \end{aligned} \tag{11.16}$$

$$\begin{aligned} U^{(\mathrm{d})}_\Pi = &-(\alpha_{\mathrm{X}}/2)R^{-4} + (\alpha_{\mathrm{X}}/64)\gamma^4[-(2z^{-1} + 3z^{-3})\psi_- \\ &-(1 + 3z^{-2})\psi_+ + 6z^{-2} + 64z^{-4} - 1152z^{-6} \\ &+ Q_\Pi(z)\exp(-z)], \end{aligned} \tag{11.17}$$

where α_X is the polarizability of atom X, $z = \gamma R$, $\psi_{\mp} = [\exp(-z)\,\mathrm{Ei}(z) \mp \exp(z)\,\mathrm{Ei}(-z)]$, $\mathrm{Ei}(z)$ is the exponential integral,

$$Q_\Sigma = 4z^{-1} + 16z^{-2} + 64z^{-3} + 224z^{-4} + 516z^{-5} + 1160z^{-6} + 1152z^{-7}, \qquad (11.18)$$

$$Q_\Pi = 16z^{-2} + 128z^{-3} + 512z^{-4} + 1152z^{-5} + 1152z^{-6}. \qquad (11.19)$$

It is seen when using the known asymptotics of the exponential integral [11.46] that for large R the functions $U_\Sigma^{(d)}$ and $U_\Pi^{(d)}$ coincide with the dispersion interactions given by (4.54, 64, 65), provided $\Delta\varepsilon_{\gamma 1}$ is neglected compared to $\Delta\varepsilon_{\gamma 2}$:

$$U_\Sigma^{(d)} = -(7/5)\,\alpha_X \langle r^2 \rangle R^{-6}, \qquad (11.20)$$

$$U_\Pi^{(d)} = -(4/5)\,\alpha_X \langle r^2 \rangle R^{-6}. \qquad (11.21)$$

Here $\langle r^2 \rangle$ is calculated with the radial function (11.14). Note, however, that these asymptotic forms are attained only for rather large distances ($R > 15$) which usually are of little interest for the collision process considered.

The core-core repulsion $U_{M^+X}^{HF}$ must be obtained solving the Hartree-Fock equations for the system M^+-X. For Na^+-He such calculations have been performed in [11.41] and for Na^+-Ne in [11.44]. In the latter case the combined statistical method considered in Sect. 4.4.1 gives results close to those of the Hartree-Fock calculations. The analytical approximations to the Hartree-Fock results which have been suggested in [11.39, 44] are used now in (11.15):

$$U_{Na^+He}^{HF} = 36.12\exp(-2.337R), \qquad (11.22)$$

$$U_{Na^+He}^{HF} = 238.6\exp(-2.75R). \qquad (11.23)$$

At distances of interest in low-energy collisions U_Π is much more sensitive to the choice of $U_{M^+X}^{HF}$ than is U_Σ. In the former case it is the competition between $U_{M^+X}^{HF}$ and $U_\Pi^{(d)}$ that fixes the potential well which can be measured experimentally.

The results of calculations using (11.14–19, 22, 23) are presented in Figs. 11.1, 2 together with some other theoretical results for He and Ne and experimental results for the $A^2\Pi$ state in the case of Ne.

For Na*–He the multiple-scattering method is seen to give results close to those obtained by the Hartree-Fock method [11.41] and by the model-potential method using the low-energy e–He scattering data [11.39]. It has been found [11.39] that these results lead to the wing profile of the Na resonance line being perturbed by He, which is in agreement with

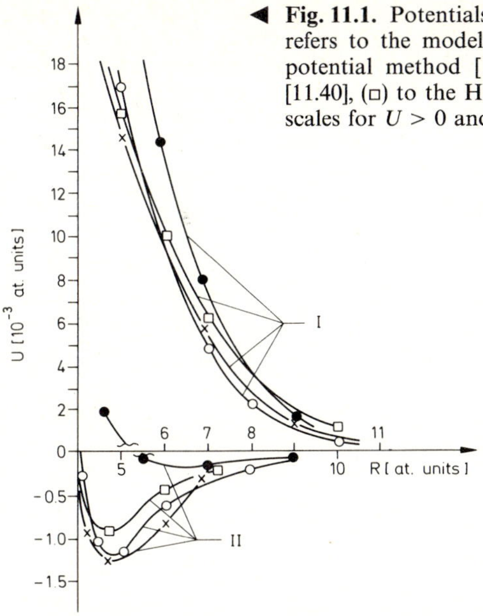

Fig. 11.1. Potentials U_Σ(I) and U_Π(II) for the Na – He system. (●) refers to the model-potential method [11.34], (×) to the model-potential method [11.39], (o) to the multiple-scattering method [11.40], (□) to the Hartree-Fock method [11.41]. Note the different scales for $U > 0$ and $U < 0$

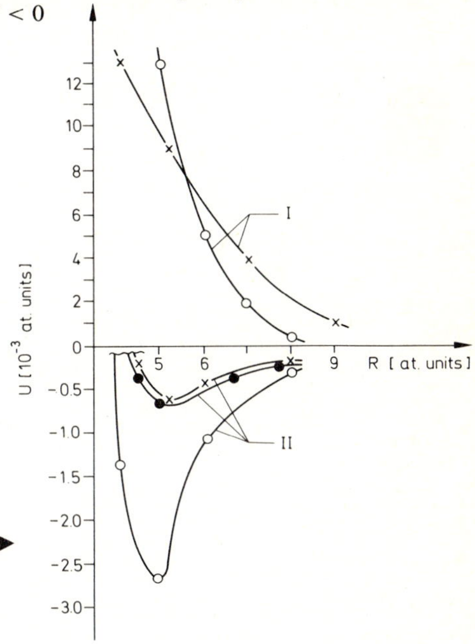

Fig. 11.2. Potentials U_Σ(I) and U_Π(II) for the Na – Ne system. (×) refers to the model-potential method [11.44], (o) to the multiple-scattering method [11.40], (●) to the experimental results [11.29]. Note the different scales for $U > 0$ and $U < 0$

the experimental data [11.33]. Thus, the analytical expressions (11.15) adequately describe the Na*–He interaction. At the same time the terms obtained in [11.34] seem to be too repulsive.

In the case of Na*–Ne (Fig. 11.2) the results of nonlocal model-potential configuration interaction calculations [11.44] (Sect. 4.3.1) agree with those available for the $A^2\Pi$ state experimental data. The multiple-scattering method gives satisfactory agreement with the results in [11.44] for the $B^2\Sigma$-state, but predicts too deep a potential well in the $A^2\Pi$ state. The reason for this probably lies in the neglect of the short-range e–Ne repulsion in the *p* state, thus omitting the additional soft exponential repulsion in the $A^2\Pi$ state. Note that in this case the local model-potential approach introduced in [11.34] and later on used in [11.37, 38, 43] fails completely, since it predicts a repulsion more than ten times stronger in the $B^2\Sigma$ state than that exhibited by the $B^2\Sigma$ terms in Fig. 11.2 and a very shallow potential well in the $A^2\Pi$ state. A conclusion that the method of [11.34] significantly overestimates the repulsion in the case of Na*–Ne has been made in [11.47] from comparison of the calculated and experimental intramultiplet-mixing cross sections.

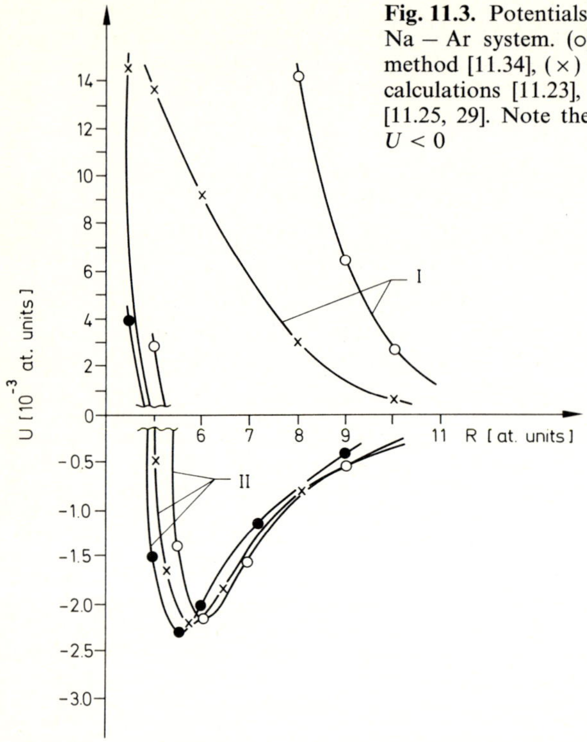

Fig. 11.3. Potentials U_Σ(I) and U_Π(II) for the Na – Ar system. (o) refers to the model-potential method [11.34], (×) to the configuration-interaction calculations [11.23], (•) to the experimental results [11.25, 29]. Note the different scales for $U > 0$ and $U < 0$

For strongly polarizable Ar for which the electron scattering length is negative, neither the multiple-scattering method, nor the model-potential method using the data for low-energy scattering are applicable (Sect. 4.3). Only the semi-empirical model-potential method proposed by Baylis [11.34] is expected to give qualitatively correct results in this case (Sect. 4.3.1). This can be seen in Fig. 11.3 where the results of the ab initio configuration interaction calculations [11.23] and the experimental data for the $A^2\Pi$ state are presented. But the $B^2\Sigma$ term obtained in [11.34] is again too repulsive. More complicated calculations [11.37, 38, 43] using the model potential suggested in [11.34] also lead to too strong repulsion in the $B^2\Sigma$ state.

Summing up, we arrive at the following conclusions. For weakly polarizable noble gas atoms the multiple-scattering method can be used in estimating U_Σ and U_Π, yielding quantitatively correct results for He and qualitatively correct results for Ne. The local model-potential method [11.34] gives considerably worse results for these cases. On the other hand, the latter method provides the only simple approach to calculating the M* – X interaction for heavy X. It gives qualitatively correct results in this case, but it can considerably overestimate the repulsion in the $B^2\Sigma$ state.

11.2 The Scattering Matrix

11.2.1 Matching Approximation

For an approximate decoupling of (11.3), we begin with an analysis of the coupling matrix presented in Table 11.1 for the case when $\Delta U_{\Sigma\Pi}$ is a positive monotonically decreasing function of R. The qualitative pattern of molecular terms for this situation is shown in Fig. 11.4 by full lines, the dot-dashed lines referring to a spinless interaction.

For this pattern of terms (11.13) gives a bell-shaped function $g(R)$ centered close to R_1 with width ΔR_1 which is defined by the characteristic divergence length of the Σ and Π terms. Close to this distance $\tan 2\vartheta$ changes from its zero asymptotic value to $(8)^{1/2}$. All matrix elements of rotational coupling containing $\cos\vartheta$, $\sin\vartheta$, $\cos 2\vartheta$ and $\sin 2\vartheta$ change correspondingly. The R dependence of all the six different coupling matrix elements in Table 11.1 is schematically presented in Fig. 11.5a. The full lines refer to the Coriolis coupling between degenerate molecular states, the dashed line, to Coriolis and radial couplings between quasi-degenerate terms (with splitting of the order $\Delta\varepsilon$ or less) and the dotted lines correspond to Coriolis coupling between states belonging to strongly split Σ and Π terms. Note that though the Coriolis interaction over extended ranges of R is represented in Fig. 11.5a by horizontal lines, it is actually stronger at a smaller distance due to the factor R^{-2}. With increasing R the Coriolis coupling decreases, with an even faster decrease of the adiabatic splitting between terms A and B. The distance at which they are equal is denoted by R_M.

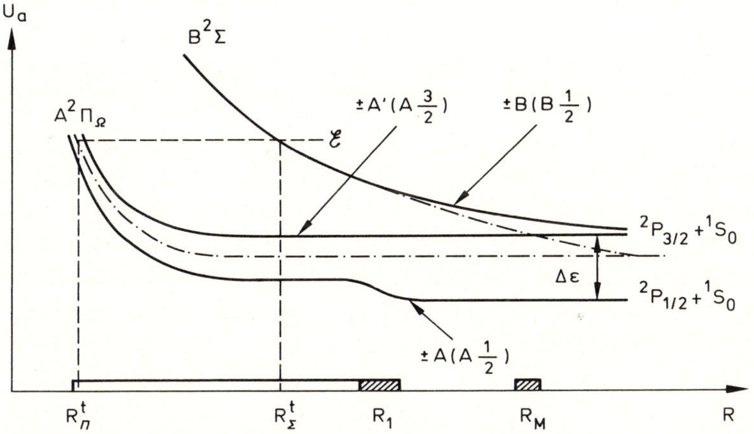

Fig. 11.4. Qualitative pattern of molecular terms for the M* – X system with low X-polarizability

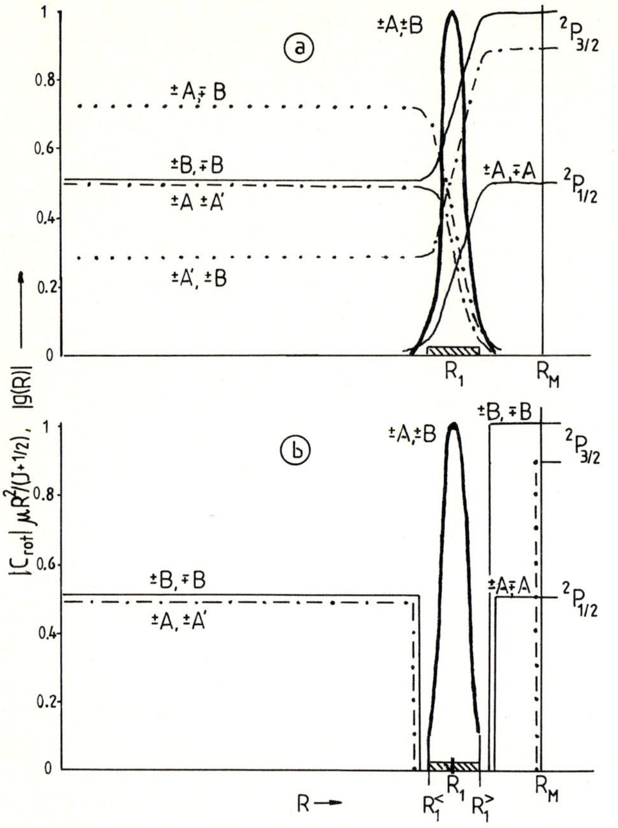

Fig. 11.5. (**a**) Qualitative pattern of the R dependence of the rotational (*full, dotted, dot-dashed lines*) and the radial (*heavy full line*) nonadiabatic couplings for the M* – X system with low X-polarizability. (**b**) Modelling of the rotational and radial couplings

Now, starting from large R we assign different Hund cases to different regions and examine the possibility of neglecting some couplings. First we introduce some simplifications which allow the matching approach to be used to construct the scattering matrix. These simplifications consist in completely neglecting certain couplings and in modeling the others and represent:

i) Omission of all Coriolis coupling in the region ΔR_1 because of the small extension of this region.
ii) Omission of the Coriolis coupling between nondegenerate states for $R < R_1$, except the couplings $+A$, $\pm A'$ and $-A$, $-A'$ as those most extensive between the quasi-degenerate terms (splitting of the order $\Delta\varepsilon$) and the only rotational couplings which can induce transition between atomic substates $^2P_{1/2}$ and $^2P_{3/2}$.
iii) Choosing the comparison equations to describe the radial coupling between states $+A$, $+B$ (and between $-A$, $-B$) within the interval ΔR_1, and the Coriolis coupling between states $+A$, $+A'$ (and between $-A$, $-A'$) for $R < R_1$.

The implication of these simplifications is seen in Fig. 11.5 b, where the rotational coupling matrix elements which were retained become zero at $R_1^<$ and at $R_1^>$, and the radial coupling lies within this interval. The sudden artificial change of the Coriolis coupling at $R_1^<$ and $R_1^>$ does not appreciably effect the transition amplitudes for degenerate states. For quasi-degenerate states this sudden change may affect the transition amplitude, so that this approximation needs further study. With the difference $R_1^> - R_1^< \ll R_1$ the radial coupling can be conceived of as occurring at one point $R = R_1$, so that in calculating the rotational coupling effect $R_1^<$ and $R_1^>$ can be identified with R_1. It is important to realize that one must first neglect the Coriolis coupling within the interval $R_1^<, R_1^>$, solve the problem of the radial coupling in this interval and then bridge the gap by going to the limit $R_1^<, R_1^> \to R_1$. This sequence of approximations gives no spurious effects, which might arise otherwise in the near-adiabatic limit because of the sudden onset of rotational perturbation.

To construct the S matrix by the matching method, the long-range Coriolis coupling between states A and B for $R > R_1$ must be removed. This will be done in the sudden approximation using (6.129). Neglecting the phase difference $\delta_{j\nu}^> - \delta_{j\Omega}^>$ we get

$$S^J_{j'\nu';j\nu} = \exp(\mathrm{i}\nu'\alpha + \mathrm{i}\nu\alpha) \sum_{\tilde{\Omega}',\tilde{\Omega}} \Delta^j_{\nu'\tilde{\Omega}'}\, {}^{\mathrm{M}}S_{j'\tilde{\Omega}';j\tilde{\Omega}}\, \Delta^j_{\nu\tilde{\Omega}}, \qquad \alpha = \arcsin(J/J_{\mathrm{M}}), \tag{11.24}$$

where J_{M} is fixed by the condition that the Coriolis coupling is equal to the splitting of the adiabatic terms A and B. The molecular scattering matrix ${}^{\mathrm{M}}S$ must now be calculated by solving the scattering equations for $R < R_{\mathrm{M}}$ which neglect the coupling between $+A$ and $+B$ (as well as between $-A$ and $-B$).

The ${}^{\mathrm{M}}S$ matrix is calculated by the matching procedure (Sect. 6.2) which requires some generalization to take the nonlocalized character of the Coriolis coupling between degenerate and quasi-degenerate states into account. To this approximation ${}^{\mathrm{M}}S$ is represented by

$${}^{\mathrm{M}}S = \lim_{R_1^<,\, R_1^> \to R_1} (r\,T\,R\,T^{\mathrm{tr}} r^{\mathrm{tr}}), \tag{11.25}$$

where the matrices r, T and R describe the following processes:

i) The matrix r accounts for the Coriolis coupling between the degenerate states $+A$, $-A$ and $+B$, $-B$ within the interval R_1, R_{M}. Accordingly the 6×6 matrix factorizes to 2×2, 2×2, 1×1, 1×1 blocks, and each can be calculated in closed form in terms of the partial adiabatic-scattering phase shifts

$$\delta_a^{>} = \int_{R_1}^{R_M} \mathscr{K}_a dR - \int_{R_1}^{R_M} \mathscr{K} dR, \qquad a = A, A', B,$$
$$\mathscr{K}_a = (2\mu)^{1/2}[\mathscr{E} - (J + 1/2)/2\mu R^2 - U_a(R)]^{1/2}, \tag{11.26}$$
$$\mathscr{K} = (2\mu)^{1/2}[\mathscr{E} - (J + 1/2)/2\mu R^2]^{1/2},$$

and in terms of the angles of rotation $2\varphi_a^{>}$ of the molecular axis for classical diatomic motion of MX in the adiabatic electronic state a from $R = R_M$ to $R = R_1$. Note that the rotation angle between two positions R_b and R_c is found from the radial action integral

$$\varphi_a(R_b, R_c) = -\frac{d}{dJ}\int_{R_b}^{R_c} \mathscr{K}_a dR. \tag{11.27}$$

The general structure of the r matrix is illustrated in Fig. 11.6.

ii) The matrix T accounts for the radial nonadiabatic coupling in the region $R_1^{<} < R < R_1^{>}$. It has the block structure shown in Fig. 11.6. In the limit $R_1^{<}, R_1^{>} \to R_1$, two 2×2 blocks of the T matrix become $\mathscr{N}$ matrices of the comparison equations for radial coupling between $+A$, $+B$ and $-A$, $-B$ states.

iii) The matrix R accounts for the Coriolis coupling between the degenerate states $+B$, $-B$ and between the quasi-degenerate states $+A$, $+A'$ and $-A$, $-A'$ for $R < R_1$. The block structure of the R matrix is presented in Fig. 11.6. The block for the $+B$, $-B$ states can be calculated explicitly, and the two blocks for the states $+A$, $+A'$ and $-A$, $-A'$ are calculated from the corresponding comparison equations. Thus, the R matrix will involve the partial adiabatic phase shifts

$$\delta_a^{<} = \int_{R_a^t}^{R_1} \mathscr{K}_a dR - \int_{R^t}^{R_1} \mathscr{K} dR, \qquad a = A, A', B, \tag{11.28}$$

with the lower integration limits corresponding to the turning points, and the rotation angle $2\varphi_B^{<}$ for motion on the B-term in the region $R < R_1$. This angle defines the elements of the block for the states $+B$, $-B$.

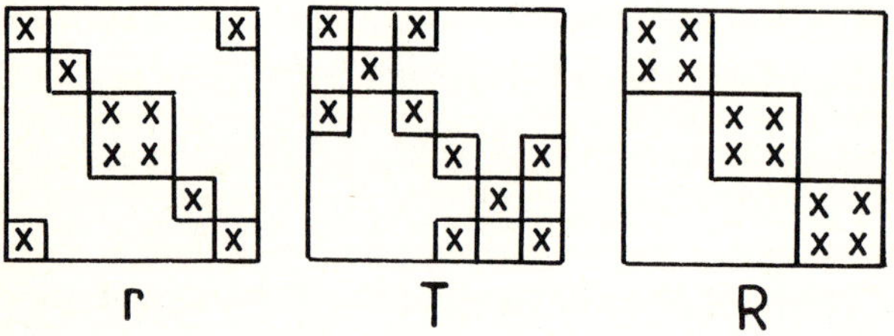

Fig. 11.6. Structures of r, T and R matrices. Nonzero elements are marked by crosses

Now note that adiabatic molecular potentials specify the partial phase shifts $\delta_{a'}^{<}, \delta_{a}^{>}$ and the classical rotation angles $2\varphi_a^{<}$ and $2\varphi_a^{>}$. The only extra information needed to construct the ${}^{M}S$ matrix consists in the solutions of two comparison equations. These solutions define the 2×2 blocks in the T and R matrices. Due to the symmetry of the coupling (see Table 11.1) the T and R matrices exhibit a symmetry that relates different blocks, thus providing a further simplification.

11.2.2 Semiclassical Comparison Equations

Consider first the radial coupling for $R_1^{<} < R < R_1^{>}$. The scattering equations for states $+A$ and $+B$ are

$$\begin{aligned} &d^2\chi_{+A}/dR^2 + 2\mu[\mathscr{E} - U_A(R) - (J+1/2)^2/2\mu R^2]\chi_{+A} \\ &\quad = 2g(R)\, d\chi_{+B}/dR, \\ &d^2\chi_{+B}/dR^2 + 2\mu[\mathscr{E} - U_B(R) - (J+1/2)^2/2\mu R^2]\chi_{+B} \\ &\quad = -2g(R)\, d\chi_{+A}/dR. \end{aligned} \tag{11.29}$$

For the states $-A$ and $-B$ the equations will differ in the sign of the rhs. According to (11.13) the function $g(R)$ with $\zeta = 2\Delta\varepsilon/3$ is

$$g(R) = -\frac{(2)^{1/2}\Delta\varepsilon(d\Delta U_{\Sigma\Pi}/dR)}{3(\Delta U_{\Sigma\Pi}^2 + 2\Delta\varepsilon\Delta U_{\Sigma\Pi}/3 + \Delta\varepsilon^2)}. \tag{11.30}$$

It follows from (11.30) that far from $R = R_1$ the function $g(R)$ vanishes because $d\Delta U_{\Sigma\Pi}/dR \to 0$ to the right of R_1 and $\Delta U_{\Sigma\Pi} \gg \Delta\varepsilon$ to the left of R_1. Define now the center of the coupling region R_s according to (ii) in Sect. 7.2 as the real part of the complex interatomic distance R_c at which the molecular terms U_A and U_B cross, i.e., $R_s = R_p = \mathrm{Re}\{R_c\}$. Now, using (11.9b) the crossing condition $\Delta U_{AB}(R_c) = 0$ transforms to

$$\Delta U_{\Sigma\Pi}(R_c) = \Delta\varepsilon \exp(\mathrm{i}2\vartheta_0), \qquad \cos 2\vartheta_0 = -1/3. \tag{11.31}$$

The solution of this equation can be found by iteration. For a zero-order approximation, take $R_c^{(0)}$ to be real and equal to R_1 defined by (11.7). Substituting $R_c = R_1 + \Delta R$ into (11.31), one can use the analytical continuation of $\Delta U_{\Sigma\Pi}(R_1 + \Delta R)$ for $\Delta R \ll R_1$. This is most readily accomplished by first approximating $\Delta U_{\Sigma\Pi}$ in a small region near R_1 by an analytical function of ΔR for real ΔR, and then making the continuation. For the monotonically decreasing function $\Delta U_{\Sigma\Pi}(R)$, a reasonable choice would be the exponential function

$$\Delta U_{\Sigma\Pi}(R_1 + \Delta R) = \Delta U_{\Sigma\Pi}(R_1)\exp(-\alpha_1 \Delta R) \tag{11.32}$$

with one free parameter α_1 for optimal simulation of the actual function $\Delta U_{\Sigma\Pi}$. Substituting (11.32) into (11.31) and using (11.7) as a definition of R_1, we get

$$R_c = R_1 \pm i2\vartheta_0/\alpha_1 . \tag{11.33}$$

It follows that $R_p = R_1$ in the exponential approximation of $\Delta U_{\Sigma\Pi}$.

It can be verified that the scattering equations (11.29) with the exponential approximation for $\Delta U_{\Sigma\Pi}$ coincide with the basic equations of the exponential model (Chap. 9) written down in the adiabatic basis for $\cos 2\vartheta_0 = -1/3$. Assume also the validity of the local common-trajectory approximation, i.e., the fulfillment of the condition

$$\Delta\varepsilon \ll \mathscr{E} - (J + 1/2)^2/2\mu R_1^2 . \tag{11.34}$$

The local common trajectory is taken to be

$$\begin{aligned} \Delta R &= v_1(t - t_1), \\ v_1 &= (2/\mu)^{1/2}[\mathscr{E} - (J + 1/2)^2/2\mu R_1^2 - U_A(R_1)/2 - U_B(R_1)/2]^{1/2}, \end{aligned} \tag{11.35}$$

where t_1 is the arrival time at R_1. Accounting for (11.35) the comparison equations (11.29) coincide with the time-dependent equations of the exponential model. The $\mathscr{N}$ matrix for this model is given by the general equation (7.20) with parameters P, φ, and ψ defined by (9.15a, b). The two parameters ξ and ξ_p of the exponential model become

$$\xi = \xi_1 = \Delta\varepsilon/\alpha_1 v_1 , \qquad \xi_p = \xi_{1p} = 2\xi_1/3 . \tag{11.36}$$

For this model the transition probability is

$$P = \exp(-2\pi\xi_1/3)\sinh(\pi\xi_1/3)/\sinh\pi\xi_1 . \tag{11.37}$$

Its limiting value for strongly nonadiabatic conditions is

$$P = 1/3 , \qquad \xi_1 \ll 1 , \tag{11.38}$$

while for a near-adiabatic case,

$$P = \exp(-4\pi\Delta\varepsilon/3\alpha_1 v_1) , \qquad \xi_1 \gg 1 . \tag{11.39}$$

In the near-adiabatic limit the first correction to the local common-trajectory approximation may be important. It can be found using the general approach discussed in connection with (9.27). When this is accomplished, and the transition probability is expressed as a function of the initial kinetic energy E_i or its radial component E_{i1} for $R = R_1$, the final result reads

$$P_{fi} = \exp[(-4\pi\Delta\varepsilon/3\alpha_1 v_{i1})(1 \pm \Delta\varepsilon/4E_{i1} + E_0/2E_{i1})],$$
$$v_{i1} = (2/\mu)^{1/2}(E_{i1})^{1/2} = (2/\mu)^{1/2}[E_i - (J+1/2)^2/2\mu R_1^2]^{1/2}. \quad (11.40)$$

The two correction terms in the brackets are assumed to be small compared to unity, but their net contribution may be notable for a large value of the Massey parameter ξ_1. The plus or minus sign in the exponent refers to up or down transitions. The last term corresponds to the last term in the exponent of (9.30), and it depends on the average potential close to R_1. For exponential repulsion in the Σ state and no appreciable interaction in the Π state, $E_0 = \Delta\varepsilon/3$ [11.48].

Turn now to the rotational coupling between the states $+A$, $+A'$ (and $-A$, $-A'$) in the region $R < R_1$. Again the common-trajectory approximation is adopted, this time for a pair of terms A and A' with the mean potential $(U_A + U_{A'})/2$, which for the model coupling is equal to U_Π. The common trajectory is defined by the differential equation

$$\mu\dot{R}^2/2 + (J+1/2)^2/2\mu R^2 + U(R) = \mathscr{E}. \quad (11.41)$$

The coupled equations for the amplitudes c_{+A} and $c_{+A'}$ are of the standard type discussed in Sect. 6.1. For the modelled Coriolis interaction these equations are coupled by the matrix element $-(J+1/2)/2\mu R^2$ up to $R = R_1$, where it suddenly becomes zero for R_1. To the same approximation the splitting $U_{A'} - U_A$ in the region $R < R_1$ is a constant, $2\Delta\varepsilon/3$. Under these conditions the comparison equations for states A and A' are

$$\mathrm{i}\dot{c}_{+A} = -[(J+1/2)/2\mu R^2]\exp(-2\mathrm{i}\Delta\varepsilon t/3)\, c_{+A'},$$
$$\mathrm{i}\dot{c}_{+A'} = -[(J+1/2)/2\mu R^2]\exp(2\mathrm{i}\Delta\varepsilon t/3)\, c_{+A}. \quad (11.42)$$

These equations must be solved for the trajectory $R(t)$ found from the differential equation (11.41) within the time interval $-t_1$ to $+t_1$. Now, the sudden change in the coupling at $t = t_1$ may cause a spurious effect due to the strongly nonadiabatic change of the interaction under the near-adiabatic condition. However, this spurious effect can be eliminated by gradually switching off the coupling for $R \gtrsim R_1$, and the final result will then depend only weakly on the switching function [11.48].

For states $+A$, $+A'$, the block of the R matrix will be

$$R = \exp(\mathrm{i}\delta_A^< + \mathrm{i}\delta_{A'}^<)\begin{pmatrix} (1-\mathscr{P})^{1/2}\exp(\mathrm{i}\gamma) & \mathrm{i}\mathscr{P}^{1/2} \\ \mathrm{i}\mathscr{P}^{1/2} & (1-\mathscr{P})^{1/2}\exp(-\mathrm{i}\gamma) \end{pmatrix}. \quad (11.43)$$

Here $\mathscr{P}$ is the probability of a transition between fine-structure components of the $^2\Pi$ state during the diatomic motion for $R < R_1$; γ comes from the additional dephasing due to the fine-structure splitting and is of the order $2\Delta\varepsilon t_1/3$.

The comparison equation can be solved numerically. However we concentrate again on two limiting cases, for small and large Massey parameters $\Delta\varepsilon t_1$. For strongly nonadiabatic conditions, $\Delta\varepsilon t_1 \ll 1$, (11.42) can be solved analytically. The transition probability in the zero-order approximation is

$$\mathscr{P} = \sin^2(\varphi_\Pi^<), \tag{11.44}$$

where $2\varphi_\Pi^<$ is the rotation angle of the molecular axis during the motion of the diatom in the region $R < R_1$. The matrix (11.43) with $\gamma = 0$ describes the spin reorientation in the body-fixed reference frame, i.e., the pure Hund case *b*.

For the near-adiabatic condition, $\Delta\varepsilon t_1 \gg 1$ the transition probability can be found using the first-order perturbation approach. In this case the main contribution to the transition amplitude comes from a region close to the turning point R^t [11.49] so that the precise form of the interaction far from the turning point has no effect on the transition probability. When the appropriate switching is accomplished by choosing the trajectory asymptotics, we get

$$\mathscr{P} = \left| \int_{-\infty}^{\infty} [(J + 1/2)/2\mu R^2] \exp(2\mathrm{i}\Delta\varepsilon t/3)\, dt \right|^2. \tag{11.45}$$

Now, near the turning point the potential U_Π is dominated by the core-core repulsion, so that a reasonable approximation to U_Π will be the shifted exponential function [11.49]

$$U_\Pi(R) = B \exp(-\alpha_2 R) - \varepsilon_0. \tag{11.46}$$

For light noble gases ε_0 is small and can be neglected. The trajectory $R(t)$ for the interaction (11.46) with $\varepsilon_0 = 0$ and for the modified wave-number approximation (Sect. 7.1) is given by (9.26) where α is replaced by α_2 and v_p by $v_2 = (2/\mu)^{1/2}[\mathscr{E} - (J + 1/2)^2/2\mu(R^t)^2]^{1/2}$. For this type of trajectory, the solution of the integral in (11.45) with the condition $\Delta\varepsilon/\alpha_2 v_2 \gg 1$ yields [11.49]

$$\mathscr{P} = a \exp(-4\pi\Delta\varepsilon/3\alpha_2 v_2). \tag{11.47}$$

Here the preexponential factor a is much lower than unity as it is proportional to the square of the rotation angle of the diatomic axis during a short time of the order of $(\alpha_2 v_2)^{-1}$. Similarly to (11.40), (11.47) can be corrected to account for the first-order term $\Delta\varepsilon/E_i$ in the exponent. The corrected exponent follows from the exponent in (9.33), viz.,

$$\mathscr{P}_{fi} = a \exp[(-4\pi \Delta\varepsilon/3\alpha_2 v_{i2})(1 \pm \Delta\varepsilon/4E_{i2})],$$
$$v_{i2} = (2/\mu)^{1/2}(E_{i2})^{1/2} = (2/\mu)^{1/2}[E_i - (J+1/2)^2/2\mu(R^t)^2]^{1/2}, \qquad (11.48)$$

where E_i is the initial kinetic energy and the plus and minus signs correspond to up and down transitions, respectively.

The central block of the matrix R (states $+B$, $-B$) describes the spin reorientation in the $^2\Sigma$ molecular state (pure Hund case b) and has the simple form

$$R = \exp(i\delta_B^<)\begin{pmatrix} \cos\varphi_B^< & i\sin\varphi_B^< \\ i\sin\varphi_B^< & \cos\varphi_B^< \end{pmatrix}. \qquad (11.49)$$

Turn now to a qualitative discussion of the effect of the long-range interaction on nonadiabatic coupling. If the polarization interaction is such that, for $R = R_1$, it is comparable to the exchange interaction, the function $g(R)$ will not be of the simple bell-shaped form. This means that the comparison equation for radial nonadiabatic coupling does not reduce to the exponential model. However, if the long-range interaction represents only a small correction to the exchange interaction, it can be accounted for approximately by substituting $\Delta\varepsilon_{\text{eff}}$ for $\Delta\varepsilon$ [11.48]. This approach follows from the general equation for $g(R)$, provided the long-range interaction close to R_1 is taken to be independent of R. The effective energy change is

$$\Delta\varepsilon_{\text{eff}} = (3\Delta\varepsilon/4)\,[(1 + 2\delta\varepsilon/3\Delta\varepsilon + \delta\varepsilon^2/\Delta\varepsilon^2)^{1/2} + 1/3 + \delta\varepsilon/\Delta\varepsilon], \qquad (11.50)$$

where $\delta\varepsilon$ is the long-range shift of $\Delta U_{\Sigma\Pi}$ at R_1. As the dispersion interaction is negative and stronger for the Σ-state, $\delta\varepsilon < 0$ and $\Delta\varepsilon_{\text{eff}} < \Delta\varepsilon$. If the dispersion interaction in the Σ- and Π-states can be described near R_1 by (11.20, 21), then

$$\delta\varepsilon = -3\alpha_X\langle r^2\rangle/5R^6. \qquad (11.51)$$

It follows that a greater polarizability of X would enhance transitions, provided the other conditions (relative velocity, range of the exchange potential, etc.) are the same.

Another effect of the strong dispersion interaction is connected with the considerable potential well for the $A^2\Pi$ term. If the well depth ε_0 is small compared to E_i, the main effect is the mutual acceleration of atoms, which can be approximately described by a slight increase in the relative velocity. However, with $\varepsilon_0 > E_i$, orbiting is possible. The spiralling approach of colliding atoms can greatly increase the transition probability. Moreover, resonant scattering would occur, but this effect cannot be described by the simplified matching method adopted here. Of course, these and other more subtle features of inelastic scattering show up when the quantum scattering equations are solved numerically, but this will not be discussed here.

11.2.3 Scattering Matrix for $^2P_{1/2}$ Substate

The matching approach which results in (11.25) enables a six-channel scattering matrix to be constructed for the doublet state ^2P. Besides the adiabatic partial phase shifts given by (11.26, 28), the rotation angles of the diatom, and the transition probabilities P and $\mathscr{P}$, the S matrix will also contain dynamic phases. These phases represent relatively small corrections to the adiabatic phases. Now, the interference pattern of the transition probability or of the differential cross section depends on phase differences. If the phase difference is not too large, the correction due to the dynamic phases is usually small, so that the interference pattern can be calculated with the dynamic phases put equal to zero. For a large phase difference the dynamic correction, though relatively small, may still be appreciable in absolute value. Then the interference pattern could not be calculated correctly ignoring the dynamic phases, but in this case the interference undulations are very fast. If averaged, the transition probability, or the differential cross section will cease to depend on the dynamic phases. These qualitative arguments suggest that in calculating either a not too rapidly oscillating interference structure of, say, the transition probability, or its average behavior in the case of rapid oscillations, all dynamic phases can be neglected. This approximation is adopted now for a presentation of the scattering matrix for the substate $^2P_{1/2}$.

This matrix actually consists of two elements,

$$\begin{aligned} S^J_{1/2,\,1/2;\,1/2,\,1/2} &= S^J_+\,, \\ S^J_{1/2,\,-1/2;\,1/2,\,-1/2} &= S^J_-\,. \end{aligned} \tag{11.52}$$

The nondiagonal elements with different v', v are zero because of parity conservation.

Equation (11.25) leads to the following expressions for the matrix elements:

$$\begin{aligned} S^J_\pm = {} & \exp(\pm\,\mathrm{i}\alpha_{1/2} \pm \mathrm{i}\varphi^>_A + 2\mathrm{i}\delta^>_A)\,[(1-P)\,(1-\mathscr{P})^{1/2} \\ & \times \exp(\mathrm{i}\delta^<_A + \mathrm{i}\delta^<_{A'} - \mathrm{i}\gamma) + P\exp(2\mathrm{i}\delta^<_B \pm \mathrm{i}\varphi^<_B)]\,. \end{aligned} \tag{11.53}$$

This is a general expression for the elements of the scattering matrix for states belonging to the $^2P_{1/2}$ component, valid for $R_1 < R_M$. The trajectory

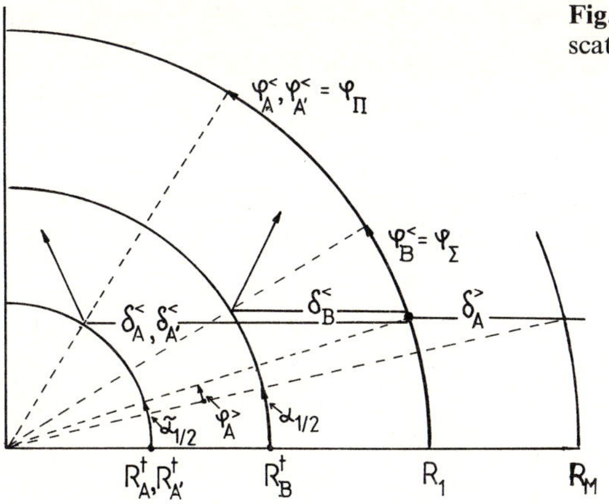

Fig. 11.7. Trajectory interpretation of the scattering matrix (11.53)

interpretation of this result is presented in Fig. 11.7 which, for simplicity, depicts trajectories specularly reflected from rigid spheres. After passing the matching radius R_M the diatom moves along a common rectilinear trajectory and reaches the distance R_1. During this motion the molecular axis rotates through an angle $\varphi_A^>$ and there appears a partial phase shift $\delta_A^>$ in the initial state. After the diatom passes R_1 two common trajectories arise: one for the adiabatic potential U_B and the other for the average potential U_Π. Hence the two terms in the squared brackets of (11.53) refer to markedly different phase factors. One coincides with that for the diagonal element of the matrix (11.43), and the other arises from a linear combination of matrix elements of the matrix (11.49). Note that as no inelastic event happens for the substate ${}^2P_{1/2}$ until the initial trajectory reaches R_1, the matching angle $\alpha_{1/2}$ naturally combines with $\varphi_A^>$ to give a new matching angle $\tilde{\alpha}_{1/2}$.

The scattering matrix (11.53) simplifies for the near adiabatic case. If there are neither $A \to B$, nor $A \to A'$ transitions, then $P = 0$, $\mathscr{P} = 0$, $\gamma = \delta_A^< - \delta_{A'}^<$. In this case

$$S_\pm^J = \exp(\pm \mathrm{i}\tilde{\alpha}_{1/2}) \exp(2\mathrm{i}\delta_A^< + 2\mathrm{i}\delta_{A'}^>), \tag{11.54}$$

which coincides with (10.22) if $\delta_A^< + \delta_A^>$ is identified with the elastic scattering phase shift for the adiabatic potential $U_A(R)$, and $\phi_{1/2}^J$ is put zero in (10.23).

The discussion of a strongly nonadiabatic case requires more attention, because for high velocities R_M may become less than R_1. We approach this case from the high-velocity limit assuming that the collision time is short compared to the inverse frequency of the fine structure splitting. Under this condition, the S matrix for intramultiplet transitions can be expressed via that for depolarizing collisions in the spinless atomic state. For the case considered, the basic equation (2.190) will read

$$S^J_{\pm} = \pm (\mathrm{i}/3) [S^{J \mp 1/2}_{10;10} - 2 S^{J \pm 1/2}_{1 \pm 1; 1 \pm 1}], \tag{11.55}$$

where $S^K_{1\nu;1\nu}$ is the scattering matrix for the P state. This matrix is calculated now by the sudden-matching procedure in terms of the phase shifts δ_Σ and δ_Π, and of the matching angle α_1. Equation (10.9) gives

$$\begin{aligned} S^{J \mp 1/2}_{10;10} &= \exp(2\mathrm{i}\delta^J_\Pi \mp \mathrm{i}\eta^J_\Pi/2), \\ S^{J \pm 1/2}_{1 \pm 1; 1 \pm 1} &= (1/2) \exp(\pm 2\mathrm{i}\alpha^J_1) \\ &\quad \times [\exp(2\mathrm{i}\delta^J_\Pi \pm \mathrm{i}\eta^J_\Pi/2) + \exp(2\mathrm{i}\delta^J_\Sigma \pm \mathrm{i}\eta^J_\Sigma/2)]. \end{aligned} \tag{11.56}$$

In this way we get

$$\begin{aligned} S^J_{\pm} &= \exp(\pm \mathrm{i}\alpha_1) (1/3) [\exp(2\mathrm{i}\delta_\Pi + \mathrm{i}\varphi_\Pi) \\ &\quad + \exp(2\mathrm{i}\delta_\Pi - \mathrm{i}\varphi_\Pi) + \exp(2\mathrm{i}\delta_\Sigma \pm \mathrm{i}\varphi_\Sigma)]. \end{aligned} \tag{11.57}$$

The two expressions for the S matrix, (11.53) and (11.57), valid for $R_1 < R_\mathrm{M}$ and $R_1 > R_\mathrm{M}$, match, provided the following equations hold in the limit $R_\mathrm{M} \to R_1$:

$$\begin{aligned} &\delta^<_A = \delta^<_{A'} = \delta_\Pi, \quad \gamma = 0, \quad \delta^<_B = \delta_\Sigma, \quad \delta^>_A = \delta^>_{A'} = 0, \\ &\alpha_1 = \alpha_{1/2}, \qquad P = 1/3, \quad \mathscr{P} = \sin^2 \varphi_\Pi. \end{aligned} \tag{11.58}$$

This is so only approximately, and a further study which seems to be beyond the simple matching procedure would be needed to bridge the gap between these two cases.

One more comment seems to be in order. Equation (11.53) shows that S^J_+ and S^J_- differ only in two phase factors. This is due to the omission of the Coriolis coupling between the degenerate states $+A$, $-A$ and $+B$, $-B$ in the region $R^<_1 < R < R^>_1$, introduced for simplification. But if the Coriolis coupling is taken into account, it can have a small but, under certain conditions, noticeable effect on the transition probability as well. Instead of approximate decoupling of six equations (11.3) to two ($+A$, $+B$) plus one ($+A'$) plus two ($-A$, $-B$) plus one ($-A'$) independent equations, the initial problem can be formulated in the total parity representation (for a quantum treatment) or in the representation of the definite symmetry character of the electronic function with respect to reflection in the collision plane (for a semiclassical treatment). These symmetry adopted electronic wave functions $|a_+\rangle$ and $|a_-\rangle$ will be linear combinations of $|+a\rangle$ and $|-a\rangle$ ($a = A, B, A'$). In this basis the scattering equations will factorize to three (A_+, B_+, A'_+) plus three (A_-, B_-, A'_-)

independent equations which are now different. This difference is due mainly to the diagonal Coriolis contribution arising from the cross terms $\langle +A|\,V_{\rm Cor}\,|-A\rangle$ and $\langle +B|\,V_{\rm Cor}\,|-B\rangle$ for states with $\Omega = 1/2$. If three equations connecting, e.g., the states A_+, B_+, A'_+ are approximately decoupled to two (A_+, B_+) plus one (A'_+), due to weak Coriolis coupling between A_+, A'_+ and B_+, A'_+, the equations for A_+, B_+ states will differ from those for the $+A$, $+B$ states.

Thus it is expected that for S^J_+ and S^J_- a more accurate treatment would give transition probabilities P_+ and P_- different from P in (11.53). A crude estimate of the effect of the diagonal Coriolis interaction on P can be obtained by replacing $\Delta\varepsilon$ by $\Delta\varepsilon + (J+1/2)/2\mu R_1^2$ or $\Delta\varepsilon - (J+1/2)/2\mu R_1^2$. It will then be seen that the difference between P_+ and P_- will be greatest when P as defined by (11.39) strongly depends on $\Delta\varepsilon$. With fixed $\Delta\varepsilon$, α_1 and v this happens for a large angular momentum J. Since P_+ and P_- bracket the probability P, the sum $P_+ + P_-$ is expected to be much closer to $2P$ than are P_+ of P_- to P. An example of this effect will be given in the next section in connection with intramultiplet mixing in Na*—He collisions.

11.3 Transition Probabilities and Cross Sections for Isotropic Collisions

The scattering matrix for the substate $^2P_{1/2}$ can be used to calculate the transition probabilities and cross sections for two inelastic events: the intramultiplet mixing $1/2 \to 3/2$, and reorientation in the substate $j = 1/2$. These will be considered now for isotropic collisions.

11.3.1 Intramultiplet Mixing

The intramultiplet mixing cross sections $\sigma_{1/2\to 3/2}$ for the transition $^2P_{1/2} \to {}^2P_{3/2}$ is defined by (2.181) via the population transfer cross section $\sigma_0^{1/2,\,3/2}$. After simple algebra we get [11.50]

$$\sigma_{1/2\to 3/2} = \frac{2\pi}{k^2}\int\limits_0^\infty J\,dJ\,(1 - |S^J_+|^2/2 - |S^J_-|^2/2). \tag{11.59}$$

The integrand in brackets defines the transition probability $P_{1/2\to 3/2}$ as a function of the total angular momentum J or of the impact parameter $b = (J+1/2)/k$. Substituting S^J_+ and S^J_- from (11.53) we get

$$\begin{aligned} P_{1/2\to 3/2} = {} & 2P(1-P) + (1-P)^2\mathscr{P} - 2P(1-P)(1-\mathscr{P})^{1/2} \\ & \times \cos(2\delta_B^< - \delta_A^< - \delta_{A'}^< + \gamma)\cos\varphi_B^< . \end{aligned} \tag{11.60}$$

For a calculation of the integral cross section the interference term in (11.60) can be neglected, and the transition probability $P_{1/2\to 3/2}$ can be replaced by the average transition probability $\bar{P}_{1/2\to 3/2}$ where

$$\bar{P}_{1/2\to 3/2} = 2P(1-P) + (1-P)^2\mathscr{P}. \tag{11.61}$$

This equation further simplifies in the near-adiabatic case:

$$\bar{P}_{1/2\to 3/2} = 2P + \mathscr{P}. \tag{11.62}$$

Under strongly nonadiabatic conditions (11.56) can be used for the S matrix. The average transition probability will then be

$$\bar{P}_{1/2\to 3/2} = \tfrac{2}{3}(1 - \tfrac{1}{3}\cos 2\varphi_{\Pi}). \tag{11.63}$$

Note that the bracketed term is the averaged probability of orientation relaxation for the P state calculated earlier, see (10.12a). For weak spin-orbit splitting this relation between $\bar{P}_{1/2\to 3/2}$ and $\bar{P}_1^{11}$ is similar to that for the integral cross sections, (2.191).

Equations (11.61, 63) match at $R_1 = R_{\mathrm{M}}$ provided (11.58) hold. This suggests a simple expression for the intramultiplet mixing cross sections,

$$\sigma_{1/2\to 3/2} = 2\pi\int_0^{b^*} \bar{P}_{1/2\to 3/2}(b^*, b)\, b\, db, \qquad b^* = \min(R_1, R_{\mathrm{M}}), \tag{11.64}$$

where $\bar{P}_{1/2\to 3/2}(R_1, b)$ is given by (11.61) and $\bar{P}_{1/2\to 3/2}(R_{\mathrm{M}}, b)$ by (11.63). For the integral cross section this approximation is similar to that for a two-state exponential model describing nonresonant charge transfer (9.58). Equation (11.64) provides a ready interpretation of the energy dependence of the intramultiplet mixing cross section. With increasing energy the cross section steeply increases in the near-adiabatic region, then reaches a broad maximum and slowly decreases. The velocity v_{m} corresponding to the maximum satisfies the equation

$$R_{\mathrm{M}}(v_{\mathrm{m}}) = R_1 \tag{11.65}$$

and the maximum cross section is

$$\sigma^{\max}_{1/2\to 3/2} = \frac{4\pi}{3}\int_0^{R_1}\left(1 - \frac{1}{3}\cos 2\varphi_{\Pi}\right) b\, db. \tag{11.66}$$

Consider now some specific cases. Take the Na* – He collisions to demonstrate the accuracy of the matching approach. Strong-coupling quantum calculations of this system for the semi-empirical potentials [11.34] at an energy $E = 0.04$ eV have been accomplished in [11.18] and will be used for comparison. At this energy R_{M} is slightly larger than R_1 so that (11.60) holds.

Turning to the matching approach, first note that for typical collisions with $b \simeq R_1/2 = 7.5$ the sequence of the Hund coupling cases is e → b → e. The case c which normally follows case e is confined for the process considered to a very narrow range.

Thus the approximate treatment [11.51] uses (11.37) for P and an appropriate version of (11.44) for $\mathscr{P}$ accounting for the nonzero value of the Massey parameter $\Delta\varepsilon R_1/v$. In calculating $P_{1/2\to 3/2}$ by (11.60) the phase difference and the rotation angle were obtained using simple analytical approximations to the Baylis potentials [11.34]. The matching-approach result is compared to the strong-coupling result presented by the full line in Fig. 11.8. The latter line represents the sum of two contributions: one from S_+^J and the other from S_-^J. The difference between the dotted and dashed curves is a measure of the diagonal Coriolis contribution to the transition probability for positive and negative states. It is seen also that though P_+ and P_- notably differ, especially in the region of the last oscillation, their sum is well approximated by simple matching.

It follows from (11.60) that the fast undulations are caused by interference of waves scattered in the fields of Σ and Π potentials, which is described by the oscillating factor containing the scattering phase shifts $\delta_B^<, \delta_A^<, \delta_{A'}^<$. The factor $\cos\varphi_B^<$ is responsible for the additional effect of slow modulation clearly seen in Fig. 11.8.

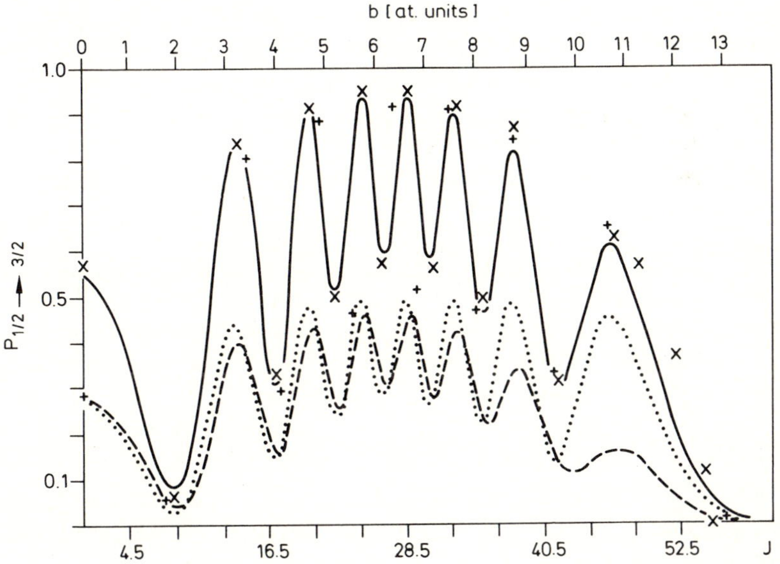

Fig. 11.8. Transition probability for Na* – He collisions. The full line refers to the close-coupling result, the dotted and dashed lines to + and − contributions to $P_{1/2\to 3/2}$. The results of semiclassical calculations are represented by (×) for the interaction calculated in [11.34] and by (+) for that in [11.43]

Another feature to be mentioned is that close to the threshold $b \simeq R_1$, both P and $\mathscr{P}$ decrease, and for $b = R_1$ the matching transition probability becomes zero. This is not the case for the quantum transition probability which extends beyond R_1. However this deficiency in the matching approximation can be readily corrected, if needed, by solving semiclassical scattering equations for large b and using the impact parameter approximation [11.52].

The matching result (11.60) provides a transparent correlation between the transition probability and the adiabatic potentials. Figure 11.8 also shows $P_{1/2 \to 3/2}$ for the potentials from [11.43]. The difference between two matching transition probabilities is interpreted by the more repulsive character of the Π term (which lowers $\mathscr{P}$) and the shallower slope of the Σ term (which lowers the undulation frequency) for the potential from [11.43] compared to the Baylis potential.

The good agreement between the essentially exact and the matching transition probabilities, except for a small threshold region, means of course that the integral cross sections are also very close. The energy dependence of the integral cross section for this case is weak, since the Na* – He collisions at a thermal energy approximately satisfy the condition (11.65) for the cross section maximum. For other alkalis, the collisions Li – X are strongly nonadiabatic, and Rb – X and Cs – X are near adiabatic.

The near-adiabatic case is particularly simple, as the transition probability is given by the sum of two terms, one for the radial and the other for the rotational nonadiabatic coupling. Accordingly, the mixing cross section is the sum of two contributions,

$$\sigma_{1/2 \to 3/2} = \sigma^{\mathrm{rad}}_{1/2 \to 3/2} + \sigma^{\mathrm{rot}}_{1/2 \to 3/2} . \tag{11.67}$$

The competition of these two contributions at decreasing energies can be inferred from (11.39, 47), taking into account that the exponents $4\pi\xi_1/3 = 4\pi\Delta\varepsilon/3\alpha_1 v_1$ and $4\pi\xi_2/3 = 4\pi\Delta\varepsilon/3\alpha_2 v_2$ are notably different (typically α_2 is approximately three times higher than α_1). If ξ_1 is not too high, the larger exponential corresponding to $\sigma^{\mathrm{rot}}_{1/2 \to 3/2}$ does not contribute to $\sigma_{1/2 \to 3/2}$ because of the small exponential coefficient a. With decreasing velocity the radial contribution $\sigma^{\mathrm{rad}}_{1/2 \to 3/2}$ decreases faster than $\sigma^{\mathrm{rot}}_{1/2 \to 3/2}$, becomes equal to the rotational contribution, and then can be neglected. This is illustrated by the collisions Rb* – He and Cs* – He that have been studied experimentally over a wide temperature range (300 – 900 K) by the sensitized fluorescence method [11.53]. The measured thermally averaged cross section $\langle\sigma_{1/2 \to 3/2}\rangle$ is related to the cross section $\sigma_{1/2 \to 3/2}$ as

$$\langle\sigma_{1/2 \to 3/2}\rangle = \int_0^\infty \sigma_{1/2 \to 3/2}(E_{\mathrm{i}}) \exp(-E_{\mathrm{i}}/kT)\, E_{\mathrm{i}}\, dE_{\mathrm{i}} (kT)^{-2} . \tag{11.68}$$

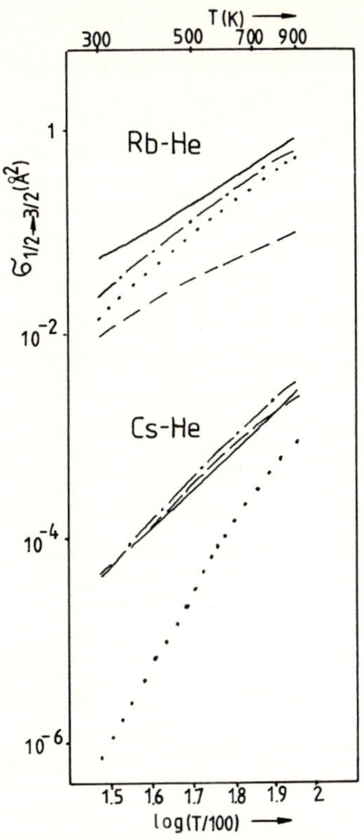

Fig. 11.9. Thermally averaged intramultiplet mixing cross sections for Rb – He and Cs – He collisions. The full lines refer to experimental results. The results of semiclassical calculations are represented by dotted (radial coupling), dashed (rotational coupling) and dot-dashed (both couplings) lines

For near adiabatic conditions this integral, with the integrand (11.67) and the transition probabilities (11.62, 40, 48), can be solved analytically [11.49] to give the temperature dependence of radial and rotational partial cross sections:

$$\begin{aligned} \sigma^{\text{rad}}_{1/2 \to 3/2} &\sim \exp[-(6\pi^2 \Delta\varepsilon^2 \mu/\alpha_1^2 k T)^{1/3}], \\ \sigma^{\text{rot}}_{1/2 \to 3/2} &\sim a \exp[-(6\pi^2 \Delta\varepsilon^2 \mu/\alpha_2^2 k T)^{1/3}]. \end{aligned} \tag{11.69}$$

We mention in connection with (11.62, 48) that condition (11.34) of the semiclassical approximation is reformulated to make the averaged cross section valid, not as $\Delta\varepsilon \ll kT$, but as a less stringent condition [11.49], since the main contribution to average cross sections comes from collisions with relative energies much higher than the average thermal energy. This condition is fulfilled for Rb* – He and Cs* – He collisions over the temperature range studied. Figure 11.9 shows the experimental cross sections (full lines) and the cross sections calculated using the Baylis potentials (dash-dotted lines) [11.49]. The theoretical cross sections represent sums of the radial (dotted line) and the rotational (dashed line) contributions which exhibit quite a different temperature dependence for each colliding pair, and radial-rotational competition in passing from one colliding pair to another.

Summing up, we note that in going from Na to Cs the intramultiplet mixing cross sections vary over a wide range, from 10^2 to 10^{-4} Å^2. It is gratifying that the simple matching approach suggests a unified and simple treatment which helps us to understand the basic features of such processes. A more sophisticated approach based on the two-trajectory approximation, but accounting for nonadiabatic coupling without any simplifications in the integration of the common-trajectory-type scattering equations [11.54], is expected to give quite reliable results.

11.3.2 Reorientation in the $^2P_{1/2}$ Substate

Consider the isotropic collisions $M^*(^2P_{1/2}) + X$ leading to reorientation, i.e., to the transitions $m = 1/2 \rightarrow m' = -1/2$. The corresponding cross section $\sigma_{1/2;-1/2}$ is expressed by (2.160) via the orientation-relaxation cross section $\sigma_1^{1/2\,1/2}$ and the population-relaxation (or population-transfer) cross section $\sigma_0^{1/2\,1/2}$. If the latter are expressed by (2.185) then the reorientation cross section has the simple form

$$\sigma_{1/2;-1/2} = \frac{\pi}{3k^2} \int_0^\infty \bar{J}\,d\bar{J}\,|S_+^{\bar{J}+1/2} + S_-^{\bar{J}-1/2}|^2 . \tag{11.70}$$

The reorientation transition probability

$$P_{1/2;-1/2}(\bar{J}) = |S_+^{\bar{J}+1/2} + S_-^{\bar{J}-1/2}|^2 \tag{11.71}$$

must be calculated using (11.53) or (11.57) in which the change of J by $\pm 1/2$ is taken into account in the partial phase shifts only (Sect. 2.1.2). Making this substitution and neglecting the interference term, we obtain the average depolarization transition probability for $R_1 < R_M$:

$$\bar{P}_{1/2;-1/2} = (P^2/3)\,(1 - \cos 4\varphi_B^<) + [(1-P)^2/3]\,(1 - \mathscr{P})\,(1 - \cos 2\varphi_A^<). \tag{11.72}$$

and for $R_1 > R_M$

$$\bar{P}_{1/2;-1/2} = \tfrac{1}{27}\,(2 - \cos 4\varphi_\Sigma - \cos 4\varphi_\Pi). \tag{11.73}$$

Again, (11.72) and (11.73) match at $R_1 = R_M$, provided (11.58) holds.

It can be seen from (11.72, 73) that the reorientation probability is close to zero when the rotation angles $\varphi_B^<$ and $\varphi_A^<$, or φ_Σ and φ_Π are small. The latter happens at large impact parameters $b \simeq R_M$ for forward scattering, and the former at very small impact parameters for backward scattering. It is also seen that the reorientation probability is smaller when effective intramultiplet mixing exists, because a large fraction of the system leaves the initial state $j = 1/2$.

In the adiabatic limit, $P = 0$ and $\mathscr{P} = 0$, and the reorientation probability is

$$\bar{P}_{1/2;\,-1/2} = \tfrac{2}{3}\sin^2\varphi_A^< . \tag{11.74}$$

This depends on the rotation angle $\varphi_A^<$ only, which must be calculated for the adiabatic motion of the diatom on the molecular term U_A. In this limit, $\bar{P}_{1/2;\,-1/2}$ is just half the probability $\bar{P}_1^{1/2\,1/2}$ for the orientation relaxation of the adiabatically isolated substate ${}^2P_{1/2}$, and because there is no interference in the adiabatic limit $\bar{P}_{1/2;\,-1/2}$ is identical with $P_{1/2;\,-1/2}$.

Equation (11.74) can now be compared with the general equations (10.25, 23), where $r^{1/2}_{1/2-1/2}$ is calculated using the matrix element between the states $+A$ and $-A$ in Table 11.1. In this way we get [11.55]

$$P_{1/2;\,-1/2} = \frac{1}{2}P_1^{1/2\,1/2} = \frac{2}{3}\sin^2\int\limits_{R_A^t}^{\infty}\frac{b(2r^{1/2}_{1/2;\,-1/2}-1)\,dR}{R^2[1-b^2/R^2-U_A(R)/E]^{1/2}}, \tag{11.75}$$

where

$$(2r^{1/2}_{1/2;\,-1/2}-1) = \frac{3}{2}\left[\frac{\Delta\varepsilon + \Delta U_{\Sigma\Pi}/3}{(\Delta\varepsilon^2 + 2\Delta U_{\Sigma\Pi}\Delta\varepsilon/3 + \Delta U_{\Sigma\Pi}^2)^{1/2}} - 1\right]. \tag{11.76}$$

When calculating the reorientation probability for this case, the matching approximation is seen to amount to a sudden change of the nondiagonal rotational-coupling matrix element from its asymptotic value $-1/2$ for $R > R_1$ to zero for $R < R_1$. For $R_1\alpha_1 \gg 1$ this approximation was found to be quite good [11.14]. Note that though in the adiabatic case the transitions from the lower molecular term A to the two upper terms A' and B are virtually prohibited, this does not mean that adiabatic mixing of free atomic states during the collision is small: for $R < R_1$ the mixing is quite strong. When the mixing is small, sometimes referred to as the situation for an energetically isolated substate $j = 1/2$, the distance R_1 is not classically accessible. Then the reorientation probability can be calculated using the perturbation approach. This approach is also applicable in the adiabatically isolated substate $j = 1/2$ for impact parameters slightly smaller than, or exceeding, R_1.

Turn now to the reorientation integral cross section $\sigma_{1/2;\,-1/2}$. Just as in (11.64), $\sigma_{1/2;\,-1/2}$ can be approximately calculated as

$$\sigma_{1/2;\,-1/2} = 2\pi\int\limits_0^{b^*}\bar{P}_{1/2;\,-1/2}(b^*, b)\,b\,db, \qquad b^* = \min(R_1, R_{\mathrm{M}}), \tag{11.77}$$

where $\bar{P}_{1/2;\,-1/2}(R_1, b)$ is given by (11.72), and $\bar{P}_{1/2;\,-1/2}(R_{\mathrm{M}}, b)$ by (11.73). This expression helps us to understand the energy dependence of $\sigma_{1/2;\,-1/2}$

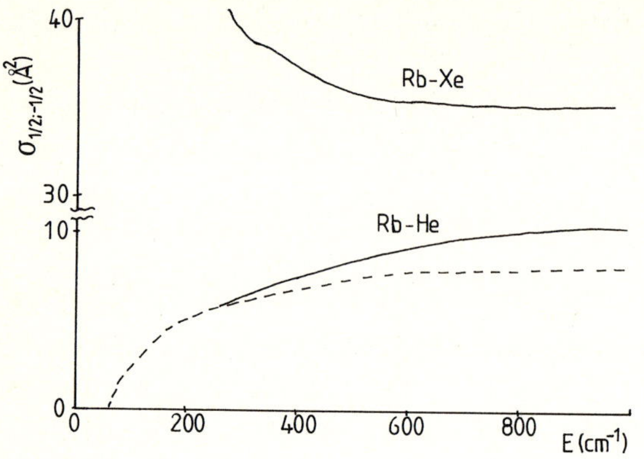

Fig. 11.10. Calculated integral reorientation cross sections for Rb – He and Rb – Xe pairs (—) from [11.14]. The dashed line refers to the rigid-sphere threshold approximation (11.78) normalized at $E = \Delta\varepsilon$

found from strong-coupling calculations for Rb* – He and Rb* – Xe with Baylis potentials [11.14] (see Fig. 11.10). In the case of He the interaction in the Σ state is essentially repulsive. Equation (11.9) for the U_A term implies that the potential U_A exhibits a step-like behavior for $R = R_1$ with a step height $E_0 = \Delta\varepsilon/3$. This means that a radial energy of at least $E_R = \Delta\varepsilon/3$ is needed to surmount the step in order to reach $R = R_1$ and approach further. Accordingly, the cross section is expected to display the characteristic threshold behavior

$$\sigma_{1/2;-1/2} \sim 1 - \Delta\varepsilon/3E. \tag{11.78}$$

This function, normalized to the calculated cross section for $E = \Delta\varepsilon$, is shown in Fig. 11.10 by the dashed line. The steeper rise of $\sigma_{1/2;-1/2}$ compared to the function (11.78) might be due to an increase of the rotation angle with decreasing R_A^t (approximately equal to the radius of the repulsive core).

In the case of Xe the potential U_A has a well with a depth $\varepsilon_0 = 210\ \mathrm{cm}^{-1}$, and the step in the potential U_A on the attractive branch does not prevent the diatom from reaching the distance R_1. In addition, the typical rotation angle $2\varphi_A^<$ is rather large. This all helps to explain the negative energy dependence of $\sigma_{1/2;-1/2}$.

Appendix

A. Quantum Theory of Angular Momentum

The quantum theory of angular momentum has been exhaustively developed and many monographs, reviews, and handbooks deal with it [A.1–13]. Nevertheless it seems to be useful to summarize in this appendix the definitions and results of angular momentum theory, using the notations and phase conditions adopted in this book.

A.1 Rotation Matrices and Spherical Functions

Consider an isolated physical system $\mathscr{A}$ in a certain space-fixed coordinate system xyz. The stationary wave functions of $\mathscr{A}$ can be taken as simultaneous eigenfunctions $|ajm\rangle$ of the total angular momentum squared $\boldsymbol{J}^2$ and its z projection J_z. These functions are such that

$$\begin{aligned} \boldsymbol{J}^2 |ajm\rangle &= j(j+1)|ajm\rangle , \\ J_z|ajm\rangle &= m|ajm\rangle , \end{aligned} \tag{A.1}$$

where j can assume integer or half-integer non-negative values and $m = -j, -j+1, \ldots, j-1, j$ (a denotes all other quantum numbers).

Now transform to another coordinate system $x_1\,y_1\,z_1$. An arbitrary orientation of this system with respect to the first one can be achieved by three successive rotations about coordinate axes. In the present-day quantum mechanical discussions of rotations these are most commonly made by

i) Rotating the system xyz through an angle α about the z axis to give the system $x'\,y'\,z'$.
ii) Rotating the system $x'\,y'\,z'$ through an angle β about the y' axis to give the system $x''\,y''\,z''$.
iii) Rotating the system $x''\,y''\,z''$ through an angle γ about the z'' axis to give the system $x_1\,y_1\,z_1$.

Note that β and α are the conventional polar and azimuthal angles which give the direction of the final z_1 axis with respect to the original coordinate system. The angles α, β, γ are called Euler angles.

The wave functions $|a j m\rangle_1$ which are simultaneous eigenfunctions of $\boldsymbol{J}^2$ and J_{z_1} can be defined in the coordinate system $x_1\, y_1\, z_1$. These functions are expressed through the functions $|a j m\rangle$ by the linear relations

$$|a j m\rangle_1 = \sum_{m'} D^j_{m' m}(\alpha, \beta, \gamma)\, |a j m'\rangle . \tag{A. 2}$$

The rotation matrices $D^j_{m m'}(\alpha, \beta, \gamma)$ (often referred to also as the D functions) can be obtained as the solutions (finite at $\beta = 0, \pi$) of the differential equations

$$-\mathrm{i}(\partial/\partial\alpha)\, D^j_{m m'}(\alpha, \beta, \gamma) = -\, m\, D^j_{m m'}(\alpha, \beta, \gamma), \tag{A. 3 a}$$

$$-\mathrm{i}(\partial/\partial\gamma)\, D^j_{m m'}(\alpha, \beta, \gamma) = -\, m'\, D^j_{m m'}(\alpha, \beta, \gamma), \tag{A. 3 b}$$

$$\begin{aligned} &-\{(\partial^2/\partial\beta^2) + \cot\beta\,(\partial/\partial\beta) + (\sin\beta)^{-2}\,[(\partial^2/\partial\alpha^2) \\ &\quad - 2\cos\beta\,(\partial^2/\partial\alpha\,\partial\gamma) + (\partial^2/\partial\gamma^2)]\}\, D^j_{m m'}(\alpha, \beta, \gamma) \\ &\quad = j(j+1)\, D^j_{m m'}(\alpha, \beta, \gamma), \end{aligned} \tag{A. 3 c}$$

which satisfy the boundary conditions

$$\begin{aligned} &D^j_{m m'}(\alpha \pm 2 n \pi, \beta, \gamma) = D^j_{m m'}(\alpha, \beta \pm 2 n \pi, \gamma) \\ &= D^j_{m m'}(\alpha, \beta, \gamma \pm 2 n \pi) = D^j_{m m'}(\alpha, \beta, \gamma), \end{aligned} \tag{A. 4}$$

where $n = 0, 1, 2, 3, \ldots$ for integer j and $n = 0, 2, 4, 6, \ldots$ for half-integer j. The normalization and phase of $D^j_{m m'}(\alpha, \beta, \gamma)$ are fixed so that

$$D^j_{m m'}(0, 0, 0) = \delta_{m m'} \tag{A. 5}$$

and

$$\begin{aligned} &\mathrm{i}(2)^{-1/2} \exp(\pm\,\mathrm{i}\,\alpha)\,[\mp \cot\beta\,(\partial/\partial\alpha) + \mathrm{i}(\partial/\partial\beta) \pm (\sin\beta)^{-1}(\partial/\partial\gamma)] \\ &\quad \times D^j_{m m'}(\alpha, \beta, \gamma) = \pm\,(2)^{-1/2}\,[j(j+1) - m(m \mp 1)]^{1/2}\, D^j_{m \mp 1\, m'}(\alpha, \beta, \gamma), \end{aligned} \tag{A.6a}$$

$$\begin{aligned} &\mathrm{i}(2)^{-1/2} \exp(\mp\,\mathrm{i}\,\gamma)\,[\pm \cot\beta\,(\partial/\partial\gamma) + \mathrm{i}(\partial/\partial\beta) \mp (\sin\beta)^{-1}(\partial/\partial\alpha)] \\ &\quad \times D^j_{m m'}(\alpha, \beta, \gamma) = \pm\,(2)^{-1/2}[j(j+1) - m'(m' \pm 1)^{1/2}\, D^j_{m\, m' \pm 1}(\alpha, \beta, \gamma). \end{aligned} \tag{A.6b}$$

The resultant matrices $D^j_{m m'}(\alpha, \beta, \gamma)$ are unitary and satisfy the conditions

$$\begin{aligned} &\int_0^{2\pi} d\alpha \int_0^{\pi} d\beta \sin\beta \int_0^{2\pi} d\gamma\, D^{j_1}_{m_1 m_1'}(\alpha, \beta, \gamma)\, D^{j_2\,*}_{m_2 m_2'}(\alpha, \beta, \gamma) \\ &\quad = \delta_{j_1 j_2}\, \delta_{m_1 m_2}\, \delta_{m_1' m_2'}\, 8\,\pi^2/(2 j_1 + 1). \end{aligned} \tag{A. 7}$$

It follows from (A. 3) that the D-functions can be represented in the form

$$D^j_{m m'}(\alpha, \beta, \gamma) = \exp(-\,\mathrm{i}\, m\, \alpha)\, d^j_{m m'}(\beta) \exp(-\,\mathrm{i}\, m'\, \gamma). \tag{A. 8}$$

The functions $d^j_{mm'}(\beta)$ possess the symmetry properties

$$\begin{aligned} d^j_{mm'}(\beta) &= (-1)^{m-m'} d^j_{-m-m'}(\beta) = (-1)^{m-m'} d^j_{m'm}(\beta) , \\ d^j_{mm'}(\pi-\beta) &= (-1)^{j-m'} d^j_{-mm'}(\beta) = (-1)^{j+m} d^j_{m-m'}(\beta) , \end{aligned} \tag{A.9}$$

and are simply expressed through $d^j_{mm'}(\pi/2)$, for which the special notation $\Delta^j_{mm'}$ is used.

For integer $j = l$, the D functions are related to the spherical functions Y_l^m (the phases of the latter are defined according to *Condon* and *Shortley* [A.14]):

$$Y_l^m(\beta,\alpha) = [(2l+1)/4\pi]^{1/2} D^{l*}_{m0}(\alpha,\beta,0) . \tag{A.10}$$

The algebraic expressions for $d^j_{mm'}(\beta)$ for $j \leqq 5$ can be found in [A. 10] and for $j \leqq 13$ in [A. 15]. Asymptotic expressions for $d^j_{mm'}$ (or Y_l^m) can be obtained for large j (or l). In particular, for $l \gg 1$, $l \gg |m|$, and $l\beta \gg 1$,

$$\begin{aligned} Y_l^m(\beta,\alpha) \simeq {} & \exp(\mathrm{i}m\alpha)[\pi(\sin\beta)^{1/2}]^{-1} \\ & \times \cos[(2l+1)(\beta/2) + (2m-1)(\pi/4)] . \end{aligned} \tag{A.11}$$

On the other hand, for $l \gg 1$ and $\beta < 1$, and when the condition

$$\beta^3 l/24 \ll 1 \tag{A.12}$$

is fulfilled,

$$Y_l^m(\beta,\alpha) \simeq (l/2\pi)^{1/2} \exp(\mathrm{i}m\alpha) J_{-m}(l\beta) \tag{A.13}$$

where J_{-m} is a Bessel function. For large l, such a range of β variations exists when the condition (A. 12) is fulfilled simultaneously with the condition $\beta l \gg 1$. In this range of β variation, an asymptotic expansion of $J_{-m}(l\beta)$ can be used (see, e.g., [A. 16]). For $|m| \ll l$ this yields an expression for $Y_l^m(\beta,\alpha)$ coinciding with (A. 11). Thus, for $l \gg 1$, the asymptotic expressions (A. 11) and (A. 13) match smoothly at some $\beta_0 < 1$. Since, in practice, the values of l are usually smaller than 10^3, typical values of β_0 are of the order 0.1.

The derivations of asymptotic expressions for Y_l^m under conditions different from those adopted above, and also for $d^j_{mm'}$ are given in [A. 17, 18] using the JWKB solution of (A. 3c), and in [A. 19] using the notion of coherent states.

A.2 Coupling of Angular Momenta, Clebsch-Gordan Coefficients and $3n-j$ Symbols

Let the system $\mathscr{A}$ now consist of N noninteracting subsystems $\mathscr{A}_1, \ldots, \mathscr{A}_N$. In this case the simultaneous eigenfunctions $|a_1 j_1 m_1\rangle |a_2 j_2 m_2\rangle \ldots |a_N j_N m_N\rangle$ of the operators $\boldsymbol{J}_1^2, \ldots, \boldsymbol{J}_N^2$ and $J_{1z}, \ldots, J_{Nz}$ can be taken as stationary wavefunctions of the total system. The simultaneous eigenfunctions of $\boldsymbol{J}^2 = (\boldsymbol{J}_1 + \ldots + \boldsymbol{J}_N)^2$ and $J_z = J_{1z} + \ldots J_{Nz}$ constructed from the above products are often of interest.

For $N = 2$ the problem of constructing such functions can be solved uniquely. The result is

$$|a_1 j_1 a_2 j_2 j m\rangle = \sum_{m_1 m_2} \begin{bmatrix} j_1 & j_2 & j \\ m_1 & m_2 & m \end{bmatrix} |a_1 j_1 m_1\rangle |a_2 j_2 m_2\rangle, \tag{A.14}$$

where $\begin{bmatrix} j_1 & j_2 & j \\ m_1 & m_2 & m \end{bmatrix}$ are the Clebsch-Gordan coefficients. These coefficients assume nonzero values only when $m_1 + m_2 = m$ and $j_1 + j_2 \geqq j \geqq |j_1 - j_2|$. They are conventionally real and satisfy the following conditions of orthogonality and completeness:

$$\sum_{jm} \begin{bmatrix} j_1 & j_2 & j \\ m_1 & m_2 & m \end{bmatrix} \begin{bmatrix} j_1 & j_2 & j \\ m_1' & m_2' & m \end{bmatrix} = \delta_{m_1 m_1'} \delta_{m_2 m_2'}, \tag{A.15}$$

$$\sum_{m_1 m_2} \begin{bmatrix} j_1 & j_2 & j \\ m_1 & m_2 & m \end{bmatrix} \begin{bmatrix} j_1 & j_2 & j' \\ m_1 & m_2 & m' \end{bmatrix} = \delta_{jj'} \delta_{mm'}. \tag{A.16}$$

The following symmetry properties of the Clebsch-Gordan coefficients are useful:

$$\begin{aligned} \begin{bmatrix} j_1 & j_2 & j \\ m_1 & m_2 & m \end{bmatrix} &= (-1)^{j_1 + j_2 - j} \begin{bmatrix} j_2 & j_1 & j \\ m_2 & m_1 & m \end{bmatrix} \\ &= (-1)^{j_1 + j_2 - j} \begin{bmatrix} j_1 & j_2 & j \\ -m_1 & -m_2 & -m \end{bmatrix}, \end{aligned} \tag{A.17}$$

$$\begin{bmatrix} j_1 & j_2 & j \\ m_1 & m_2 & m \end{bmatrix} = (-1)^{j_2 + m_2} [(2j+1)/(2j_1+1)]^{1/2} \begin{bmatrix} j_2 & j & j_1 \\ -m_2 & m & m_1 \end{bmatrix}, \tag{A.18}$$

$$\begin{bmatrix} j_1 & j_2 & j \\ m_1 & m_2 & m \end{bmatrix} = (-1)^{j_1 - m_1} [(2j+1)/(2j_2+1)]^{1/2} \begin{bmatrix} j & j_1 & j_2 \\ m & -m_1 & m_2 \end{bmatrix}. \tag{A.19}$$

More symmetrical $3-jm$ symbols $\begin{pmatrix} j_1 & j_2 & j_3 \\ m_1 & m_2 & m_3 \end{pmatrix}$ are often used in scientific literature, instead of the Clebsch-Gordan coefficients, and are connected with the latter by the relations

$$\begin{pmatrix} j_1 & j_2 & j_3 \\ m_1 & m_2 & m_3 \end{pmatrix} = (-1)^{j_1-j_2-m_3}(2j_3+1)^{-1/2} \begin{bmatrix} j_1 & j_2 & j_3 \\ m_1 & m_2 & -m_3 \end{bmatrix}. \tag{A.20}$$

Algebraic and numerical tables of Clebsch-Gordan coefficients and $3-jm$ symbols can be found in [A. 10, 12, 20–22]. The analytical asymptotic expressions for these coefficients hold when two or the three values in the coefficient are high. In particular, when $j_1, j \gg j_2$,

$$\begin{bmatrix} j_1 & j_2 & j \\ m_1 & m_2 & m \end{bmatrix} = \delta_{m-m_1;m_2} d^{j_2}_{m_2 \nu}(\beta), \tag{A.21}$$

where $\nu = j - j_1$ and

$$\cos\beta = m/(j+1/2). \tag{A.22}$$

The derivation of (A. 21) is given in [A. 17]. The asymptotic expressions for the Clebsch-Gordan coefficients for $j_1, j_2, j \gg 1$ have been derived in [A. 18, 23, 24].

When $N > 2$ there arise several linearly independent wavefunctions corresponding to the same values of j and m. Several equivalent sets of such functions can be obtained by adopting different addition sequences for the subsystem angular momenta. These sets are related by the unitary transformations independent of the magnetic quantum numbers. The transformation matrices are expressed through the familiar $3n-j$ symbols, where $n = N-1$. The $6-j$ and $9-j$ symbols are most often encountered.

Three significantly different schemes for angular momenta addition are possible for $N = 3$:

(i) $j_1 + j_2 \to j_{12}, \quad j_{12} + j_3 \to j$;

(ii) $j_1 + j_3 \to j_{13}, \quad j_{13} + j_2 \to j$;

(iii) $j_3 + j_2 \to j_{32}, \quad j_{32} + j_1 \to j$.

The matrix of the transformation between, e.g., schemes (i) and (iii) is

$$\begin{aligned} &\langle j_1 j_2 j_{12} j_3 j | j_3 j_2 j_{32} j \rangle \\ &= (2j+1)^{-1} \sum_{\substack{m_1 m_2 m_3 \\ m_{12} m_{32} m}} \begin{bmatrix} j_1 & j_2 & j_{12} \\ m_1 & m_2 & m_{12} \end{bmatrix} \begin{bmatrix} j_{12} & j_3 & j \\ m_{12} & m_3 & m \end{bmatrix} \\ &\times \begin{bmatrix} j_3 & j_2 & j_{32} \\ m_3 & m_2 & m_{32} \end{bmatrix} \begin{bmatrix} j_{32} & j_1 & j \\ m_{32} & m_1 & m \end{bmatrix} = (-1)^{2j}[(2j_{12}+1)(2j_{32}+1)]^{1/2} \\ &\times \begin{Bmatrix} j_1 & j_2 & j_{12} \\ j_3 & j & j_{32} \end{Bmatrix}. \end{aligned} \tag{A.23}$$

Thus the $6-j$ symbol $\begin{Bmatrix} j_1 & j_2 & j_{12} \\ j_3 & j & j_{32} \end{Bmatrix}$ is expressed as the sum of products of four Clebsch-Gordan coefficients. Note that, due to the properties of the latter, the summation in (A. 23) is actually performed over two indices only. It follows from (A. 23) that the $6-j$ symbol $\begin{Bmatrix} j_1 & j_2 & j_3 \\ j_4 & j_5 & j_6 \end{Bmatrix}$ is zero if the triangular conditions in (j_1, j_2, j_3), (j_1, j_5, j_6), (j_4, j_2, j_6) and (j_4, j_5, j_3) are not simultaneously satisfied. The $6-j$ symbols possess several symmetry properties. In particular they are invariant under interchange of any two columns and under simultaneous interchange of upper and lower arguments in any two columns.

The so-called Racah coefficients $W(j_1 j_2 j_5 j_4; j_3 j_6)$ are sometimes used in the literature. These are related to the $6-j$ symbols by

$$W(j_1 j_2 j_5 j_4; j_3 j_6) = (-1)^{j_1+j_2+j_4+j_5} \begin{Bmatrix} j_1 & j_2 & j_3 \\ j_4 & j_5 & j_6 \end{Bmatrix}. \tag{A. 24}$$

Algebraic expressions and numerical values of $6-j$ symbols are tabulated in [A. 10, 12, 20, 21]. Various asymptotic expressions for $6-j$ symbols with several large arguments have been derived [A. 2, 17, 18, 23, 24]. For instance, when the $6-j$ symbols can be expressed in the form $\begin{Bmatrix} a & b & c \\ d+l & e+l & f+l \end{Bmatrix}$ where $l \gg 1$ and a, b, c, d, e, f are arbitrary, then

$$\begin{Bmatrix} a & b & c \\ d+l & e+l & f+l \end{Bmatrix} \simeq [2l(2c+1)]^{-1/2}(-1)^{2l+a+b+d+e} \begin{bmatrix} a & b & c \\ \nu_1 & \nu_2 & \nu_3 \end{bmatrix}, \tag{A. 25}$$

where $\nu_1 = f-e$, $\nu_2 = d-f$, $\nu_3 = d-e$.

The $9-j$ symbols are connected with the transformation matrices between different schemes of addition of four angular momenta, e.g.,

$$\langle j_1 j_2 j_{12}, j_3 j_4 j_{34} j | j_1 j_3 j_{13}, j_2 j_4 j_{24}, j \rangle = \sum_{m_i m_{kl} m} (2j+1)^{-1}$$

$$\begin{bmatrix} j_1 & j_2 & j_{12} \\ m_1 & m_2 & m_{12} \end{bmatrix} \begin{bmatrix} j_3 & j_4 & j_{34} \\ m_3 & m_4 & m_{34} \end{bmatrix} \begin{bmatrix} j_{12} & j_{34} & j \\ m_{12} & m_{34} & m \end{bmatrix} \begin{bmatrix} j_1 & j_3 & j_{13} \\ m_1 & m_3 & m_{13} \end{bmatrix}$$

$$\begin{bmatrix} j_2 & j_4 & j_{24} \\ m_2 & m_4 & m_{24} \end{bmatrix} \begin{bmatrix} j_{13} & j_{24} & j \\ m_{13} & m_{24} & m \end{bmatrix} = [(2j_{12}+1)(2j_{34}+1)$$

$$\times (2j_{13}+1)(2j_{24}+1)]^{1/2} \begin{Bmatrix} j_1 & j_2 & j_{12} \\ j_3 & j_4 & j_{34} \\ j_{13} & j_{24} & j \end{Bmatrix}. \tag{A. 26}$$

It follows from (A. 26) that the $9-j$ symbol $\{j_1 j_2 j_{12}; j_3 j_4 j_{34}; j_{13} j_{24} j\}$ is zero if the triangular conditions in (j_1, j_2, j_{12}), (j_3, j_4, j_{34}), (j_{12}, j_{34}, j), (j_1, j_3, j_{13}), (j_2, j_4, j_{24}), (j_{13}, j_{24}, j) are not fulfilled simultaneously. The algebraic expressions and numerical values of the $9-j$ symbols are tabulated in [A. 10, 12, 20, 25]. The following expression for the $9-j$ symbols in terms of the $6-j$ symbols is useful for practical calculations of the former:

$$\begin{Bmatrix} j_1 & j_2 & j_{12} \\ j_3 & j_4 & j_{34} \\ j_{13} & j_{24} & j \end{Bmatrix} = \sum_{\varkappa} (-1)^{\varkappa} (2\varkappa + 1) \begin{Bmatrix} j_1 & j_3 & j_{13} \\ j_{24} & j & \varkappa \end{Bmatrix} \begin{Bmatrix} j_2 & j_4 & j_{24} \\ j_3 & \varkappa & j_{34} \end{Bmatrix} \begin{Bmatrix} j_{12} & j_{34} & j \\ \varkappa & j_1 & j_2 \end{Bmatrix}. \tag{A. 27}$$

Here the summation is performed over all values of $\varkappa$ for which the product of the $6-j$ symbols is not zero.

A. 3 Matrix Elements of the Irreducible Tensor Operators

The irreducible tensor operators $T_{\varkappa}^{q}$ of rank $\varkappa$ transform under rotations of the coordinate system similarly to the eigenfunctions of $\boldsymbol{J}^2$ and J_z with $j = \varkappa$ and $m = q$. For $\varkappa = 1$, these operators are the spherical components of three-dimensional vectors, and

$$T_1^{(0)} = T_z, \quad T_1^{(\pm 1)} = \mp (2)^{-1/2} (T_x \pm \mathrm{i}\, T_y). \tag{A. 28}$$

The operators T_2^q are the spherical components of the traceless symmetric tensors. The scalar corresponds to $\varkappa = 0$.

The irreducible tensor operator $T_{\varkappa}^{q}(\mathscr{A}_1, \mathscr{A}_2)$ can be constructed from the products of two commuting operators $T_{\varkappa}^{q_1}(\mathscr{A}_1)$ and $T_{\varkappa}^{q_2}(\mathscr{A}_2)$, which depend on the coordinates of two different subsystems, using the angular momenta coupling expression

$$T_{\varkappa}^{q}(\mathscr{A}_1, \mathscr{A}_2) = \sum_{q_1 q_2} \begin{bmatrix} \varkappa_1 & \varkappa_2 & \varkappa \\ q_1 & q_2 & q \end{bmatrix} T_{\varkappa_1}^{q_2}(\mathscr{A}_1)\, T_{\varkappa_2}^{q_2}(\mathscr{A}_2). \tag{A. 29}$$

The scalar product of two tensor operators of the same rank $T_{\varkappa}(\mathscr{A}_1) \cdot T_{\varkappa}(\mathscr{A}_2)$ is

$$T_{\varkappa}(\mathscr{A}_1) \cdot T_{\varkappa}(\mathscr{A}_2) = \sum_{q} (-1)^q\, T_{\varkappa}^{q}(\mathscr{A}_1)\, T_{\varkappa}^{-q}(\mathscr{A}_2). \tag{A. 30}$$

The scalar product $T_{\varkappa}(\mathscr{A}_1) \cdot T_{\varkappa}(\mathscr{A}_2)$ differs from $T_0^0(\mathscr{A}_1, \mathscr{A}_2)$ defined by (A. 29) by a factor $(-1)^{\varkappa}(2\varkappa + 1)^{-1/2}$. For $\varkappa = 1$, (A. 30) gives the familiar scalar product of two vectors.

According to the Wigner-Eckart theorem,

$$\langle a j m | T_\varkappa^q | a' j' m' \rangle = \begin{bmatrix} j' & \varkappa & j \\ m' & q & m \end{bmatrix} f(aj, a'j', \varkappa). \tag{A. 31}$$

The definition of the reduced matrix elements $\langle aj \| T_\varkappa \| a'j' \rangle$ adopted in this book coincides with that used in the theory of atomic spectra:

$$\langle a j m | T_\varkappa^q | a' j' m' \rangle = (-1)^{2\varkappa} (2j+1)^{-1/2} \times \begin{bmatrix} j' & \varkappa & j \\ m' & q & m \end{bmatrix} \langle aj \| T_\varkappa \| a'j' \rangle. \tag{A. 32}$$

The reduced matrix element for the operator of the total angular momentum J_1^q is

$$\langle aj \| J_1 \| a'j' \rangle = \delta_{aa'} \delta_{jj'} [(2j+1)(j+1)j]^{1/2}. \tag{A. 33}$$

The spherical function $Y_\varkappa^q$ is obviously an irreducible tensor operator of rank $\varkappa$ and the reduced matrix element $\langle l \| Y_\varkappa \| l' \rangle$ (the wave functions $|lm\rangle$ and $|l'm'\rangle$ are the spherical functions Y_l^m and $Y_{l'}^{m'}$) is

$$\langle l \| Y_\varkappa \| l' \rangle = (-1)^\varkappa (2)^{-1} [(2l+1)(2\varkappa+1)/\pi]^{1/2} \begin{bmatrix} l & \varkappa & l' \\ 0 & 0 & 0 \end{bmatrix}. \tag{A. 34}$$

When there are two subsystems $\mathscr{A}_1$ and $\mathscr{A}_2$, the reduced matrix element of $T_\varkappa^q(\mathscr{A}_1)$ between the wave functions given by (A. 14), $\langle a_1 j_1 a_2 j_2 j \| T_\varkappa(\mathscr{A}_1) \| a_1' j_1' a_2' j_2' j' \rangle$ is expressed through the reduced matrix elements $\langle a_1 j_1 \| T_\varkappa(\mathscr{A}_1) \| a_1' j_1' \rangle$ as

$$\begin{aligned} \langle a_1 j_1 a_2 j_2 j \| T_\varkappa(\mathscr{A}_1) \| a_1' j_1' a_2' j_2' j' \rangle &= \delta_{a_2 a_2'} \delta_{j_2 j_2'} \\ &\times (-1)^{j_1 + j_2 + j' + \varkappa} [(2j+1)(2j'+1)]^{1/2} \\ &\times \begin{Bmatrix} j_1 & j & j_2 \\ j' & j_1' & \varkappa \end{Bmatrix} \langle a_1 j_1 \| T_\varkappa(\mathscr{A}_1) \| a_1' j_1' \rangle. \end{aligned} \tag{A. 35}$$

The expression for the matrix element of the scalar product $T_\varkappa(\mathscr{A}_1) \cdot T_\varkappa(\mathscr{A}_2)$ is

$$\begin{aligned} &\langle a_1 j_1 a_2 j_2 j m | T_\varkappa(\mathscr{A}_1) \cdot T_\varkappa(\mathscr{A}_2 | a_1' j_1' a_2' j_2' j' m' \rangle \\ &= \delta_{mm'} \delta_{jj'} (-1)^{j + j_2 + j_1'} \begin{Bmatrix} j & j_2 & j_1 \\ \varkappa & j_1' & j_2' \end{Bmatrix} \\ &\quad \times \langle a_1 j_1 \| T_\varkappa(\mathscr{A}_1) \| a_1' j_1' \rangle \langle a_2 j_2 \| T_\varkappa(\mathscr{A}_2) \| a_2' j_2' \rangle. \end{aligned} \tag{A. 36}$$

References

Chapter 1

1.1 N.F. Mott, H.S.W. Massey: *The Theory of Atomic Collisions* (Clarendon, Oxford 1965)
1.2 V. de Alfaro, T. Regge: *Potential Scattering* (North-Holland, Amsterdam 1965)
1.3 A.I. Baz', Ya.B. Zel'dovich, A.M. Perelomov: *Rasseyanie, Reaktsii i Raspady v Nerelyativistskoi Kvantovoi Mekhanike* (Nauka, Moscow 1971)
1.4 M.L. Goldberger, K.M. Watson: *Collision Theory* (Wiley, New York 1964)
1.5 R.G. Newton: *Scattering Theory of Waves and Particles*, 2nd ed., Texts and Monographs in Phys. (Springer, Berlin, Heidelberg, New York 1982)
1.6 J.R. Taylor: *Scattering Theory. The Quantum Theory of Nonrelativistic Collisions* (Wiley, New York 1972)
1.7 T.Y. Wu, T. Ohmura: *Quantum Theory of Scattering* (Prentice Hall, New York 1962)
1.8 V.P. Zhigunov, B.N. Zakhar'ev: *Methody Sil'noi Svyazi Kanalov v Kvantovoi Teorii Rasseyaniya* (Atomizdat, Moscow 1974)
1.9 M.S. Child: *Molecular Collision Theory* (Academic, New York 1974)
1.10 S. Geltman: *Topics in Atomic Collision Theory* (Academic, New York 1969)
1.11 H.S.W. Massey, E.H.S. Burhop, H.B. Gilbody: *Slow Collisions of Heavy Particles*, Electronic and Ionic Impact Phenomena, Vol.3, 2nd ed. (Clarendon, Oxford 1971)
1.12 M.R.C. Mc Dowell, J.P. Coleman: *Introduction to the Theory of Ion-Atom Collisions* (North-Holland, Amsterdam 1970)
1.13 E.E. Nikitin: *Theory of Elementary Atomic and Molecular Processes in Gases* (Clarendon, Oxford 1974)
1.14 K. Smith: *The Calculation of Atomic Collision Processes* (Wiley Interscience, New York 1971);
R.E. Johnson: *Introduction to Atomic and Molecular Collisions* (Plenum, New York 1982);
H.O. Lutz, J.S. Briggs, H. Kleinpoppen (eds.): *Fundamental Processes in Energetic Atomic Collisions*, NATO ASI Series B, Physics, Vol.103 (Plenum, New York 1983)
1.15 B.M. Smirnov: *Atomnye Stolknoveniya i Elementarnye Protsessy v Plazme* (Atomizdat, Moscow 1968)
1.16 B.M. Smirnov: *Asimptoticheskie Metody v Teorii Atomnykh Stoknovenii* (Atomizdat, Moscow 1973)
1.17 I.I. Sobelman: *Atomic Spectra and Radiative Transitions*, Springer Ser. Chem. Phys., Vol.1 (Springer, Berlin, Heidelberg, New York 1979)
1.18 I.I. Sobelman, L.A. Vainshtein, E.A. Yukov: *Excitation of Atoms and Broadening of Spectral Lines*, Springer Ser. Chem. Phys., Vol.7 (Springer, Berlin, Heidelberg, New York 1981)
1.19 A.P. Yutsis, A.A. Bandzaitis: *Teoriya Momenta Kolichestva Dvizheniya v Kvantovoi Mekhanike* (Mintis, Vilnius 1965)

Chapter 2

2.1 D.G. Truhlar, A.C. Mead, A.M. Brandt: Adv. Chem. Phys. **33**, 295 (1975)

2.2 E.U. Condon, G.H. Shortley: *Theory of Atomic Spectra* (Cambridge University Press, London 1951)
2.3 R.G. Newton: *Scattering Theory of Waves and Particles*, 2nd ed., Texts and Monographs in Phys. (Springer, Berlin, Heidelberg, New York 1982)
2.4 M.L. Goldberger, K.M. Watson: *Collision Theory* (Wiley, New York 1964)
2.5 M. Jacob, G.C. Wick: Ann. Phys. **7**, 404 (1959)
2.6 B.H. Choi, R.T. Poe, K.T. Tang: J. Chem. Phys. **69**, 411 (1977)
2.7 A. Bárány: J. Phys. B**12**, 2841 (1979)
2.8 R.J. Cross, Jr.: J. Chem. Phys. **51**, 5163 (1969)
2.9 G.V. Dubrovskii: Zh. Eksp. Teor. Fiz. **58**, 1075 (1970) [English transl.: Sov. Phys. JETP **31**(58), 577 (1970)]
2.10 T.A. Green, R.E. Johnson: Phys. Rev. **152**, 9 (1966)
2.11 R.P. Marchi: Phys. Rev. **183**, 185 (1969)
2.12 R. McCarroll, A. Salin: J. Phys. B**1**, 163 (1968); R.E. Turner: Phys. Rev. A**27**, 841 (1983); R.E. Turner, J.S. Dahler: J. Phys. A**16**, 975 (1983)
2.13 M.E. Riley: Phys. Rev. A**8**, 742 (1973)
2.14 A. Russek: Phys. Rev. A**20**, 113 (1979); E.E. Nikitin: In *New Horizons of Quantum Chemistry*, ed. by P.O. Löwdin, P. Pullman (Reidel, Dordrecht 1983) p.117
2.15 F.T. Smith: Phys. Lett. **10**, 290 (1964)
2.16 F.T. Smith: In *Physics of the One- and Two-Electron Atoms*, ed. by H. Kleinpoppen (North-Holland, Amsterdam 1969) p.755
2.17 F.T. Smith: In *Atomic Collision Processes*, ed. by S. Geltman, K.T. Mahantappa, W.E. Brittin (Gordon and Breach, New York 1969) p.95
2.18 F.T. Smith: In *Atomic Physics*, ed. by F. Bederson, V.W. Cohen, F.M.J. Pichanick (Plenum, New York 1969) p.353
2.19 M.S. Child: In *Dynamics of Molecular Collisions*, Part B, ed. by W.H. Miller (Plenum, New York 1976) p.171
2.20 M.S. Child: Adv. Atom. Mol. Phys. **14**, 225 (1978)
2.21 W.H. Miller: Adv. Chem. Phys. **25**, 69 (1974)
2.22 W.H. Miller: Adv. Chem. Phys. **30**, 77 (1975)
2.23 W. Kolos, L. Wolniewicz: Rev. Mod. Phys. **35**, 473 (1963)
2.24 L.D. Landau, E.M. Lifshitz: *Quantum Mechanics*, 3rd rev. ed. (Pergamon, Oxford 1977)
2.25 M. Mizushima: *The Theory of Rotating Diatomic Molecules* (Wiley, New York 1975)
2.26 R.T. Pack, J.O. Hirschfelder: J. Chem. Phys. **49**, 4009 (1968)
2.27 J.H. Van Vleck: Rev. Mod. Phys. **23**, 213 (1951)
2.28 M.S. Child: *Molecular Collision Theory* (Academic, New York 1974)
2.29 D.C.S. Allison, P.G. Burke: J. Phys. B**2**, 941 (1969)
2.30 C.G. Carrington, D.N. Stacey, J. Cooper: J. Phys. B**6**, 417 (1973)
2.31 K. Smith: *The Calculation of Atomic Collision Processes* (Wiley Interscience, New York 1971)
2.32 D.R. Bates: Comments At. Mol. Phys. **1**, 127 (1969/70)
2.33 D.R. Bates, R. McCarroll: Proc. Roy. Soc. London A**245**, 175 (1958)
2.34 J.S. Briggs: Rep. Prog. Phys. **39**, 217 (1976)
2.35 J.C.Y. Chen, G.J. Hatton, T. Ishihara, K.M. Watson: Phys. Rev. A**12**, 1281 (1975)
2.36 J.C.Y. Chen, V.H. Ponce, K.M. Watson: J. Phys. B**6**, 965 (1973)
2.37 Yu.N. Demkov, V.N. Ostrovskii: Zh. Eksp. Teor. Fiz. **69**, 1582 (1975) [English transl.: Sov. Phys. JETP **42**(69), 806 (1975)]
2.38 L.I. Ponomarev, S.I. Vinitsky, F.R. Vukailovich: J. Phys. B**13**, 847 (1980); S.I. Vinitsky, L.I. Ponomarev: Fiz. Elem. Chastits At. Yadra **13**, 1336 (1982)
2.39 A. Riera, A. Salin: J. Phys. B**9**, 2877 (1976); A. Salin: Comments At. Mol. Phys. **9**, 165 (1980)
2.40 S.B. Schneiderman, A. Russek: Phys. Rev. **181**, 311 (1969); J.B. Delos: Rev. Mod. Phys. **53**, 287 (1981);

J.B. Delos: Comments At. Mol. Phys. **10**, 211 (1981)
2.41 R. Albat, N. Gruen: J. Phys. B**9**, L463 (1976)
2.42 J. Hahn: J. Phys. B**11**, 3221 (1978)
2.43 U. Heinz, W. Greiner, B. Müller: Phys. Rev. A**23**, 562 (1981)
2.44 V.H. Ponce: J. Phys. B**12**, 3731 (1979);
D.S.F. Crothers, N.R. Todd: J. Phys. B**14**, 2233 (1981);
V.H. Ponce: J. Phys. B**14**, 2823 (1981);
J. Vaaben, K. Taulbjerg: J. Phys. B**14**, 1815 (1981);
L.F. Errea, L. Mêndez, A. Riera: J. Phys. B**15**, 101 (1982);
V. Sidis, C. Kubach: Phys. Rev. A**27**, 2431 (1983)
2.45 G.B. Schmid: Phys. Rev. A**15**, 1459 (1977)
2.46 G.B. Schmid: J. Phys. B**12**, 3909 (1979)
2.47 W.R. Thorson, J.B. Delos: Phys. Rev. A**18**, 117 (1978)
2.48 J.B. Delos: Phys. Rev. A**23**, 2301 (1981)
2.49 T.A. Green: Phys. Rev. A**23**, 519 (1981)
2.50 M.H. Mittleman, H. Tai: Phys. Rev. A**8**, 1880 (1973)
2.51 T.A. Green: Phys. Rev. A**23**, 532 (1981);
A.S. Dickinson, R. McCarrol: J. Phys. B**16**, 459 (1983);
A. Salin: J. Phys. B**16**, L661 (1983)
2.52 W.R. Thorson, J.B. Delos: Phys. Rev. A**18**, 135 (1978)
2.53 E.I. Dashevskaya, E.A. Kobzeva: Opt. Spektrosk. **30**, 807 (1971)
2.54 M.I. D'yakonov, V.I. Perel': Zh. Eksp. Teor. Fiz. **48**, 345 (1965) [English transl.: Sov. Phys. JETP **21**(48), 227 (1965)]
2.55 E.E. Nikitin: Comments At. Mol. Phys. **3**, 7 (1971)
2.56 A. Omont: J. Phys. Paris **26**, 26 (1965)
2.57 W. Baylis: In *Progress in Atomic Spectroscopy*, Part B, ed. by W. Hanle, H. Kleinpoppen (Plenum, New York 1978) p.1227
2.58 A. Omont: Prog. Quantum Electron. **5**, 69 (1977)
2.59 G. Nienhuis: J. Phys. B**9**, 167 (1976)
2.60 L. Waldman: In *Fundamental Problems in Statistical Mechanics*, ed. by E.G. Cohen (North-Holland, Amsterdam 1968) p.276
2.61 U. Fano: Rev. Mod. Phys. **29**, 74 (1957)
2.62 G. Grawert: Z. Phys. **225**, 283 (1969)
2.63 R.H.G. Reid: J. Phys. B**8**, 2255 (1975)
2.64 V.N. Rebane, T.K. Rebane: Opt. Spektrosk. **33**, 405 (1972) [English transl.: Opt. Spectrosc. USSR **33**, 219 (1972)]
2.65 A.P. Yutsis, A.A. Bandzaitis: *Teoriya Momenta Kolichestva Dvizheniya v Kvantovoi Mekhanike* (Mintis, Vilnius 1965)
2.66 A.P. Yutsis, J.B. Levinson, V.V. Vanagas: *Mathematical Apparatus of the Theory of Angular Momentum* (Israel Program for Scientific Translation, Jerusalem 1962)
2.67 D.A. Coombe, R.F. Snider, B.C. Sanctuary: J. Chem. Phys. **63**, 3015 (1975)
2.68 R.H.G. Reid: J. Phys. B**6**, 2018 (1973)

Chapter 3

3.1 H.A. Bethe, E.E. Salpeter: *Quantum Mechanics of One- and Two-Electron Atoms* (Academic, New York 1957)
3.2 E.U. Condon, G.H. Shortley: *Theory of Atomic Spectra* (Cambridge University Press, London 1951)
3.3 V. Heine: *Group Theory in Quantum Mechanics* (Pergamon, London 1960)
3.4 J.C. Slater: *Quantum Theory of Atomic Structure*, Vols.1,2 (McGraw-Hill, New York 1960)
3.5 I.I. Sobelman: *Atomic Spectra and Radiative Transitions*, Springer Ser. Chem. Phys., Vol.1 (Springer, Berlin, Heidelberg, New York 1979)
3.6 E. Wigner: *Group Theory and Its Application to the Quantum Mechanics of Atomic Spectra* (Academic, New York 1959)

3.7 L.D. Landau, E.M. Lifshitz: *Quantum Mechanics*, 3rd ed. (Pergamon, Oxford 1977)
3.8 A.I. Baz', Ya.B. Zel'dovich, A.M. Perelomov: *Rasseyanie, Reaktsii i Raspady v Nerelyativistskoi Kvantovoi Mekhanike* (Nauka, Moscow 1971)
3.9 M. Barat, W. Lichten: Phys. Rev. A**6**, 211 (1972)
3.10 C.W. Kern, M. Karplus: J. Chem. Phys. **40**, 1374 (1964)
3.11 R. McWeeny, B.T. Sutcliffe: *Methods of Molecular Quantum Mechanics* (Academic, London 1969)
3.12 O. Sinanoglu: Adv. Chem. Phys. **6**, 315 (1964); R.J. Boyd: Chem. Phys. Lett. **44**, 363 (1976)
3.13 E. Wigner, E.E. Witmer: Z. Phys. **51**, 859 (1928)
3.14 E.E. Nikitin, S.Ya. Umanskii: Theor. Chim. Acta **13**, 91 (1969)
3.15 Ying-Nan Chin: J. Chem. Phys. **58**, 722 (1973)
3.16 E.E. Nikitin: Opt. Spektrosk. **10**, 443 (1961)
3.17 R.S. Mulliken: Rev. Mod. Phys. **4**, 1 (1932)
3.18 J.C. Slater: *Quantum Theory of Molecules and Solids. Electronic Structure of Molecules*, Vol.1 (McGraw-Hill, New York 1963)
3.19 E.E. Nikitin, S.Ya. Umanskii: Theor. Chim. Acta **28**, 121 (1973)
3.20 T.E.M. Walker: Mol. Phys. **23**, 489 (1972)
3.21 J. Raftery, P.R. Scott, W.G. Richards: J. Phys. B**7**, 1293 (1972)
3.22 D.M. Bishop: Adv. Quant. Chem. **3**, 25 (1967)
3.23 W. Byers-Brown, J.D. Power: Proc. Roy. Soc. London A**317**, 545 (1970)
3.24 R. Anholt, W.E. Mayerhof, A. Salin: Phys. Rev. A**16**, 951 (1977)
3.25 J. Eichler, U. Wille: Phys. Rev. A**11**, 1973 (1975)
3.26 I.V. Komarov, L.I. Ponomarev, S.Yu. Slavyanov: *Sferoidal'nye i Kulonovskie Sferoidal'nye Funktsii* (Nauka, Moscow 1976)
3.27 W.A. Bingel: J. Chem. Phys. **49**, 1931 (1968)
3.28 C.A. Coulson: *Valence* (Oxford University Press, New York 1961)
3.29 J. von Neumann, E.P. Wigner: Phys. Z. **30**, 467 (1929)
3.30 E. Teller: J. Phys. Chem. **41**, 109 (1937)
3.31 G.J. Hatton: Phys. Rev. A**14**, 901 (1976)
3.32 L. Lathouwers: Phys. Lett. A**63**, 228 (1977)
3.33 K.R. Naqvi, W.B. Brown: Int. J. Quantum Chem. **6**, 271 (1972)
3.34 K.R. Naqvi: Phys. Rev. A**15**, 1807 (1977)
3.35 T. Carrington: J. Chem. Phys. **70**, 2958 (1979)
3.36 G.J. Hatton, W.L. Lichten, N. Ostrove: Chem. Phys. Lett. **40**, 437 (1976)
3.37 G.J. Hatton, W.L. Lichten, N. Ostrove: J. Chem. Phys. **67**, 2169 (1977)
3.38 J.S. Cohen, W.R. Wadt, J.P. Hay: J. Chem. Phys. **71**, 2955 (1979)
3.39 H. Hellman, J.K. Syrkin: Acta Physicochim. URSS **2**, 433 (1935)
3.40 L. Landau: Phys. Z. Sowietunion **1**, 88 (1932)
3.41 C. Zener: Proc. Roy. Soc. London A**137**, 696 (1932)
3.42 U. Fano, W. Lichten: Phys. Rev. Lett. **14**, 627 (1965)
3.43 W. Lichten: Phys. Rev. **131**, 229 (1963)
3.44 W. Lichten: Phys. Rev. **164**, 131 (1967)
3.45 W. Lichten: In *Atomic Physics*, Vol.4, ed. by G. zu Putlitz, E.W. Weber, A. Winnacker (Plenum, New York 1975) p.249
3.46 W. Lichten: J. Phys. Chem. **84**, 2102 (1980)
3.47 B. Andersen, S.E. Nielsen: Mol. Phys. **21**, 523 (1971)
3.48 J.S. Briggs: Rep. Prog. Phys. **39**, 217 (1976)
3.49 J.S. Cohen: Phys. Rev. A**13**, 86 (1976)
3.50 J. Eichler, U. Wille: In *Electronic and Atomic Collisions*. Proc. of the 10th Int. Conf. on the Physics of Electronic and Atomic Collisions, ed. by G. Watel (North-Holland, Amsterdam 1978) p.331
3.51 J.P. Gauyacq: In *Electronic and Atomic Collisions*. Proc. of the 10th Int. Conf. on the Physics of Electronic and Atomic Collisions, ed. by G. Watel (North-Holland, Amsterdam 1978) p.431
3.52 V. Sidis: In *Quantum Theory of Chemical Reactions*, ed. by R. Daudel, A. Pullman, S. Salem, A. Viellard, Vol.1 (Reidel, Dordrecht 1979) p.7; J.S. Briggs, K. Taulbjerg: "Theory of Inelastic Atom-Atom Collisions",

in *Structure and Collisions of Ions and Atoms*, ed. by I.A. Sellin, Topic Curr. Phys., Vol.5 (Springer, Berlin, Heidelberg, New York 1978) p.105; C. Falcon, A. Maciâs, A. Riera, A. Salin: J. Phys. B**14**, 1983 (1981); C. Courbin-Gaussorgues, V. Sidis, J. Vaaben: J. Phys. B**16**, 2817 (1983)
3.53 V. Sidis, H. Lefebvre-Brion: J. Phys. B**4**, 1040 (1971)
3.54 R. Cimiraglia, M. Persico: Mol. Phys. **38**, 1707 (1979)
3.55 J.B. Delos, W.R. Thorson: J. Chem. Phys. **70**, 1774 (1979)
3.56 H. Gabriel, K. Taulbjerg: Phys. Rev. A**10**, 741 (1974)
3.57 T.G. Heil, A. Dalgarno: J. Phys. B**12**, L557 (1979)
3.58 B.R. Johnson: Chem. Phys. Lett. **27**, 289 (1974)
3.59 H. Klar, U. Fano: Phys. Rev. Lett. **37**, 1132 (1976); J. Hendekovic: Chem. Phys. Lett. **90**, 193 (1982); C.A. Mead, D.G. Truhlar: J. Chem. Phys. **77**, 6090 (1982); J.A. Olson, D.A. Micha: Int. J. Quant. Chem. **22**, 971 (1982)
3.60 T.F. O'Malley: Phys. Rev. **162**, 98 (1967)
3.61 T.F. O'Malley: Adv. At. Mol. Phys. **7**, 223 (1971)
3.62 F.T. Smith: Phys. Rev. **179**, 111 (1969)
3.63 G. Herzberg: *Spectra of Diatomic Molecules* (Van Nostrand, New York 1950)
3.64 M. Mizushima: *The Theory of Rotating Diatomic Molecules* (Wiley, New York 1975)
3.65 E.E. Nikitin: J. Chem. Phys. **43**, 744 (1965)
3.66 V. Aquilanti, P. Casavecchia, G. Grossi, A. Lagana: J. Chem. Phys. **73**, 1173 (1980)
3.67 V. Aquilanti, G. Grossi: J. Chem. Phys. **73**, 1165 (1980); V. Aquilanti, G. Grossi, A. Logana: Nuovo Cimento B**63**, 7 (1981)
3.68 E.I. Dashevskaya, F. Masnou, R. McCarroll, E.E. Nikitin: Opt. Spektrosk. **37**, 209 (1974) [English transl.: Opt. Spectrosc. USSR **37**, 119 (1974)]
3.69 H. Laue: J. Chem. Phys. **46**, 3034 (1967)
3.70 F. Masnou-Seeuws, R. McCarroll: J. Phys. B**7**, 2230 (1974)
3.71 F.H. Mies: Phys. Rev. A**7**, 942 (1973); F.H. Mies: Mol. Phys. **41**, 953 (1980); F.H. Mies: Mol. Phys. **41**, 973 (1980); J. Grosser: J. Phys. B**14**, 1149 (1981); J.L. Le Gonët, P.R. Berman: Phys. Rev. A**24**, 1831 (1981)
3.72 E.E. Nikitin: In *Atomic Physics*, Vol.4, Proc. of the 4th Int. Congr. on Atomic Physics, ed. by G. zu Putlitz (Plenum, New York 1975) p.529
3.73 E.E. Nikitin: Adv. Chem. Phys. **28**, 317 (1975)
3.74 R.T. Pack, J.O. Hirschfelder: J. Chem. Phys. **52**, 521 (1970)
3.75 R.T. Pack, J.O. Hirschfelder: J. Chem. Phys. **52**, 4198 (1970)
3.76 E.S. Chang, U. Fano: Phys. Rev. A**6**, 173 (1972)
3.77 U. Fano: Comments At. Mol. Phys. **2**, 47 (1970)
3.78 R.S. Mulliken: Rev. Mod. Phys. **2**, 60 (1930)
3.79 J. Eichler, U. Wille, B. Fastrup, K. Taulbjerg: Phys. Rev. A**14**, 707 (1976)
3.80 J.D. Power: Phil. Trans. Roy. Soc. A**274**, 663 (1973)
3.81 S.S. Gershtein, V.D. Krivchenkov: Zh. Eksp. Teor. Fiz. **40**, 1491 (1961) [English transl.: Sov. Phys. JETP **13**(40), 1044 (1961)]
3.82 V.M. Lavrov: ICPEAC X, Abstracts of Papers (ICPEAC, Paris 1977) p.10
3.83 J. Aguilar, H. Nakamura: Chem. Phys. **32**, 115 (1978)
3.84 M. Aubert, N. Bessis, G. Bessis: Phys. Rev. A**10**, 51 (1974)
3.85 M. Aubert, N. Bessis, G. Bessis: Phys. Rev. A**10**, 61 (1974)
3.86 M. Aubert, C. Le Sech: Phys. Rev. A**13**, 632 (1976)
3.87 V.K. Nikulin, N.A. Gushchina: J. Phys. B**11**, 3553 (1978)
3.88 E.E. Nikitin, B.M. Smirnov: Usp. Fiz. Nauk **124**, 201 (1978) [English transl.: Sov. Phys. Usp. **21**, 95 (1978)]
3.89 B.M. Smirnov: *Asimptoticheskie Metody v Teorii Atomnykh Stolknovenii* (Atomizdat, Moscow 1973)
3.90 V. Sidis: J. Chem. Phys. **55**, 5838 (1971)
3.91 M.H. Mittleman: Phys. Rev. A**9**, 704 (1974)

Chapter 4

4.1 J.C. Browne: Adv. At. Mol. Phys. **7**, 47 (1971)
4.2 J.C. Browne, F.A. Matsen: Adv. Chem. Phys. **23**, 161 (1973)
4.3 J. Goodisman: *Diatomic Interaction Potential Theory*, Vols.1,2 (Academic, New York 1973)
4.4 A.C. Hurley: *Introduction to the Electron Theory of Small Molecules* (Academic, New York 1976)
4.5 A.C. Hurley: *Electron Correlation in Small Molecules* (Academic, New York 1976)
4.6 J.J. Kaufman: Adv. Chem. Phys. **28**, 113 (1975)
4.7 R. McWeeny, B.T. Sutcliffe: *Methods of Molecular Quantum Mechanics* (Academic, New York 1969)
4.8 H.F. Schaefer (ed.): *Modern Theoretical Chemistry. Methods of Electronic Structure Theory*
4.9 N.G. Rambidi, N.F. Stepanov, A.I. Dement'ev: *Kvantovomekhanicheskie Raschety Dvukhatomnykh Molekul* (VINITI, Moscow 1979)
4.10 H.F. Schaefer: *The Electronic Structure of Atoms and Molecules. A Survey of Rigorous Quantum Mechanical Results* (Addison-Wesley, Reading, MA 1972)
4.11 R.K. Janev: Adv. At. Mol. Phys. **11**, 3 (1976)
4.12 B.M. Smirnov: *Asimptoticheskie Metody v Teorii Atomnykh Stolknovenii* (Atomizdat, Moscow 1973)
4.13 D.R. Hartree: *The Calculation of Atomic Structures* (Wiley, New York 1957);
M. Cohen, R.P. Eachran: Adv. At. Mol. Phys. **16**, 1 (1980)
4.14 J.C. Slater: *Quantum Theory of Atomic Structure*, Vols.1,2 (McGraw-Hill, New York 1960)
4.15 V.F. Bratsev: *Tablitsy Atomnykh Funktsii* (Nauka, Leningrad 1966)
4.16 V.F. Bratsev: *Tablitsy Atomnykh Volnovykh Funktsii* (Nauka, Leningrad 1971)
4.17 E. Clementi, C. Roetti: At. Data Nucl. Data Tables **14**, 177 (1974)
4.18 P. Gombas, T. Szondy: *Solutions of the Simplified Self-Consistent Field for All Atoms of the Periodic System of Elements from Z = 2 to Z = 92* (Hilger, London 1970)
4.19 F. Herman, S. Skillman: *Atomic Structure Calculations* (Prentice Hall, Englewood Cliffs, NJ 1963)
4.20 P. Gombas: *Die statistische Theorie des Atoms und ihre Anwendungen* (Springer, Wien 1949)
4.21 D.A. Kirzhnits, Yu.E. Lozovik, G.V. Shpikovskaya: Usp. Fiz. Nauk **117**, 1 (1975) [English transl.: Sov. Phys. Usp. **18**, 649 (1975)];
J. Schwinger: Phys. Rev. A**22**, 1827 (1980);
J. Schwinger: Phys. Rev. A**24**, 2353 (1981)
4.22 D. Shalitin: Phys. Rev. A**140**, 1857 (1965)
4.23 J.C. Stewart, M. Rotenberg: Phys. Rev. A**140**, 1508 (1965)
4.24 M. Berrondo, J.P. Daudey, O. Goschinski: Chem. Phys. Lett. **62**, 34 (1979)
4.25 R. Gaspar, J. Szabo, J. Szanozlō: Acta Phys. Acad. Sci. Hung. **39**, 261 (1975)
4.26 A.E.S. Green: Adv. Quantum Chem. **7**, 161 (1973)
4.27 A.E.S. Green, R.H. Garvey, C.H. Jackman: Int. J. Quantum Chem. Symp. **9**, 43 (1975)
4.28 A.E.S. Green, D.L. Sellin, A.S. Zachor: Phys. Rev. **184**, 1 (1969)
4.29 S.T. Manson, J.E. Purcell: Phys. Rev. A**15**, 1319 (1977);
N. Bessis, G. Bessis: J. Chem. Phys. **74**, 3628 (1981)
4.30 P.P. Szydlik, A.E.S. Green: Phys. Rev. A**9**, 1885 (1974)
4.31 C.A. Coulson: *Valence* (Oxford University Press, New York 1961)
4.32 J. Sakui, T. Anno: J. Chem. Phys. **60**, 620 (1974)
4.33 M.E. Schwartz: Chem. Phys. Lett. **43**, 291 (1976)
4.34 R. Ahlrichs: J. Math. Phys. **14**, 1860 (1973)
4.35 R. Ahlrichs: J. Chem. Phys. **64**, 2706 (1976)

4.36 E.L. Slaggie, E.H. Wichman: J. Math. Phys. **3**, 946 (1962)
4.37 M. Hoffman-Ostenhoff, T. Hoffman-Ostenhoff: Phys. Rev. A**16**, 1782 (1977)
4.38 B.M. Smirnov: *Atomnye Stolknoveniya i Elementarnye Protsessy v Plazme* (Atomizdat, Moscow 1968)
4.39 R. Ahlrichs, T. Hoffman-Ostenhoff, M. Hoffman-Ostenhoff, J.D. Morgan: Phys. Rev. A**23**, 2106 (1981)
4.40 J. Tal: Phys. Rev. A**18**, 1781 (1978);
F.W. King: J. Chem. Phys. **78**, 2459 (1983)
4.41 A.V. Evseev, A.A. Radtsig, B.M. Smirnov: Opt. Spektrosk. **44**, 833 (1978) [English transl.: Opt. Spectrosc. USSR **44**, 495 (1978)]
4.42 A.A. Radtsig, B.M. Smirnov: *Spravochnik po Atomnoi i Molekulyarnoi Fizike* (Atomizdat, Moscow 1980)
4.43 G.S. Handler, D.W. Smith, H.J. Silverstone: J. Chem. Phys. **73**, 3936 (1980)
4.44 N.C. Handy, M.T. Marron, H.J. Silverstone: Phys. Rev. **180**, 45 (1969)
4.45 H. Hellman: Acta Physicochim. URSS **1**, 913 (1935)
4.46 P. Gombas: *Pseudopotentiale* (Springer, New York 1967)
4.47 J.N. Bardsley: In *Case Studies in Atomic and Molecular Physics*, Vol.4, ed. by E.W. McDaniel (North-Holland, Amsterdam 1975) p.302
4.48 V. Bonifacic, S. Huzinaga: J. Chem. Phys. **60**, 2779 (1974);
P. Fuentealba, H. Stoll, L. von Szentpaly, P. Schwerdtfeger, H. Preuss: J. Phys. B**16**, L323 (1983)
4.49 C. Bottcher, A. Dalgarno: Proc. Roy. Soc. London A**340**, 187 (1974)
4.50 P. Cavaliere, G. Ferrante, R. Geracitani: Nuovo Cimento B**9**, 96 (1972)
4.51 T.C. Chang, P. Habitz, B. Pittel, W.H.E. Schwarz: Theor. Chim. Acta **34**, 263 (1974)
4.52 G. Csanak, H.S. Taylor, R. Yaris: Adv. At. Mol. Phys. **7**, 287 (1971)
4.53 P. Durand, J.C. Barthelat: Theor. Chim. Acta **38**, 283 (1975)
4.54 J. Goodisman: Int. J. Quantum Chem. **10**, 341 (1976);
G.H. Jeung, J.P. Malrien, J.P. Daudey: J. Chem. Phys. **77**, 3571 (1982)
4.55 V. Heine: Solid State Phys. **24**, 1 (1970)
4.56 L.R. Kahn, P. Baybutt, D.G. Truhlar: J. Chem. Phys. **65**, 3826 (1976)
4.57 B.J. Laurenzi: J. Chem. Phys. **69**, 4838 (1978)
4.58 C.F. Metius, W.A. Goddard: Phys. Rev. A**10**, 1528 (1974)
4.59 J.C. Phillips, L. Kleinmann: Phys. Rev. **116**, 287 (1959)
4.60 L. Szasz: Z. Naturforsch. A**32**, 829 (1977)
4.61 L. Szasz, L. Brown: J. Chem. Phys. **65**, 1393 (1976)
4.62 J.C. Tully: Phys. Rev. **181**, 7 (1969)
4.63 S.W. Wang, H.S. Taylor, R. Yaris: Chem. Phys. **14**, 53 (1976)
4.64 J.D. Weeks, A. Hazi, S.A. Rice: Adv. Chem. Phys. **16**, 283 (1969)
4.65 B. Schneider: Phys. Rev. A**7**, 557 (1973)
4.66 B.S. Yarlagadda, G. Csanak, H.S. Taylor, B. Schneider, R. Yaris: Phys. Rev. A**7**, 146 (1973)
4.67 L. Szasz, G. McGinn: J. Chem. Phys. **47**, 3495 (1967)
4.68 D.R. Hartree, W. Hartree: Proc. Roy. Soc. London A**193**, 299 (1948)
4.69 I.V. Avilova, M.I. Podlubnii: Opt. Spektrosk. **38**, 1059 (1975) [English transl.: Opt. Spectrosc. USSR **38**, 613 (1975)]
4.70 V. Bonifacic, S. Huzinaga: J. Chem. Phys. **65**, 2322 (1976)
4.71 N.L. Manakov, V.D. Ovsyannikov: J. Phys. B**10**, 569 (1977)
4.72 I. Martin, G. Simons: Mol. Phys. **32**, 1017 (1976)
4.73 H.A. Pohl, D.R. Fowler: Int. J. Quantum Chem. **8**, 435 (1974)
4.74 G.A. Hart, P.L. Goodfriend: J. Chem. Phys. **53**, 448 (1970)
4.75 G. Simons: J. Chem. Phys. **55**, 756 (1971)
4.76 R.G. Parsons, V.F. Weisskopf: Z. Phys. **202**, 492 (1967)
4.77 M.V. Basilevsky, M.M. Berenfeld: Int. J. Quantum Chem. **6**, 23 (1972)
4.78 D.M. Chipman, J.D. Bowan, J.A. Hirschfelder: J. Chem. Phys. **59**, 2830 (1973)
4.79 L. Jansen: Phys. Rev. **162**, 63 (1967)

4.80 I.G. Kaplan, O.B. Rodimova: Usp. Fiz. Nauk **126**, 403 (1978) [English transl.: Sov. Phys. Usp. **21**, 918 (1978)]
4.81 J.N. Murrel, M. Randić, D.R. Williams: Proc. Roy. Soc. London A**284**, 566 (1965)
4.82 J.N. Murrel, G. Shaw: J. Chem. Phys. **46**, 1768 (1961)
4.83 J. Musher, A. Amos: Phys. Rev. **164**, 31 (1967); A.T. Amos, C.S. Van den Berghe: Mol. Phys. **47**, 897 (1982)
4.84 A.B. Ritchie: J. Chem. Phys. **49**, 2167 (1968)
4.85 A.A. Rumyantsev: Zh. Eksp. Teor. Fiz. **65**, 926 (1973) [English transl.: Sov. Phys. JETP **38**(65), 459 (1974)]
4.86 A. Van der Avoird: J. Chem. Phys. **47**, 3649 (1967)
4.87 P.R. Certain, J.O. Hirschfelder, W. Kolos, L. Wolniewicz: J. Chem. Phys. **49**, 24 (1968)
4.88 G. Chalasinski, B. Jeziorski: Mol. Phys. **32**, 81 (1976); G. Chalasinski: Mol. Phys. **49**, 1353 (1983)
4.89 W. Kolos: Int. J. Quantum Chem. Symp. **8**, 241 (1974)
4.90 L.P. Gor'kov, L.P. Pitaevskii: Dokl. Akad. Nauk SSSR **151**, 822 (1963)
4.91 C. Herring: Rev. Mod. Phys. **34**, 631 (1962)
4.92 C. Herring: In *Magnetism*, Vol.2B, ed. by G.T. Rado, H. Suhl (Academic, New York 1966) p.1
4.93 C. Herring, M. Flicker: Phys. Rev. A**134**, 362 (1964)
4.94 E.L. Duman, B.M. Smirnov: Opt. Spektrosk. **29**, 425 (1970)
4.95 E.E. Nikitin, S.Ya. Umanskii: Theor. Chim. Acta **13**, 91 (1969)
4.96 E.E. Nikitin, S.Ya. Umanskii: Theor. Chim. Acta **28**, 121 (1973)
4.97 H. Margenau: Phys. Rev. **56**, 1000 (1939)
4.98 J.F. Bazet, C. Harel, R. McCarroll, A. Riera: Astron. Astrophys. **43**, 223 (1975)
4.99 E.I. Dashevskaya, E.E. Nikitin, A.I. Voronin, A.A. Zembekov: Can. J. Phys. **48**, 981 (1970)
4.100 E.I. Dashevskaya, A.I. Voronin, E.E. Nikitin: Can. J. Phys. **47**, 1237 (1969)
4.101 E.P. Gordeev, S.Ya. Umanskii, A.I. Voronin: Chem. Phys. Lett. **23**, 524 (1973)
4.102 E.P. Gordeev, S.Ya. Umanskii, A.I. Voronin: Chem. Phys. Lett. **44**, 36 (1976)
4.103 T.M. Kereselidze, M.I. Chibisov: Teor. Eksp. Khim. **14**, 147 (1978) [English transl.: Theor. Exp. Chem. USSR **14**, 113 (1978)]
4.104 Ng Kin-Chue, W.J. Meath, A.R. Allnat: Mol. Phys. **37**, 237 (1979)
4.105 J.N. Murrel, A.J.C. Varandas: Mol. Phys. **30**, 223 (1975)
4.106 L.A. Nadeikina, S.Ya. Umanskii: Teor. Eksp. Khim. **14**, 828 (1978)
4.107 A.I. Reznikov, S.Ya. Umanskii: Teor. Eksp. Khim. **7**, 585 (1971)
4.108 K.T. Tang, J.P. Toennies: J. Chem. Phys. **66**, 1496 (1977)
4.109 A.I. Voronin, E.P. Gordeev, S.Ya. Umanskii: Astron. Zh. **48**, 275 (1971)
4.110 R. Block: Physica **70**, 397 (1973)
4.111 R. Block, L. Jansen: Physica BC**98**, 39 (1979)
4.112 P.D. Dacre, R. McWeeny: Proc. Roy. Soc. London A**317**, 435 (1970)
4.113 J. Gerrat: Proc. Roy. Soc. London A**350**, 363 (1976)
4.114 Yu. Narushis, I. Batarunas: Opt. Spektrosk. **33**, 450 (1972) [English transl.: Opt. Spectrosc. USSR **33**, 243 (1972)]
4.115 W.E. Baylis: In *Progress in Atomic Spectroscopy*, Part A, ed. by W. Hanle, H. Kleinpoppen (Plenum, New York 1978) p.207
4.116 A. Dalgarno: Adv. Chem. Phys. **12**, 143 (1967)
4.117 A. Dalgarno, W.D. Davidson: Adv. At. Mol. Phys. **2**, 1 (1966)
4.118 J.O. Hirschfelder, W.J. Meath: Adv. Chem. Phys. **12**, 3 (1967)
4.119 H. Margenau, N.R. Kestner: *Theory of Intermolecular Forces* (Pergamon, Oxford 1971)
4.120 Tai Yup Chang: Rev. Mod. Phys. **39**, 911 (1967)
4.121 B.C. Carlson, G.S. Rushbrooke: Proc. Cambridge Philos. Soc. **46**, 626 (1950)

4.122 C.G. Gray: Can. J. Phys. **46** 135 (1968)
4.123 M.E. Rose: J. Math. Phys. **37**, 215 (1957)
4.124 I.I. Sobelman: *Atomic Spectra and Radiative Transitions*, Springer Ser. Chem. Phys., Vol.1 (Springer, Berlin, Heidelberg, New York 1979)
4.125 M. Movre, G. Pichler: J. Phys. B**10**, 2631 (1977)
4.126 W.F. Meggers, C.H. Corliss, B.F. Schribner: *Tables of Spectral Line Intensities*, Parts 1,2 (Government Printing Office, Washington 1975)
4.127 W.L. Wiese: Nucl. Instrum. Meth. **90**, 25 (1970)
4.128 R.J.S. Crossley, Adv. At. Mol. Phys. **5**, 237 (1969)
4.129 U. Fano, J.W. Cooper: Rev. Mod. Phys. **40**, 441 (1968)
4.130 A. Hibbert: Phys. Scr. **16**, 7 (1977)
4.131 A.L. Stewart: In *Atomic Processes and Applications*, ed. by P.G. Burke, B.L. Moiseiwitsch (North-Holland, Amsterdam 1976) p.251
4.132 W.L. Wiese: Nucl. Instrum. Meth. **90**, 121 (1970)
4.133 W.L. Wiese: In *Progress in Atomic Spectroscopy*, Part B, ed. by W. Hanle, H. Kleinpoppen (Plenum, New York 1979) p.1101
4.134 C. Mavroyannis, M.J. Stephen: Mol. Phys. **5**, 629 (1962)
4.135 D.C.S. Allison, P.G. Burke, W.D. Robb: J. Phys. B**5**, 1431 (1972)
4.136 M.B. Doran: J. Phys. B**7**, 558 (1974)
4.137 H.P. Kelly: Int. J. Quantum Chem. Symp. **3**, 349 (1970)
4.138 M.N. Adamov, N.P. Borisova: In *Kvantovaya Mekhanika*, Problemy Teoreticheskoi Fiziki, Vol.1, ed. by M.G. Veselov, Yu.V. Novozhilov, P.P. Pavinskii (Leningrad University, Leningrad 1974) p.117
4.139 G.P. Arrighini, C. Guidotti: Mol. Phys. **28**, 273 (1974)
4.140 V. Magnasco, M. Battezzati, R. Austi: Chem. Phys. Lett. **51**, 375 (1977)
4.141 H.L. Kramer: J. Chem. Phys. **53**, 2783 (1970)
4.142 P.W. Langhoff, M. Karplus: J. Chem. Phys. **53**, 233 (1970)
4.143 P.W. Langhoff, M. Karplus: In *The Padé Approximant in Theoretical Physics*, ed. by G.A. Baker, J.L. Gammel (Academic, New York 1970) p.41
4.144 A.V. Tulub: In *Problemy Teoreticheskoi Fiziki. Kvantovaya Mekhanika*, Vol.1, ed. by M.G. Veselov, Yu.V. Novozhilov, P.P. Pavinskii (Leningrad University, Leningrad 1974) p.301
4.145 R. Luyckx, F. Delbaen, Ph. Coulon, H.N.W. Lekkerkerker: Phys. Rev. A**19**, 324 (1979)
4.146 K.T. Tang, J.M. Norbeck, P.R. Certain: J. Chem. Phys. **64**, 3063 (1976)
4.147 H.L. Kramer, D.R. Herschbach: J. Chem. Phys. **53**, 2792 (1970)
4.148 K.T. Tang: Phys. Rev. **177**, 108 (1969);
M. Diaz Pena, C. Pondo, J.A.R. Renuncio: J. Chem. Phys. **72**, 5269 (1980)
4.149 H.P. Kelly: Phys. Lett. A**29**, 30 (1969)
4.150 J.H. Miller, H.P. Kelly: Phys. Rev. A**5**, 516 (1972);
D.B. Neuman, M. Krauss: J. Chem. Phys. **75**, 315 (1981);
V.D. Ovsyannikov: Opt. Spektrosk. **53**, 600 (1982)
4.151 J.C. Slater, J.G. Kirkwood: Phys. Rev. **37**, 682 (1931)
4.152 S. Fraga, J. Karwowski, K.M.S. Saxena: At. Data Nucl. Data Tables **12**, 467 (1973)
4.153 T.M. Miller, B. Bederson: Adv. At. Mol. Phys. **13**, 1 (1977)
4.154 R.R. Teachout, R.T. Pack: At. Data **3**, 195 (1971)
4.155 J. Callaway, E. Bauer: Phys. Rev. A**140**, 1072 (1965)
4.156 W.E. Baylis: J. Chem. Phys. **51**, 2665 (1969)
4.157 G.D. Mahan: J. Chem. Phys. **50**, 2755 (1969)
4.158 T.R. Proctor, W.C. Stwalley: J. Chem. Phys. **66**, 2063 (1977)
4.159 A.M. Ratner, O.Ya. Shanfarova: Ukr. Fiz. Zh. **21**, 773 (1976)
4.160 J.C. Barthelat, P. Durand, A. Serafini: Mol. Phys. **33**, 159 (1977)
4.161 V. Bonifacic, S. Huzinaga: J. Chem. Phys. **62**, 1509 (1975)
4.162 E.A. Andreev: Theor. Chim. Acta **28**, 235 (1973)
4.163 B.M. Smirnov, M.I. Chibisov: Zh. Eksp. Teor. Fiz. **48**, 939 (1965) [English transl.: Sov. Phys. JETP **21**(48), 624 (1965)]
4.164 S.Ya. Umanskii, A.I. Voronin: Theor. Chim. Acta **12**, 166 (1968)
4.165 H.F. Schaefer, F.E. Harris: J. Chem. Phys. **48**, 4946 (1968)

4.166 E.E. Shpil'rain, A.Ya. Polishchuk: Teplofiz. Vys. Temp. **17**, 285 (1979) [English transl.: High Temp. USSR **17**, 240 (1979)]
4.167 T.C. Chang, P. Habitz, W.H.E. Schwarz: Theor. Chim. Acta **44**, 61 (1977)
4.168 G. Das, A.C. Wahl: J. Chem. Phys. **69**, 53 (1978)
4.169 R.N. Dixon, I.L. Robertson: Mol. Phys. **37**, 1223 (1979)
4.170 W.C. Ermler, Y.S. Lee, K.S. Pitzer, N.W. Winter: J. Chem. Phys. **69**, 976 (1978)
4.171 P.J. Hay, W.R. Wadt, L.R. Kahn, F.W. Bobrowicz: J. Chem. Phys. **69**, 984 (1979)
4.172 J.N. Murrel, I.G. Vincent: J. Chem. Soc. Faraday Trans. II, **73**, 973 (1977)
4.173 E. Nagy-Felsobuki, J.B. Peel: Austr. J. Chem. **31**, 2571 (1978)
4.174 J.G. Snyders, E.J. Bearends: Mol. Phys. **33**, 1651 (1977); S. Runge: Chem. Phys. Lett. **101**, 167 (1983)
4.175 E. Fermi. Nuovo Cimento **11**, 157 (1934)
4.176 G. Csanak, H.S. Taylor: Phys. Rev. A**6**, 1843 (1972)
4.177 C. Bottcher: Chem. Phys. Lett. **18**, 457 (1973)
4.178 C. Bottcher, T.C. Cravens, A. Dalgarno: Proc. Roy. Soc. London A**346**, 157 (1975)
4.179 C. Bottcher, A. Dalgarno, E.L. Wright: Phys. Rev. A**7**, 1606 (1973)
4.180 E. Czuchaj, J. Sienkiewicz: Z. Naturforsch. A**34**, 694 (1979)
4.181 R. Düren: J. Phys. B**10**, 3467 (1977)
4.182 J. Hanssen, R. McCarroll, P. Valiron: J. Phys. B**12**, 899 (1979)
4.183 A.R. Malvern: J. Phys. B**11**, 831 (1978)
4.184 F. Masnou-Seeuws, M. Philippe, P. Valiron: Phys. Rev. Lett. **41**, 395 (1978)
4.185 J. Pascale, J. Vandeplanque: J. Chem. Phys. **60**, 2278 (1974)
4.186 G. Peach: J. Phys. B**11**, 2107 (1978)
4.187 M. Philippe, F. Masnou-Seeuws, P. Valiron: J. Phys. B**12**, 2493 (1979)
4.188 P.E. Siska: J. Chem. Phys. **71**, 3931 (1979)
4.189 P.E. Siska: J. Chem. Phys. **73**, 2372 (1980)
4.190 P. Valiron, R. Gayet, R. McCarroll, F. Masnou-Seeuws, M. Philippe: J. Phys. B**12**, 53 (1979)
4.191 K.J. Wahlstrand, R.W. Munrich, J.S. Dahler, S.E. Nielsen: J. Phys. B**10**, 1687 (1977)
4.192 N.F. Mott, H.S.W. Massey: *The Theory of Atomic Collisions* (Clarendon, Oxford 1965)
4.193 L.D. Landau, E.M. Lifshitz: *Quantum Mechanics*, 3rd rev. ed., (Pergamon, Oxford 1977)
4.194 S.T. Epstein: J. Chem. Phys. **44**, 836 (1966)
4.195 H. Eissa, U. Opik: Proc. Phys. Soc. London **92**, 556 (1967)
4.196 R.H. Young, W.J. Deal: Int. J. Quantum Chem. **9**, 835 (1975); D.W. Norcross: Phys. Rev. A**28**, 3095 (1983)
4.197 R. Düren, G. Moritz: J. Chem. Phys. **73**, 5155 (1980)
4.198 F.I. Dalidchik, G.K. Ivanov: Teor. Eksp. Khim. **8**, 9 (1972)
4.199 G.K. Ivanov: Opt. Spektrosk. **37**, 636 (1974) [English transl.: Opt. Spectrosc. USSR **37**, 361 (1974)]
4.200 G.K. Ivanov: Opt. Spektrosk. **39**, 834 (1975) [English transl.: Opt. Spectrosc. USSR **39**, 474 (1975)]
4.201 G.K. Ivanov: Teor. Eksp. Khim. **14**, 610 (1978)
4.202 A. Omont: J. Phys. Paris **38**, 1343 (1977)
4.203 M.L. Goldberger, K.M. Watson: *Collision Theory* (Wiley, New York 1964)
4.204 R.G. Newton: *Scattering Theory of Waves and Particles*, 2nd ed., Texts and Monographs in Phys. (Springer, Berlin, Heidelberg, New York 1982)
4.205 Yu.N. Demkov, V.N. Ostrovskii: *Metod Potentsialov Nulevogo Radiusa v Atomnoi Fizike* (Leningrad University 1975)
4.206 G.F. Drukarev: Adv. Quantum Chem. **11**, 251 (1978)
4.207 A.I. Baz', Ya.B. Zel'dovich, A.M. Perelomov: *Rasseyanie, Reaktsii i Raspady v Nerelyativistskoi Kvantovoi Mekhanike* (Nauka, Moscow 1971)
4.208 I.V. Komarov, G.A. Pogorelyi, A.S. Tabilov: Opt. Spektrosk. **27**, 198 (1969)

4.209 A. Maquet: Phys. Rev. A**15**, 1088 (1977)
4.210 T.M. Kereselidze, M.I. Chibisov: Zh. Eksp. Teor. Fiz. **68**, 12 (1975) [English transl.: Sov. Phys. JETP **41**(68), 6 (1975)]
4.211 M.Ya. Ovchinnikova: Zh. Eksp. Teor. Fiz. **49**, 275 (1965) [English transl.: Sov. Phys. JETP **22**(49), 194 (1966)]
4.212 V.A. Alekseev, I.I. Sobelman: Zh. Eksp. Teor. Fiz. **49**, 1274 (1965) [English transl.: Sov. Phys. JETP **22**(49), 882 (1966)]
4.213 M.H. Mittleman: Phys. Rev. **162**, 81 (1967)
4.214 G.K. Ivanov: Opt. Spektrosk. **40**, 965 (1976) [English transl.: Opt. Spectrosc. USSR **40**, 554 (1976)]
4.215 O. Hinckelman, L. Spruch: Phys. Rev. A**3**, 642 (1971)
4.216 G.K. Ivanov: Teor. Eksp. Khim. **15**, 644 (1979) [English transl.: Theor. Exp. Chem. USSR **15**, 502 (1979)]
4.217 J.S. Briggs: Rep. Prog. Phys. **39**, 217 (1976); A. Maciâs, A. Riera: Phys. Rep. **90**, 299 (1982)
4.218 J.C. Slater: *Quantum Theory of Molecules and Solids*, Vol.1 (McGraw-Hill, New York 1963)
4.219 J.S. Briggs, M.R. Hayns: J. Phys. B**6**, 514 (1973)
4.220 A. Macias, A. Riera: J. Phys. B**9**, 497 (1976)
4.221 F. Hund: Z. Phys. **77**, 12 (1932)
4.222 A. Recknagel: Z. Phys. **87**, 375 (1934)
4.223 J. Eichler, U. Wille: Phys. Rev. Lett. **33**, 56 (1974)
4.224 J. Eichler, U. Wille: Phys. Rev. A**11**, 1973 (1975)
4.225 J. Eichler, U. Wille: In *Electronic and Atomic Collisions*. Proc. of the 10th Int. Conf. on the Physics of Electronic and Atomic Collisions, ed. by G. Watel (North-Holland, Amsterdam 1978) p.331
4.226 J. Eichler, U. Wille, B. Fastrup, K. Taulbjerg: Phys. Rev. A**14**, 707 (1976)
4.227 E. Gross, R.M. Dreizler: Phys. Lett. A**57**, 131 (1976)
4.228 E. Gross, R.M. Dreizler: Phys. Rev. A**20**, 1798 (1979); G.L. Plindov, S.K. Pogrebnya: Chem. Phys. Lett. **90**, 317 (1982); G.L. Plindov, S.K. Pogrebnya: Phys. Lett. A**91**, 219 (1982)
4.229 E. Gross, A. Toepfer, R.M. Dreizler: Phys. Rev. A**20**, 1808 (1979)
4.230 E. Teller: Rev. Mod. Phys. **34**, 627 (1962)
4.231 J.S. Cohen, R.T. Pack: J. Chem. Phys. **61**, 2372 (1974)
4.232 N.H. March: Adv. Phys. **6**, 1 (1957); R. Benguria, H. Brezis, E.H. Lieb: Commun. Math. Phys. **79**, 167 (1981); E.H. Lieb: Rev. Mod. Phys. **53**, 603 (1981)
4.233 V.K. Nikulin, Yu.N. Tsarev: Chem. Phys. **10**, 433 (1975)
4.234 F.T. Smith: In *Physics of Electronic and Atomic Collisions*, Invited Papers to VII ICPEAC (North-Holland, Amsterdam 1971) p.1; H.J. Böhm, R. Ahlrichs: J. Chem. Phys. **77**, 2028 (1982)
4.235 I.M. Torrens: *Interatomic Potentials* (Academic, New York 1972)
4.236 J.W. Sheldon: Phys. Rev. **99**, 1291 (1955)
4.237 J.R. Townsend, G.S. Hundler: J. Chem. Phys. **36**, 3325 (1962)
4.238 K. Yonei, J. Goodisman: J. Chem. Phys. **66**, 4551 (1977)
4.239 O.B. Firsov: Zh. Eksp. Teor. Fiz. **32**, 1464 (1957) [English transl.: Sov. Phys. JETP **5**(32), 1192 (1957)]
4.240 K. Günter: Ann. Phys. **14**, 296 (1964)
4.241 B.R. Junker, M.G. Menendez: J. Chem. Phys. **63**, 1243 (1975)
4.242 O.B. Firsov: Zh. Eksp. Teor. Fiz. **33**, 696 (1957) [English transl.: Sov. Phys. JETP **6**(33), 534 (1957)]
4.243 R.S. Mulliken: Chem. Phys. Lett. **14**, 137 (1972)
4.244 J.S. Cohen, B. Schneider: J. Chem. Phys. **61**, 3230 (1974)
4.245 R.G. Gordon, J.S. Kim: J. Chem. Phys. **56**, 3122 (1972)
4.246 T.L. Gilbert, A.C. Wahl: J. Chem. Phys. **47**, 3425 (1967)
4.247 M. Barat, D. Dhuicg, R. Francois, V. Sidis: J. Phys. B**6**, 2072 (1973)
4.248 K. Fuchs: Proc. Roy. Soc. London A**151**, 585 (1935)
4.249 H.S.W. Massey, D.W. Sida: Philos. Mag. **46**, 190 (1955)

4.250 D.W. Sida: Philos. Mag. **2**, 761 (1957)
4.251 G. Brual, S.M. Rothstein: J. Chem. Phys. **69**, 1177 (1978)
4.252 G. Brual, S.M. Rothstein: Chem. Phys. Lett. **61**, 167 (1979)
4.253 M.J. Clugston, R.G. Gordon: J. Chem. Phys. **66**, 239 (1977)
4.254 V.I. Gaydaenko, V.K. Nikulin: Chem. Phys. Lett. **7**, 360 (1970)
4.255 F.A. Gianturco: J. Chem. Phys. **64**, 1973 (1976)
4.256 T. Ishikawa, S. Kito, H. Inouye: J. Phys. Soc. Jpn. **40**, 1446 (1976)
4.257 J.Y. Kim, R.G. Gordon: J. Chem. Phys. **60**, 1842 (1974)
4.258 J. Lloyd, D. Pugh: J. Chem. Soc. Faraday Trans. I **74**, 32 (1978)
4.259 B. Schneider: J. Chem. Phys. **58**, 447 (1973)
4.260 R. Ahlrichs, P. Penco, G. Scoles: Chem. Phys. **19**, 119 (1977)
4.261 J.S. Griffith: *The Irreducible Tensor Method for Molecular Symmetry Groups* (Prentice Hall, New York 1962)
4.262 T.E.H. Walker: Mol. Phys. **23**, 489 (1972)
4.263 S.R. Langhoff, C.W. Kern: In *Modern Theoretical Chemistry*, Vol.4, ed. by H.F. Schaefer (Plenum, New York 1977) p.381;
W.G. Richards, H.P. Triverdi, D.L. Cooper: *Spin-Orbit Coupling in Molecules* (Clarendon, Oxford 1981)
4.264 J.H. van Vleck: Rev. Mod. Phys. **23**, 213 (1951)
4.265 S.R. Kinnersly, J.N. Murrel: J. Chem. Soc. Faraday Trans. II **74**, 600 (1978)
4.266 S.R. Langhoff: J. Chem. Phys. **73**, 2379 (1980)
4.267 H.P. Trivedi, W.G. Richards: J. Chem. Phys. **72**, 3438 (1980);
W.J. Stevens, M. Krauss: J. Chem. Phys. **76**, 3834 (1982)
4.268 W.H. Moores, R. McWeeny: Proc. Roy. Soc. London A**332**, 365 (1973)
4.269 M. Blume, R.E. Watson: Proc. Roy. Soc. London A**270**, 127 (1962)
4.270 M. Blume, R.E. Watson: Proc. Roy. Soc. London A**271**, 565 (1963)
4.271 H.A. Bethe, E.E. Salpeter: *Quantum Mechanics of One- and Two-Electron Atoms* (Academic, New York 1957)
4.272 J.S. Cohen, W.R. Wadt, J.P. Hay: J. Chem. Phys. **71**, 2955 (1979)
4.273 E. Ishiguro, M. Kobori: J. Phys. Soc. Jpn. **22**, 263 (1967)
4.274 R. Anholt, W.E. Meyerhof, A. Salin: Phys. Rev. A**16**, 951 (1977)
4.275 P.R. Fontana: Phys. Rev. **125**, 1597 (9161)
4.276 E.E. Nikitin: Opt. Spektrosk. **10**, 443 (1961)
4.277 P.J. Hay, T.H. Dunning: J. Chem. Phys. **66**, 1306 (1977)
4.278 F.H. Mies: Phys. Rev. A**7**, 942 (1973)
4.279 W.R. Wadt: J. Chem. Phys. **68**, 402 (1978)
4.280 V. Sidis: J. Chem. Phys. **55**, 5838 (1971)
4.281 S.E. Butler: Phys. Rev. A**23**, 1 (1981)
4.282 A. Macias, A. Riera: J. Phys. B**10**, 861 (1977)
4.283 A. Riera, A. Salin: J. Chem. Phys. **62**, 1195 (1975);
A. Riera: Comments At. Mol. Phys. **12**, 95 (1982)
4.284 T.A. Green, J.C. Browne, H.H. Michels, M.M. Madsen: J. Chem. Phys. **61**, 5198 (1974)
4.285 A. Maciás, A. Riera: J. Phys. B**11**, 1077 (1978);
A. Maciás, A. Riera, M. Yanez: Phys. Rev. A**27**, 206 (1983)
4.286 R.E. Olson, E.J. Shipsey, J.C. Browne: J. Phys. B**11**, 699 (1978);
I. Demetropoulos, K.P. Lawley: J. Phys. B**14**, 4787 (1981);
L.T. Redmon: Phys. Rev. A**25**, 2453 (1982)
4.287 P. Habitz, C. Votava: J. Chem. Phys. **72**, 5532 (1980)
4.288 S.T. Epstein: *The Variation Method in Quantum Chemistry* (Academic, New York 1974)
4.289 S.E. Butler: Phys. Rev. A**20**, 2317 (1979)
4.290 M. Oppenheimer: J. Chem. Phys. **57**, 3899 (1972)
4.291 M.B. Faist, R.B. Bernstein: J. Chem. Phys. **64**, 2971 (1976)
4.292 M.B. Faist, R.D. Levine: J. Chem. Phys. **64**, 2953 (1976)

Chapter 5

5.1 R.B. Bernstein: Adv. Chem. Phys. **10**, 75 (1966)
5.2 M.V. Berry, K.E. Mount: Rep. Prog. Phys. **35**, 315 (1972)
5.3 U. Buck: Adv. Chem. Phys. **30**, 313 (1975)
5.4 J.N.L. Connor: In *Semiclassical Methods in Molecular Scattering and Spectroscopy*, ed. by M.S. Child (Reidel, Dordrecht 1980) p.45
5.5 B.J.B. Crowley: Phys. Rep. **57**, 47 (1980)
5.6 K.W. Ford, J.A. Wheeler: Ann. Phys. **7**, 259 (1959)
5.7 K.W. Ford, J.A. Wheeler: Ann. Phys. **7**, 287 (1959)
5.8 R. McCarroll: In *Atoms, Molecules and Lasers*, Lectures presented at the Int. Winter Colloquium in Trieste 1973 (Intern. Atomic Energy Agency, Vienna 1974) p.471
5.9 E.E. Nikitin, M.Ya. Ovchinnikova: Usp. Fiz. Nauk **104**, 379 (1971) [English transl.: Sov. Phys. Usp. **14**(104), 394 (1972)]
5.10 H. Pauly: In *Atom Molecule Collision Theory*, ed. by R.B. Bernstein (Plenum, New York 1979) p.111;
J.N.L. Connor, D. Farrely: J. Chem. Phys. **75**, 2831 (1981)
5.11 H. Pauly, J.P. Toennies: Adv. At. Mol. Phys. **1**, 195 (1965)
5.12 P.G. Burke: *Potential Scattering in Atomic Physics* (Plenum, New York (1977)
5.13 M.S. Child: *Molecular Collision Theory* (Academic, New York 1974)
5.14 M.L. Goldberger, K.M. Watson: *Collision Theory* (Wiley, New York 1964)
5.15 N.F. Mott, H.S.W. Massey: *The Theory of Atomic Collisions* (Clarendon, Oxford 1965)
5.16 R.G. Newton: *Scattering Theory of Waves and Particles*, 2nd ed., Texts and Monographs in Phys. (Springer, Berlin, Heidelberg, New York 1982)
5.17 L.D. Landau, E.M. Lifshitz: *Quantum Mechanics*, 3rd rev. ed. (Pergamon, Oxford 1977)
5.18 M. Rosen, D.R. Yennie: J. Math. Phys. **5**, 1505 (1964)
5.19 F.T. Smith: J. Chem. Phys. **42**, 2419 (1965)
5.20 R.J. Gordon: J. Chem. Phys. **63**, 3109 (1975)
5.21 D.M. Brink, N. Takigawa: Nucl. Phys. A**279**, 159 (1977)
5.22 N. Fröman, P.O. Fröman: *JWKB Approximation: Contribution to the Theory* (North-Holland, Amsterdam 1965)
5.23 J. Heading: *An Introduction to Phase-Integral Methods* (Methuen's Monographs on Physical Subjects, London 1962)
5.24 V.P. Maslov: *Théorie des Perturbation et Méthodes Asymptotiques* (Dunod, Paris 1972)
5.25 V.P. Maslov, M.V. Fedoryuk: *Kvasiklassischeskoe Priblizhenie dlya Uravnenii Kvantovoi Mekhaniki* (Nauka, Moscow 1976);
A.Ya. Povzner: Teor. Mat. Fiz. **51**, 54 (1982)
5.26 D.S.F. Crothers: J. Phys. B**9**, L513 (1976)
5.27 K.W. Ford, D.L. Hill, M. Wakano, J.A. Wheeler: Ann. Phys. **7**, 239 (1959)
5.28 N. Fröman: In *Semiclassical Methods in Molecular Scattering and Spectroscopy*, ed. by M.S. Child (Reidel, Dordrecht 1980) p.1
5.29 N. Fröman, P.O. Fröman: Ann. Phys. **83**, 103 (1974)
5.30 N. Fröman, P.O. Fröman: Nuovo Cimento B**20**, 121 (1974)
5.31 M.G. Barwell, R.J. Le Roy, P. Pajunen, M.S. Child: J. Chem. Phys. **71**, 2618 (1979)
5.32 P. Pajunen: Mol. Phys. **40**, 605 (1980);
J. Luppi, P. Pajunen: J. Chem. Phys. **77**, 1505 (1982);
P. Pajunen, R.J. LeRoy: J. Chem. Phys. **77**, 3527 (1982)
5.33 F. Karlsson: Phys. Rev. D**11**, 2120 (1975)
5.34 J.N.L. Connor: Mol. Phys. **15**, 621 (1968)
5.35 J.N.L. Connor: Mol. Phys. **23**, 717 (1972);
J.N.L. Connor, A.D. Smith: Mol. Phys. **43**, 397 (1981);
J.N.L. Connor, A.D. Smith: Mol. Phys. **45**, 149 (1982);
J. Luppi, P. Pajunen: J. Chem. Phys. **78**, 4451 (1983)

5.36 R.R. Herm: J. Chem. Phys. **47**, 4290 (1967)
5.37 P.M. Livingston: J. Chem. Phys. **45**, 601 (1966)
5.38 J.N.L. Connor: Mol. Phys. **25**, 1469 (1973)
5.39 G.V. Dubrovskii, M.Yu. Sumetskii: Teor. Mat. Fiz. **32**, 380 (1977) [English transl.: Theor. Math. Phys. USSR **32**, 808 (1977)]
5.40 S. Tani, M. Inokuti: J. Chem. Phys. **61**, 4422 (1974)
5.41 V.S. Masterov, A.A. Seregin: Yad. Fiz. **31**, 280 (1980)
5.42 W.N. Whitton, J.N.L. Connor: Mol. Phys. **26**, 1511 (1973); H.K. Shepard: Phys. Rev. D**27**, 1288 (1983)
5.43 V.S. Masterov, A.A. Seregin: Yad. Fiz. **28**, 1195 (1978); M. Yu. Sumetskii: Yad. Fiz. **36**, 130 (1982)
5.44 J. Grosser: Physica **64**, 550 (1973)
5.45 H. Goldstein: *Classical Mechanics* (Addison-Wesley, Cambridge, MA 1950)
5.46 L.D. Landau, E.M. Lifshitz: *Mechanics*, 3rd rev. ed. (Pergamon, Oxford 1976)
5.47 E.A. Gislason: J. Chem. Phys. **56**, 2493 (1972)
5.48 H.M. Mott-Smith: Phys. Fluids **3**, 721 (1960)
5.49 S.I. Rubinow, T.T. Wu: J. Appl. Phys. **27**, 1032 (1956)
5.50 P.E. Siska: J. Chem. Phys. **59**, 3429 (1973)
5.51 M.E. Riley, P.E. Siska: J. Chem. Phys. **61**, 2485 (1974)
5.52 R.B. Bernstein, R.A. La Budde: J. Chem. Phys. **58**, 1109 (1973)
5.53 H. Pauly: In *Atomic Physics 2*, Proc. 2nd Int. Conf. (Oxford 1970) p.155; E.A. Mason, C. Nyeland, J.J.H. Van den Biesen, C.J.N. Van den Meijdenberg: Physica A**116**, 133 (1982)
5.54 A. Schutte, D. Bassi, F. Tommasini, G. Scoles: J. Chem. Phys. **62**, 600 (1975)
5.55 W.C. Stwalley, A. Niehaus, D.R. Herschbach: J. Chem. Phys. **63**, 3081 (1975)
5.56 G.E. Zahr, W.H. Miller: Mol. Phys. **30**, 951 (1975)
5.57 H.J. Korsch, D. Leissing: J. Phys. B**9**, 1857 (1976)
5.58 J. Knoll, R. Schaeffer: Ann. Phys. **97**, 307 (1976)
5.59 S.H. Sheen, J.G. Skofronick, C.R. Müller: Int. J. Quantum Chem. **9**, 817 (1975);
H.J. Korsch, K.E. Thylwe: J. Phys. B**16**, 793 (1983)
5.60 H.C.W. Beijerinck, P.M.A. van der Kam, W.J.G. Thijssen, N.F. Verster: Chem. Phys. **45**, 225 (1980);
J.N.L. Connor, D. Farrely, D.C. Mackay: J. Chem. Phys. **74**, 3278 (1981)
5.61 R.W. Anderson: J. Chem. Phys. **60**, 2680 (1974)
5.62 M.V. Berry: J. Phys. B**2**, 381 (1969)
5.63 P. Pechukas: Phys. Rev. **181**, 166 (1969)
5.64 Yu.N. Demkov, J. Los: Phys. Lett. A**46**, 13 (1973)
5.65 V.N. Ostrovskii: J. Phys. B**8**, 2979 (1975)

Chapter 6

6.1 J.A. Gaunt: Proc. Cambridge Philos. Soc. **23**, 732 (1927)
6.2 N.F. Mott: Proc. Cambridge Philos. Soc. **27**, 553 (1931)
6.3 D.R. Bates: In *Atomic and Molecular Processes*, ed. by D.R. Bates (Academic, New York 1962) p.550
6.4 D.R. Bates: Comments At. Mol. Phys. **3**, 23 (1971)
6.5 B.H. Bransden: *Atomic Collision Theory* (Benjamin, New York 1970)
6.6 H.S.W. Massey, E.H.S. Burhop, H.B. Gilbody: *Slow Collisions of Heavy Particles*, 2nd ed., Electronic and Ionic Impact Phenomena, Vol.3 (Clarendon, Oxford 1971)
6.7 M.R.C. McDowell, J.P. Coleman: *Introduction to the Theory of Ion-Atom Collisions* (North-Holland, Amsterdam 1970)
6.8 N.F. Mott, H.S.W. Massey: *The Theory of Atomic Collisions* (Clarendon, Oxford 1965)

6.9 B.M. Smirnov: *Atomnye Stolknoveniya i Elementarnye Protesessy v Plazme* (Atomizdat, Moscow 1968)
6.10 B.M. Smirnov: *Asimptoticheskie Metody v Teorii Atomnykh Stolknovenii* (Atomizdat, Moscow 1973)
6.11 I.I. Sobelman, L.A. Vainshtein, E.A. Yukov: *Excitation of Atoms and Broadening of Spectral Lines*, Springer Ser. Chem. Phys., Vol.7 (Springer, Berlin, Heidelberg, New York 1981)
6.12 R.B. Bernstein (ed.): *Atom-Molecule Collision Theory. A Guide for Experimentalists* (Plenum, New York 1979)
6.13 V.N. Kondratiev, E.E. Nikitin: *Gas-Phase Reactions. Kinetics and Mechanisms* (Springer, Berlin, Heidelberg, New York 1981)
6.14 E.E. Nikitin: *Theory of Elementary Atomic and Molecular Processes in Gases* (Clarendon, Oxford 1974)
6.15 G. Breit, R.L. Gluckstern: In *Handbuch der Physik*, Vol.41, ed. by S. Flügge (Springer, Berlin, Heidelberg, New York 1959) p.496
6.16 W. Nörenberg, H.A. Weidenmüller: *Introduction to the Theory of Heavy-Ion Collisions*, 2nd ed., Lecture Notes Phys., Vol.51 (Springer, Berlin, Heidelberg, New York 1980)
6.17 J.W. Frame: Proc. Cambridge Philos. Soc. **27**, 511 (1931)
6.18 B.L. Moiseiwitsch: Proc. Phys. Soc. London **87**, 885 (1966)
6.19 A.V. Vinogradov: Opt. Spektrosk. **22**, 663 (1967)
6.20 F.W. Byron: Phys. Rev. A**4**, 1907 (1971)
6.21 D. Campos, H. Krüger: J. Phys. B**11**, 687 (1978)
6.22 J.C.Y. Chen, T. Ishihara, V.H. Ponce, K.M. Watson: Phys. Rev. A**8**, 1334 (1973)
6.23 J.C.Y. Chen, C.J. Joachain, K.M. Watson: Phys. Rev. A**5**, 1268 (1972)
6.24 J.C.Y. Chen, C.S. Wang, K.M. Watson: Phys. Rev. A**1**, 1150 (1970)
6.25 J.C.Y. Chen, K.M. Watson: Phys. Rev. **174**, 152 (1968)
6.26 M.R. Flannery, K.J. McCann: Phys. Rev. A**8**, 2915 (1973)
6.27 R. McCarroll: In *Atoms, Molecules and Lasers*, Lectures presented at the Int. Winter Colloquium in Trieste 1973 (Intern. Atomic Energy Agency, Vienna 1974) p.471
6.28 M.E. Riley: Phys. Rev. A**7**, 626 (1973);
R.E. Turner: Phys. Rev. A**26**, 3155 (1982)
6.29 D.R. Bates, D.S.F. Crothers: Proc. Roy. Soc. London A**315**, 465 (1970)
6.30 I.Ya. Berson: Izv. Akad. Nauk Latv. SSSR, Ser. Fiz. Tekh. Nauk **4**, 47 (1968)
6.31 M.S. Child: Mol. Phys. **20**, 171 (1971)
6.32 R.J. Cross, Jr.: J. Chem. Phys. **51**, 5163 (1969)
6.33 J.B. Delos, W.R. Thorson: Phys. Rev. A**6**, 720 (1972)
6.34 J.B. Delos, W.R. Thorson, S.K. Knudson: Phys. Rev. A**6**, 709 (1972)
6.35 G.V. Dubrovskii: Zh. Eksp. Teor. Fiz. **58**, 1075 (1970) [English transl.: Sov. Phys. JETP **31**(58), 577 (1970)]
6.36 Ch. Gaussorgues, C. Le Sech, F. Masnou-Seeuws, R. McCarroll, A. Riera: J. Phys. B**8**, 240 (1975)
6.37 K.P. Lawley, J. Ross: J. Chem. Phys. **43**, 2930 (1965)
6.38 R. McCarroll: In *Atomic Processes and Applications*, ed. by P.G. Burke, B.L. Moiseiwitsch (North-Holland, Amsterdam 1976) p.469
6.39 B.L. Moisweiwitsch: Rep. Prog. Phys. **40**, 843 (1977)
6.40 M.E. Riley: Phys. Rev. A**8**, 742 (1973)
6.41 S.D. Augustin, H. Rabitz: J. Chem. Phys. **69**, 4195 (1978)
6.42 M.H. Mittleman: Phys. Rev. **122**, 499 (1961)
6.43 P. Pechukas: Phys. Rev. **181**, 166 (1969)
6.44 P. Pechukas: Phys. Rev. **181**, 174 (1969)
6.45 A.P. Penner, R. Wallace: Phys. Rev. A**7**, 1007 (1973)
6.46 B. Corrigall, B. Kupers, R. Wallace: Phys. Rev. A**4**, 977 (1971)
6.47 T.A. Green: Phys. Rev. A**23**, 532 (1981)
6.48 R.E. Turner, J.S. Dahler: J. Phys. B**13**, 161 (1980)
6.49 E. Gerjuoy: Case Stud. At. Phys. **3**, 1 (1972)

6.50 R. Shakershaft: J. Phys. B**5**, 559 (1972)
6.51 R. Shakershaft, J. Macek, E. Gerjuoy: J. Phys. B**6**, 794 (1973)
6.52 R.J. Glauber: In *Lectures in Theoretical Physics*, Vol.1, ed. by W.E. Brittin, L.G. Dunham (Interscience, New York 1959) p.315
6.53 M.L. Goldberger, K.M. Watson: *Collision Theory* (Wiley, New York 1964)
6.54 R.G. Newton: *Scattering Theory of Waves and Particles*, 2nd ed., Texts and Monographs in Phys. (Springer, Berlin, Heidelberg, New York 1982)
6.55 E. Gerjuoy, B.C. Thomas: Rep. Prog. Phys. **37**, 1345 (1974)
6.56 C.J. Joachim, C. Quigg: Rev. Mod. Phys. **46**, 279 (1974)
6.57 L.D. Landau, E.M. Lifshitz: *Quantum Mechanics*, 3rd rev. ed. (Pergamon, Oxford 1977)
6.58 N. Fröman, P.O. Fröman: *JWKB Approximation: Contribution to the Theory* (North-Holland, Amsterdam 1965)
6.59 A.H. Nayfeh: *Perturbation Methods* (Wiley, New York 1973)
6.60 B.C. Eu, N. Zaritsky: J. Chem. Phys. **70**, 4986 (1979)
6.61 T.A. Green, M.E. Riley: Phys. Rev. A**8**, 2938 (1973)
6.62 B.R. Johnson: Chem. Phys. **2**, 381 (1973);
D. Campos: J. Phys. B**14**, 2585 (1981);
M.F. Herman: J. Chem. Phys. **76**, 2949 (1982);
H. Nakamura: Phys. Rev. A**26**, 3125 (1982);
M.F. Herman: J. Chem. Phys. **79**, 2771 (1983)
6.63 E.J. Shipsey: J. Chem. Phys. **59**, 5109 (1973)
6.64 E.J. Shipsey: J. Chem. Phys. **60**, 1589 (1974)
6.65 E.J. Shipsey: Mol. Phys. **35**, 81 (1978)
6.66 E.J. Shipsey: J. Chem. Phys. **70**, 5281 (1979)
6.67 L. Landau: Phys. T. Sowietunion **2**, 46 (1932)
6.68 A.M. Dykhne: Zh. Eksp. Teor. Fiz. **41**, 1324 (1961) [English transl.: Sov. Phys. JETP **14**(41), 941 (1962)]
6.69 M. Matsuzawa: J. Phys. Soc. Jpn. **23**, 1383 (1967)
6.70 J.P. Davis, P. Pechukas: J. Chem. Phys. **64**, 3129 (1976)
6.71 Yu.N. Demkov, V.N. Ostrovskii, E.D. Solov'ev: Phys. Rev. A**18**, 2089 (1978)
6.72 E.A. Solov'ev: Vestn. Leningr. Univ. Fiz. Khim. **4**, 10 (1976)
6.73 J.T. Hwang, P. Pechukas: J. Chem. Phys. **67**, 4640 (1977)
6.74 F.J. McLafferty, T.F. George: J. Chem. Phys. **63**, 2609 (1975)
6.75 E.P. Gordeev, E.E. Niktin, A.I. Shushin: Mol. Phys. **33**, 1611 (1977)
6.76 E.A. Solov'ev: Teor. Mat. Fiz. **28**, 115 (1976) [English transl.: Theor. Math. Phys. USSR **28**, 669 (1976)]
6.77 T.F. George, K. Morokuma: Chem. Phys. **2**, 129 (1973)
6.78 R.L. Jaffe, A.D. Isaacson, J.R. Laing, K. Morokuma, T.F. George: Theor. Chim. Acta **40**, 189 (1975)
6.79 R.L. Jaffe, T.F. George, K. Morokuma: Mol. Phys. **28**, 1489 (1974)
6.80 K.D. Jordan: Chem. Phys. **9**, 199 (1975)
6.81 K.D. Jordan: Int. J. Quantum Chem. Symp. **9**, 325 (1975)
6.82 R.K. Preston, C. Sloane, W.H. Miller: J. Chem. Phys. **60**, 4961 (1974)
6.83 P. Pechukas, T.F. George, K. Morokuma, F.J. McLafferty, J.R. Laing: J. Chem. Phys. **64**, 1099 (1976)
6.84 W.N. Whitton: Mol. Phys. **36**, 941 (1978)
6.85 R.M. Wilcox: J. Math. Phys. **8**, 962 (1967)
6.86 A.I. Shushin: Chem. Phys. Lett. **55**, 535 (1978)
6.87 A.I. Shushin: Opt. Spektrosk. **49**, 103 (1980) [English transl.: Opt. Spectrosc. USSR **49**, 54 (1980)]
6.88 A.M. Dykhne, G.L. Yudin: Usp. Fiz. Nauk **125**, 377 (1978) [English transl. Sov. Phys. Usp. **21**, 549 (1978)]
6.89 E.I. Dashevskaya: Chem. Phys. Lett. **11**, 184 (1971)
6.90 E.I. Dashevskaya, N.A. Mokhova: Opt. Spektrosk. **33**, 817 (1972) [English transl.: Opt. Spectrosc. USSR **33**, 452 (1972)]
6.91 B.C. Eu: J. Chem. Phys. **56**, 2507 (1972)
6.92 B.C. Eu: J. Chem. Phys. **58**, 472 (1973)
6.93 B.C. Eu: J. Chem. Phys. **59**, 4705 (1973)

6.94 B.C. Eu, T.P. Tsien: Chem. Phys. Lett. **17**, 256 (1972)
6.95 T. Uzer, M.S. Child: Mol. Phys. **41**, 1177 (1980);
R.J. Bieniek: Phys. Rev. A**23**, 2826 (1981)
6.96 S. Bosanac: J. Math. Phys. **20**, 396 (1979)
6.97 R.D. Levine: Mol. Phys. **22**, 497 (1971)
6.98 U.I. Cho, B.C. Eu: Mol. Phys. **32**, 19 (1976)
6.99 B.C. Eu, T.P. Tsien: Phys. Rev. A**7**, 648 (1973);
H.J. Korsch: Mol. Phys. **49**, 325 (1983);
B.C. Eu: *Semiclassical Theories of Molecular Scattering*, Springer Ser. Chem. Phys., Vol.26 (Springer, Berlin, Heidelberg, New York, Tokyo 1984)
6.100 C.H. Wang, W.J. Tomlison: Phys. Rev. **181**, 115 (1969)
6.101 M. Mori: J. Phys. Soc. Jpn. **23**, 1086 (1967)
6.102 P. Pechukas, J.C. Light: J. Chem. Phys. **44**, 3897 (1966)
6.103 D. Richards: J. Phys. B**14**, 1465 (1981)
6.104 J. Eichler: Phys. Rev. A**15**, 1856 (1977);
R.E. Turner: J. Phys. A**13**, 2079 (1980)

Chapter 7

7.1 N.F. Mott, H.S.W. Massey: *The Theory of Atomic Collisions* (Clarendon, Oxford 1965)
7.2 R. Cimiraglia, M. Persico: Mol. Phys. **38**, 1707 (1979)
7.3 M. Desouter-Lecomte, J.C. Lorquet: J. Chem. Phys. **66**, 4006 (1977)
7.4 T.G. Heil, A. Dalgarno: J. Phys. B**12**, L557 (1979)
7.5 R.D. Levine, B.R. Johnson, R.B. Bernstein: J. Chem. Phys. **50**, 1694 (1969)
7.6 F.T. Smith: Phys. Rev. **179**, 111 (1969)
7.7 K. Takayanagi: Adv. At. Mol. Phys. **1**, 149 (1965)
7.8 B.R. Johnson: Chem. Phys. Lett. **27**, 373 (1974)
7.9 M.S. Child: In *Molecular Spectroscopy*, Specialist Periodical Report, Vol.2, ed. by R.F. Barrow, D.A. Long, D.J. Miller (Chemical Society, London 1974) p.466
7.10 R. Colle: J. Chem. Phys. **74**, 2910 (1981);
H. Nakamura: Chem. Phys. **78**, 235 (1983)
7.11 M.S. Child: J. Phys. B**13**, 2557 (1980)
7.12 J.B. Delos, W.R. Thorson: Phys. Rev. A**6**, 720 (1972)
7.13 J.B. Delos, W.R. Thorson: Phys. Rev. Lett. **28**, 647 (1972)
7.14 V.K. Bykhovskii, E.E. Nikitin, M.Ya. Ovchinnikova: Zh. Eksp. Teor. Fiz. **47**, 750 (1964) [English transl.: Sov. Phys. JETP **20**(47), 500 (1965)]
7.15 J.B. Delos, W.R. Thorson: Phys. Rev. A**6**, 728 (1972)
7.16 M. Desouter-Lecomte, J.C. Lorquet: J. Chem. Phys. **71**, 4391 (1979)
7.17 Yu.N. Demkov, V.N. Ostrovskii, E.D. Solov'ev: Phys. Rev. A**18**, 2089 (1978)
7.18 N. Rosen, C. Zener: Phys. Rev. **40**, 502 (1932)
7.19 D.S.F. Crothers: J. Phys. B**6**, 1418 (1973)
7.20 E.F. Gurnee, J.H. Magee: J. Chem. Phys. **26**, 1237 (1957)
7.21 J. Callaway, J.Q. Bartling: Phys. Rev. **150**, 69 (1966)
7.22 J. Callaway, E. Bauer: Phys. Rev. A**140**, 1072 (1965)
7.23 P. Pechukas, J.C. Light: J. Chem. Phys. **44**, 3897 (1966)
7.24 J.C. Light: J. Chem. Phys. **66**, 5241 (1977)
7.25 L.A. Vainshtein, L.P. Presnyakov, I.I. Sobelman: Zh. Eksp. Teor. Fiz. **43**, 518 (1962) [English transl.: Sov. Phys. JETP **16**(43), 370 (1963)]
7.26 I.I. Sobelman, L.A. Vainshtein, E.A. Yukov: *Excitation of Atoms and Broadening of Spectral Lines*, Springer Ser. Chem. Phys., Vol.7 (Springer, Berlin, Heidelberg, New York 1981)
7.27 D. Arora, J.E. Turner: Phys. Rev. A**17**, 1614 (1978)
7.28 M.G. Paype, M.H. Nayfeh: Phys. Rev. A**13**, 595 (1976)
7.29 S.C. Mehrotra, J.E. Boggs: J. Chem. Phys. **64**, 2796 (1976)
7.30 M. Namiki, H. Yagisawa, H. Nakamura: J. Phys. B**13**, 743 (1980)

7.31 H. Yagisawa: J. Phys. B**9**, 2757 (1976);
H.A. Hyman: J. Phys. B**15**, L773 (1982);
R.T. Robiscoe: Phys. Rev. A**27**, 1365 (1983);
G.F. Thomas: Phys. Rev. A**27**, 2744 (1983)
7.32 B.G. Skinner: Proc. Phys. Soc. London **77**, 551 (1961)
7.33 M. Mori, H. Fujita: J. Phys. Soc. Jpn. **20**, 432 (1965)
7.34 N. Nakamura: Mol. Phys. **25**, 577 (1973)
7.35 J.C. Gay, A. Omont: J. Phys. Paris **35**, 9 (1974)
7.36 H.J. Korsch, H. Krüger: Mol. Phys. **39**, 51 (1980)
7.37 B. Brutschy, H. Haberland: Phys. Rev. A**19**, 2232 (1979)
7.38 R.M. Jordan, P.E. Siska: J. Chem. Phys. **69**, 4634 (1978)

Chapter 8

8.1 L. Landau: Phys. Z. Sowietunion **1**, 88 (1932)
8.2 L. Landau: Phys. Z. Sowietunion **2**, 46 (1932)
8.3 C. Zener: Proc. Roy. Soc. London A**137**, 696 (1932)
8.4 E.C.G. Stueckelberg: Helv. Phys. Acta **5**, 369 (1932)
8.5 Z.Z. Latypov: Zh. Tekh. Fiz. **42**, 1329 (1972) [English transl.: Sov. Phys. Tech. Phys. USSR **17**, 1061 (1973)]
8.6 V.M. Lavrov: In *Physics of Ionized Gases*, ed. by V. Vujnovič (Institute of Physics of the University of Zagreb 1974) p.199
8.7 E.E. Niktin, M.Ya. Ovchinnikova: Usp. Fiz. Nauk **104**, 379 (1971) [English transl.: Sov. Phys. Usp. **14**(104), 394 (1972)]
8.8 F.T. Smith: In *Physics of the One- and Two-Electron Atoms*, ed. by H. Kleinpoppen (North-Holland, Amsterdam 1969) p.755
8.9 F.T. Smith: In *Atomic Collision Processes*, ed. by S. Geltman, K.T. Mahantappa, W.E. Brittin (Gordon and Breach, New York 1969) p.95
8.10 D.R. Bates: Proc. Roy. Soc. London A**257**, 22 (1960)
8.11 D.S.F. Crothers: Adv. Phys. **20**, 405 (1971);
D.S.F. Crothers: Adv. At. Mol. Phys. **17**, 55 (1981)
8.12 G.V. Dubrovskii: Zh. Eksp. Teor. Fiz. **46**, 863 (1964) [English transl.: Sov. Phys. JEPT **19**(46), 591 (1964)]
8.13 M.S. Child: *Molecular Collision Theory* (Academic, London 1974);
E.E. Nikitin: In *Chemische Elementarprozesse*, ed. by H. Hartman (Springer, Berlin, Heidelberg, New York 1968) p.43
8.14 J.B. Delos, W.R. Thorson: Phys. Rev. A**6**, 728 (1972)
8.15 W.R. Thorson, J.B. Delos, S.A. Boorstein: Phys. Rev. A**4**, 1052 (1971)
8.16 V.K. Bykhovskii, E.E. Nikitin, M.Ya. Ovchinnikova: Zh. Eksp. Teor. Fiz. **47**, 750 (1964) [English transl.: Sov. Phys. JETP **20**(47), 500 (1965)]
8.17 M.S. Child: In *Molecular Spectroscopy*, Specialist Periodical Report, Vol.2, ed. by R.F. Barrow, D.A. Long, D.J. Miller (Chemical Society, London 1974) p.466
8.18 M.S. Child: In *Semiclassical Methods in Molecular Scattering and Spectroscopy*, ed. by M.S. Child (Reidel, Dordrecht 1980) p.127
8.19 A.M. Dykhne: Zh. Eksp. Teor. Fiz. **41**, 1324 (1961) [English transl.: Sov. Phys. JETP **14**(41), 941 (1962)]
8.20 M.S. Child: Mol. Phys. **16**, 313 (1969)
8.21 M.S. Child: Mol. Phys. **20**, 171 (1971)
8.22 L.P. Kotova: Zh. Eksp. Teor. Fiz. **55**, 1375 (1968) [English transl.: Sov. Phys. JETP **28**(55), 719 (1969)]
8.23 J. Heading: *An Introduction to Phase-Integral Methods* (Methuen's Monographs on Physical Subjects, London 1962)
8.24 G.V. Dubrovskii: Vestn. Leningr. Univ. **16**, 24 (1967)
8.25 G.V. Dubrovskii, I. Fischer-Hjalmars: J. Phys. B**7**, 892 (1974)
8.26 G.V. Dubrovskii, I. Fischer-Hjalmars: Phys. Scr. **19**, 318 (1979)
8.27 E.E. Nikitin: Opt. Spektrosk. **11**, 452 (1961) [English transl.: Opt. Spectrosc. USSR **11**, 246 (1961)]

8.28 J.B. Delos, W.R. Thorson: Phys. Rev. A**6**, 720 (1972)
8.29 M.Ya. Ovchinnikova: Opt. Spektrosk. **17**, 822 (1964) [English transl.: Opt. Spectrosc. USSR **17**, 447 (1964)]
8.30 D.S.F. Crothers, J.G. Hughes: J. Phys. B**10**, L557 (1977)
8.31 A.D. Bandrauk, W.H. Miller: Mol. Phys. **38**, 1893 (1979)
8.32 A. Bárány: J. Phys. B**11**, L399 (1978)
8.33 M.Ya. Ovchinnikova: Dokl. Akad. Nauk SSSR **161**, 641 (1965)
8.34 M. Baer, M.S. Child: Mol. Phys. **36**, 1449 (1978)
8.35 J.R. Laing, J.M. Yuan, I.H. Zimmerman, P.L. De Vries, T.F. George: J. Chem. Phys. **66**, 2801 (1977)
8.36 J. Heinrichs: Phys. Rev. **176**, 141 (1968)
8.37 J. Grosser, A.E. de Vries: Chem. Phys. **3**, 180 (1974)
8.38 J. Grosser, A.E. de Vries: Chem. Phys. **10**, 229 (1975)
8.39 A.V. Chaplik: Zh. Eksp. Teor. Fiz. **43**, 462 (1962) [English transl.: Sov. Phys. JETP **16**(43), 331 (1963)]
8.40 A. Bárány: J. Phys. B**13**, 147 (1980)
8.41 M.S. Child: Mol. Phys. **28**, 495 (1974)
8.42 H.J. Korsch, H. Krüger: Mol. Phys. **39**, 51 (1980)
8.43 E.E. Nikitin, A.I. Reznikov: Phys. Rev. A**6**, 522 (1972)
8.44 A.I. Shushin: Chem. Phys. Lett. **43**, 110 (1976)
8.45 J.B. Delos: Phys. Rev. A**9**, 1626 (1974)
8.46 A.K. Belyaev, A.Z. Devdariani: Opt. Spektrosk. **45**, 448 (1978)
8.47 M.B. Faist, R.B. Bernstein: J. Chem. Phys. **64**, 3924 (1976)
8.48 S.D. Augustin, W.H. Miller, P.K. Pearson, H.F. Schaefer: J. Chem. Phys. **58**, 2845 (1973)
8.49 R.E. Olson, F.T. Smith: Phys. Rev. A**3**, 1607 (1971)
8.50 A.Z. Devdariani, S.V. Bobashev: Zh. Eksp. Teor. Fiz. **60**, 485 (1971) [English transl.: Sov. Phys. JETP **33**(60), 260 (1971)]
8.51 J.B. Delos, W.R. Thorson: Phys. Rev. Lett. **28**, 647 (1972)
8.52 M.Ya. Ovchinnikova: Zh. Eksp. Teor. Fiz. **64**, 129 (1973) [English transl.: Sov. Phys. JETP **37**(64), 68 (1973)]
8.53 E.E. Nikitin, M.Ya. Ovchinnikova, B. Andersen, A.E. de Vries: Chem. Phys. **14**, 121 (1976)
8.54 T.A. Vartanyan, S.T. Przhibel'skii: Opt. Spektrosk. **45**, 433 (1978)
8.55 A.M. Chang, D.E. Pritchard: J. Chem. Phys. **70**, 4524 (1979)
8.56 A. Russek: Phys. Rev. A**4**, 1918 (1971)
8.57 M.S. Child, R.B. Gerber: Mol. Phys. **38**, 421 (1979)
8.58 V.V. Afrosimov, Yu.S. Gordeev, V.K. Nikulin, A.M. Polyanskii, A.P. Shergin: Zh. Eksp. Teor. Fiz. **62**, 848 (1972) [English transl.: Sov. Phys. JETP **35**(62), 449 (1972)]
8.59 M.Ya. Ovchinnikova: Zh. Eksp. Teor. Fiz. **59**, 1795 (1970) [English transl.: Sov. Phys. JETP **32**(59), 974 (1971)]
8.60 F.A. Elisenkov, V.K. Nikulin: Zh. Eksp. Teor. Fiz. **64**, 122 (1973) [English transl.: Sov. Phys. JETP **37**(64), 65 (1973)]
8.61 J. Boudon, M. Barat, M. Abignoli: J. Phys. B**3**, 207 (1970)
8.62 D. Coffey, D.C. Lorents, F.T. Smith: Phys. Rev. **187**, 201 (1969)
8.63 U.I. Cho, B.C. Eu: Mol. Phys. **32**, 19 (1976)
8.64 B.C. Eu, T.P. Tsien: Phys. Rev. A**7**, 648 (1973)
8.65 Ch. Gaussorgues: J. Phys. B**6**, 675 (1973)
8.66 L.P. Kotova, M.Ya. Ovchinnikova: Zh. Eksp. Teor. Fiz. **60**, 2026 (1971) [English transl.: Sov. Phys. JETP **33**(60), 1092 (1971)]
8.67 R.P. Saxon, R.E. Olson: Phys. Rev. A**12**, 830 (1975)
8.68 G.A.L. Delvigne, J. Los: Physica **67**, 166 (1973)

Chapter 9

9.1 N. Rosen, C. Zener: Phys. Rev. **40**, 502 (1932)
9.2 E.E. Nikitin: Opt. Spektrosk. **13**, 761 (1962) [English transl.: Opt. Spectrosc. USSR **13**, 431 (1962)]

9.3 E.E. Nikitin: Discuss. Faraday Soc. **33**, 14 (1962)
9.4 E.E. Nikitin: Izv. Akad. Nauk SSSR Ser. Fiz. **27**, 996 (1963)
9.5 Yu.N. Demkov: Zh. Eksp. Teor. Fiz. **45**, 195 (1963) [English transl.: Sov. Phys. JETP **18**(45), 138 (1964)]
9.6 E.G. Boving: J. Phys. B**10**, L63 (1977)
9.7 W. Fritsch, U. Wille: J. Phys. B**12**, L645 (1979)
9.8 W.E. Meyerhof: Comments At. Mol. Phys. **5**, 33 (1975)
9.9 W.E. Meyerhof, R. Anholt, J. Eichler, A. Salop: Phys. Rev. A**17**, 108 (1978);
N. Stolterfoht: J. Phys. B**16**, 2385 (1983)
9.10 N.P.F.B. van Asselt, J.G. Maas, J. Los: Chem. Phys. **16**, 81 (1976)
9.11 E.E. Nikitin: Adv. Quantum Chem. **5**, 135 (1970)
9.12 J.E. Bayfield, E.E. Nikitin, A.I. Reznikov: Chem. Phys. Lett. **19**, 471 (1973); **21**, 212 (1973);
E.E. Nikitin, A.I. Reznikov: J. Phys. B**11**, L659 (1978)
9.13 A. Bárány, D.S.F. Crothers: Proc. Roy. Soc. London A385, 129 (1983);
D.S.F. Crothers: J. Phys. B**11**, 1025 (1978);
D.S.F. Crothers, N.R. Todd: J. Phys. B**11**, L663 (1978)
9.14 J.M. Jackson, N.F. Mott: Proc. Roy. Soc. London A**137**, 703 (1932)
9.15 F.H. Mies: J. Chem. Phys. **40**, 523 (1964)
9.16 L.D. Landau, E.M. Lifshitz: *Quantum Mechanics*, 3rd rev. ed. (Pergamon, Oxford 1977)
9.17 R.Z. Vitlina, A.V. Chaplik, M.V. Entin: Zh. Eksp. Teor. Fiz. **67**, 1667 (1974) [Enlgish transl.: Sov. Phys. JETP **40**(67), 829 (1974)]
9.18 J. Grosser, A.E. de Vries: Chem. Phys. **3**, 180 (1974)
9.19 V.M. Galitskii, E.E. Nikitin, B.M. Smirnov: *Teoriya Stolknovenii Atomnykh Chastits* (Nauka, Moscow 1980)
9.20 B.M. Smirnov: *Asimptoticheskie Metody v Teorii Atomnykh Stolknovenii* (Atomizdat, Moscow 1973)
9.21 T.R. Dinterman, J.B. Delos: Phys. Rev. A**15**, 463 (1977)
9.22 O.B. Firsov: Zh. Eksp. Teor. Fiz. **21**, 1001 (1951)
9.23 D.S.F. Crothers, N.R. Todd: J. Phys. B**13**, 547 (1980)
9.24 C. Bottcher: J. Phys. B**9**, 3099 (1976)
9.25 T.R. Dinterman, J.B. Delos: Phys. Rev. A**15**, 475 (1977)
9.26 J. Grosser: Chem. Phys. **30**, 187 (1978)
9.27 R.W. Wijnaendts van Resandt, C. de Vreugd, R.L. Champion, J. Los: Chem. Phys. **29**, 151 (1978)
9.28 S.V. Bobashev: Adv. At. Mol. Phys. **14**, 341 (1978)
9.29 V. Kempter: Adv. Chem. Phys. **30**, 417 (1975)
9.30 V. Kempter: In *The Physics of Electronic and Atomic Collisions*, ed. by J. Risley (University of Washington Press 1976) p.327
9.31 H. Rosenthal, H.M. Foley: Phys. Rev. Lett. **23**, 1480 (1969)
9.32 V.A. Ankundinov, S.V. Bobashev, V.I. Perel': Zh. Eksp. Teor. Fiz. **60**, 906 (1971) [English transl.: Sov. Phys. JETP **33**(60), 490 (1971)]
9.33 R.K. Janev: Adv. At. Mol. Phys. **11**, 3 (1976);
F. Borondo, A. Maciás, A. Riera: J. Chem. Phys. **74**, 6126 (1981);
F. Borondo, A. Maciás, A. Riera: Chem. Phys. **81**, 303 (1983)
9.34 V.K. Bykhovskii, E.E. Nikitin: Zh. Eksp. Teor. Fiz. **48**, 1499 (1965) [English transl.: Sov. Phys. JETP **21**(48), 1003 (1965)]
9.35 M.B. Faist, R.D. Levine: J. Chem. Phys. **64**, 2953 (1976)
9.36 M.S. Child: Mol. Phys. **23**, 469 (1972)
9.37 A.I. Shushin: Chem. Phys. Lett. **43**, 110 (1976)
9.38 M.Ya. Ovchinnikova: Zh. Eksp. Teor. Fiz. **64**, 129 (1973) [English transl.: Sov. Phys. JETP **37**(64), 68 (1973)]
9.39 Yu.N. Demkov, M. Kunike: Vestn. Leningr. Univ. **16**, 39 (1969);
A. Bambini, P.R. Berman: Phys. Rev. A**23**, 2496 (1981)
9.40 E.E. Nikitin: Chem. Phys. Lett. **2**, 402 (1968)
9.41 E.C.G. Stueckelberg: Helv. Phys. Acta **5**, 369 (1932)
9.42 D.R. Bates: Discuss. Faraday Soc. **33**, 7 (1962)

9.43 A. Gallagher: Phys. Rev. **179**, 105 (1969)
9.44 J. Callaway, J.Q. Bartling: Phys. Rev. **150**, 69 (1966)
9.45 A.D. Bandrauk: Mol. Phys. **24**, 661 (1972)
9.46 Yu.N. Demkov, G.V. Dubrovskii, A.M. Ermolaev: Abstracts of Papers of 5th ICPEAC (Leningrad 1967) p.186
9.47 Yu.N. Demkov, C.V. Kunacz, V.N. Ostrovskii: Phys. Rev. A**18**, 2097 (1978)
9.48 E.A. Solov'ev: Vestn. Leningr. Univ. **4**, 10 (1976)
9.49 A.I. Voronin, S.P. Karkach, V.I. Osherov, V.G. Ushakov: Zh. Eksp. Teor. Fiz. **71**, 884 (1976) [English transl.: Sov. Phys. JETP **44**(71), 465 (1976)]
9.50 U. Wille: Phys. Lett. A**82**, 389 (1981); U. Wille: Z. Phys. A**308**, 3 (1982)
9.51 E.L. Duman, B.M. Smirnov: Zh. Eksp. Teor. Fiz. **69**, 59 (1975) [English transl.: Sov. Phys. JETP **42**(69), 29 (1975)]
9.52 J.S. Briggs: Rep. Prog. Phys. **39**, 217 (1976)
9.53 S.K. Knudson, W.R. Thorson: Can. J. Phys. **48**, 313 (1970)
9.54 K. Taulbjerg, J.S. Briggs, J. Vaaben: J. Phys. B**9**, 1351 (1976)
9.55 D.S.F. Crothers, J.G. Hughes: J. Phys. B**12**, 3063 (1979)
9.56 R. Anholt, W.E. Meyerhof, A. Salin: Phys. Rev. A**16**, 951 (1977)

Chapter 10

10.1 J.P. Faroux, J. Brossel: C. R. Acad. Sci. Paris B**263**, 612 (1966)
10.2 V.N. Rebane: Opt. Spektrosk. **24**, 296 (1968)
10.3 E.P. Gordeev, E.E. Nikitin, M.Ya. Ovchinnikova: Opt. Spektrosk. **30**, 189 (1971)
10.4 A.I. Okunevich, V.I. Perel': Zh. Eksp. Teor. Fiz. **58**, 666 (1970) [English transl.: Sov. Phys. JETP **31**(58), 356 (1970)]
10.5 A.G. Petrashen', V.N. Rebane, T.K. Rebane: Opt. Spektrosk. **35**, 408 (1973) [English transl.: Opt. Spectrosc. USSR **35**, 240 (1973)]
10.6 E.I. Dashevskaya, N.A. Mokhova: Opt. Spektrosk. **33**, 817 (1972) [English transl.: Opt. Spectrosc. USSR **33**, 452 (1972)]
10.7 J. Grosser: J. Phys. B**14**, 1449 (1981)
10.8 E.I. Dashevskaya: Chem. Phys. Lett. **11**, 184 (1971)
10.9 Yu.N. Demkov, B.H. Rebane, T.K. Rebane: In *Problemy Teoreticheskoi Fiziki. Kvantovaya Mekhanika*, Vol.1, ed. by M.G. Veselov, Yu.V. Novozhilov, P.P. Pavinskii (Leningrad University 1974) p.263
10.10 E.E. Nikitin, M.Ya. Ovchinnikova: In Abstracts of Papers of 5th ICPEAC (Nauka, Leningrad 1967) p.226
10.11 E.I. Dashevskaya, E.A. Kobzeva: Opt. Spektrosk. **30**, 807 (1971)
10.12 A.P. Kazantsev: Zh. Eksp. Teor. Fiz. **51**, 1751 (1966) [English transl.: Sov. Phys. JETP **24**(51), 1183 (1967)]
10.13 Yu.A. Vdovin, N.A. Dobrodeev: Zh. Eksp. Teor. Fiz. **55**, 1047 (1968) [English transl.: Sov. Phys. JETP **28**(55), 544 (1969)]
10.14 T. Watanabe: Phys. Rev. A**138**, 1573 (1965)
10.15 C.G. Carrington, D.N. Stacey, J. Cooper: J. Phys. B**6**, 417 (1973)
10.16 E.E. Nikitin: Adv. Chem. Phys. **28**, 317 (1975)
10.17 Yu.A. Vdovin, V.M. Galitskii: Zh. Eksp. Teor. Fiz. **69**, 103 (1975) [English transl.: Sov. Phys. JETP **42**(69), 51 (1975)]
10.18 J.C. Gay: J. Phys. Paris **35**, 813 (1974)
10.19 Yu.N. Demkov, V.N. Ostrovskii, E.A. Solov'ev: Zh. Eksp. Teor. Fiz. **66**, 125 (1974) [English transl.: Sov. Phys. JETP **39**(66), 57 (1974)]
10.20 V.N. Ostrovskii, E.A. Solov'ev: Zh. Eksp. Teor. Fiz. **66**, 1590 (1974) [English transl.: Sov. Phys. JETP **39**(66), 1590 (1974)]
10.21 N. Toshima: J. Phys. Soc. Jpn. **42**, 633 (1977)
10.22 N. Toshima: J. Phys. Soc. Jpn. **43**, 605 (1977)
10.23 Yu.N. Demkov, B.S. Monozon, V.N. Ostrovskii: Zh. Eksp. Teor. Fiz. **57**, 1431 (1969) [English transl.: Sov. Phys. JETP **30**(57), 775 (1970)]
10.24 V.I. Kogan, V.M. Galitsky: *Problems in Quantum Mechanics* (Prentice Hall, New York 1963)

10.25 L.D. Landau, E.M. Lifshitz: *Quantum Mechanics*, 3rd rev. ed. (Pergamon, Oxford 1977)
10.26 W.H. Miller: Adv. Chem. Phys. **25**, 69 (1974)
10.27 W.H. Miller: Adv. Chem. Phys. **30**, 77 (1975)
10.28 A.P. Clark: J. Phys. B**6**, 1153 (1973)
10.29 A.P. Clark, A.S. Dickinson: J. Phys. B**4**, L112 (1971)
10.30 A.P. Clark, A.S. Dickinson, D. Richards: Adv. Chem. Phys. **36**, 63 (1977)
10.31 D. Richards: J. Phys. B**5**, L53 (1972)
10.32 I.L. Beigman, L.A. Vainshtein, I.I. Sobelman: Zh. Eksp. Teor. Fiz. **57**, 1703 (1969) [English transl.: Sov. Phys. JETP **30**(57), 920 (1970)]
10.33 I.C. Percival, D. Richards: J. Phys. B**3**, 1035 (1970)
10.34 L.P. Presnyakov, A.M. Urnov: J. Phys. B**3**, 1267 (1970)
10.35 I.L. Beigman: Zh. Eksp. Teor. Fiz. **73**, 1729 (1977) [English transl.: Sov. Phys. JETP **46**(73), 908 (1977)]
10.36 I.C. Percival, D. Richards: Adv. At. Mol. Phys. **11**, 1 (1975)
10.37 I.C. Percival, D. Richards: J. Phys. B**10**, 1497 (1977)
10.38 M. Matsuzawa: J. Phys. B**12**, 3743 (1979)
10.39 M. Matsuzawa: In *Electronic and Atomic Collisions*, ed. by N. Oda, K. Takayanagy (North-Holland, Amsterdam 1980) p.493
10.40 M.Ya. Ovchinnikova: *Nekotorye Voptrosy Teorii Atomnykh Stolknovenii*, Ph. D. Thesis (Institute of Chemical Physics, Academy of Sciences USSR 1965)
10.41 V.I. Osherov: Zh. Eksp. Teor. Fiz. **49**, 1157 (1965) [English transl.: Sov. Phys. JETP **22**(49), 804 (1966)]
10.42 V.I. Osherov: Fiz. Tverd. Tela **10**, 30 (1968)
10.43 Yu.N. Demkov: Dokl. Akad. Nauk SSSR **166**, 1076 (1966)
10.44 M.Ya. Ovchinnikova: Opt. Spektrosk. **17**, 822 (1964) [English transl.: Opt. Spectrosc. USSR **17**, 447 (1964)]
10.45 Yu.N. Demkov, V.I. Osherov: Zh. Eksp. Teor. Fiz. **53**, 1584 (1967) [English transl.: Sov. Phys. JETP **26**(53), 916 (1968)]
10.46 V.I. Osherov: Dokl. Akad. Nauk SSSR **168**, 1291 (1966)
10.47 Yu.N. Demkov, V.N. Ostrovskii: *Metod Potentsialov Nulevogo Radiusa v Atomnoi Fizike* (Leningrad University 1975)
10.48 R.J. Bieniek: J. Phys. B**13**, 4405 (1980)
10.49 J.P. Gauyacq: J. Phys. B**13**, 4417 (1980); R.D. Taylor, J.B. Delos: Proc. Roy. Soc. London A**379**, 179 (1982)
10.50 A.Z. Devdariani, V.N. Ostrovskii, Yu.N. Sebyakin: Zh. Eksp. Teor. Fiz. **73**, 412 (1977) English transl.: Sov. Phys. JETP **46**(73), 215 (1977)
10.51 R.B. Barker, H.W. Berry: Phys. Rev. **151**, 14 (1966)
10.52 J.S. Cohen: Phys. Rev. A**13**, 99 (1976)
10.53 W.H. Miller: J. Chem. Phys. **52**, 3563 (1970)
10.54 W.H. Miller, H. Morgner: J. Chem. Phys. **67**, 4923 (1977)
10.55 V.A. Basylev, N.K. Zhevago: Zh. Eksp. Teor. Fiz. **69**, 853 (1975) [English transl.: Sov. Phys. JETP **42**(69), 436 (1975)]
10.56 A.Z. Devdariani, V.N. Ostrovskii, Yu.N. Sebyakin: Zh. Eksp. Teor. Fiz. **71**, 909 (1976) [English transl.: Sov. Phys. JETP **44**(71), 477 (1976)]
10.57 V.A. Basylev, N.K. Zhevago, M.I. Chibisov: Zh. Eksp. Teor. Fiz. **71**, 1286 (1976) English transl.: Sov. Phys. JETP **44**(71), 671 (1976)
10.58 V.P. Krainov, A.V. Kruglikov: Opt. Spektrosk. **50**, 826 (1981)
10.59 A.Z. Devdariani, V.N. Ostrovskii, Yu.N. Sebyakin: Zh. Eksp. Teor. Fiz. **76**, 529 (1979) [English transl.: Sov. Phys. JETP **49**(76), 266 (1979)]

Chapter 11

11.1 R.W. Anderson, T.P. Goddard, C. Parravano, J. Warner: J. Chem. Phys. **64**, 4037 (1976)
11.2 J. Apt, D.E. Pritchard: J. Phys. B**12**, 83 (1979)
11.3 G.M. Carter, D.E. Pritchard, K. Martin, D. Theodore: Phys. Rev. Lett. **35**, 1144 (1975)

11.4 J. Cuvellier, J. Berlande, C. Benoit, M.Y. Perrin, J.M. Mestdagh, J. de Mesmay: J. Phys. B**12**, L461 (1979)
11.5 M. Elbel: In *Progress in Atomic Spectroscopy*, Part B, ed. by W. Hanle, H. Kleinpoppen (Plenum, New York 1979) p.1299
11.6 I.V. Hertel, W. Stoll: Adv. At. Mol. Phys. **13**, 113 (1977)
11.7 L. Krause: Adv. Chem. Phys. **28**, 267 (1975)
11.8 E.L. Lewis: Phys. Rep. **1**, 1 (1980); N. Allard, J. Kielkopf: Rev. Mod. Phys. **54**, 1103 (1982)
11.9 P.L. Lijnse: *Review of Literature on Quenching, Excitation and Mixing Collision Cross Sections for the First Resonance Doublets of the Alkalics*, Report No.398 (Rijksuniversiteit, Utrecht 1972)
11.10 J.M. Mestaagh, J. Cuvelier, J. Berlande, A. Binet, P. de Pujo: J. Phys. B**13**, 4589 (1980)
11.11 P. Münster, J. Marek: J. Phys. B**14**, 1009 (1981); R. Boggy, F.A. Franz: Phys. Rev. A**25**, 1887 (1982)
11.12 W.D. Phillips, C.L. Glaser, D. Kleppner: Phys. Rev. Lett. **38**, 1018 (1977)
11.13 W.D. Phillips, J.A. Serri, D.J. Ely, D.E. Pritchard, K.R. Way, J.L. Kinsey: Phys. Rev. Lett. **41**, 937 (1978)
11.14 B. Brouillaud, R. Gayet: J. Phys. B**10**, 2143 (1977)
11.15 F. Masnou-Seeuws, M. Philippe, E. Roueff, A. Spielfiedel: J. Phys. B**12**, 4065 (1979)
11.16 R.E. Olson: Chem. Phys. Lett. **33**, 250 (1975)
11.17 J. Pascale, R.E. Olson: J. Chem. Phys. **64**, 3538 (1976)
11.18 R.H.G. Reid: J. Phys. B**6**, 2018 (1973)
11.19 R.H.G. Reid: J. Phys. B**8**, 2255 (1975)
11.20 R.H.G. Reid: J. Phys. B**8**, L493 (1975)
11.21 R.H.G. Reid, A. Dalgarno: Chem. Phys. Lett. **6**, 85 (1970); R.W. Anderson: J. Chem. Phys. **77**, 5426 (1982); D.E. Fitz, D.J. Kouri: J. Chem. Phys. **73**, 5115 (1980); P.L. De Vries: J. Chem. Phys. **80**, 186 (1984)
11.22 R.H.G. Reid, R.F. Rankin: J. Phys. B**11**, 55 (1978)
11.23 R.P. Saxon, R.E. Olson, B. Liu: J. Chem. Phys. **67**, 2692 (1977)
11.24 Y. Shimoni, A.D. Wilson: Chem. Phys. **11**, 289 (1975)
11.25 R.E. Smalley, D.A. Auerbach, P.S.H. Fitch, D.H. Levy, L. Warton: J. Chem. Phys. **66**, 3778 (1977); L.C. Balling, J.J. Wright, M.D. Havey: Phys. Rev. A**26**, 1426 (1982); G. Aepfelbach, A. Nunnemann, D. Zimmermann: Chem. Phys. Lett. **96**, 311 (1983)
11.26 J. Tellinghuisen, A. Ragone, M.S. Kim, J. Auerbach, R.E. Smalley, L. Warton, D.H. Levy: J. Chem. Phys. **71**, 1283 (1979)
11.27 R. Ahmad-Bitar, W.P. Lapatovich, D.E. Pritchard, I. Renhorn: Phys. Rev. Lett. **39**, 1657 (1977)
11.28 R. Düren, W. Gröger: Chem. Phys. Lett. **61**, 6 (1979); R. Düren, E. Hasselbrink, G. Moritz: Z. Phys. A**307**, 1 (1982); R. Düren, E. Hasselbrink, H. Tischer: J. Chem. Phys. **77**, 3286 (1982); R. Düren, E. Hasselbrink, H. Tischer, S. Milosevic, G. Pichler: Chem. Phys. Lett. **89**, 218 (1982); J.M.M. Van Deventer, F. Van Der Valk: Physica BC**112**, 119 (1982)
11.29 J.H. Goble, J.S. Winn: J. Chem. Phys. **70**, 2051 (1979)
11.30 J.F. Kielkopf: J. Chem. Phys. **61**, 4733 (1974)
11.31 W.P. Lapatovich, R. Ahmad-Bitar, P.E. Moskowitz, I. Renhorn, R.A. Gottscho, D.E. Pritchard: J. Chem. Phys. **73**, 5419 (1980)
11.32 G. Smith: J. Phys. B**8**, 2273 (1975)
11.33 G. York, R. Scheps, A. Gallagher: J. Chem. Phys. **63**, 1052 (1975)
11.34 W.E. Baylis: J. Chem. Phys. **51**, 2665 (1969)
11.35 C. Bottcher: Chem. Phys. Lett. **18**, 457 (1973)
11.36 C. Bottcher, T.C. Cravens, A. Dalgarno: Proc. Roy. Soc. London A**346**, 157 (1975)

11.37 E. Czuchaj, J. Sienkiewicz: Z. Naturforsch. A**34**, 694 (1979)
11.38 R. Düren, G. Moritz: J. Chem. Phys. **73**, 5155 (1980);
R. Düren: Adv. At. Mol. Phys. **16**, 55 (1980)
11.39 J. Hanssen, R. McCarroll, P. Valiron: J. Phys. B**12**, 899 (1979)
11.40 G.K. Ivanov: Teoret. Eksp. Khim. **14**, 610 (1978)
11.41 M. Krauss, P. Maldonado, A.C. Wahl: J. Chem. Phys. **54**, 4944 (1971)
11.42 F. Masnou-Seeuws, M. Philippe, P. Valiron: Phys. Rev. Lett. **41**, 395 (1978)
11.43 J. Pascale, J. Vandeplanque: J. Chem. Phys. **60**, 2278 (1974)
11.44 M. Philippe, F. Masnou-Seeuws, P. Valiron: J. Phys. B**12**, 2493 (1979)
11.45 K.J. Wahlstrand, R.W. Munrich, J.S. Dahler, S.E. Nielsen: J. Phys. B**10**, 1687 (1977);
J. Pascale: Phys. Rev. A**28**, 632 (1983);
E.M. Pontius, K.M. Sando: Phys. Rev. A**28**, 3117 (1983)
11.46 M. Abramowitz, I.A. Stegun (eds.): *Handbook of Mathematical Functions* (National Bureau of Standards, Washington 1964)
11.47 A.I. Reznikov: Chem. Phys. Lett. **44**, 41 (1976)
11.48 E.E. Nikitin: Opt. Spektrosk. **22**, 689 (1967)
11.49 E.I. Dashevskaya, E.E. Nikitin, A.I. Reznikov: J. Chem. Phys. **53**, 1175 (1970)
11.50 E.I. Dashevskaya, E.E. Nikitin: Can. J. Phys. **54**, 709 (1976)
11.51 E.I. Dashevskaya, A.I. Reznikov: Opt. Spektrosk. **48**, 644 (1980) [English transl.: Opt. Spectrosc. USSR **48**, 355 (1980)]
11.52 E.I. Dashevskaya, F. Masnou, R. McCarroll, E.E. Nikitin: Opt. Spektrosk. **37**, 209 (1974) [English transl.: Opt. Spectrosc. USSR **37**, 119 (1974)]
11.53 A. Gallagher: Phys. Rev. **172**, 88 (1968)
11.54 Ch. Gaussorgues, F. Masnou-Seeuws: J. Phys. B**10**, 2125 (1977)
11.55 E.P. Gordeev, E.E. Nikitin, M.Ya. Ovchinnikova: Can. J. Phys. **47**, 1819 (1969)

Appendix

A.1 D.M. Brink, G.R. Satchler: *Angular Momentum* (Clarendon, Oxford 1968)
A.2 A.R. Edmonds: *Angular Momentum in Quantum Mechanics* (Princeton Press, New Jersey 1957)
A.3 E. El Baz, B. Castel: *Graphical Methods of Spin Algebras* (Deccer, New York 1972)
A.4 U. Fano, G. Racah: *Irreducible Tensorial Sets* (Academic, New York 1959)
A.5 V. Heine: *Group Theory in Quantum Mechanics* (Pergamon, Oxford 1960)
A.6 L.D. Landau, E.M. Lifshitz: *Quantum Mechanics*, 3rd rev. ed. (Pergamon, Oxford 1977)
A.7 L.C. Biedenharn, H. van Dam (eds.): *Quantum Theory of Angular Momentum* (Academic, New York 1965)
A.8 M.E. Rose: *Elementary Theory of Angular Momentum* (Wiley, New York; Chapman Hall, London 1957)
A.9 I.I. Sobelman: *Atomic Spectra and Radiative Transitions*, Springer Ser. Chem. Phys., Vol.1 (Springer, Berlin, Heidelberg, New York 1979)
A.10 D.A. Varshalovich, A.N. Moskalev, V.K. Khersonskii: *Kvantovaya Teoriya Uglovogo Momenta* (Nauka, Leningrad 1975)
A.11 E. Wigner: *Group Theory and Its Application to the Quantum Mechanics of Atomic Spectra* (Academic, New York 1959)
A.12 A.P. Yutsis, A.A. Bandzaitis: *Teoriya Momenta Kolichestva Dvizheniya v Kvantovoi Mekhanike* (Mintis, Vilnius 1965)

Subject Index

Volume 17

Inelastic Particle-Surface Collisions

Proceedings of the Third International Workshop on Inelastic Ion-Surface Collisions
Feldkirchen-Westerham, Federal Republic of Germany, September 17-19, 1980
Editors: **E. Taglauer, W. Heiland**
1981. 194 figures. VIII, 329 pages. ISBN 3-540-10898-X

Volume 18

Modelling of Chemical Reaction Systems

Proceedings of an International Workshop, Heidelberg, Federal Republic of Germany, September 1-5, 1980
Editors: **K. H. Ebert, P. Deuflhard, W. Jäger**
1981. 163 figures. X, 389 pages. ISBN 3-540-10983-8

Volume 19

Secondary Ion Mass Spectrometry SIMS III

Proceedings of the Third International Conference, Technical University, Budapest, Hungary, August 30 - September 5, 1981
Editors: **A. Benninghoven, J. Giber, J. László, M. Riedel, H. W. Werner**
1982. 289 figures. XI, 444 pages. ISBN 3-540-11372-X

Volume 20

Chemistry and Physics of Solid Surfaces IV

Editors: **R. Vanselow, R. Howe**
1982. 247 figures. XIII, 496 pages
ISBN 3-540-11397-5

Volume 21

Dynamics of Gas-Surface Interaction

Proceedings of the International School on Material Science and Technology, Erice, Italy, July 1-15, 1981
Editors: **G. Benedek, U. Valbusa**
1982. 132 figures. XI, 282 pages. ISBN 3-540-11693-1

Volume 22
V. S. Letokhov

Nonlinear Laser Chemistry
Multiple-Photon Excitation
1983. 152 figures. XIV, 417 pages. ISBN 3-540-11705-9

Volume 23

Picosecond Phenomena III

Proceedings of the Third International Conference on Picosecond Phenomena Garmisch-Partenkirchen, Federal Republic of Germany, June 16-18, 1982
Editors: **K. B. Eisenthal, R. M. Hochstrasser, W. Kaiser, A. Laubereau**
1982. 288 figures. XIII, 401 pages. ISBN 3-540-11912-4

Volume 24

Desorption Induced by Electronic Transitions, DIET I

Proceedings of the First International Workshop, Williamsburg, Virginia, USA, May 12-14, 1982
Editors: **N. H. Tolk, M. M. Traum, J. C. Tully, T. E. Madley**
1983. 112 figures. XI, 269 pages. ISBN 3-540-12127-7

Volume 25

Ion Formation from Organic Solids

Proceedings of the Second International Conference, Münster, Federal Republic of Germany, September 7-9, 1982
Editor: **A. Benninghoven**
1983. 170 figures. IX, 269 pages. ISBN 3-540-12244-3

Volume 26
B. C. Eu

Semiclassical Theories of Molecular Scattering

1984. 17 figures. XII, 229 pages. ISBN 3-540-12410-1

Volume 27

EXAFS and Near Edge Structures

Proceedings of the International Conference, Frascati, Italy, September 13-17, 1982
Editors: **A. Bianconi, L. Incoccia, S. Stipcich**
1983. 316 figures. XII, 420 pages. ISBN 3-540-12411-X

Volume 33

Surface Studies with Lasers

Proceedings of the International Conference, Mauterndorf, Austria, March 9-11, 1983
Editors: **F. R. Aussenegg, A. Leitner, M. E. Lippitsch**
1983. 146 figures. IX, 241 pages. ISBN 3-540-12598-1

Volume 34

Inert Gases
Potentials, Dynamics, and Energy Transfer in Doped Crystals
Editor: **M. L. Klein**
With contributions by R. A. Aziz, S. S. Cohen, H. Dubost, M. L. Klein
1984. 89 figures. XI, 266 pages. ISBN 3-540-13128-0

Volume 35

Chemistry and Physics of Solid Surfaces V

Editors: **R. Vanselow, R. Howe**
1984. 303 figures. XXI, 554 pages. ISBN 3-540-13315-1

Volume 36

Secondary Ion Mass Spectrometry SIMS IV

Proceedings of the Fourth International Conference, Osaka, Japan, November 13-19, 1983
Editors: **A. Benninghoven, J. Okano, R. Shimizu, H. W. Werner**
1984. 415 figures. XV, 503 pages. ISBN 3-540-13316-X

Springer-Verlag
Berlin
Heidelberg
New York
Tokyo

DATE DUE

81195			